Walter Nance

THE PRINCIPLES OF HUMAN BIOCHEMICAL GENETICS

NORTH-HOLLAND RESEARCH MONOGRAPHS

FRONTIERS OF BIOLOGY

VOLUME 19*

Under the General Editorship of

A. NEUBERGER

London

and

E. L. TATUM

New York

NORTH-HOLLAND PUBLISHING COMPANY
AMSTERDAM · OXFORD

THE PRINCIPLES
OF HUMAN
BIOCHEMICAL GENETICS

Second revised and enlarged edition

HARRY HARRIS

Galton Laboratory, University College, London

1975

NORTH-HOLLAND PUBLISHING COMPANY
AMSTERDAM · OXFORD

Library of Congress Catalog Card Number: 74–83271
ISBN North-Holland for this series 0 7204 7100 1
ISBN North-Holland for this volume 0 7204 7119 2
ISBN American Elsevier 0 444 10012 1

PUBLISHERS:

NORTH-HOLLAND PUBLISHING COMPANY – AMSTERDAM
NORTH-HOLLAND PUBLISHING COMPANY, LTD. – OXFORD

SOLE DISTRIBUTORS FOR THE U.S.A. AND CANADA:

AMERICAN ELSEVIER PUBLISHING COMPANY, INC.
52 VANDERBILT AVENUE, NEW YORK, N.Y. 10017

This book is a completely revised edition of Volume 19, published by North-Holland Publishing Co in 1970.

This book·is also published in a paperback student edition

PRINTED IN THE NETHERLANDS

Editors' preface

The aim of the publication of this series of monographs, known under the collective title of '*Frontiers of Biology*', is to present coherent and up-to-date views of the fundamental concepts which dominate modern biology.

Biology in its widest sense has made very great advances during the past decade, and the rate of progress has been steadily accelerating. Undoubtedly important factors in this acceleration have been the effective use by biologists of new techniques, including electron microscopy, isotopic labels, and a great variety of physical and chemical techniques, especially those with varying degrees of automation. In addition, scientists with partly physical or chemical backgrounds have become interested in the great variety of problems presented by living organisms. Most significant, however, increasing interest in and understanding of the biology of the cell, especially in regard to the molecular events involved in genetic phenomena and in metabolism and its control, have led to the recognition of patterns common to all forms of life from bacteria to man. These factors and unifying concepts have led to a situation in which the sharp boundaries between the various classical biological disciplines are rapidly disappearing.

Thus, while scientists are becoming increasingly specialized in their techniques, to an increasing extent they need an intellectual and conceptual approach on a wide and non-specialized basis. It is with these considerations and needs in mind that this series of monographs, '*Frontiers of Biology*' has been conceived.

The advances in various areas of biology, including microbiology, biochemistry, genetics, cytology, and cell structure and function in general will be presented by authors who have themselves contributed significantly to these developments. They will have, in this series, the opportunity of bringing together, from diverse sources, theories and experimental data, and of integrating these into a more general conceptual framework. It is unavoidable, and probably even desirable, that the special bias of the individual authors will become evident in their contributions. Scope will also be given for presentation of new and challenging ideas and hypotheses for which complete evidence is at present lacking. However, the main emphasis will be on fairly complete and objective presentation of the more important and more rapidly advancing aspects of biology. The level will be advanced, directed primarily to the needs of the graduate student and research worker.

Most monographs in this series will be in the range of 200–300 pages, but on occasion a collective work of major importance may be included exceeding this figure. The intent of the publishers is to bring out these books promptly and in fairly quick succession.

It is on the basis of all these various considerations that we welcome the opportunity of supporting the publication of the series '*Frontiers of Biology*' by North-Holland Publishing Company.

E. L. TATUM
A. NEUBERGER, *Editors*

Preface

This book has grown out of a course of lectures given at the Galton Laboratory, which was intended not only for students and research workers in human genetics, but also for biochemistry, biology and medical students as well as for interested research workers in related fields. I was concerned to explain the principal concepts which underlie modern ideas in human biochemical genetics, to present a picture of the extraordinary degree of inherited biochemical diversity which recent research has shown to be a characteristic feature of human populations, and to show how the detailed analysis of genetically determined biochemical differences between individual members of our species could throw new light on fundamental problems not only in genetics, but also in medicine and more generally in human biology.

Just over ten years ago I wrote an account of the subject (Human Biochemical Genetics, Cambridge University Press, 1959) covering most of the information which was then available, in what seemed at the time a logical order. Since then, however, research in this field has expanded almost explosively and in preparing the present book, it became very obvious that one could not be content with simply trying to update the earlier text. The many advances now called for a very different arrangement if present knowledge and concepts were to be presented in a coherent manner. This is not merely because a great deal more is now known about the particular topics that were dealt with in the earlier book, but because whole new areas of the subject have been opened up in a manner which could hardly have been envisaged only a few years ago. One of the important consequences of these developments has been the greater unity they have given to the subject as a whole. The interrelationships between what at one time seemed very different and unconnected types of phenomena such as the inborn errors of metabolism, the blood group antigenic differences, the haemoglobin diseases and the enzyme and protein polymorphisms, can now be thought about within a consistent theoretical framework in a way that was hardly possible previously. This of course gives one an opportunity to try and present the subject in a more systematic and analytical manner. So the present work differs considerably in its approach and arrangement from my earlier one, and of course much of the material discussed is new.

One of the difficulties in writing this kind of text is deciding what examples should be used to illustrate the various points in the argument, and in how much detail they should be presented. Also one has to decide what must be left out, if the overall length is to be kept to a manageable size, and the argument not be obscured by an excessive amount of descriptive material. Since a book of this sort may also be useful as a source of reference, one is often placed in something of a quandary. I have tried to resolve this difficulty by giving key references to much material which is not described in detail in the text, and a great deal of this has been arranged in the form of tables or appendices so that the appropriate references can be extracted more easily. Nevertheless the scope of the subject is now so very extensive and the literature so vast and distributed over such a wide range of journals, that reference to many topics must inevitably have been omitted. I hope nevertheless that enough has been included so that the book may serve not only as an introductory text in which the main principles of the subject are formulated, but also as guide to further reading on specific topics.

Galton Laboratory,   HARRY HARRIS, M.D., F.R.S.
University College, London   July, 1969

Preface to 2nd edition

In the past five years, work in human biochemical genetics has continued to advance very rapidly. Consequently in revising the text for this edition much new material has been included, though the general arrangement of the book has been retained.

Galton Laboratory,
University College London

Harry Harris
March 1974

Harry Harris

March 1974

Acknowledgements

It is a great pleasure to thank Mrs. N. Parry-Jones for preparing the figures, Mrs. J. Barrie for her secretarial assistance and Dr. D. A. Hopkinson for much helpful discussion and advice while the book was being written.

Many of the illustrations come from other publications and the permission for their reproduction is gratefully acknowledged. The original sources are given in the captions and bibliography.

Contents

CHAPTER 1

Gene mutations and single aminoacid substitutions

1.1. Introduction: genes, DNA and proteins

Human beings are exceedingly diverse. They differ from one another in their normal physical, physiological and mental attributes. They also differ in whether they suffer from particular diseases or other abnormalities. These variations are caused in part by differences in the environmental conditions in which they live. But they also depend on inborn differences. Indeed it is very probable that no two individuals with the exception of monozygotic twins are exactly alike in their inherited constitutions. The analysis in molecular terms of the nature and effects of such genetically determined differences forms the subject matter of human biochemical genetics.

Classical genetics led to the concept of the gene as the fundamental biological unit of heredity and postulated that it must possess three basic properties. It had to have a specific function in the cell, and hence in the organism as a whole. It had to be capable of exact self-replication so that its functional specificity would be preserved from one cell generation to the next. Finally, although usually an extremely stable entity, it had to be susceptible to occasional sudden change or mutation, which could result in the appearance of a new unit or allele differing functionally from the original one but self-replicating in its new form. It was shown how such units are arranged in linear order in the chromosomes, each gene having its own characteristic position or locus; how they are transmitted to an individual from his parents via the ovum and sperm, so that they are usually present in pairs, one member of a pair being derived from one parent and one from the other; and how because of mutational changes in previous generations multiple allelic forms of a gene can occupy a particular gene locus, so that individual members of a natural population may differ from one another in their characteristics according to the specific nature of the alleles that they happened to have received from their parents.

Four major advances made it possible to begin to understand the nature of

1

genetical diversity in molecular terms. The first was the discovery that the particular chemical substance which endows a gene with its characteristic properties is deoxyribosenucleic acid or DNA. The second was the elucidation of the molecular structure of this substance. The third was the recognition that the primary biochemical role of DNA in the cells of an organism is to direct the synthesis of enzymes and other proteins. The fourth was the unravelling of the genetic code. That is the details of the relationships between the structure of nucleic acids and the structure of proteins.

The main features of the molecular architecture of DNA were first formulated by Watson and Crick in 1953, who at the same time pointed out how the proposed structure would account for the three basic attributes of genetic material; gene specificity, gene replication and gene mutation.

The molecule is made up of two very long polynucleotide chains coiled round a common axis to form a double helix. The backbone of each chain consists of a regular alternation of phosphate and sugar (deoxyribose) groups. To each sugar group and projecting inwards from the chain is attached a nitrogenous base. This may be one of four different types; adenine or guanine which are purines, thymine or cytosine which are pyrimidines. The two chains are held together by hydrogen bonding between pairs of bases projecting at the same level from each chain, so that the whole structure may be likened to a spiral staircase, the pairs of bases representing the steps. There are certain restrictions on which bases can constitute a pair. In any one pair one base must be a purine and the other a pyrimidine, and of the possible combinations only two can occur; adenine with thymine, and guanine with cytosine. A given pair may be either way round. Thymine for example can occur in either chain, but when it does its partner on the other chain must be adenine.

A gene can be regarded as being represented by a length of DNA containing several hundred or thousand base pairs. While the phosphate–sugar backbones of the two chains which form the double helix are quite regular, the base pairs may occur in any sequence. A great many different permutations are therefore possible, and so each gene can have its own unique structure, from which is derived its functional specificity. The precise sequence of base pairs in a particular gene carries as it were in coded form a specific piece of genetical information.

Since the nature of one base fixes the nature of the other member of the pair, the two polynucleotide chains which make up the molecule, though qualitatively different, are exactly complementary. The sequence of bases in one chain fixes the sequence of bases in the other. Replication can occur by the unwinding and separation of the chains and the reformation on each chain of its appropriate companion from an available pool of nucleotides. Each chain may thus act as a template for the formation of the other, so that from one molecule two precise replicates are produced each with exactly the same sequence of base pairs as the original.

A gene mutation can be envisaged as the consequence of some kind of event which results in an alteration of the base pair sequence of the particular gene. Many and perhaps most mutations probably represent no more than the change of one base for another at some point in the sequence. Others however involve more drastic changes such as the duplication or deletion of part of the sequence or some other kind of rearrangement. In general the new gene structure once formed will then be conserved in subsequent cell divisions by the ordinary process of DNA replication.

A great variety of different enzymes and other proteins are synthesised in the cells of a single organism. They each have their own distinctive properties and functions and together they define and control the complex pattern of metabolic and developmental processes which characterise the species and the individual. Proteins are composed of one or more polypeptide chains which are made up of long strings of aminoacids linked by peptide bonds in a specific linear order. Twenty different aminoacids may be present and typical polypeptide chains have sequences 100–500 aminoacids long, so as with DNA the number of possible structures is enormous. Furthermore the three-dimensional arrangement and hence the characteristic properties and functional activity of any given protein ultimately depends on the precise sequence of aminoacids in its constituent polypeptide chains.

The fundamental idea relating DNA structure to protein structure is that the sequence of base pairs in a given gene determines the sequence of aminoacids in a corresponding polypeptide chain. So the structures and hence the properties of all the enzymes and proteins an individual can make are thought to be defined by the base pair sequences of his genes.

The details of the genetic code – that is the relationship of base sequence to aminoacid sequence – have largely been worked out by experimental studies on microorganisms. But there is little doubt that in their main features they also apply in higher organisms including man. Each aminoacid is specified by a sequence of three bases. The base triplets occur consecutively and do not overlap. That is to say a triplet specifying one aminoacid is immediately followed by a separate triplet specifying the next aminoacid and so on. Thus the two sequences are colinear. The four characteristic bases of a DNA chain can occur in 64 different triplet sequences, and 61 of these triplets each specify one of the twenty aminoacids, so that a particular aminoacid may be coded by two or more different base triplets (see fig. 1.5, p. 14). There are also three so-called 'nonsense' triplets which do not specify aminoacids but probably designate chain termination.

The series of processes by which the sequence of bases in the DNA of a gene is translated into the corresponding sequence of aminoacids in a polypeptide chain are complex and involve as intermediaries certain types of ribosenucleic acid (RNA) molecules. The first step involves the separation of the two polynucleotide chains of the DNA, so that one of them may serve as a template for the synthesis from available ribonucleotides of a complementary RNA chain. In this process the same base pairing rules as in DNA apply, except that uracil, which occurs in RNA instead of thymine, pairs with adenine. Thus a strand of RNA carrying the same genetic information as the DNA strand is formed, but it is coded in a complementary base sequence. This RNA strand, known as messenger or mRNA, then separates from the DNA and passes out of the cell nucleus to the ribosomes in the cytoplasm, which are the site of protein synthesis. The strands of mRNA attached to ribosomes then serve as templates for the formation of the polypeptide chains. Aminoacids come to the mRNA template attached to another species of RNA molecule, known as transfer RNA or tRNA. The tRNA molecules are relatively small (about 80 nucleotides) and occur as a series of distinct molecular types. Each is specific for a particular aminoacid which can be attached to one end of the tRNA molecule, and each contains within its polynucleotide sequence a characteristic base triplet complementary to a base triplet in mRNA which codes for that aminoacid. The attached aminoacid can thus be placed in the correct position for the synthesis of the polypeptide chain defined by the coded sequence of the mRNA.

The polypeptide chain is made sequentially, one aminoacid being added at a time, starting from the amino-terminal end. A key feature of the processes leading to polypeptide synthesis is that each aminoacid in the sequence is designated by a trinucleotide or base triplet in three types of molecule: DNA, mRNA and tRNA. The triplet in mRNA is complementary to that in DNA and also to that in tRNA, so that although the actual bases differ the same aminoacid is specified.

Thus the inherited information coded in the genes can be regarded as a kind of blueprint which defines the structures of all the enzymes and other proteins which an individual makes. But genes not only determine the structures of proteins, they are also apparently concerned in regulating their synthesis. The nature of the molecular relationships involved in these regulatory functions are still however very obscure, and so far no satisfactory general theory, at least for multicellular organisms, has as yet been developed (see ch. 4, p. 126). Nevertheless, one can say that the enzyme and protein makeup of each individual must in a very direct sense be a reflexion of his genetical constitution. Furthermore, one may anticipate that inherited differences between individual members of the species, whether they are expressed as differences in normal physical, physiological or mental characteristics, or as differences in the development of particular abnormalities, are likely to be a consequence of differences in enzyme or protein synthesis.

1.2. The haemoglobin variants

The first direct evidence that a gene mutation can result in the synthesis of an altered protein came from work on haemoglobin in the condition called sickle-cell disease or sickle-cell anaemia.

It has been known for many years that the red blood cells of certain individuals have the peculiar property of undergoing reversible alterations in shape when subjected to changes in the partial pressure of oxygen (Herrick 1910, Hahn and Gillespie 1927). When oxygenated, the cells are biconcave discs like the red cells of normal individuals. However, when they are deoxygenated, they become elongated, filamentous and sickle-shaped. Most people whose red cells show the so-called 'sickling phenomenon' are quite healthy. They are said to have the sickle-cell trait. In some individuals, however, 'sickling' is associated with a severe and characteristic form of anaemia, which is commonly fatal in childhood or adolescence. This condition is usually called sickle-cell disease. The sickling phenomenon is common in Central Africa where in many areas 20% or more of the population may have the sickle-cell trait, and a significant fraction of individuals (1–2%) may be expected to die of sickle-cell anaemia in early life. It is also found not infrequently among Negro populations living in other parts of the world, such as the U.S.A. But it is rare or absent in most other human populations.

The peculiarity is an inherited characteristic, and in 1949 Neel and Beet working independently showed that the pedigrees could be most simply explained by the hypothesis that individuals with the sickle-cell trait and those with sickle-cell anaemia are respectively heterozygous and homozygous for a particular abnormal gene located on one of the autosomal chromosomes. Thus individuals who receive the abnormal gene from one parent but its normal allele from the other would be expected to have the sickle-cell trait, while those who receive the abnormal gene from each parent would develop sickle-cell anaemia. On this hypothesis one expects that in matings between a normal individual and one with the sickle-cell trait, half the children should on the average be normal and half should have the sickle-cell trait. From matings between two individuals with the sickle-cell trait, there should be normal children, children with the sickle-cell trait, and children with sickle-cell anaemia, and these should occur on average in the Mendelian ratio of 1:2:1. Furthermore both parents of a patient with sickle-cell anaemia should exhibit the sickling phenomenon. The family data (Neel 1951) were shown to be in good agreement with these and other expectations, and sub-sequent studies fully confirmed the hypothesis.

At about the same time Pauling and his colleagues (1949) made the crucial discovery that the haemoglobin present in the red cells in sickle-cell anaemia is qualitatively different from the haemoglobin present in normal red cells. They showed that the two proteins differed in their physical properties and presumably therefore in their structures, since they could be separated by electrophoresis. Furthermore they demonstrated that the red cells from individuals with the sickle-cell trait contained both normal haemoglobin (Hb A) and also the abnormal sickle-cell haemoglobin (Hb S). Thus there appeared to be a direct correspondence between the genetic constitution of an individual and the haemoglobins that were synthesised. Individuals homozygous for the sickle-cell gene formed Hb S, individuals homozygous for the normal allele formed Hb A, and heterozygous individuals who had inherited the sickle-cell gene from one parent and its normal allele from the other formed both types of haemoglobin.

Shortly afterwards another abnormal haemoglobin, called Hb C, was discovered (Itano and Neel 1950). Like Hb S this also appeared from family studies to be determined by a single gene. Heterozygotes for this gene and its normal allele form both Hb A and Hb C in their red cells. They are quite healthy and are said to have the haemoglobin C trait. Homozygotes for the gene form Hb C but no Hb A and may show a moderate or mild degree of anaemia. This condition, known as haemoglobin C disease, is clinically much less severe than sickle-cell disease. A third condition of intermediate severity also occurs. This is known as sickle-cell–haemoglobin C disease, and affected individuals have both Hb S and Hb C in their red cells. Family studies indi-cate that they are heterozygous for both the gene determining Hb C and also for the sickle-cell gene, having inherited one of these abnormal genes from one parent and one from the other. These various conditions can be readily distinguished from each other by electrophoresis. Typical separations are illustrated in fig. 1.1.

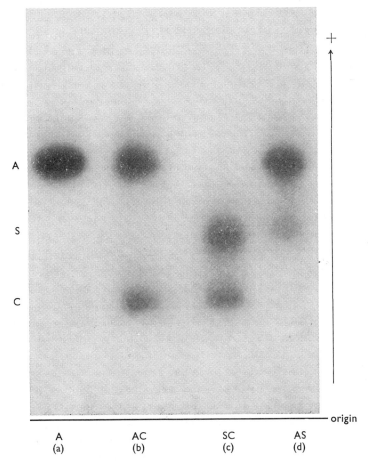

Fig. 1.1. Electrophoresis of haemoglobins in (a) normal adult, (b) haemoglobin C trait, (c) sickle-cell–haemoglobin C disease, and (d) sickle-cell trait. (Electrophoresis carried out at *p*H 8.6 in starch gel.)

In sickle-cell–haemoglobin C disease no Hb A is found. This suggests that the genes determining Hb S and Hb C are allelic, and that when both occur the normal allele presumed to be necessary for the formation of Hb A cannot be present. Studies of families in which one parent had sickle-cell–haemoglobin C disease while the other was normal proved to be consistent with this

idea. The offspring were found to include individuals with either the haemoglobin C trait (AC) or the sickle-cell trait (AS), but not individuals with
sickle-cell–haemoglobin C disease (SC) or with normal haemoglobin (A).
As will be seen later the conclusion that the genes determining Hb S and
Hb C are in fact allelic is also supported by a quite different line of evidence.

The discovery of these abnormalities initiated a search for other variant
forms of haemoglobin, and in the following years a wide variety of different
forms of anaemia were investigated. Also extensive electrophoretic surveys
of haemoglobins obtained from randomly selected individuals in many
populations were carried out. As a result of these studies more than a
hundred and thirty different genetically determined variants of human
haemoglobin have now been identified (Lehmann and Carrell 1969, Stamatoyannopoulos 1972). The majority of these are very rare and have only
been seen in the heterozygous state, occurring together with normal haemoglobin. However there are some which, like sickle-cell haemoglobin, have
quite an appreciable incidence in certain parts of the world. Hb E, for
example, is found to occur with relatively high frequencies in a number of
populations in South East Asia, Hb C is common in West Africa, and Hb D
Punjab is found with an appreciable frequency in North West India.

Many individuals who are heterozygous for one of the genes determining
a variant haemoglobin and for its normal allele are quite healthy. But there
are important exceptions in which the heterozygous state is found to be
associated with a characteristic form of chronic haemolytic disease or other
abnormality (see pp. 20–28).

It has only been possible to study the homozygous state in a relatively
small number of these different mutant genes that determine variant haemoglobins. In some cases like sickle-cell disease chronic haemolytic anaemia is
a regular feature. But there are evidently also other cases where the homozygote for a gene determining a particular haemoglobin variant may be quite
healthy (e.g. Hb G Accra, Edington and Lehmann 1954).

Chronic haemolytic disease can also be a characteristic feature of certain
heterozygous states like sickle-cell–haemoglobin C disease where two different abnormal alleles are present. It is also often seen in individuals who are
heterozygous for one of the genes determining a particular variant haemoglobin, and also heterozygous for one of the genes to be discussed later
(ch. 4) which result in a specific defect in haemoglobin synthesis (the so-called
thalassaemia genes). Thus a variety of different haematological abnormalities
which had previously not been clearly differentiated from one another can
now be defined in terms of specific abnormalities in haemoglobin formation.

1.3. The structure of the variant haemoglobins

With the discovery of the many genetically determined variants of haemoglobin came the problem as to the precise way these proteins differ in molecular structure from normal haemoglobin.

A protein consists of one or more polypeptide chains each of which is composed of aminoacids linked by peptide bonds and arranged in a definite sequence. Where there are two or more polypeptide chains in the structure of a particular protein molecule, these may be identical in sequence or non-identical. Haemoglobin A has four polypeptide chains. They are of two different types, each with a characteristic aminoacid sequence and each represented twice in the molecule. These different polypeptides are called the α- and the β-chains, so the haemoglobin A molecule may be said to have the structure $\alpha_2\beta_2$. The α-chain contains 141 amino acid residues and the β-chain 146 amino acid residues, and their precise sequences (Braunitzer et al. 1964) have been established (see fig. 2.8, p. 44).

Each of the polypeptide chains in a protein is coiled and folded in a characteristic manner so that the whole molecule has a complex three-dimensional arrangement. The spatial configuration taken up by each of the polypeptide chains and the consequent overall three-dimensional arrangement of the protein molecule is thought to be determined primarily by the sequence of aminoacids in the constituent polypeptide chains, the so-called primary structure. In the case of haemoglobin, the detailed manner in which the four chains are coiled and folded and the way they fit together to form the globular protein molecule has been worked out by Perutz and his colleagues (1968) by X-ray crystallographic analysis.

Many proteins have attached to them some additional and usually relatively small groupings not made up of aminoacids. These are called prosthetic groups and they may have a very specific role in the functioning of the molecule as a whole. The prosthetic group in haemoglobin is haem. This is a porphyrin ring structure with an iron atom at its centre. There are four haem groups in the haemoglobin molecule, one being attached to each of the polypeptide chains by a linkage between the iron atom and a particular histidine residue in each of the chains. In the three-dimensional model of haeomoglobin the haem groups are seen to lie in four separate pockets formed by folds in the corresponding polypeptide chains. When oxygen combines with haemoglobin it attaches to the iron atoms in the haems, and this is associated with subtle changes in the three-dimensional conformation of the whole molecule.

The fact that Hb S differed in its electrophoretic properties from Hb A implied that the two molecules differed in structure, and it was soon shown that the difference did not lie in the haem groups but was present in the protein proper. The nature of the difference was elucidated by Ingram (1957, 1959) who found that a particular position in the aminoacid sequence of the β-polypeptide chain which is occupied by a glutamic acid residue in Hb A, is occupied by a valine residue in Hb S.

The method originally used to demonstrate this difference in the proteins is now generally referred to as 'finger printing' and it has been widely applied in the study of the primary structures of many different proteins and their genetically determined variants. The protein is first digested with a specific proteolytic enzyme such as trypsin or chymotrypsin, so that the polypeptide chains are split at a number of separate points to give a mixture of different smaller peptides. These peptides are then separated two-dimensionally on filter paper, first by electrophoresis and then by chromatography. The pattern of peptide 'spots' so produced is in general characteristic for the particular protein. In the case of haemoglobin, Ingram found that after digestion with trypsin which splits lysyl or arginyl peptide bonds, a complex pattern of peptides was obtained (fig. 1.2). The peptide pattern obtained

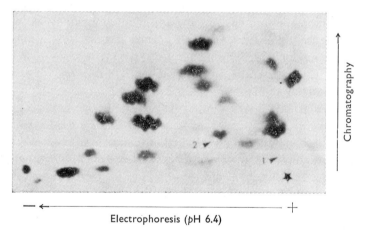

Electrophoresis (pH 6.4)

Fig. 1.2. 'Fingerprint' of Hb S. The peptides obtained by digestion of the haemoglobin have been separated on filter paper first by electrophoresis (horizontal dimension) and then by chromatography (vertical dimension). The peptide 'spots' were then developed with ninhydrin. *Key:* ★: point of application of peptide mixture to sheet of filter paper. **1**: position of peptide present in Hb A but not in Hb S. **2**: peptide present in Hb S but not in Hb A. (From Lehmann and Huntsman 1966.)

from Hb A appeared in most respects to be identical with that from Hb S. However there was one peptide present in Hb A which was not present in the Hb S pattern, and one peptide present in Hb S which was not seen in Hb A. These two peptides were isolated and the sequences of aminoacids in them determined. They each contained eight aminoacid residues, and their sequences were the same except that in the Hb A peptide the sixth residue from the amino end of the chain was glutamic acid, while in Hb S it was valine (fig. 1.3). Subsequently it was shown that all the other peptides

```
        1   2    3    4    5    6    7    8
HbA    Val. His. Leu. Thr. Pro. Glu. Glu. Lys.....
                                  ↑
HbS    Val. His. Leu. Thr. Pro. Val. Glu. Lys.....
                                  ↑
```

Fig. 1.3. Aminoacid sequences of the peptide present in tryptic digests of Hb A but not Hb S, and of the peptide present in tryptic digests of Hb S but not Hb A. The sequences represent the first eight aminoacids of the β-polypeptide chains of the two haemoglobins. They differ only at position 6. The arrows indicate the site of cleavage by trypsin. (For aminoacid abbreviations see caption fig. 1.5, p. 14.)

obtained from the two haemoglobins were the same as one another. It was also found that the peptides in the Hb A and Hb S which differed, represented the sequences at the amino-terminal end of the β-chains. Thus it could be concluded that the two haemoglobins differed only in a single aminoacid residue at the sixth position from the amino-terminal end of the β-chain, and that the sequence of the rest of the β-chain and the sequence of the whole of the α-chain was identical in the two molecules. Conventionally the aminoacid residues in a polypeptide chain are numbered from the amino-terminal end. If then the structure of Hb A is written as $\alpha_2\beta_2$, the structure of Hb S may be written as $\alpha_2\beta_2^{6\ Glu\rightarrow Val}$.

Hb S was the first example of a genetically determined protein variant in which the structural peculiarity was precisely identified. The result obtained was remarkable for its simplicity. A single gene difference such as that between the sickle-cell gene and its normal allele could be presumed to be the result of a single mutational step. This is the smallest unit of inherited variation. It could evidently lead to the smallest unit difference in the primary structure of a specific protein, namely the substitution of a single aminoacid residue by another. Subsequent work has shown that this is a general rule. Stamatoyannopoulos (1972) lists more than one hundred and twenty other variant forms of haemoglobin which have now been shown to differ from the normal by a single aminoacid substitution, occurring at one position or another in either the α- or the β-chains (fig. 1.4 illustrates some

β		β		β	
1 Val:		51 Pro:		101 Glu:	
2 His:	Tyr	52 Asp:	Asn	102 Asn:	Thr^b, Lys
3 Leu:		53 Ala:		103 Phe:	
4 Thr:		54 Val:		104 Arg:	
5 Pro:		55 Met:		105 Leu:	
6 Glu:	Val, Lys, Ala	56 Gly:	Asp	106 Leu:	
7 Glu:	Gly, Lys	57 Asn:		107 Gly:	
8 Lys:		58 Pro:	Arg	108 Asn:	Asp^b
9 Ser:	Cys	59 Lys:	Glu, Thr	109 Val:	
10 Ala:		60 Val:		110 Leu:	
11 Val:		61 Lys:	Asn, Glu	111 Val:	Phe^b
12 Thr:		62 Ala:		112 Cys:	
13 Ala:		63 His:	Tyr^c, Arg^{a,b}	113 Val:	Glu
14 Leu:	Arg	64 Gly:		114 Leu:	
15 Try:		65 Lys:		115 Ala:	
16 Gly:	Asp, Arg	66 Lys:	Glu^a	116 His:	
17 Lys:	Glu	67 Val:	Glu^c, Asp^{a,b}, Ala^a	117 His:	Arg
18 Val:		68 Leu:		118 Phe:	
19 Asn:		69 Gly:	Asp	119 Gly:	
20 Val:	Met^b	70 Ala:	Asp^{a,b}	120 Lys:	Glu
21 Asp:		71 Phe:	Ser^a	121 Glu:	Lys, Gln
22 Glu:	Lys, Ala, Gly	72 Ser:		122 Phe:	
23 Val:		73 Asp:	Asn	123 Thr:	
24 Gly:	Arg^a, Val^a	74 Gly:	Asp^{a,b}	124 Pro:	Arg^a
25 Gly:	Arg$_b$	75 Leu:		125 Pro:	
26 Glu:	Lys$_b$	76 Ala:		126 Val:	Glu
27 Ala:		77 His:	Asp	127 Gln:	
28 Leu:	Pro^a	78 Leu:		128 Ala:	
29 Gly:		79 Asp:	Asn, Gly	129 Ala:	Asp
30 Arg:	Ser^a	80 Asn:	Lys	130 Tyr:	Asp^a
31 Leu:		81 Leu:		131 Gln:	
32 Leu:		82 Lys:		132 Lys:	Gln
33 Val:		83 Gly:	Cys	133 Val:	
34 Val:		84 Thr:		134 Val:	
35 Tyr:	Phe^a	85 Phe:		135 Ala:	
36 Pro:		86 Ala:		136 Gly:	Asp
37 Try:	Ser^b	87 Thr:	Lys	137 Val:	
38 Thr:		88 Leu:	Pro, Arg^a	138 Ala:	
39 Gln:		89 Ser:		139 Asn:	
40 Arg:		90 Glu:	Lys	140 Ala:	
41 Phe:		91 Leu:	Pro^a	141 Leu:	Arg^a
42 Phe:	Ser^{a,b}, Leu^{a,b}	92 His:	Tyr^c	142 Ala:	
43 Glu:	Ala	93 Cys:		143 His:	
44 Ser:		94 Asp:	Asn	144 Lys:	
45 Phe:		95 Lys:	Asp, Glu	145 Tyr:	His^b, Cys^b
46 Gly:	Glu	96 Leu:		146 His:	Asp
47 Asp:	Asn	97 His:	Gln^b		
48 Leu:		98 Val:	Met^{a,b}		
49 Ser:		99 Asp:	His^b, Asn^b, Tyr^b		
50 Thr:		100 Pro:			

Fig. 1.4. Single aminoacid substitutions in 84 β-chain haemoglobin variants. Based on the tabulations given by McKusick (1971) and Stamatoyannopoulos (1972). The full numbered aminoacid sequence of the normal haemoglobin β-chain is given on the left of each column, and the various substitutions in the different positions are indicated. The aminoacid abbreviations are as listed in caption to fig. 1.5, p. 14. Key: *a* unstable haemoglobin (p. 24); *b* altered oxygen affinity (p. 26); *c* methaemoglobinaemia (p. 20).

of the β-chain substitutions). The same has also been found to be the case for a variety of other altered proteins attributable to single gene mutations both in man and in other species.

1.4. Single aminoacid substitutions and the genetic code

The 64 base triplets which may occur in mRNA and which of course correspond to 64 complementary triplets in DNA, are shown in fig. 1.5, together with the various aminoacids which they are believed to specify in polypeptide synthesis. This is the so-called genetic code, and it is probably much the same for all species. Many aminoacids are designated by two different triplets and some by four or six. There are also three so-called 'nonsense' triplets whose position in the polynucleotide chain is thought to specify the termination of polypeptide chain synthesis.

It is instructive to examine the single aminoacid substitutions of the different haemoglobin variants in terms of these base triplets or codons. The mutational change from Hb A to Hb C for example has caused the replacement of a glutamic acid residue in the sixth position of the β-chain by a lysine residue. From fig. 1.5 one finds that the nucleotide triplet which codes for glutamic acid is either GAA or GAG, and the nucleotide triplet that codes for lysine is either AAA or AAG. So the mutation need have involved the change of only a single base in the triplet (i.e. GAA→AAA or GAG→AAG). All the other single aminoacid substitutions found in the abnormal haemoglobins may be considered in the same way. It turns out that in each case a single mutation need have involved no more than a single base change in a particular triplet. A similar conclusion is reached when mutations resulting in single aminoacid substitutions in other proteins and in other quite different species are considered. Taken as a whole the data which are now very extensive provide strong support for the view that most gene mutations represent simply a change of a single base in the whole base sequence of the DNA of a particular gene.

The normal β-polypeptide chain of human haemoglobin which contains 146 aminoacid residues, is presumably represented in the corresponding gene by a specific sequence of 438 bases (146 × 3). The mutation from the normal gene to the gene determining Hb C results in a single aminoacid substitution in the sixth position in the polypeptide sequence. Presumably therefore it occurred in the sixth base triplet in the corresponding stretch of DNA, and from the code we may infer that the mutation actually involved a change in the sixteenth base of the whole sequence. Similarly we can infer that the mutation from the normal gene to the sickle-cell gene involved the adjacent base in the same triplet (GAA→GUA or GAG→GUG). This would represent the seventeenth base in the full sequence. Similar precise derivations may be made for most of the other variant haemoglobins where the nature of the aminoacid substitution has been determined.

It is convenient (Crick 1967) to display the various single aminoacid substitutions detected in mutants of a given protein by an arrow on the diagram of the genetic code as shown for the haemoglobin variants in fig. 1.6. If the substitution is due to single base change in the first position of the triplet, the arrow marking the change will be vertical and will begin and end in the same relative position within a square on the figure. If the change is in

Second base

		U	C	A	G	
First base	U	UUU }Phe UUC } UUA }Leu UUG }	UCU } UCC }Ser UCA } UCG }	UAU }Tyr UAC } UAA term. UAG term.	UGU }Cys UGC } UGA term. UGG Try	U C A G
	C	CUU } CUC }Leu CUA } CUG }	CCU } CCC }Pro CCA } CCG }	CAU }His CAC } CAA }Gln CAG }	CGU } CGC }Arg CGA } CGG }	U C A G
	A	AUU } AUC }Ile AUA } AUG Met	ACU } ACC }Thr ACA } ACG }	AAU }Asn AAC } AAA }Lys AAG }	AGU }Ser AGC } AGA }Arg AGG }	U C A G
	G	GUU } GUC }Val GUA } GUG }	GCU } GCC }Ala GCA } GCG }	GAU }Asp GAC } GAA }Glu GAG }	GGU } GGC }Gly GGA } GGG }	U C A G

Third base

Fig. 1.5. The genetic code. [For detailed references to derivation of the code see Cold Spring Harbor Symp. Quant. Biol. *31* (1966), Crick (1967), Woese (1967) and Yčas (1969).] *Bases:* U: uracil, C: cytosine, A: adenine, G: guanine.

Aminoacids:

Ala:	alanine	Gly:	glycine	Pro:	proline
Arg:	arginine	His:	histidine	Ser:	serine
Asn:	asparagine	Ile:	isoleucine	Thr:	threonine
Asp:	aspartic acid	Leu:	leucine	Try:	tryptophan
Cys:	cysteine	Lys:	lysine	Tyr:	tyrosine
Gln:	glutamine	Met:	methionine	Val:	valine
Glu:	glutamic acid	Phe:	phenylalanine		

term. = 'nonsense' triplet – chain termination.

the second base, the arrow will be horizontal. If the change is in the third base, the arrow will be vertical but will begin and end in the same square. Thus if an aminoacid substitution requires at a minimum only a single base change in a coding triplet then the arrow will be either horizontal or vertical, and this is seen to be so in all cases. A diagonal arrow would imply that at least two bases had been changed, and this would have been expected in a considerable fraction of the cases if the changes in aminoacids were quite arbitrary and at random.

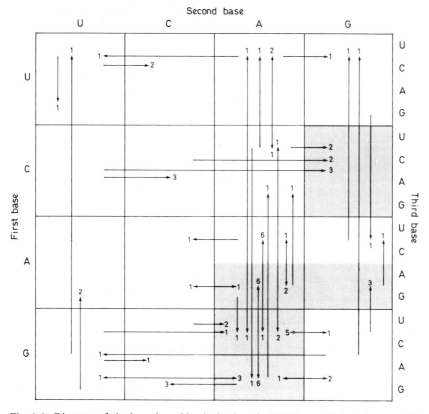

Fig. 1.6. Diagram of single aminoacid substitutions in 84 different β-chain haemoglobin variants (after Crick 1967). The code is as arranged in fig. 1.5. Each arrow represents a single aminoacid substitution, and the number attached to an arrow gives the number of different variants with that particular aminoacid substitution. The shaded areas of the diagram represent charged aminoacids (i.e. the basic aminoacids lysine and arginine *or* the acidic aminoacids aspartic acid and glutamic acid). The aminoacid substitutions shown are those given in fig. 1.4, p. 12.

It will be seen that the majority of the arrows are localised to particular areas of the diagram. This arises because of the method by which most of the variant haemoglobins were discovered. They were mainly detected because they showed an electrophoretic difference from normal haemoglobin A, and consequently the aminoacid change was likely to involve a change in charge. The areas of the diagram occupied by charged aminoacids are shaded, while the remainder is occupied by neutral aminoacids. As expected from the mode of discovery a large proportion of the arrows start or end in a shaded region (61 out of 84). But the widespread use of electrophoresis in the search for variants of haemoglobin has largely been a matter of technical convenience. One may therefore anticipate that there exist many other variants of haemoglobin not involving a change in charge, and therefore remaining undetected. The arrows representing such mutations would presumably occur in the unfilled areas of the diagram shown in fig. 1.6. No doubt these 'concealed' mutations will be progressively uncovered as other techniques are more widely applied in the search for variants.

Furthermore, not all mutations involving a single base change in a gene determining a particular polypeptide chain will result in an altered or variant protein. This is because the code is degenerate and the majority of aminoacids are represented by two different nucleotide triplets, and some by four or six. It follows that many single base changes can occur which will result in a new triplet which codes for the same aminoacid as the original triplet. Thus the primary structure of the protein concerned will not be altered. It is possible, however, as will be discussed later, that such mutations may on occasion result in changes in the rate of synthesis of the protein.

One may also note that a further class of mutations having quite a different effect on protein structure may be expected. These would involve base changes which alter a triplet coding for a particular aminoacid to one which codes for chain termination. This would result in the synthesis of an abbreviated polypeptide which would lack the sequence of aminoacids from the point affected by the mutation to the carboxyl terminal end of the normal polypeptide. Such a shortened polypeptide is in most cases unlikely to assume a stable three-dimensional conformation and would in general be expected to differ profoundly in properties from its normal counterpart. Such mutations, although they have been demonstrated under rather special circumstances in certain microorganisms, are likely to be very difficult to identify with certainty in man and may perhaps in most instances be manifest simply as a protein or enzyme deficiency.

Clearly a very considerable number of different single base alterations to the DNA sequence of a gene coding for a particular polypeptide chain could arise as a result of separate mutations. Their effects on the structure of the polypeptide will vary according to the position in the sequence at which they occur and the specific nature of the base alteration. A general picture of the relative frequencies of the different kinds of effects that may ensue can be obtained by considering the sequence of 438 bases which code for 146 aminoacids in the β-polypeptide chain of haemoglobin (see p. 279). Of all the possible single base changes that could occur, about 23% would cause no alteration in the aminoacid sequence, about 4% would result in shortening of the polypeptide by premature chain termination, and about 73% would be expected to result in a single aminoacid substitution in the polypeptide. Of those that result in single aminoacid substitutions, only about one-third would be expected to cause an alteration in charge which might allow the variant protein to be detected electrophoretically.

1.5. The effects of single aminoacid substitutions

We have seen that in any particular protein separate mutations can result in a considerable number of different single aminoacid substitutions. Some of these may cause little or no change in the structural integrity of the protein molecule or in its other properties, so that the effects of the mutation on function may be very slight and perhaps undetectable. In other cases however the protein may be altered in such a way as to give rise to a variety of secondary pathological and clinical consequences.

The specific change in the properties of a protein produced by a particular aminoacid substitution will of course depend on the chemical nature of the aminoacid substituted. Whether for example the substitution involves a change from a hydrophobic to a hydrophilic residue, a change in ionisation, or a marked change in the physical dimensions of the side chain. Furthermore it will depend on the particular aminoacid site in the three-dimensional structure of the protein which is substituted. In the case of haemoglobin it appears that substitutions occurring at sites on the external surface of the molecule are in general less likely to result in significant effects than changes elsewhere. The most critical sites appear to be those which involve areas of contact between the different polypeptide chains in the tetramers or between the polypeptide chains and the haem groups, and also other internal sites which happen in one way or another to be critical for maintaining the conformational integrity of the molecule (Perutz and Lehmann 1968).

An alteration in protein structure of this sort is probably the specific underlying biochemical cause of many different forms of inherited disease. The characteristic features of such a disease will be determined by the nature and function of the protein involved, and the manner in which its properties have been changed. In such conditions therefore it should eventually be possible to trace in detail the precise way in which the alteration in the primary structure of the protein leads to functional changes in the molecule, and how these in turn give rise to the pathological disturbances that are manifest (also see Ch. 9 p. 340). So far however a satisfactory insight into this causal sequence has been achieved in relatively few conditions. However the general nature of the problem is well illustrated by the various and in many cases quite strikingly different disorders due to structural abnormalities in haemoglobin. They serve as models for the kind of analysis of inherited diseases which should become increasingly possible as the structures of other proteins (particularly enzyme proteins) and of their variants are elucidated.

1.5.1. Sickle-cell disease

Hb S was first distinguished from Hb A because of the difference in electrophoretic mobility. Shortly afterwards however another difference but in a quite different property was discovered. It was found (Perutz and Mitchison 1950) that deoxygenated sickle-cell haemoglobin is very much less soluble than deoxygenated normal haemoglobin, whereas in the oxygenated state the two haemoglobins are equally soluble (fig. 1.7). This finding immediately suggested a simple explanation for the sickling phenomenon which occurs when red cells containing Hb S are subjected to low oxygen tension. When deoxygenated, Hb S which is present in high concentrations in the red cell would tend to come out of free solution and cause a characteristic deformation of the cell shape. It can indeed be shown that concentrated sickle-cell haemoglobin solutions free of red cell stroma become increasingly viscous as the oxygen tension is reduced and eventually assume a semi-solid gel-like state, in which spindle-shaped birefringent bodies, $1-15\ \mu$ in length, can be seen microscopically (Harris 1950). These are remarkably similar in shape to the elongated sickle-like forms that intact red cells containing Hb S take up at low oxygen tension.

The aminoacid substitution in sickle-cell haemoglobin involves a replacement of a glutamic acid in haemoglobin A by a valine, i.e. the substitution of a polar by a non-polar residue. The site at which this occurs, the sixth position in the β-polypeptide chain, has been shown to lie at the surface of

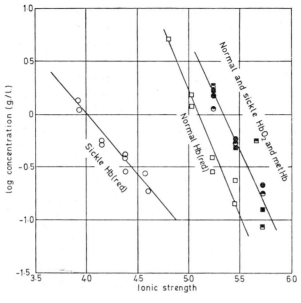

Key

○ sickle haemoglobin (reduced)	□ normal haemoglobin (reduced)
● sickle haemoglobin (oxygenated)	■ normal haemoglobin (oxygenated)
◕ sickle methaemoglobin	▫ normal methaemoglobin

Fig. 1.7. Solubilities (g/l) of normal and sickle-cell haemoglobins plotted against ionic strength of buffer solution. (From Perutz and Mitchison 1950.)

the molecule, and Perutz and Lehmann (1968) have suggested that in the deoxygenated state the non-polar valine residue in this position is able to adhere to a complementary site on a neighbouring haemoglobin molecule. The substituted valine will of course be represented twice in the sickle-cell haemoglobin molecule, once on each of the two symmetrically arranged β-chains. Similarly the complementary site will be represented twice. So there will be a tendency in concentrated solutions of sickle-cell haemoglobin for the formation of long linear aggregates of haemoglobin molecules, which would explain the low solubility. Linear aggregates of this sort in deoxygenated haemoglobin S have indeed been observed under the electron microscope (Stetson 1966, Edelstein et al. 1973, Finch et al. 1973). To account for the marked difference in solubility between oxygenated and deoxygenated sickle-cell haemoglobin it is suggested (Perutz and Lehmann 1968) that the postulated complementary site on the surface of the molecule to which the valine residue adheres, is created by the conformational change in the tetrameric molecules which has been shown to occur when oxyhaemoglobin is deoxygenated. The nature of the complementary site is still unknown, but it is probably present in both deoxygenated Hb A as well as in Hb S.

When the intracellular concentration of Hb S is high, as is the case in sickle-cell homozygotes, sickling of red cells will tend to occur *in vivo* on the venous side of the circulation where the oxygen tension is reduced. This results in increased viscosity of the blood and this is particularly likely to impede the circulation in the smaller veins, and the venous side of the capillaries. This will tend to cause further deoxygenation and increased sickling and so a vicious circle is likely to be set up accentuating the effect. The sickled cells may also block the smaller blood vessels by forming thrombi, and these probably lead to the multiple scattered foci of tissue destruction which are a characteristic feature of the disease and which can result in a wide variety of symptoms. The deformed red cells also tend to be broken down at a much increased rate. Thus one can see at least in a general way, how both the haemolytic anaemia and the other diverse pathological manifestations which appear in the disease may be traced back to the altera- tion in the solubility of the haemoglobin produced by the single specific aminoacid substitution.

It is of interest to contrast sickle-cell anaemia with haemoglobin C disease. Sickle-cell anaemia is a severe disorder which is often fatal in childhood or adolescence. Haemoglobin C disease is by comparison benign. The degree of anaemia that occurs is relatively slight and in many cases the affected individ- uals live a normal and active life with no obvious disability. Yet the two conditions are due to the substitution of the same aminoacid residue in the haemoglobin molecule. This is the glutamic acid in the sixth position on the β-chains, which is replaced by valine in Hb S and by lysine in Hb C. But while the valine substitution leads to the sickling phenomenon from which most of the deleterious consequences ensue, the lysine substitution does not. Indeed it is still not clear exactly how the lysine substitution in Hb C causes even the mild degree of haemolytic disease which is observed.

1.5.2. Hereditary methaemoglobinaemias

Another illustration of how the site of an aminoacid substitution is important in determining the properties of an altered protein is provided by the abnor- mal haemoglobins found in the group of conditions known as the hereditary methaemoglobinaemias.

Methaemoglobin is the oxidised derivative of haemoglobin in which the iron of the haem group is changed to the ferric state from the usual ferrous state. It is unable to combine reversibly with oxygen and so fulfil its normal function in oxygen transport. In the normal individual only a very small proportion (less than 0.5%) of the total haemoglobin present in the circulating red cells occurs as methaemoglobin, because although methaemoglobin is constantly being formed by oxidation of haemoglobin, there are powerful enzymic reducing systems present in the red cell which reconvert the haem iron back to the ferrous state. The result is an equilibrium which may be written:

$$\text{Hb (Fe}_4{}^{++})\rightleftarrows\text{Hb (Fe}_3{}^{++}\text{Fe}^{+++})\rightleftarrows\text{Hb (Fe}_2{}^{++}\text{Fe}_2{}^{+++})\rightleftarrows\text{Hb (Fe}^{++}\text{Fe}_3{}^{+++}\rightleftarrows\text{Hb (Fe}_4{}^{+++})$$

Haemoglobin   Methaemoglobin

and which is normally kept well over to the left. A number of rare inherited abnormalities are known however in which a markedly increased proportion of the haem iron is in the ferric state. These disorders are known as hereditary methaemoglobinaemias and affected individuals are characterised by the blue cyanotic appearance which occurs when a significant proportion of the haemoglobin in the circulating red cells is not combined with oxygen.

Some cases of hereditary methaemoglobinaemia are due to a genetically determined defect of an enzyme (NADH diaphorase, see p. 356) which is normally concerned with maintaining haem iron in the ferrous state. Other cases are due to the presence of an abnormal haemoglobin in which the ferric state is unusually stable and so is not readily reduced by the normal enzyme systems. Several different haemoglobin variants are known in which this effect occurs and they each involve single aminoacid substitutions in the polypeptide chains in the immediate neighbourhood of the affected haems.

The four haem groups of haemoglobin lie in separate pockets on the surface of the haemoglobin molecule formed by folds in each of the four polypeptide chains. Each haem is attached to its polypeptide chain by a co-ordinate linkage from the iron to a specific histidine residue (fig. 1.8). This is at position 87 in the α-chain and position 92 in the β-chain. The iron is also linked through an oxygen molecule in oxygenated haemoglobin or a water molecule in reduced haemoglobin to another histidine residue on the opposite side of the fold (position 58 in the α-chain and position 63 in the β-chain).

One of the variant haemoglobins which gives rise to methaemoglobinaemia is known as Hb M Boston, and it has a substititution at position 58 in the α-chain where the normal histidine residue is replaced by a tyrosine residue (Gerald and Efron 1961). This position is spatially very close to the α-chain haem iron. When this haem iron in a given molecule is in the ferrous state

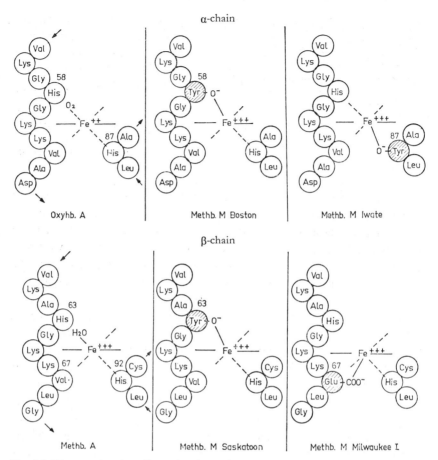

Fig. 1.8. Diagram showing effects of aminoacid substitutions in different types of Hb M. The substitution in each case involves a position close to the haem iron, so that an internal complex can be formed between the iron in the ferric state, and the phenolic side chain of a substituted tyrosine or the negatively charged side chain of glutamic acid (Tönz 1968).

nothing probably happens, but when, as will occur sooner or later, it becomes oxidised to the ferric state, the phenol side chain of the substituted tyrosine residue at position 58 is in just the right position to bond to the ferric iron. This forms a very stable complex which is not readily reduced by the normal methaemoglobin reductase system present in the red cell. Consequently the irons of the haem groups associated with the two α-chains of the molecule remain in the ferric state and are incapable of combining with and therefore transporting oxygen. The haems of the two β-chains are not directly affected, but because of stereochemical changes in the three-dimensional organisation of the haemoglobin molecule, the oxygen affinity of the normal β-subunits is reduced (Pulsinelli et al. 1973).

It is interesting that another haemoglobin variant is known, Hb Norfolk (Ager et al. 1958), in which the aminoacid substitution, a replacement of a glycine residue by an aspartic acid residue, occurs at position 57 of the α-chain (Baglioni 1962b), immediately adjacent to the position affected in Hb M Boston. In this case, however, methaemoglobinaemia does not occur. This is presumably because the carboxyl side chain of the aspartic acid residue, although very close to the haem, is nevertheless not quite in the right orientation to form a stable complex with the ferric iron. This illustrates rather clearly the very precise specificity which is involved in producing functional abnormality.

Other haemoglobin variants which result in methaemoglobinaemia are listed in table 1.1 and illustrated diagrammatically in fig. 1.8. In each case a

TABLE 1.1

Haemoglobin variants which result in methaemoglobinaemia.

Haemoglobin	Aminoacid substitution		Reference
	α-chain	β-chain	
M Boston	58 His → Tyr	—	Gerald and Efron (1961)
M Iwate	87 His → Tyr	—	Miyaji et al. (1963)
M Saskatoon	—	63 His → Tyr	Gerald and Efron (1961)
M Hyde Park	—	92 His → Tyr	Shibata et al. (1967)
M Milwaukee	—	67 Val → Glu	Gerald and Efron (1961)

substitution occurs which is such as to form a stable and not easily reduced complex with the closely adjacent haem iron when it is in the ferric state. Four of the cases represent substitutions of one of the haem-linked histidines by tyrosine.

These different methaemoglobinaemia variants are all rare and have so far only been seen in heterozygotes. In affected individuals both the normal and the abnormal haemoglobin are present in the red cells, and in the abnormal haemoglobin only either the haems of the α-chains or the haems of the β-chains are kept in the ferric state. This is nevertheless sufficient to produce marked methaemoglobinaemia and obvious cyanosis.

One interesting indication of the intramolecular changes that occur in these abnormal methaemoglobins is a characteristic alteration in their absorption spectra (Tönz 1968). The spectral difference involves a shift of the normal absorption maximum at 632 mμ to a lower wavelength (around 600 mμ). The extent of this shift, however, varies in the different abnormal methaemoglobins (fig. 1.9) and presumably depends on the characteristic distortion of the three-dimensional structure resulting from the different aminoacid substitutions.

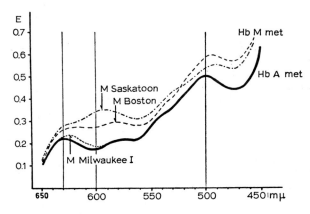

Fig. 1.9. Methaemoglobin spectra of haemolysates from patients with different Hb M types (Tönz 1968).

1.5.3. *Unstable haemoglobins*

One would expect that there will be some aminoacid substitutions which, because of the particular chemical properties and size of the aminoacid involved and the particular site of the substitution in the polypeptide chain, will so alter the three-dimensional conformation of the protein as to render it much less stable than its normal counterpart. Such variants would tend to be denatured more readily than normal haemoglobin, and their half-life *in vivo* greatly reduced. Several haemoglobin variants in which this appears to occur are listed in table 1.2.

Characteristically individuals who are heterozygous for a gene determining one of these so-called 'unstable' haemoglobins show a significant degree of chronic haemolytic anaemia, a situation not usually seen in simple heterozygotes for the genes determining most of the other known haemoglobin variants. The heterozygotes synthesise both normal Hb A and the abnormal

TABLE 1.2

Some unstable haemoglobins. Variants with aminoacid substitution in same position are bracketed. Some variants have been shown to have altered oxygen affinities as indicated: * increased oxygen affinity; ** reduced oxygen affinity.

Haemoglobin	Aminoacid substitution		References
	α-chain	β-chain	
Torino	43 Phe → Val		Beretta et al. (1968), Prato et al. (1970)
L Ferrara	⎰ 47 Asp → Gly		Nagel et al. (1969)
Hasharon	⎱ 47 Asp → His		Halbrecht et al. (1967)
Ann Arbor	80 Leu → Arg		Rucknagel et al. (1971)
Bibba	136 Leu → Pro		Kleihauer et al. (1968)
Riverdale-Bronx	—	⎰ 24 Gly → Arg	Ranney et al. (1968)
Savannah	—	⎱ 24 Gly → Val	Huisman et al. (1971)
Genova	—	28 Leu → Pro	Sansone et al. (1967), Sansone and Pik (1965)
Tacoma	—	30 Arg → Ser	Brimhall et al. (1969)
Philly	—	35 Tyr → Phe	Rieder et al. (1969)
Hammersmith	—	⎰ 42 Phe → Ser**	Dacie et al. (1967), Morimoto et al. (1971)
Bucaresti	—	⎱ 42 Phe → Leu**	Bratu et al. (1971), Morimoto et al. (1971)
Zürich	—	63 His → Arg*	Muller and Kingma (1961), Frick et al. (1962), Winterhalter et al. (1969)
I Toulouse	—	66 Lys → Glu	Labie et al. (1971)
Sydney	—	⎰ 67 Val → Ala	Carrell et al. (1967)
Bristol	—	⎱ 67 Val → Asp**	Steadman et al. (1970)
Seattle	—	70 Ala → Asp**	Stamatoyannopoulos et al. (1969), Kurachi et al. (1973), Anderson et al. (1973)
Christchurch	—	71 Phe → Ser	Carrell and Owen (1971)
Shepherds Bush	—	74 Gly → Asp*	White et al. (1970)
Santa Ana	—	⎰ 88 Leu → Pro	Opfell et al. (1968)
Boras	—	⎱ 88 Leu → Arg	Hollender et al. (1969)
Sabine	—	91 Leu → Pro	Schneider et al. (1969)
Köln	—	98 Val → Met*	Carrell et al. (1966), Bellingham and Huehns (1968)

Khartoum	—	124 Pro → Arg	Clegg et al. (1969)
Wein	—	130 Tyr → Asp	Perutz and Lehmann (1968)
Olmstead	—	141 Leu → Arg	Lorkin et al. (1970)

haemoglobin, but the abnormal haemoglobin fraction because of its inherent instability tends to be denatured and become functionally inactive relatively rapidly. A typical feature of the anaemia that ensues is the appearance of so-called inclusion bodies (Heinz bodies) in the red cells, which are particularly prominent in patients who have been splenectomised. These appear to be composed of denatured haemoglobin, and they apparently render the red cells abnormally susceptible to premature destruction in the spleen.

The 'unstable' haemoglobin variants listed in table 1.2 mainly involve substitutions at internal sites in the molecular configuration which are normally occupied by non-polar aminoacids. The manner in which substitutions of this sort tend to distort the three-dimensional structure of the haemoglobin and so lead to molecular instability is discussed in detail by Perutz and Lehmann (1968).

1.5.4. Altered oxygen affinity

The principal functional effect of certain aminoacid substitutions (table 1.3) is an alteration in the oxygen affinity of the haemoglobin. Since these mutants are rare they have been studied only in heterozygotes.

In some cases the oxygen affinity is elevated. The proportion of oxygenated haemoglobin, at any given partial pressure of oxygen, is increased so that

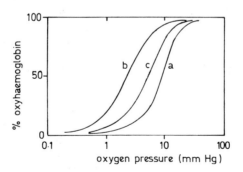

Fig. 1.10. Oxygen dissociation curves (at 20 °C, *p*H 7.0). a) Hb A, b) Hb Hiroshima, and c) Unfractionated haemolysate from heterozygote containing equal proportions of Hb A and Hb Hiroshima. (From Hamilton et al. 1969.)

the oxygen dissociation curve is shifted to the left (fig. 1.10). As a result oxygen is given up to the tissues less readily, and this has the secondary effect of stimulating the erythropoetic system which leads to increased red cell production and polycythaemia. In other cases the oxygen affinity is reduced which enhances oxygen delivery to the tissues and this in turn leads to a compensatory reduction in red cell and haemoglobin production.

Thus the polycythaemia which occurs in heterozygotes with high affinity variants, and the anaemia in those with low affinity variants represent physiological compensatory changes in red cell mass which tend to normalise the oxygen tension in the tissues (Stamatoyannopoulos et al. 1971). Sometimes, as in the case of Hb Kansas (table 1.3), the oxygen affinity of the haemoglobin may be so reduced that the oxygen saturation of the arterial blood

TABLE 1.3

Some haemoglobins with altered oxygen affinity: * increased oxygen affinity; ** reduced oxygen affinity. Variants with amino acid substitutions in same position are bracketed.

Haemoglobin	Aminoacid substitution		References
	α-chain	β-chain	
J. Capetown	92 Arg → Gln*	—	Lines et al. (1967)
Chesapeake	92 Arg → Leu*	—	Charache et al. (1966)
Olympia	—	20 Val → Met*	Nute et al. (1972)
E	—	26 Gly → Lys**	Bellingham et al. (1968)
Hirose	—	37 Try → Ser*	Yamaoka et al. (1971)
Malmö	—	97 His → Gln*	Boyer et al. (1972)
Yakima	—	99 Asp → His*	Novy et al. (1967)
Kempsey	—	99 Asp → Asn*	Reed et al. (1968)
Ypsi	—	99 Asp → Tyr*	Rucknagel, D. cited by Stamatoyannopoulos (1972)
Kansas	—	102 Asn → Thr**	Bonaventura and Riggs (1968)
Yoshizuka	—	108 Asn → Asp**	Imamura et al. (1969)
Peterborough	—	111 Val → Phe**	King et al. (1972)
Rainier	—	145 Tyr → Cys*	Hayashi et al. (1971)
Bethesda	—	145 Tyr → His*	Hayashi et al. (1971)
Hiroshima	—	146 His → Asp*	Hamilton et al. (1969), Perutz et al. (1971)

may be as low as 60–70%, and the amount of deoxygenated haemoglobin gives the patient a blue cyanotic appearance.

Marimoto, Lehmann and Perutz (1971) have discussed the different ways in which oxygen affinity may be altered by aminoacid substitutions. It may be due to changes in conformation in the neighbourhood of the haem groups; alterations in the binding of hydrogen or of diphosphoglycerate ions; or shifts in the allosteric equilibrium between the three-dimensional oxy- and deoxy-conformations. Quite small stereochemical alterations in the molecule may produce drastic and sometimes strikingly different changes in function. For example the substitutions in Hb Yakima and Hb Kansas (table 1.3) are only three residues apart. In both cases interaction between the four subunits of the molecule is abolished so that the oxygen dissociation curves are hyperbolic rather than sigmoid, but in Hb Yakima the oxygen affinity is markedly increased (Novy et al. 1967), while in Hb Kansas it is markedly decreased (Bonaventura and Riggs 1968).

Some of the unstable haemoglobins discussed earlier (p. 24) also have altered oxygen affinities (table 1.2), and corresponding functional compensatory effects occur. In patients with unstable haemoglobin disease, the red cell mass may be normal or even slightly elevated if the oxygen affinity of the abnormal haemoglobin is increased. Whereas the anaemia may appear more severe than it really is if the unstable haemoglobin has reduced oxygen affinity (Stamatoyannopoulos et al. 1971).

1.6. Two separate aminoacid substitutions in a single polypeptide chain

Although mutations are very rare events, occasionally two separate mutations may come to be represented in a single allele. The two alterations will then be reflected in the structure of the polypeptide it determines.

An example of this is provided by the rare haemoglobin variant Hb Harlem (Bookchin et al. 1967) which was originally found along with normal Hb A in an otherwise healthy Negro. He was presumably heterozygous for an abnormal gene at the β-locus and for the normal β-allele, because his red cells, besides containing the variant haemoglobin, also contained Hb A, and because six of his nine children showed the same combination of haemoglobins. The β-chain of the variant haemoglobin has a substitution at position 6 of valine for glutamic acid, the same as occurs in sickle-cell haemoglobin; and a substitution at position 73 of asparagine for aspartic acid which is the same as has been found by itself in another rare variant – Hb Korle-Bu

(Konotey-Ahulu et al. 1968). The structure of Hb C Harlem may therefore be written: $\alpha_2 \beta_2^{6 \text{ Glu} \rightarrow \text{Val, 73 Asp} \rightarrow \text{Asn}}$. As in sickle-cell heterozygotes, red cells with the variant haemoglobin and Hb A showed the sickling phenomenon, and the variant like Hb S was found to be very insoluble in the deoxygenated state.

It is very unlikely that the allele coding for the unusual β polypeptide chain in Hb C Harlem originated from a single mutational event which led to two such well-separated single base alterations in the DNA sequence of the gene. Most probably it is the product of two quite distinct mutations occurring in different individuals living at quite different times. Since the sickle-cell allele is so relatively common in Negroes, the change at the $\beta6$ position was probably the earliest of the two mutations to occur. The other mutation is likely to have occurred in a much more recent generation.

It is of interest to note that the separate mutational alterations could have come to be present together in a single allele, in two somewhat different ways. The Hb C Harlem allele might have originated in an individual carrying the sickle-cell allele on one chromosome and the normal β allele on the other, by a mutation in the sickle-cell allele at the seventy-third base triplet. Alternatively it could have arisen in an individual who carried the allele determining Hb S ($\alpha_2 \beta_2^{6 \text{ Glu} \rightarrow \text{Val}}$) on one chromosome and the allele determining Hb Korle-Bu ($\alpha_2 \beta_2^{73 \text{ Asp} \rightarrow \text{Asn}}$) on the other, as a result of crossing-over somewhere between the sixth base triplet and the seventy-third base triplet during normal chromosome pairing at meiosis. Such a cross-over would have resulted in one of the recombination products coding for the abnormal β-chain of Hb C Harlem, and the other coding for the normal β-chain.

CHAPTER 2

One gene–one polypeptide chain

2.1. 'Hybrid' proteins in heterozygotes

2.1.1. Haemoglobin

When haemolysates from individuals with the sickle-cell trait are examined electrophoretically, the two characteristic forms of haemoglobin seen are Hb A, $\alpha_2\beta_2$, and Hb S, $\alpha_2\beta_2^{6\ \mathrm{Glu}\rightarrow\mathrm{Val}}$ (conveniently written $\alpha_2\beta_2^{S}$). One does not observe the other possible kind of haemoglobin $\alpha_2\beta\beta^{S}$, which might be expected to occur in individuals who are synthesising both the β and β^{S} polypeptide chains. This curious and for a long time very puzzling situation also obtains in heterozygotes for other alleles determining haemoglobin variants. For example in the haemoglobin C trait one finds Hb A, $\alpha_2\beta_2$, and Hb C, $\alpha_2\beta_2^{C}$, but not $\alpha_2\beta\beta^{C}$, and in sickle-cell–haemoglobin C disease one finds Hb S, $\alpha_2\beta_2^{S}$, and Hb C, $\alpha_2\beta_2^{C}$, but not $\alpha_2\beta^{S}\beta^{C}$.

It has in fact been shown (Guidotti et al. 1963, Benesch et al. 1966) that hybrid molecules of the general type $\alpha_2\beta\beta^{S}$ in sickle-cell trait, and $\alpha_2\beta^{S}\beta^{C}$ in sickle-cell–haemoglobin C disease, are almost certainly present in quantity in the red cells of such heterozygotes. They are not however usually detected because the identification of the individual protein components in a mixture of haemoglobins requires in the ordinary way their physical separation by methods such as electrophoresis or column chromatography, and also because haemoglobin very readily dissociates into half molecules.

It seems that in solution the haemoglobin tetramer $\alpha_2\beta_2$ partially dissociates to give dimers of structure $\alpha\beta$, and there is a rapid dissociation–association equilibrium which may be written $\alpha_2\beta_2 \rightleftarrows 2\alpha\beta$. Similarly for Hb S one has $\alpha_2\beta_2^{S} \rightleftarrows 2\alpha\beta^{S}$. In a solution containing both $\alpha_2\beta_2$ and $\alpha_2\beta_2^{S}$, the dimers $\alpha\beta$ and $\alpha\beta^{S}$ are of course present and there is also the third form of the tetramer $\alpha_2\beta\beta^{S}$. The two dimers and the three tetramers are in equilibrium and the whole situation may be represented by the scheme shown in fig. 2.1. When such a solution is subjected to electrophoresis or chromatography this

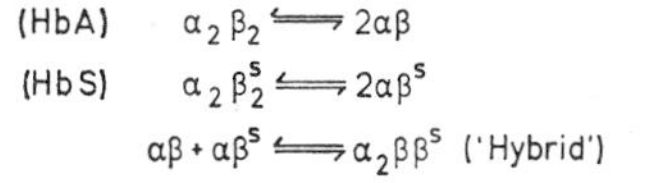

(HbA) $\alpha_2\beta_2 \rightleftharpoons 2\alpha\beta$

(HbS) $\alpha_2\beta_2^S \rightleftharpoons 2\alpha\beta^S$

          $\alpha\beta + \alpha\beta^S \rightleftharpoons \alpha_2\beta\beta^S$ ('Hybrid')

Fig. 2.1. Equilibria of haemoglobin tetramers and dimers in red cells of sickle-cell heterozygote (sickle-cell trait).

equilibrium is immediately disturbed by the preferential removal of $\alpha_2\beta_2$ and $\alpha_2\beta_2^S$ at the two extreme ends of the separation. As the most rapidly moving tetramer, say $\alpha_2\beta_2$, separates out of the mixture this effectively removes $\alpha\beta$ dimers and leads in turn to further dissociation of $\alpha_2\beta\beta^S$. Thus as the separation proceeds the amount of $\alpha_2\beta\beta^S$ is progressively reduced and by the time there is a complete separation of $\alpha_2\beta_2$ from $\alpha_2\beta_2^S$ no molecules of $\alpha_2\beta\beta^S$ remain (fig. 2.2). In consequence when haemolysates from sickle-cell

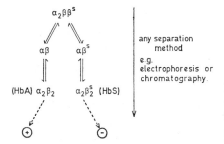

any separation method

e.g. electrophoresis or chromatography.

Fig. 2.2. Diagram showing how separation of haemoglobins in sickle-cell heterozygote haemolysate leads to the disappearance of the hybrid form, $\alpha_2\beta\beta^S$ (Benesch et al. 1966).

trait individuals are examined by these standard procedures only Hb A and Hb S are seen. Thus the inability to detect this type of haemoglobin hybrid molecule in heterozygotes by electrophoresis or chromatography depends on the fact that the rate of attainment of equilibrium between the tetramers and dimers under ordinary conditions is more rapid than the separation process. This appears to be the general rule in heterozygotes for haemoglobin alleles, and a very large number of different allelic combinations have indeed been examined. However very occasionally an exception may occur.

One such exceptional case has been reported by Efremov et al. (1969) and concerns individuals heterozygous for a rare allele determining a variant β-polypeptide chain ($\beta^{Richmond}$) containing a substitution of asparagine at position 102 by lysine. In heterozygous combination with the normal β allele, the hybrid form $\alpha_2\beta\beta^{Richmond}$ as well as $\alpha_2\beta_2$ and $\alpha_2\beta_2^{Richmond}$ could be readily separated by electrophoresis, and similarly the form $\alpha_2\beta^S\beta^{Richmond}$ was demonstrable in individuals heterozygous for the sickle-cell allele and the allele determining $\beta^{Richmond}$. In this very unusual case it appears that the particular aminoacid substitution is so sited in the molecule as to render the process of subunit exchange between the haemoglobin tetramers slower than the rate of separation of the molecules by electrophoresis.

The formation of 'hybrid' proteins in heterozygotes, that is the occurrence of a single protein molecule containing each of the structurally different polypeptide chains coded by the two alleles, appears to be a common and widespread phenomenon. The phenomenon is of particular interest, because by its very nature a 'hybrid' molecule of this sort cannot be present in individuals homozygous for either allele. It therefore represents a special molecular form peculiar to the heterozygous state. Also since a heterozygote receives one allele from one parent and one from the other, this particular molecular form can occur in individuals neither of whose parents possess it.

2.1.2. 'Hybrid' enzymes in heterozygotes

It is the facility with which the haemoglobin tetramer dissociates into half molecules which results in the failure to demonstrate by electrophoresis the hybrid forms which occur in heterozygotes, at least in the great majority of cases. However haemoglobin appears to be unusual in this respect. Most enzymes and other proteins which contain two or more polypeptide chains are much less readily dissociated into their subunits and consequently such hybrid molecules as occur in heterozygotes are more readily detected. In fact electrophoretic studies of allelic variants of many different enzymes have led to the discovery of numerous examples of such hybrid molecules in heterozygotes.

A simple illustration is provided by the enzyme known as peptidase A (Lewis and Harris 1967, Lewis et al. 1968, Lewis 1973). This is a dipeptidase which hydrolyses a variety of dipeptides to their constituent aminoacids, and it can be detected after electrophoresis by the use of a specific staining technique. In the course of electrophoretic surveys of the enzyme in red cells from many different individuals a number of genetically determined variant types of peptidase A were discovered. Fig. 2.3 shows the electrophoretic patterns in three such types, Pep A 1, Pep A 2-1 and Pep A 2. Pep A 1 is the most common type in all populations which have been studied. Pep A 2-1 and Pep A 2 have quite appreciable frequencies in Black populations, around 15% and 1% respectively, but they appear to be very rare in other populations. Family studies have shown that these three types are determined by two autosomal alleles (*PEP A*[1] and *PEP A*[2]), types Pep A 1 and Pep A 2 representing the homozygotes and Pep A 2-1 the heterozygote.

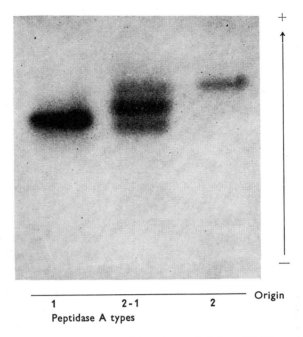

Fig. 2.3. Electrophoretic patterns of peptidase A types 1, 2-1, and 2. Electrophoresis and staining carried out as described by Lewis and Harris (1967).

Most of the activity in Pep A 1 and Pep A 2 occurs in a single main electrophoretic zone, though this differs in mobility in the two types (fig. 2.3). In the heterozygote however three main zones of activity are seen, two corresponding in mobility to the single zones seen in the homozygotes, whereas the third is exactly intermediate. This rather characteristic type of triple-banded electrophoretic pattern has also been observed in heterozygotes for alleles determining variant forms of a number of other enzymes (Shaw 1965). It is most simply accounted for by the hypothesis that in the homozygous state the enzyme protein contains two identical polypeptide subunits. In the case of peptidase A for example, one may suppose that the allele $PEP\ A^1$ codes for a polypeptide α^1, and the allele $PEP\ A^2$ codes for a polypeptide α^2, which since the allelic difference is likely to represent a single mutational step, possibly only differs from α^1 in a single aminoacid substitution. In the heterozygote where both alleles are present, one expects both types of polypeptide (α^1 and α^2) to be formed, and if they combine at random to give dimers, then three distinct enzyme proteins will occur (fig. 2.4). Two of these with subunit structures $\alpha^1\alpha^1$ and $\alpha^2\alpha^2$ will correspond to the forms seen separately in the homozygotes. The third will be a hybrid isozyme, $\alpha^1\alpha^2$, peculiar to the heterozygous state.

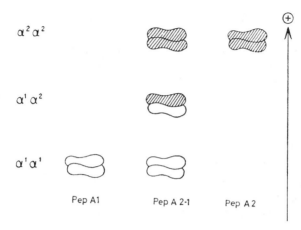

Fig. 2.4. Diagram showing postulated subunit composition of peptidase A types 1, 2-1 and 2.

This interpretation in the case of peptidase A is supported by the results of 'hybridisation' experiments *in vitro*. If solutions of Pep A 1 and Pep A 2 preparations are simply mixed and then subjected to electrophoresis, only the two zones corresponding to those seen in the separate types are observed. However if the mixture is treated under appropriate conditions with urea in the presence of mercaptoethanol, and then after dialysis subjected to electrophoresis, a triple-banded pattern essentially the same as that seen in the heterozygous type Pep A 2-1 is obtained (Lewis and Harris 1969a). The treatment evidently results in the dissociation of the enzyme proteins into their subunits (α^1 and α^2) which subsequently recombine at random to give the hybrid ($\alpha^1\alpha^2$) as well as the original forms ($\alpha^1\alpha^1$ and $\alpha^2\alpha^2$).

In general, any enzyme or protein which contains multiple polypeptide chains at least two of which are identical in homozygotes, may be expected to show 'hybrid' forms in heterozygotes. The simplest situation is when the enzyme is dimeric and is made up of two polypeptide chains whose amino-acid sequences are determined at a single gene locus, and so are identical in the homozygous state. Under these circumstances three different components are expected in heterozygotes, one of which is a 'hybrid'. Peptidase A is a typical example. More complex patterns of 'hybrid' formation in hetero-zygotes will occur if the enzyme protein in the homozygous state contains more than two identical polypeptide chains. For instance if it is a trimer, consisting of three identical subunits in the homozygote, then at least four distinct components will be expected in heterozygotes, two of which are 'hybrids'. Thus if one allele determined the polypeptide α^x and another α^y such that the molecular structure of the enzyme in one type of homozygote is α^x_3 and in the other α^y_3, then the heterozygote should have four molecular forms – α^x_3, $\alpha^x_2\alpha^y$, $\alpha^x\alpha^y_2$ and α^y_3. The enzyme purine nucleoside phosphorylase provides an example of this (see p. 87). If the enzyme is a tetramer and in the homozygote is made up of four identical polypeptides, then at least five distinct molecular forms are expected in a heterozygote, three of which are hybrids. In this case if the two alleles determine polypeptides α^x and α^y respectively then the subunit structures of the five molecular forms expected in heterozygotes would be – α^x_4, $\alpha^x_3\alpha^y$, $\alpha^x_2\alpha^y_2$, $\alpha^x\alpha^y_3$ and α^y_4. Examples of this are the lactate dehydrogenase heterozygotes (see pp. 56–58).

Electrophoresis has been the most widely used technique for separating the different molecular forms of enzymes in heterozygotes. Its effectiveness depends in general on there being a difference in electric charge between the polypeptide products of the two different alleles present in the particular heterozygote, so that enzyme molecules with different polypeptide compositions will differ in their electrophoretic mobilities. In these circumstances characteristic electrophoretic patterns are observed in heterozygotes, according to the subunit structure of the particular enzyme protein under consideration. Fig. 2.5 illustrates diagrammatically the main features of the patterns expected for enzyme proteins which are respectively, monomers, dimers, trimers and tetramers. In the case of monomers where the enzyme contains only a single polypeptide chain, the electrophoretic pattern in heterozygotes represents a simple mixture of the molecular forms seen in the two homozygotes, and of course no 'hybrid' form occurs. In the case of a dimer there is a triple banded pattern, in the case of a trimer, a four banded pattern, and in the case of a tetramer, a five banded pattern. In each case the outermost bands correspond to the two molecular forms present separately in the two corresponding homozygotes, and the bands with intermediate mobilities represent the 'hybrids'.

Electrophoretic patterns occurring in heterozygotes have now been examined in a wide variety of enzymes, and in general the patterns observed fit one or other of the situations illustrated diagrammatically in fig. 2.5. However the number and the pattern of components actually observed may often be more complex than those suggested in the diagram. This is commonly due to the occurrence of so-called 'secondary' isozymes (see p. 78) which are

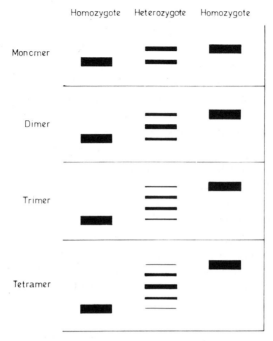

Fig. 2.5. Diagram of characteristic isozyme patterns expected in heterozygotes in the case of enzymes which are monomers, dimers, trimers or tetramers.

additional molecular forms generated from the main components after their primary synthesis. Variations in the symmetry of the patterns may also be observed (see 2.1.3 p. 37).

It will be apparent that the demonstration of 'hybrid' components in heterozygotes gives information about the subunit structure of the protein concerned. If such 'hybrids' are demonstrable then the protein must presumably be made up of two or more polypeptide subunits, at least some of which are identical in homozygotes. Enzymes or proteins which contain only a single polypeptide chain will not be expected to show 'hybrid' formation of this kind and in such cases the electrophoretic pattern in the heterozygote will represent a simple mixture of the molecular forms seen in the respective homozygotes. However it should be noted that the failure to observe 'hybrids' in heterozygotes cannot be taken by itself as sufficient evidence that the enzyme is a monomer. The absence or apparent absence of hybrids in heterozygotes even though the enzyme is polymeric could be due to several different causes. The enzyme protein may be so readily dissociated into its subunits as occurs for example with haemoglobin, that in the course of the electrophoretic or chromatographic separation of the mixture in heterozygotes the hybrid forms are lost. Alternatively there may be some special structural feature of the alternative polypeptide chains which prevents or limits their association in the same multimeric molecule. A further possible cause is that the polypeptide products of the two alleles although they are both synthesised in the organism may only be formed in different cells so they do not have the opportunity to combine together in a single protein molecule. This has been shown to be the case for the enzyme glucose-6-phosphate dehydrogenase which is determined by a gene locus on the X chromosome. Here hybrid formation has been demonstrated *in vitro* (Yoshida et al. 1967b) and also in somatic cell hybrids (see p. 86) but is not seen in female heterozygotes. This is thought to occur because one or other of the two alleles present in any single cell of the female is not functional, being present on the so-called 'inactivated' X chromosome (pp. 174–175). One may anticipate that the same will be generally true for other multimeric enzymes whose polypeptide chains are coded by gene loci on the X chromosome.

2.1.3. Symmetric and asymmetric electrophoretic patterns in heterozygotes

The relative activities of the several enzyme components observed in heterozygotes depend on the rates of synthesis and on the specific properties of the polypeptide products of the two alleles which are present. If the two polypeptides are synthesised at the same rate, have similar stabilities and contribute equally to enzyme activity, and if once synthesised they associate at random, then a characteristic pattern of activity will be observed. These patterns are indicated diagrammatically in fig. 2.5. In the case of a dimer, where the three enzyme components can be separated electrophoretically, the relative activities of the three bands observed in the heterozygote will be in the ratio 1:2:1. The two outermost bands which represent the homo-dimeric components will each account for 25% of the total activity, while the middle band which represents the 'hybrid' or heterodimeric component will account for 50% of the total activity. In the case of a trimer, the relative activities of the four components in the heterozygotes will be in the ratio 1:3:3:1, the two inner components with the highest activities representing the 'hybrids'. In the case of a tetramer the ratio of activities of the five components in heterozygotes will be 1:4:6:4:1. Thus in the case of a dimer half the total activity will arise from the single 'hybrid' component, in the case of a trimer three quarters of the total activity will come from the 'hybrids', and in the case of a tetramer, seven eighths. In each case the pattern will appear symmetrical, with the two outermost components which correspond to the homomeric forms seen alone in the homozygotes for the two separate alleles, contributing the least activity.

Numerous examples of such symmetric enzyme patterns in heterozygotes have indeed been observed in electrophoretic studies where the various enzyme components are detected by specific staining techniques. However in other cases, using the same methods, asymmetric patterns are observed, one of the outermost bands showing distinctly less activity than the other, and a similar asymmetry being manifest in the inner hybrid bands. Such asymmetric patterns occur when the polypeptide product of one of the two alleles contributes less to the total enzyme activity than the other. This may be because it is synthesised at a slower rate; or is significantly less stable so that it is broken down more rapidly than its counterpart before association into the multimeric enzyme molecules can occur; or because it renders the enzyme molecules in which it is present less stable than the other enzyme molecules so that they are more rapidly degraded; or because its structure so modifies the catalytic activity of the enzyme molecules in which it occurs that they show reduced enzyme activity. If for example in the case of a dimer, the polypeptide chains produced by the two separate alleles are formed in the cell at different rates in the ratio of 2:1, and if they are similar in their stabilities and in their catalytic properties, then the ratio of the three components seen in the heterozygote will be 4:4:1. If the relative rates of synthesis are 3:1, then the three components will be 9:6:1. Similar ratios can be estimated for trimers and tetramers. However it should be noted that where the asymmetry is due to differences in stability or to catalytic differences, the relative activities of the several enzyme components in the heterozygote are less easily predictable because the effects of instability or catalytic differences will not necessarily affect the activities of components with differing polypeptide compositions in a strictly proportionate manner.

However, there is usually in practice little difficulty in distinguishing between symmetric and asymmetric isozyme patterns in individuals heterozygous for particular allelic combinations. This is useful because it provides, in a simple way, information about the relative contributions of the two alleles to the total enzyme activity. And comparison of the patterns in different heterozygous combinations of alleles at the same gene locus, allows one to assess the relative effects on the enzyme activity of the different members of the allelic series.

The point can be illustrated by the enzyme known as placental alkaline phosphatase. This is a phosphatase with a high pH optimum which is found in relatively large amounts in the placenta and is apparently peculiar to this organ since it differs in a variety of respects (thermostability, inhibition characteristics, immunologically, etc.) from the alkaline phosphatases present in other tissues such as liver, bone, intestine and kidney. It also differs from these other human alkaline phosphatases in showing a remarkable degree of genetically determined variation.

Electrophoretic studies of the alkaline phosphatase in placentae from a large number of different individuals have led to the discovery of a considerable variety of separate phenotypes which can be distinguished one from another by their electrophoretic patterns (Boyer 1961, Harris and Robson 1965, 1967). Detailed analysis of the findings in pairs of placentae from dizygotic twins indicate that these various phenotypes are determined by different combinations of an extensive series of alleles at a single gene locus. As many as 18 different alleles have now been identified in the course of studies on a few thousand placentae (Harris and Robson 1967, Donald and Robson 1973). The twin studies have also shown that the alkaline phosphatase phenotype in any particular placenta is determined by the foetal rather than the maternal genotype.

The three most common alleles in European populations are now for convenience referred to as PL^1, PL^2 and PL^3 (Donald and Robson 1973). Previously they were designated as PL^{S_1}, PL^{F_1} and PL^{I_1}. They give rise to the six common phenotypes shown in fig. 2.6. Three of these phenotypes represent the homozygotes for each of the alleles, and the other three represent the different heterozygous combinations. To obtain optimal discrimination between these phenotypes electrophoretic separations at two different pHs are required, and generally pH 8.6 and pH 6.0 are used. The reason for this is that at the higher pH, the product of the PL^3 allele has a mobility very close to that of the PL^2 allele, so that in the heterozygote PL^2PL^3 the isozymes are not easily resolved; and at the lower pH the product of the PL^3 allele has the same mobility as that of the PL^1 allele, so that the isozymes present in the heterozygote PL^1PL^3 cannot be resolved. Thus resolution of the isozyme patterns in the heterozygotes PL^1PL^2 and PL^1PL^3 is obtained at pH 8.6, and resolution of the isozyme patterns in the heterozygotes PL^1PL^2 and PL^2PL^3 is obtained at pH 6.0.

Under these conditions it is found that each of the heterozygotes gives a triple banded pattern typical of a dimer. However while the heterozygote PL^1PL^2 shows a symmetric or near symmetric pattern, the heterozygotes

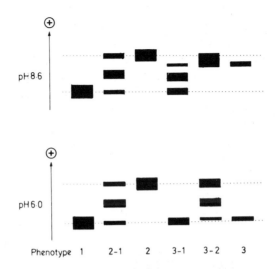

Fig. 2.6. Diagram of electrophoretic patterns of placental alkaline phosphatase in six common phenotypes at *p*H 8.6 and *p*H 6.0. Only the main 'fast moving' components are shown. For further details of the patterns and of other components which may be observed see Robson and Harris (1967) and Beratis et al. (1970, 1971)

PL^1PL^3 and PL^2PL^3 both show asymmetric patterns and in each case it is the isozyme corresponding to that seen alone in the homozygote PL^3PL^3 which has relatively weak activity. Thus one may infer that while the alleles PL^1 and PL^2 contribute about equally to the total enzyme activity in heterozygotes, the allele PL^3 contributes very much less to the total activity. This conclusion finds support in the results obtained by quantitative assay of the total alkaline phosphatase activity in extracts from a series of different placentae. On average, placentae from homozygotes or heterozygotes with only the PL^1 or PL^2 alleles show appreciably greater activities than heterozygotes with the PL^3 allele, and the homozygote PL^3PL^3 shows the least activity (Beckman 1970). Studies on the catalytic properties of the placental alkaline phosphatase of the various types have revealed no significant differences (Byers et al. 1972), which suggests that the reduced activity attributable to the PL^3 allele is due either to a reduced rate of synthesis of its polypeptide product, or to reduced stability of the polypeptide itself or of the dimers in which it occurs. The homodimeric product of the PL^3 allele has been shown to be more thermolabile than that of either the PL^1 or PL^2 alleles (Thomas and Harris 1973), but the differences are only apparent at relatively high temperatures and it is not at present clear whether such differences in stability are significant *in vivo*.

Most of the other alleles at this gene locus are relatively uncommon so that they are mainly found in heterozygous combination with either PL^1, PL^2 or PL^3. Such heterozygotes show typical triple banded isozyme patterns, which in some cases are symmetric and in others asymmetric. By a comparison of the patterns in the extensive series of different heterozygous combinations which have now been observed it has proved possible to arrange the alleles in a consistent order in terms of their relative contributions to the total enzyme activity (Harris and Robson 1967, Donald and Robson 1973). The findings also indicate that there is quite a wide variation in activities from allele to allele.

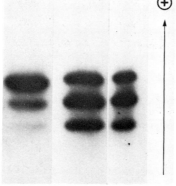

a   b   c

Fig. 2.7. Peptidase A isozyme patterns observed in different cells of an individual of genotype *Pep A⁸ Pep A⁶*. (Sinha et al. 1970, Lewis 1973). Key: a) red cells, b) leucocytes, c) fibroblasts in tissue culture.

It is of some interest that in certain cases the electrophoretic pattern of a particular enzyme in a single heterozygous individual may be asymmetric in certain cells and symmetric in others. An illustration of this is shown in fig. 2.7. It involves peptidase A and shows the pattern observed in red cells, leucocytes and fibroblasts grown in tissue culture, from a heterozygous individual of genotype $PEP\ A^8\ PEP\ A^6$. The allele $PEP\ A^8$ is relatively common and gives rise to a polypeptide product which in red cells contributes much less activity than the product of the commonest allele $PEP\ A^1$ (see p. 32) or certain very rare alleles such as $PEP\ A^6$ (Sinha et al. 1970, Lewis 1973). In other cells such as leucocytes or fibroblasts, the peptidase activities attributable to these different alleles are however very similar to one another. It appears that peptidase A molecules containing the polypeptide derived from $PEP\ A^8$ are relatively unstable in the red cell and decay quickly so that an asymmetric pattern is produced. In other cells this effect is much less marked, if indeed it occurs at all, and so symmetrical patterns are observed. The difference is presumably a consequence of the fact that the red cell is relatively long lived but loses its nucleus and its capacity for enzyme synthesis at a very early stage in its life span. Other cells tend to have shorter lives and retain their capacity for enzyme synthesis throughout.

Another illustration of the same effect is seen in the case of heterozygotes for certain alleles determining the enzyme phosphogluconate dehydrogenase (Parr 1966). Here again the enzyme is a dimer, and the red cells show an asymmetric triple banded pattern, while leucocytes show a symmetric pattern.

In other cases the effect may be even more extreme with the result that while enzyme components attributable to the products of both alleles in the heterozygotes are clearly seen in extracts of leucocytes, other tissues and fibroblasts grown in tissue culture, the circulating red cells are peculiar and show only the enzyme product attributable to one of the alleles, presumably because the other decayed too rapidly. Clear examples of this effect are seen in certain heterozygotes for alleles determining variants of the soluble form of isocitric dehydrogenase (Turner et al. 1973) and also the enzyme known as peptidase C (Povey et al. 1972).

2.2. *Multiple gene loci*

So far we have been mainly concerned with the effects on protein and enzyme structure of mutations at single gene loci. We now consider the complexities that arise when two or more separate gene loci are involved in defining the structure of a protein or a closely related set of proteins. Three rather different examples will be considered in some detail in order to illustrate various aspects of the matter. They are haemoglobin lactate dehydrogenase (p. 52) and phosphoglucomutase (p. 60).

2.2.1. *The haemoglobin loci*

In the normal adult the principal type of haemoglobin present is Hb A, which has already been discussed. In the red cells of the foetus, however, another form predominates. This is known as foetal haemoglobin, or Hb F. Although it differs in its detailed structure from Hb A it resembles it very closely in its three-dimensional conformation and in most of its properties. In the newborn about 70–80% of the haemoglobin present is Hb F, and nearly all the remainder Hb A. This proportion rapidly changes during the first few months after birth, so that by six or twelve months Hb F is present in no more than trace amounts and Hb A now predominates.

Yet a further distinctive molecular form of haemoglobin known as Hb A_2 is also consistently found in normal individuals. It occurs along with Hb A in the red cells of the adult, but it constitutes only a small fraction (about 2.5 %) of the total haemoglobin present. It is apparently synthesised synchronously with Hb A but at a much slower rate.

Hb A_2 and Hb F, like Hb A, both contain two distinct types of polypeptide, each of which is represented twice in the molecule. One of these is an α-chain identical with that in Hb A. They differ from Hb A and also from each other in the nature of the other chain, which in Hb A_2 is called the δ-chain and in Hb F the γ-chain. So these three different forms of normal haemoglobin may be written as follows:

Hb A $\alpha_2\beta_2$
Hb A_2 $\alpha_2\delta_2$
Hb F $\alpha_2\gamma_2$

As in the β-chain there are 146 aminoacid residues in the δ- and the γ-chains, and the sequences of the three chains show many similarities (fig. 2.8). The β-chain and the δ-chain sequences have identical aminoacid residues in 136 positions and differ in only ten positions. The γ-chain shows more differences but still shares identical residues in 107 positions with the β-chain, and in 105 positions with the δ-chain. There are also many similarities between the sequences of the β-, δ- and γ-chains on the one hand and the α-chain on the other. Exact comparisons are complicated because the latter chain has five fewer aminoacid residues, but making reasonable assumptions about the positions of the 'gaps', it has been estimated that there is a correspondence of perhaps 46 % in the aminoacid sequences of the α- and the β-chains. The remarkable degree of homology in the aminoacid sequences of these different polypeptide chains suggests that they have all been derived during the course of evolution from a single ancestral form. This point is considered in more detail later (ch. 3, pp. 101–103).

Since the α-, β-, γ- and δ-chains all occur in normal individuals, and since they differ from each other in varying degrees in their aminoacid sequences, one must suppose that they are determined at separate gene loci on the chromosomes. Thus at least four gene loci will be concerned in determining the structures of the three normal haemoglobins A, A_2 and F, and the situation may be represented as shown in fig. 2.9. An important conclusion follows. A mutant allele at the locus determining the α-chain will be expected to result in the occurrence of variant forms of all three haemoglobins A, A_2 and F. On the other hand a mutant allele at one or other of the loci determining the β-, δ- or γ-chains should only result in a variant form of the haemoglobin

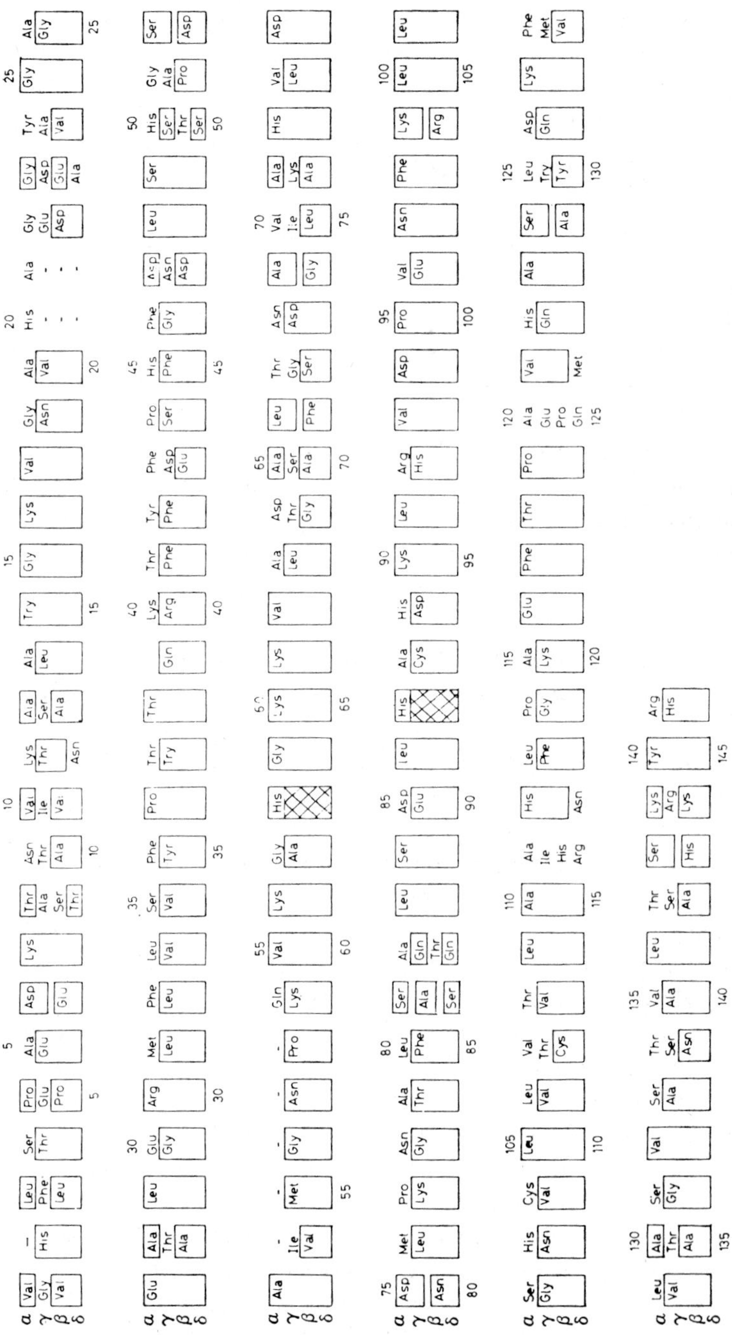

One gene–one polypeptide chain

Fig. 2.8. The aminoacid sequences of the α, β, γ and δ polypeptide chains of haemoglobin. Identical aminoacids are boxed. The haem linked histidine residues (p. 20) are hatched. The 141 aminoacids in the α-chain are numbered at the top, and the 146 aminoacids in the β-, γ- and δ-chains are numbered at the bottom. The 'gaps' are arranged to secure maximum homology in the aminoacid sequences. (From Lehmann and Carrell (1969). Aminoacid abbreviations as in fig. 1.5 p. 14).

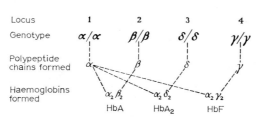

Fig. 2.9. Formation of haemoglobins A, A₂ and F in the normal individual.

which normally contains the corresponding chain. A number of variants of Hb A_2 and Hb F have indeed been identified, and the findings are fully consistent with these expectations. In individuals heterozygous for a mutant allele at either the β-, δ- or γ-locus, only a variant form of the haemoglobin normally containing that particular polypeptide chain is produced. Thus in the sickle-cell trait where the β-locus is involved, haemoglobin A_2 ($\alpha_2\delta_2$) and haemoglobin F ($\alpha_2\gamma_2$) are normal. On the other hand, an individual heterozygous for a mutant allele at the α-locus synthesises variant forms as well as the normal forms of each of the three kinds of haemoglobin, Hb A ($\alpha_2\beta_2$), Hb A_2 ($\alpha_2\delta_2$) and Hb F ($\alpha_2\gamma_2$).

In early embryonic and foetal life at least two other types of haemoglobin chain have been found to occur. One is referred to as the epsilon (ε)-chain (Huehns et al. 1964a, b) and the other as the zeta (ζ)-chain (Capp et al. 1970). Their sequences have not yet been determined in detail but they evidently differ from one another and also from those of the α-, β-, γ- or δ-chains. Several so-called embryonic haemoglobins with structures thought to be $\alpha_2\varepsilon_2$, $\gamma_2\zeta_2$ and ε_4 have been found in significant quantities in early embryos and have also been detected in very small quantities in later foetal life and occasionally in neonates (for review see Lorkin 1973). Such haemoglobins are of course peculiarly difficult to study and so far no variants have been detected. However it seems very probable that the ε-chain and the ζ-chain are determined by separate gene loci distinct from the loci which code for the α-, β-, γ- and δ-chains.

Furthermore it has also been shown that two slightly different sorts of γ-chain are formed by the normal organism. In one of these, position 136 in the aminoacid sequence is occupied by glycine (as shown in fig. 2.8) while in the other this position is occupied by alanine (Schroeder et al. 1968). The two chains are referred to as $^G\gamma$ and $^A\gamma$ respectively. There seems no doubt that they are coded by separate gene loci, since in the analyses of variants of Hb F where the abnormal γ-chain must be the product of a single gene it has been found that the variant has either all alanine or all glycine at position 136 and not a mixture. At birth the ratio of $^G\gamma$ to $^A\gamma$ chains being formed is about 3:1 (Schroeder et al. 1972) but during the period immediately after birth it changes in favour of $^A\gamma$ chains (Schroeder et al. 1971, 1972).

There is also evidence for the occurrence of two separate gene loci coding for α-chains which are either very similar in structure or perhaps identical. Hollan et al. (1972)· described a Hungarian family in which two alleles determining the rare α-chain variants Hb J-Buda and Hb G-Pest were segregating (fig. 2.14, p. 51). Three brothers were found to have both the abnormal α-chain haemoglobins and also Hb A ($\alpha_2\beta_2$), suggesting that they carried at least three different genes determining α-chains. Other members of the family had either Hb A and Hb J-Buda, or Hb A and Hb J-Pest. Bernini et al. (1970) have also described an individual with two different α-chain variants as well as Hb A. In addition certain individuals thought to be homozygous for the gene determining the very unusual α-chain variant Hb Constant Spring (see pp. 115, 144) have been found to have normal Hb A as well (Lie-Injo et al. 1974). These various findings suggest the occurrence of two α-loci.

The observation that in heterozygotes for single α-chain variants, the amount of the variant present is usually only about half that found in heterozygotes for β-chain variants has also been considered as evidence for the occurrence of two α-chain loci (Lehmann and Carrell 1968). In addition it is known that certain other animal species have more than one α-globin locus. On the other hand evidence against the occurrence of two α-loci comes from the finding that in individuals who are homozygous for a gene which determines the α-chain variant Hb Tongariki and which has an appreciable incidence in Melanesia, no Hb A occurs (Abramson et al. 1970; Beaven et al. 1972). It is possible therefore that the human species as a whole or possibly only certain populations may be polymorphic for chromosomes bearing duplicated and single α-gene loci. The evidence has been reviewed by Wasi (1973).

Thus it now seems that there are at least seven or eight separate gene loci determining the structure of haemoglobin: the loci coding for the β-, $^G\gamma$-, $^A\gamma$-, δ-, ε- and ζ-haemoglobin chains, and probably two loci coding for α-chains.

The general hypothesis that each of the different haemoglobin polypeptide chains is determined by a separate gene locus accounts rather nicely for the biochemical findings in individuals who are simultaneously heterozygous at a locus which determines the α-chain and also at one of the other loci. Under such circumstances a particularly complex mixture of haemoglobins is produced (Baglioni and Ingram 1961, McCurdy et al. 1961). As an example (Weatherall et al. 1962) one may consider an individual who is heterozygous at the α-locus for the mutant allele which determines the variant polypeptide chain α^G ($\alpha^{68\ \mathrm{Asn}\ \to\ \mathrm{Lys}}$), and is also heterozygous at the β-locus for the Hb C gene which determines the variant polypeptide chain β^C ($\beta^{6\ \mathrm{Glu}\ \to\ \mathrm{Lys}}$). Two sorts of α-chain are synthesised and two sorts of β-chain, and as a result four different kinds of haemoglobins containing α- and β-type chains are demonstrable by electrophoresis (fig. 2.10). These are normal Hb A ($\alpha_2\beta_2$), Hb C ($\alpha_2\beta_2^C$), Hb G ($\alpha_2^G\beta_2$), and a new haemoglobin, $\alpha_2^G\beta_2^C$, which contains both the variant α-chain and also the variant β-chain.

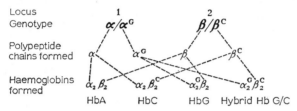

Fig. 2.10. Formation of haemoglobins A, C, G and G/C in an individual heterozygous at the α locus for an allele which determines the variant polypeptide α^G, and heterozygous at the β locus for an allele which determines the polypeptide β^C.

Also, of course, in such an individual a small fraction of the haemoglobin present in adult life will consist of Hb A$_2$ ($\alpha_2\delta_2$) and the variant Hb A$_2^G$ ($\alpha_2^G\delta_2$); while in foetal life and immediately after birth most of the haemoglobin formed will consist of Hb F ($\alpha_2\gamma_2$), and the variant Hb F^G ($\alpha_2^G\gamma_2$), and since there are two sorts of γ-chain there will presumably be two sorts of each of these kinds of molecule. Besides this, various 'hybrid' forms of haemoglobin molecules, such as $\alpha_2\beta\beta^C$ and $\alpha\alpha^G\beta_2$, will also be present in the red cells of such individuals, though they will not be demonstrable in ordinary electrophoretic or chromatographic separation of haemolysates for the reasons already discussed.

Thus one can see that the multiple gene loci determining the several distinct haemoglobin polypeptide chains allow the generation of a quite complex set of different haemoglobin molecules in normal individuals, and that this complexity is greatly enhanced if an individual is heterozygous at one or more of these loci.

When it is found that more than one gene locus is concerned in determining the structure of a particular protein it becomes of some interest to find out where the loci are positioned on the chromosomes, and in particular whether they are close together on a single chromosome (i.e. closely linked) or whether they are well separated on the same chromosome, or occur on different chromosomes (see also pp. 59, 69, 77, 122).

In the case of haemoglobin, information about the linkage relationships of the α-, β- and δ-loci has been obtained from pedigree studies. There is also some less direct information about the γ-loci.

α- and β-loci: Several families have been found in which a gene determining an α-chain variant and also a gene determining a β-chain variant are segregating. An example (Smith and Torbert 1958, Itano and Robinson 1960, Bradley et al. 1961, Charache et al. 1971) is shown in fig. 2.11, where the α-chain variant occurs in an unusual haemoglobin known as Hopkins-2 (α 112 His → Asp) and the β-chain variant is sickle cell haemoglobin.

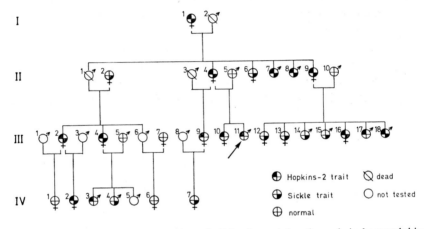

Fig. 2.11. Pedigree showing segregation of alleles determining the α-chain haemoglobin variant Hopkins-2, and the β-chain variant Hb S (Smith and Torbert 1958, Bradley et al. 1961.)

Several individuals in the pedigree are heterozygous for the gene determining Hopkins-2 and its allele which determines the normal α-chain, and also for the sickle-cell gene and its allele which determines the normal β-chain. The segregation of these genes in the pedigree suggests that the two loci determining the α- and β-chains cannot be very closely linked.

Thus if one considers the seven children of the mating between II$_9$ who is heterozygous at both the 'α' and the 'β' loci, and II$_{10}$ who is homozygous for the normal allele at these loci, one finds that two of the children are heterozygous at both loci like II$_9$, while five are heterozygous at one locus or the other but not at both. If II$_9$ carried the genes for Hb Hopkins-2 and Hb S in coupling on the same chromosome this would imply that there are five recombinants among the seven children. If II$_9$ carried the abnormal genes in repulsion (i.e. on different chromosomes) then there would be two recombinants and five non-recombinants. In either case some cross-overs must have occurred, and this makes it unlikely that the two loci are very closely linked. This conclusion is supported by data from other families (Raper et al. 1960, Hall Craggs et al. 1964, Rothman and Ranney 1971) and detailed analysis of the findings as a whole indicates that the loci are either on different chromosomes, or if they are on the same chromosome they are sufficiently well separated for crossingover between them to be a not infrequent event.

β- and δ-loci: A number of families have also been found which enable one to examine the linkage relation between the locus determining the β-chain and the locus determining the δ-chain. Here a quite different result has been obtained. No definite example of a cross-over has yet been observed, though more than sixty children of matings where one parent is doubly heterozygous have been studied (Mishu and Nance 1969).

A typical pedigree (Horton and Huisman 1963) is shown in fig. 2.12.

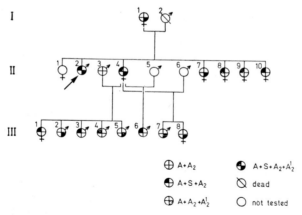

Fig. 2.12. Pedigree showing segregation of alleles determining the β-chain variant Hb S and the δ-chain variant Hb A$_2^1$ (Horton and Huisman 1963).

Here II$_4$ is heterozygous at the β-locus for the sickle-cell gene, and is also heterozygous at the δ-locus for the gene which determines an Hb A$_2$ variant referred to as Hb A$_2^1$. Of her eight children by three fathers, four show only

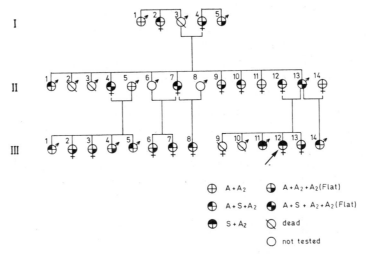

Fig. 2.13. Pedigree showing segregation of alleles determining the haemoglobin β-chain variant Hb S and the δ-chain variant Hb A₂ Flatbush (Ranney et al. 1963).

the sickle-cell trait (i.e. are heterozygous for the β-locus, but not for the δ-locus), and four show only the Hb A_2^1 trait (i.e. are heterozygous at the δ-locus, but not the β-locus). There were no children who were either homozygous or heterozygous at both loci. The pedigree thus supports the idea that the two loci are linked and suggests that II$_4$ carried the β^S- and the δ^1-genes in repulsion. She presumably received the β^S-gene from her father and the δ^1-gene from her mother. Fig. 2.13 shows another pedigree (Ranney et al. 1963) illustrating the same phenomenon, but in this case with a different mutant allele at the δ-locus. Taken together these and the other pedigrees of families where mutant alleles at both the β-locus and the δ-locus were found to be segregating provide convincing evidence that these two loci are very closely linked (see also Lepore haemoglobins pp. 104–107).

At present, then, one can say that the loci determining the β- and δ-polypeptide chains of haemoglobin lie close together on the same chromosome, possibly in immediate juxtaposition, and that the locus determining the α-polypeptide chain lies either on a quite different chromosome or, if it is on the same chromosome as the β- and δ-loci, it must be some distance away from them.

Other haemoglobin loci: Fig. 2.14 shows the pedigree of the family described by Hollan et al. (1972) in which two different α-chain variants were segre-

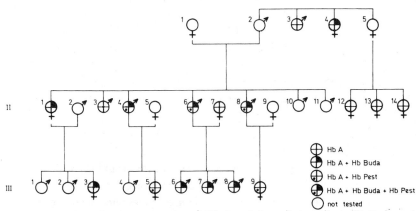

Fig. 2.14. Pedigree showing segregation of Hb A and the α-chain variants Hb Buda and Hb Pest. Note that individuals II 4, II 6 and II 8 have all three haemoglobins (Hollan et al. 1972).

gating and certain individuals had both of these as well as normal Hb A. The findings suggest that two α-gene loci occur (p. 46, p. 144). The pedigree is consistent with the two loci being closely linked, but too few opportunities for recombination exist in the pedigree to be certain about this conclusion.

Because of the difficulties of studying variants of foetal (or indeed embryonic) haemoglobin in more than one generation of a family, direct pedigree information about the linkage relations of the γ-loci has not been obtained. However the evidence from studies on Hb Kenya which contains a fused γ-β gene which probably arose by unequal crossing over, supports the idea that there is close linkage between the β- and γ-loci.

Studies using the technique of *in situ* hybridisation of labelled haemoglobin messenger RNA applied to human chromosome preparation have led to the claim that there are two chromosomal regions where haemoglobin loci occur, one on chromosome No. 2 and the other on chromosome No. 4 or No. 5 (Price et al. 1972). Presumably, one represents the α- and the other the β-, δ- and γ-loci.

***

More is known about the structure and genetical determination of haemoglobin than any other human protein. But for a number of other proteins there is now good, though less complete evidence that two or more gene loci are concerned in determining their structures. The phenomenon, though common, is not universal, since there are also many proteins which appear to contain only one type of polypeptide chain apparently defined by just a single gene locus.

Where an enzyme or protein is determined by two or more gene loci each coding for the aminoacid sequence of a distinctive polypeptide chain, several molecular forms of the protein will often be demonstrable in the normal individual. Their number, properties and characteristics will depend on whether or not the different polypeptide products of the several loci combine in a single molecule, and on the possible sorts of combination that can occur. The general point is illustrated by the two enzymes discussed below: lactate dehydrogenase *and* phosphoglucomutase. In each case more than one gene locus is involved in determining the structure of the enzyme protein, and multiple forms (often referred to as isozymes – see pp. 70 ff) occur.

2.2.2. Lactate dehydrogenase

Lactate dehydrogenase is a key enzyme in carbohydrate metabolism and occurs in virtually all tissues. It catalyses the interconversion of lactate and pyruvate with the concomitant oxidation and reduction of the coenzyme NAD:

$$CH_3, CO.COOH + NADH_2 \rightleftarrows CH_3, CHOH.COOH + NAD$$
pyruvic acid $\qquad\qquad\qquad\qquad$ lactic acid

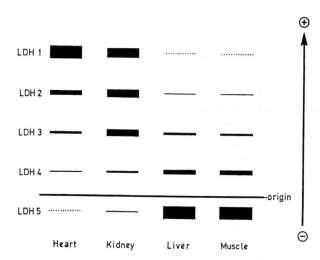

Fig. 2.15. Lactate dehydrogenase isozymes in heart muscle, kidney, liver, and skeletal muscle.

When extracts from various tissues are examined by electrophoresis, at least five distinct proteins (isozymes) with lactate dehydrogenase activity can be separated. The relative amount of the several isozymes is found to vary markedly from tissue to tissue. This is illustrated in fig. 2.15 where it can be seen, for instance, that the electrophoretic pattern of lactate dehydrogenase observed in heart muscle and in kidney is strikingly different from that observed in liver or in skeletal muscle. The five isozymes are referred to as LDH 1, LDH 2, LDH 3, etc. in the order of their electrophoretic mobilities towards the anode at alkaline pH. In heart muscle and in kidney LDH 1 predominates. The other isozymes are present but in much smaller amounts, and their relative activities decrease progressively from LDH 2 to LDH 5. In skeletal muscle and in liver, LDH 5 predominates, and the pattern of relative activities is effectively the opposite to that seen in heart.

It has been shown that each of these lactate dehydrogenase proteins are tetramers composed of four polypeptide subunits (Appella and Markert 1961, Markert 1963, 1968), and the polypeptide subunits can be of two different kinds which are usually referred to as A and B (or M and H). Although the detailed structures are not yet known, it appears from comparisons of their aminoacid compositions and their peptide fingerprints that they probably differ quite extensively in aminoacid sequence, though they are evidently similar in molecular size. Extensive evidence has been obtained to show that the five characteristic isozymes have subunit structures which may be represented as follows: LDH 1 = B_4, LDH 2 = AB_3, LDH 3 = A_2B_2, LDH 4 = A_3B and LDH 5 = A_4.

The LDH isozyme patterns characteristic of different tissues can be largely accounted for, if it is supposed that although both the A and B subunits are usually formed they occur in widely different amounts in different tissues, and the relative quantities of the several isozymes which are observed are the consequence of random combinations of the subunits present. Thus in skeletal muscle subunit A is evidently in great excess compared with subunit B, so that LDH 5 is the predominant form and the others are present in decreasing amounts in the order LDH 4 > LDH 3 > LDH 2 > LDH 1. Heart muscle by contrast has a great excess of the B subunit, so that here LDH 1 > LDH 2 > LDH 3 etc. These tissue differences have usually been attributed entirely to differences in the relative rates of synthesis of the A and B polypeptides in different cell types. However differences in their rates of degradation, or in the rates at which the various isozymes containing them are degraded, may also play a significant part in determining the electrophoretic patterns in different tissues (Fritz et al. 1969, 1971).

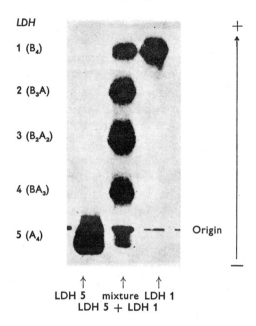

Fig. 2.16. Experiment showing the dissociation and recombination of lactate dehydrogenase subunits (Markert 1963a). Preparations of LDH 1 (B₄) and LDH 5 (A₄), separately and mixed together in equal proportions, were placed in 1 M NaCl and frozen overnight. The electrophoretic patterns obtained after thawing the samples are shown. In the mixed preparation the five isozymes LDH 1 (B₄), LDH 2 (B₃A), LDH 3 (B₂A₂), LDH 4 (BA₃) and LDH 5 (A₄) were present in the approximate ratio 1:4:6:4:1.

It is possible experimentally (Markert 1963a) to dissociate the separate isozyme proteins into their constituent subunits and then allow them to recombine to give enzymically active products (fig. 2.16). Applying this to artificial mixtures of purified LDH 1 (B_4) and LDH 5 (A_4) in different proportions, isozyme patterns corresponding to those observed in different tissues can be generated. From such experiments (Vessell 1965a) it appears that the ratio of A to B subunits in human skeletal muscle and liver cells may be of the order 10:1. For heart muscle on the other hand the ratio appears to be close to 1:20, and for kidney 1:10.

The characteristic asymmetric lactate dehydrogenase patterns described above for heart, muscle and liver are consistently found both in adults and in children. However in early foetal life a much more symmetric pattern is found, indicating that both A and B polypeptides are present in roughly equal amounts. During the course of foetal development the patterns change

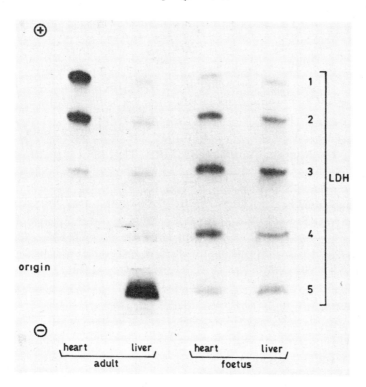

Fig. 2.17. Foetal and adult lactate dehydrogenase isozyme patterns in heart and liver.

(fig. 2.17). In heart the B polypeptides progressively increase relative to the A polypeptides, while in liver the opposite occurs. The changes are most rapid during the third trimester of foetal life (Werthamer et al. 1973).

Numerous investigations have been directed to comparing the properties of the various isozymes. Significant differences between LDH 1 (B_4) and LDH 5 (A_4) have been demonstrated in K_m's, in reactivity with various analogues of the coenzyme NAD, in inhibition by substrate excess, in thermostability and in other respects. These differences are presumably reflected functionally, and one would expect the pattern of isozymes present in a given tissue to be related to its particular metabolic characteristics. However, although certain general correlations have been noted, for example the preponderance of A subunits in tissues which tend to be subjected to periods of oxygen lack (Cahn et al. 1962, Wilson et al. 1963), the full details of such possible functional relationships remain to be elucidated (Vessell 1968, Kaplan et al. 1968, Wuntch et al. 1970a, b).

One may anticipate that the two distinct polypeptide subunits A and B are determined by separate gene loci. If so, a mutation would be expected to produce quite different effects on the LDH isozymes according to whether it occurred at the locus determining the A polypeptide, or at the locus determining the B polypeptide. Thus a mutant allele at the '*A*' locus should not affect the LDH 1 protein because this does not contain an A subunit. It should however affect in different degrees LDH 2, LDH 3, LDH 4 and LDH 5, which contain respectively one, two, three and four A type subunits. Similarly a mutant allele at the '*B*' locus should not affect LDH 5 but should affect the four other isozymes.

It is apparent that very complex sets of isozymes may be generated in heterozygotes, and it is possible to predict in some detail the general pattern of variant proteins to be expected in different cases. Thus an individual heterozygous at the '*A*' locus but homozygous for the normal allele at the '*B*' locus would be expected to form the fifteen isozymes shown in column (c) of table 2.1, while an individual heterozygous at the '*B*' locus but homozygous at the '*A*' locus would be expected to form the isozymes shown in column (d) of the table. The relative proportions of the normal and variant isozymes formed will be determined by the amounts of the normal and variant polypeptides available to form tetramers. If for example in the case of an '*A*' locus mutation, equal amounts of the variant '*A*' polypeptide and the normal A polypeptide were formed, and if association to give tetramers occurred at random then the two isozymes corresponding to LDH 2 should occur in the ratio 1:1, the three isozymes corresponding to LDH 3 in the ratio 1:2:1, the four isozymes corresponding to LDH 4 in the ratio 1:3:3:1, and the five isozymes corresponding to LDH 5 in the ratio 1:4:6:4:1. These ratios would be consistent from tissue to tissue, but the overall pattern observed in any given tissue would of course be correlated with the relative amounts of the five isozymes normally present, that is the relative activities attributable to the two loci ('*A*' and '*B*').

A number of different genetically determined variants of lactate dehydrogenase have indeed been discovered (Boyer et al. 1963, Kraus and Neely 1964, Nance et al. 1963, Davidson et al. 1965, Vessell 1965b, Ananthakrishnan et al. 1970, Blake et al. 1969, Das et al. 1972). They are each relatively uncommon and they have mainly been found during the course of routine electrophoretic surveys in various populations. None of those so far discovered appears to be specifically associated with any clinical abnormality. In each of the variants a complex but quite characteristic electrophoretic pattern of many isozymes is seen. This runs true to type in the particular family in which it is found and the familial distribution indicates that the individuals showing the peculiarity must be heterozygous at an autosomal locus for a rare mutant gene and its normal allele.

From the electrophoretic patterns it is possible to classify these different

TABLE 2.1

Postulated subunit constitutions of LDH isozymes in normal individuals and in individuals heterozygous for mutant genes at either the '*A*' or '*B*' loci.

(a) Isozyme	(b) Normal subunit constitution	(c) Subunit constitution of the fifteen possible isozymes in a heterozygote for an '*A*' locus variant	(d) Subunit constitution of the fifteen possible isozymes in a heterozygote for a '*B*' locus variant
LDH 1	B_4	B_4	B_4 B_3B^* $B_2B^*_2$ BB^*_3 B^*_4
LDH 2	B_3A	B_3A B_3A^*	B_3A B_2B^*A BB^*_2A B^*_3A
LDH 3	B_2A_2	B_2A_2 B_2AA^* $B_2A^*_2$	B_2A_2 BB^*A_2 $B^*_2A_2$
LDH 4	BA_3	BA_3 BA_2A^* BAA^*_2 BA^*_3	BA_3 B^*A_3
LDH 5	A_4	A_4 A_3A^* $A_2A^*_2$ AA^*_3 A^*_4	A_4

A and B represent the normal subunits.
A* and B* represent the variant subunits.

heterozygous types quite readily into distinct classes, according to whether the rare mutant allele is at the '*A*' locus or at the '*B*' locus, and so determines a variant form of the A polypeptide or the B polypeptide. The point is illustrated in fig. 2.18, which shows the electrophoretic patterns observed in

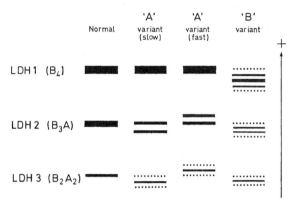

Fig. 2.18. Diagram of variant electrophoretic patterns of lactate dehydrogenase as seen in red cells. Two of the variants are from individuals heterozygous for different alleles at the gene locus determining the A subunit, and the other is from an individual heterozygous at the gene locus determining the B subunit.

red cell lysates from different types of heterozygotes. In normal red cells the three main isozymes found are LDH 1, LDH 2 and LDH 3. It will be seen that the patterns observed in one of the variant types corresponds exactly to that expected in a heterozygote for a '*B*' locus mutant, and that the other two agree with the expectations for '*A*' locus mutants.

An interesting variant has been reported by Kitamura et al. (1971). The mutation evidently affected the '*B*' locus in such a way as lead to the absence of an enzymatically active product. In an individual thought to be homozygous for this allele, the red cells were found to have only about 5% of the normal level of lactate dehydrogenase activity and this was entirely attributable to LDH 5 (i.e. A_4). No LDH 1 (B_4), LDH 2 (B_3A) or LDH 3 (B_2A_2) isozymes which are the principal ones seen in normal red cells could be detected. The findings in serum and saliva were similar. The affected individual was a 64 year old man with mild diabetes, but there was no indication that the lactate dehydrogenase activity was in itself the cause of the metabolic abnormality. In his five children, all presumably heterozygous, it appeared from the electrophoretic patterns and from assays of lactate dehydrogenase activity that only about half the normal amount of B subunits were being produced.

Besides the gene loci giving rise to the characteristic A and B subunits of LDH, there is evidence for yet a third locus determining a distinctive subunit (Goldberg 1963, 1965, Bianco and Zinkham 1963, Zinkham 1968). This so-called C subunit leads to the formation of a further type of tetrameric isozyme (usually referred to as the X band), which migrates electrophoretically between LDH 3 and LDH 4. The X band is readily detected in sperm and in extracts of postpubertial testis, but is not found in pre-pubertial testis or in other tissues. It accounts for some 80–90% of the total lactate dehydrogenase activity in sperm. The C isozyme has been observed in man and in various other species, but inherited variants have so far only been demonstrated in pigeons (Bianco et al. 1964). Apparently the C subunit is formed only in the primary spermatocyte and then only for a short time. Although isozymic tetramers containing the C subunit as well as the A or B subunits may be generated *in vitro* by appropriate dissociation and recombination experiments, they do not occur *in vivo*. This is presumably because the A or B subunits do not occur together with the C subunits when it is being synthesised in the primary spermatocyte.

Chromosomal assignments of LDH A- and B-loci: The linkage relations and chromosomal locations of the loci determining the A- and B-polypeptide chains of lactate dehydrogenase have been studied using the technique of interspecies somatic cell hybridisation.

This technique (Ruddle 1972) involves the formation in tissue culture of hybrid cells from two different species (e.g. mouse and man, or chinese hamster and man) and the subsequent study of clones derived from single hybrid cells. Very often the enzyme products of particular homologous loci in the two species can be identified in the hybrid cells and distinguished from one another by electrophoresis. However the clones have usually lost a number of the chromosomes originally present when the primary fusion to form the hybrid cell occurred, and of course only the products of loci on chromosomes retained by the cells will be present. Furthermore in hybrid cells derived from fusions between permanent animal cell lines and human fibroblasts or leucocytes, it appears to be a general rule that only the human chromosomes are lost. But the loss of different chromosomes occurs in an apparently irregular manner, so that a series of independently derived clones will vary from one to another in the particular human chromosomes which have been retained. Thus if an extensive series of such clones is examined for the presence of the enzyme products of two different human loci it is possible to discover whether the products of the two loci are retained or lost together, or whether they are retained or lost independently. If they are consistently retained or lost together, this implies that the two loci are located on the same chromosome. But if their retention or loss is uncorrelated then they are presumably located on different chromosomes. Furthermore by identifying cytologically the particular human chromosomes retained in the same series of clones, it is possible to correlate the retention or loss of the enzyme product of a particular locus with the retention or loss of a particular chromosome. In this way specific loci can be assigned to particular chromosomes. The principles and practical problems involved in this kind of analysis have been extensively reviewed (e.g. Ruddle 1972, 1973).

In the case of lactate dehydrogenase it was found that the human A and B polypeptides appeared to be retained or lost independently in clones from mouse–human somatic cell hybrids (Boone and Ruddle 1969, Nabholz et al. 1969). Subsequently it was shown that the LDH_A locus is located on chromosome No. 11 (Boone et al. 1972) and that the LDH_B locus is located on chromosome No. 12 (Chen et al. 1973).

Linkage studies by pedigree analysis such as those described on pp. 48–50, have not been possible in this particular case because such studies require the investigation of the children of individuals who are heterozygous for alleles at both loci, and since allelic variants at each of these loci are quite rare, informative families are hard to find.

Linkage information about the LDH_C locus is at present lacking in man, because the C-polypeptide cannot be detected in tissue cultured cells. Also no human LDH C variants have been found. However it is of interest that in pigeons the LDH_B and LDH_C loci appear to be closely linked (Zinkham et al. 1969). Here both B- and C-polypeptide variants occur and the segregation of the genes determining them was studied in breeding experiments.

2.2.3. Phosphoglucomutase

Phosphoglucomutase is a phosphotransferase enzyme which catalyses the interconversion of glucose-1-phosphate and glucose-6-phosphate. It has an important role in carbohydrate metabolism and is found in most tissues. Electrophoretic studies have shown that multiple molecular forms of the enzyme occur and it has been established that at least three gene loci are involved in determining their structures. However unlike the situation with lactate dehydrogenase, it seems that in this case the characteristic polypeptide products of the different gene loci do not associate together in single isozymes, because the individual isozymes are monomers.

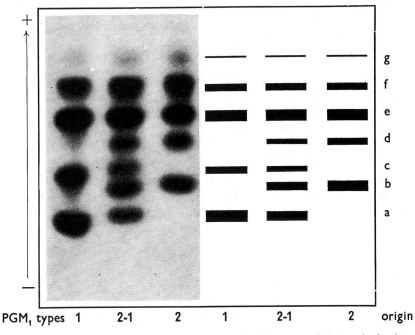

Fig. 2.19. Photograph and diagram of red cell phosphoglucomutase isozymes in the three common PGM₁ types: PGM₁ 1, PGM₁ 2-1 and PGM₁ 2. Electrophoresis in starch gel at pH 7.4 (Spencer et al. 1964a).

The complexity of the phosphoglucomutase isozymes first became apparent in studies on human red cells obtained from different individuals (Spencer et al. 1964a). When red cell lysates were subjected to starch gel electrophoresis and a staining procedure which specifically detected phosphoglucomutase activity was applied, multiple zones of activity were demonstrable, and it was found that there were clear cut and consistent person to person differences in the electrophoretic pattern. Three common phenotypes were readily recognised (fig. 2.19) and they are now referred to as $PGM_1 1$, PGM_1 2–1 and PGM_1 2. Isozymes *a* and *c* occur in PGM_1 1 and PGM_1 2–1, but not in PGM_1 2. Isozymes *b* and *d* are absent in PGM_1 1 but occur in PGM_1 2–1 and PGM_1 2. Isozymes *e*, *f* and *g* are present in all three types. In the English population about 58% of people are PGM_1 1, about 36% PGM_1 2–1 and about 6% PGM_1 2. These three common types also occur in many other human populations, so this polymorphism is evidently widespread throughout the species.

Studies of these phenotypes among the members of a very large number of different families (table 2.2) show that the segregation pattern conforms very

TABLE 2.2

The distribution of phosphoglucomutase types (PGM_1) in 537 families (Harris et al. 1968).

Parents	Number of matings	Children			Total
		1	2–1	2	
1×1	199	392	—	—	392
1×2–1	203	207	215	—	422
1×2	34	—	71	—	71
2–1×2–1	77	35	81	41	157
2–1×2	21	—	24	31	55
2×2	3	—	—	13	13
Totals	537	634	391	85	1110

closely to Mendelian expectations on the simple hypothesis that the polymorphism is determined by two common autosomal alleles (PGM_1^1 and PGM_1^2). Phenotypes PGM$_1$ 1 and PGM$_1$ 2 represent the homozygous genotypes $PGM_1^1 PGM_1^1$ and $PGM_1^2 PGM_1^2$, and phenotype PGM$_1$ 2–1 the heterozygous genotype $PGM_1^1 PGM_1^2$. Thus the isozymes *a* and *c* appear to be determined by one allele (PGM_1^1), and the isozymes *b* and *d* by the other (PGM_1^2). In the heterozygote where both alleles are present all four isozymes occur, but there is no evidence in this case of 'hybrid' isozyme formation.

Following the discovery of these common phosphoglucomutase types, blood samples from several thousand different individuals were examined, and in the course of this work a number of other isozyme patterns each characteristic of the individual, but each very rare, were discovered (Hopkinson and Harris 1966). Pedigree studies show that many of these can be attributed to the heterozygous combination of either PGM_1^1 or PGM_1^2 with one or another of a series of rare alleles at the same gene locus. Some examples of such isozyme patterns are shown in fig. 2.20. In each case there are two unusual isozymes presumably determined by the particular rare allele, and

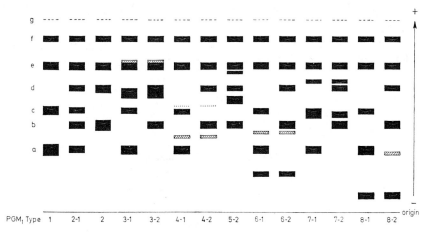

Fig. 2.20. Diagram showing red cell phosphoglucomutase isozymes observed in 14 different PGM₁ types (Hopkinson and Harris 1966).

also either the *a* and *c* isozymes determined by PGM_1^1, or the *b* and *d* isozymes determined by PGM_1^2.

The fact that isozymes *e, f* and *g* are not apparently affected by the common or rare alleles mentioned so far, suggests that they are determined by a second gene locus (PGM_2). Direct evidence for this (Hopkinson and Harris 1965, 1966) came from the recognition of several rare variant types which appeared to involve *e, f* and *g* and to occur independently of the variants attributable to the first locus (PGM_1). Several examples of these PGM_2 variant isozyme patterns are illustrated in fig. 2.21. They are shown together with the *a, b, c* and *d* isozymes characteristic of PGM₁ 2–1, but they may also be associated with *a* and *c* alone (i.e. with PGM₁ 1) or with *b* and *d* alone (i.e. with PGM₁ 2).

From pedigree studies one may infer that these different variant *e, f, g* patterns occur in individuals heterozygous for one or another of several uncommon alleles at a second gene locus (PGM_2). The usual allele at this locus (PGM_2^1) for which most people are homozygous, is thought to determine isozymes *e, f* and *g*, and the different mutant alleles to result in altered forms of these isozymes with characteristic and consistent changes of electrophoretic mobility. The patterns seen in the different heterozygotes can be interpreted as due to a simple mixture of the isozymes determined by the two alleles present. In some cases (e.g. PGM₂ 3–1) the new isozymes determined by the mutant allele are clearly separated from isozymes *e, f* and *g* determined by the usual allele (PGM_2^1). But in other cases the two groups may

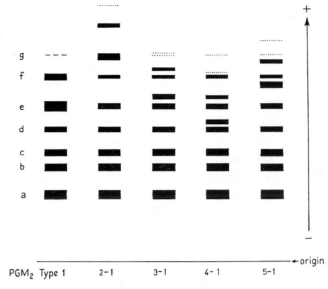

Fig. 2.21. Diagram showing red cell phosphoglucomutase isozyme patterns in five different PGM₂ types. In each case the PGM₁ type shown is PGM₁ 2-1 (Hopkinson and Harris 1966; Parrington et al. 1968.)

overlap to some extent. For example in PGM_2 2–1, the slowest isozyme of the variant group appears to have about the same electrophoretic mobility as g, the fastest of the three isozymes determined by PGM_2^1.

Although the phosphoglucomutase isozymes were first identified in red cells, it was subsequently found that they also occur in their characteristic types in other tissues, such as liver, kidney, muscle, brain and placenta. Also the isozymes can be demonstrated in fibroblasts grown *in vitro* in tissue cultures derived from small explants of skin, and these have been shown to retain after quite a number of cell generations the characteristic phenotypes of the original donors. In many of these tissues the total phosphoglucomutase activity is considerably greater than in red cells and this led to the discovery (Hopkinson and Harris 1968) of a further set of isozymes determined by yet a third gene locus (PGM_3). These additional isozymes usually account for only a very small proportion of the total phosphoglucomutase activity present in a given tissue, and so can only be conveniently studied where the total activity is relatively high. In practice placenta has proved to be an extremely useful source of material for this purpose.

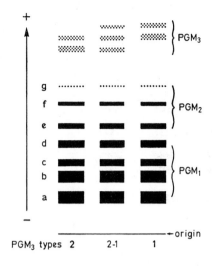

Fig. 2.22. Diagram of three types of PGM$_3$ isozymes, PGM$_3$ 1, PGM$_3$ 2-1 and PGM$_3$ 2, as seen in placental extracts. In each case the PGM$_1$ isozymes are PGM$_1$ 2-1, and the PGM$_2$ isozymes are PGM$_2$ 1 (Hopkinson and Harris 1968).

This third set of isozymes migrate more rapidly towards the anode than the others, and the typical electrophoretic patterns that are observed are shown in fig. 2.22. Three distinct phenotypes have been recognised (PGM$_3$ 1, PGM$_3$ 2–1 and PGM$_3$ 2) and genetical analysis indicates that they are determined by two common alleles PGM_3^1 and PGM_3^2, the phenotype PGM$_3$ 2–1 representing the heterozygote. The types occur independently of those determined by the alleles at the other loci.

Thus evidence for three different gene loci separately determining a distinct set of phosphoglucomutase isozymes has been obtained. At each of these loci multiple alleles occur. The sets of isozymes thought to be attributable to the various alleles may be conveniently displayed as in fig. 2.23. Individuals homozygous at all three loci show at least eight phosphoglucomutase isozymes. In individuals heterozygous at one or more loci the isozyme patterns are more complex, and it is apparent that a very large number of different combinations of phosphoglucomutase isozymes are possible in different individuals. Their relative incidence will of course depend on the frequencies of the different alleles in the particular population, and most of the possible allelic combinations are very rare. However there are three common PGM$_1$ types and also three common PGM$_3$ types so one can classify people in the English population for example, into at least nine distinct types, each of which has an appreciable frequency in the general population (table 2.3). Thus quite a high degree of individual differentiation occurs.

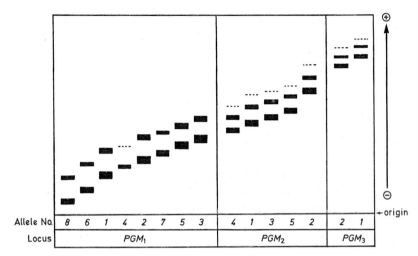

Fig. 2.23. Diagram of phosphoglucomutase isozymes determined by eight different alleles at locus PGM_1, five alleles at locus PGM_2, and two alleles at locus PGM_1 (Hopkinson and Harris 1969). At least one additional, relatively weak, fast moving isozyme probably also occurs consistently in the PGM_1 series, but is generally obscured by the other isozymes present in the different genotypes.

Studies of the relative contributions of the isozyme products of the three loci to the total phosphoglucomutase activity in different tissues indicate that usually most of the activity (80–95%) is attributable to PGM_1 and the remainder is mainly due to PGM_2 (McAlpine et al. 1970). Red cells however are exceptional since here half or more of the total phosphoglucomutase is

TABLE 2.3

Estimates of the percentage incidence in the English population of individuals with various combinations of PGM₁ and PGM₃ phenotypes.

| | | PGM₁ phenotype | | | |
		PGM₁ 1	PGM₁ 2-1	PGM₁ 2	Total
PGM₃	PGM₃ 1	31.8	19.4	3.0	54.2
phenotype	PGM₃ 2-1	22.7	13.9	2.1	38.7
	PGM₃ 2	4.0	2.5	0.4	6.9
	Total	58.5	35.8	5.5	99.8

due to PGM_2. In all tissues PGM_3 contributes at most only a small fraction to the total activity and in some tissues (e.g. red cells and muscle) it is not detectable.

Important differences in the catalytic properties of the various isozymes were discovered by the development of staining methods for detecting phosphoribomutase activity using ribose-1-phosphate or ribose-5-phosphate as substrates (Quick, Fisher and Harris 1972). It was found that the PGM_2 isozymes gave strong phosphoribomutase activity, whereas the other isozymes showed only trace activity or none at all.

The possible metabolic significance of the enzyme products of the three loci was further examined by kinetic studies using glucose-1-phosphate and ribose-1-phosphate as substrates in the presence of varying concentrations of the coenzyme glucose-1, 6-diphosphate (Quick et al. 1974). The main results are summarised in table 2.4. It will be seen that at very low concentrations of the coenzyme, the PGM_1 isozymes are much more effective in catalysing the phosphoglucomutase reaction than either the PGM_2 or PGM_3 isozymes. However as the coenzyme concentration is increased, the $K_{m\ G-1-P}$ for the PGM_2 isozymes decreases markedly while no such change occurs in the $K_{m\ G-1-P}$ for either PGM_1 or PGM_3 isozymes. At coenzyme concentrations of between 250 and 340 μM, the $K_{m\ G-1-P}$ for the PGM_2 isozymes is virtually the same as for the PGM_1 isozymes.

With ribose-1-phosphate as substrate the findings are quite different. The PGM_2 isozymes are evidently much more effective than the PGM_1 isozymes in catalysing the phosphoribomutase reaction, and the $K_{m\ R-1-P}$ for either the PGM_1 or PGM_2 isozymes is not appreciably affected by increasing the coenzyme concentration. The $K_{m\ R-1-P}$ for the PGM_3 isozymes could not be determined because of the very low activities. Inhibition studies showed that ribose-1-phosphate inhibited the phosphoglucomutase activity of the isozymes from all three loci, and was apparently competing with glucose-1-phosphate for the same sites. However the relative affinity of ribose-1-phosphate is evidently very much greater for the PGM_2 isozymes than for the PGM_1 or PGM_3 isozymes.

The values obtained for the Michaelis constants (K_m) in these various experiments suggest that in vivo the PGM_1 isozymes function primarily as phosphoglucomutases. The PGM_2 isozymes however may function as phosphoribomutases or phosphoglucomutases, the latter depending on the intracellular concentration of glucose-1,6-diphosphate. In fact estimates of the concentrations of this substance in different tissues indicate that they often lie in the critical range in which marked changes in the $K_{m\ G-1-P}$ for

TABLE 2.4

Estimated K_m values of PGM_1, PGM_2 and PGM_3 for glucose-1-phosphate and ribose-1-phosphate with increasing concentrations of glucose-1,6-diphosphate (Quick et al. 1974).

Substrate	Concentration of G-1,6-diP (μM)	K_m (μM)		
		PGM_1	PGM_2	PGM_3
Glucose-1-phosphate	10	49	1600	31,000
	190	53	490	17,000
	250	45	110	36,000
	340	34	19	18,000
Ribose-1-phosphate	10	12,000	120	—
	190	11,000	130	—
	340	—	150	—

the PGM_2 isozymes occur, so the actual concentration may exert some kind of regulatory effect.

The function of the PGM_3 isozymes remains obscure. The estimates of the kinetic parameters with either glucose-1-phosphate or ribose-1-phosphate do not suggest that these isozymes would operate effectively as either phosphoglucomutases or phosphoribomutases *in vivo*. Possibly some other phosphate ester is the true substrate.

The isozymes derived from the three *PGM* loci have also been found to differ in certain other characteristics. The PGM_2 isozymes appear to be appreciably larger in molecular size (about 10–20%) than either the PGM_1 or PGM_3 isozymes (McAlpine et al. 1969b, Monn 1969a, Santachiara and Modiano 1969). However the individual isozymes determined by the alleles at any one gene locus (e.g. a, b, c and d from PGM_1) are all of the same size.

It has also been shown by thermostability studies that the PGM_2 isozymes are significantly more stable than either the PGM_1 or PGM_3 isozymes (McAlpine et al. 1969c). The isozymes of the three loci have also been found to differ in their reactivities with various sulphydryl reagents (Fisher and Harris 1972).

Each of the alleles at the different loci appears to determine a set of two or more separate isozymes (fig. 2.23), the strongest activity being seen in the slowest moving band (towards the anode), with the others showing decreasing activity with increasing anodal mobility. The relative activities of the isozymes in each set determined by any particular allele varies however with the average age of the cell population from which they come. For example the proportion of the total activity contributed by the slowest moving isozyme in each set is much greater in reticulocytes and young red cell populations than in mature red cells (Monn 1969b). It is also much greater in leucocytes than in mature red cells (Monn 1969b). Similarly the slowest migrating isozyme is much more prominent in extracts of tissues such as placenta and also in lymphoid or fibroblast cells growing *in vitro* in tissue culture, than in red cells (Fisher and Harris 1972). Red cells in contrast to leucocytes, cells grown in tissue culture and cells in tissues like placenta, are relatively long lived (about 120 days) and lose their nuclei and hence their capacity for further enzyme synthesis at an early stage of their development. These findings, which are consistent for the sets of isozymes of all the alleles which have been studied, suggest that in each set the slowest migrating isozyme is the primary form synthesised and that the others are generated from this by secondary structural modifications in the course of the life span of the cell. The precise nature of these secondary modifications to the enzyme protein structure in this case is not known, but they might perhaps be due to deamidation of certain asparagine or glutamine residues in the molecule. The occurrence of such so-called secondary isozymes is not an uncommon phenomenon (see pp. 78–84).

Linkage relations and chromosomal assignment of PGM loci: That the pairs of loci PGM_1 and PGM_2, PGM_1 and PGM_3 and PGM_2 and PGM_3 are not closely linked was first demonstrated by pedigree studies (Hopkinson and Harris 1965, 1968, Parrington et al. 1968, Lamm 1970). This was subsequently confirmed using the technique of somatic cell hybridisation (p. 59), and these studies led to the localisation of the PGM_1 locus to the short arm of chromosome No. 1 (Van Cong et al. 1971, Douglas et al. 1973, Burgerhout et al. 1973, Jongsma et al. 1973) and the assignment of the PGM_2 locus to chromosome No. 4 (McAlpine 1974) and the PGM_3 locus to chromosome No. 6 (Jongsma et al. 1973).

2.3. Genes and isozymes

Lactate dehydrogenase and phosphoglucomutase illustrate how in the cells of a single individual a number of distinct and separable proteins may exhibit the same or very similar enzyme activities. Such multiple molecular forms of an enzyme occurring in a single organism are usually referred to as isozymes (or isoenzymes), a term first introduced by Markert and Møller in 1959 and now widely found in the biochemical literature. It is a convenient term operationally, because it implies no specific type of structural relationship between the several protein species which may be observed to have similar enzyme activities. Indeed, as more and more examples of the phenomenon have been studied, it has become clear that different sorts of molecular relationship are likely to be involved in different cases, and they may be brought about in a variety of ways.

The rapid discovery of many examples of isozyme systems in recent years has largely been a product of the extensive use of relatively simple methods of zone electrophoresis in such supporting media as starch gel or acrylamide, combined with the development of sensitive and specific staining methods for demonstrating the particular zones of enzyme activity in what is usually a very complex protein mixture. This general procedure first developed for esterases (Hunter and Markert 1957) and dehydrogenases (Markert and Møller 1959) has since been extended to many types of enzymes, and has proved to be a peculiarly powerful experimental tool. Often it can be applied directly to crude homogenates of fresh tissues, so that possible *in vitro* changes of the enzyme components of the mixtures can be minimised. The procedure can of course also be applied at successive stages of the separation and purification of the isozyme components so that any alterations produced by particular biochemical manipulations may be readily assessed. Furthermore only relatively small quantities of material are generally required, and this is of particular value in genetical studies where it is usually necessary to examine the isozymes in a given tissue from a large number of different individuals.

The widespread application of this general procedure, combined of course in different cases with other techniques, has led to the identification of a considerable variety of isozyme systems in many different species, and the subject has been extensively reviewed (see, for example, 'Multiple molecular forms of enzymes', *Annals New York Acad. Sci. 151*, 1–689, 1968, Harris 1969, Hopkinson and Harris 1971). It has become clear that the phenomenon is a very general one and quite probably it is exhibited by most enzymes at least in some degree. Furthermore, many non-enzymic proteins have been found to show essentially the same kind of phenomenon, so that it seems that a multiplicity of molecular forms with similar functional activities is a regular feature of most enzymes and proteins.

But such multiple forms of a given enzyme can evidently be generated in a variety of different ways and a combination of both biochemical and genetical methods of analysis are usually required to elucidate any particular case.

The different possible causes of isozymes or, more generally, of the multiple molecular forms of functionally similar proteins can be classified for convenience into three distinct categories (Harris 1969). These are 1) multiple gene loci; 2) multiple allelism at a single locus; and 3) secondary or post-translational changes.

Although these different types of cause are considered separately in what follows, it is important to remember that in the case of any particular enzyme more than one of them may be operating to generate the multiple molecular forms which are observed. For example in the case of phosphoglucomutase (p. 60), there are three separate loci each determining a distinctive set of isozymes; at each locus multiple allelism occurs, each allele determining a variant of the particular isozyme set; and finally each set of isozymes determined by any given allele appears to consist of a 'primary' isozyme which is the form thought to be synthesised initially, and so-called 'secondary' isozymes which are generated from it in the course of the life span of the cell.

2.3.1. Multiple loci determining isozymes

Both lactate dehydrogenase (p. 52) and phosphoglucomutase (p. 60) are examples of situations where three separate loci are concerned in determining the various isozymes which are seen. But the formation of the isozymes differ markedly in the two examples. The lactate dehydrogenase are tetramers and in most cells where both the '*A*' and '*B*' genes are active, isozymes representing various tetrameric combinations of the two different polypeptide products of the loci are found, though their relative proportions vary from tissue to tissue. Hybrid isozymes containing the polypeptide product of the '*C*' gene locus are however not observed. This is apparently because the '*C*' locus produces its polypeptide product only in a particular cell type, the primary spermatocyte, and then evidently only at a time when the A and B polypeptides are not being formed. Thus the pattern of isozyme formation depends both on the subunit structure (in this case tetrameric) and also on the tissue site and timing of the action of the separate gene loci. The phosphoglucomutase isozymes in contrast, appear to be monomers, so that although the products of separate loci are present in most cells, they do not associate to produce hybrid molecules. Also one may note that although tissue differences in isozyme pattern certainly occur with phosphoglucomutase, they are much less marked than with lactate dehydrogenase.

Table 2.5 lists a series of other enzymes in which two or more loci appear to be concerned in isozyme formation. They illustrate in different ways various features of the phenomenon. Amylase, for instance is an example of marked tissue differentiation. The isozyme products of one locus (AMY_S) are only formed in cells of the salivary glands and the products of the other locus (AMY_P) only in cells of the pancreas. The same general phenomenon is illustrated by the three loci ('A' 'B' and 'C') of aldolase (p. 199). The A polypeptide is the predominant form in muscle though it also occurs in most other tissues. The B polypeptide predominates in liver and the C polypeptide is characteristic of brain.

In some cases marked changes in isozyme pattern occur during the course of development of a single tissue, apparently due to progressive changes in the relative activities of the several loci. One example of this is lactate dehydrogenase in the development of heart muscle and liver (p. 54). In early foetal life similar amounts of the A and B lactate dehydrogenase polypeptides are present in both tissues, but as foetal development proceeds there is a progressive shift in their relative proportions so that eventually B polypeptides predominate in heart and A polypeptides in liver.

The sequential changes which occur in the isozyme pattern of alcohol dehydrogenase in liver during development provide another illustration of the phenomenon (Smith, Hopkinson and Harris 1971, 1972, 1973). They are illustrated in fig. 2.24. The alcohol dehydrogenase isozymes are dimers and may contain α-polypeptide subunits derived from locus ADH_1, β-subunits from ADH_2 or γ-subunits derived from ADH_3. In foetal liver before about the 20th week of gestation the ADH_1 locus activity predominates so that the only isozyme seen has the subunit structure αα. Later in foetal life there is a progressive increase in the amount of β polypeptide derived from the ADH_2 locus, and isozymes with the subunit structures αβ and ββ appear. At birth αα, αβ and ββ isozymes are all present. Sometime later, during the first year or two of post-natal life the γ polypeptides determined by the ADH_3 locus appear as well and so one finds in addition αγ, βγ and γγ isozymes. This situation persists into adult life and in adult liver at least six types of ADH isozymes are observed, three of which are homodimeric

TABLE 2.5

Some enzymes determined by two or more gene loci.

Enzyme	Loci	Polypeptide subunits	Tissue activity distribution	Isozyme subunit structure	Interloci hybrids in tissues	References
Lactate dehydrogenase	LDH_A LDH_B LDH_C	A B C	widespread testis and sperm	tetramers	A and B, not A & C, or B & C	see pp. 52–60
Phosphoglucomutase	PGM_1 PGM_2 PGM_3	'a' 'e' 'h'	widespread most tissues, but not red cells or skeletal muscle	monomers	—	see pp. 60–69
Alcohol dehydrogenase	ADH_1 ADH_2 ADH_3	α β γ	liver: α, β, γ lung: β stomach: γ kidney: γ, β	dimers	α, β & γ in liver; not in lung, stomach or kidney	Smith et al. 1971, 1972, 1973
Aldolase	ALD_A ALD_B ALD_C	A B C	widespread liver, kidney brain	tetramers	A, B & C where they occur together	see pp. 199–203

TABLE 2.5 (continued)

Enzyme	Loci	Polypeptide subunits	Tissue activity distribution	Isozyme subunit structure	Interloci hybrids in tissues	References
Amylase	AMY_S AMY_P	S P	salivary glands pancreas	? ?	—	Kamaryt and Laxova 1965, Merritt et al. 1971, 1973, Boetcher and De la Lande 1969, De la Lande and Boetcher 1969, Karn et al. 1973
Pyruvate kinase	PK_1 PK_3		red cells, liver widespread, but not red cells	? probably tetramer	?	see pp. 203–204
Carbonic anhydrase	$CA\ I$ $CA\ II$	B C	red cells, kidney medulla, gall bladder mucosa widespread except heart and testis	monomers	—	Tashian et al. 1968, Tashian 1969, Moore et al. 1971, 1973, Andersson et al. 1972, Henderson et al. 1973, Funakoshi and Deutsch 1971
Superoxide dismutase	SOD_A SOD_B	A B	widespread	dimer tetramer	—	Brewer 1967, Beauchamp and Fridovich 1971, Beckman 1973, Ruddle 1973
Acid phosphatase	ACP_1 ACP_2 ACP_3	'red cell' β a	widespread	monomer dimers	— a & β in some tissues e.g. placenta	see pp. 179–186, Swallow et al. 1973, Beckman and Beckman 1967, Swallow and Harris 1972

Enzyme	Gene	Tissue distribution	Structure	α and β	References
Glycerol-3-phosphate dehydrogenase	GPD_1 α GPD_2 β	muscle α & β, liver α; α & β widespread in other tissues but weak activity	dimers	α and β	Kompf et al. 1971, 1972, Hopkinson et al. 1974
Glutamate-oxalate transaminase (soluble (S) and mitochondrial (M) forms)	GOT_S GOT_M	widespread	dimers	—	Bodansky et al. 1966, Davidson et al. 1970, Hackel et al. 1972
Isocitrate dehydrogenase (soluble (S) and mitochondrial (M) forms)	ICD_S ICD_M	widespread	dimer ?	—	Chen et al. 1972, Turner et al. 1974b, Turner et al. 1974a
Malate dehydrogenase (NAD requiring) (soluble (S) and mitochondrial (M) forms)	MOR_S MOR_M	widespread	dimers	—	Davidson and Cortner 1967a, Davison and Cortner 1967b
Malic enzyme (NADP requiring) (soluble (S) and mitochondrial (M) forms)	MOD_S MOD_M	widespread	tetramers	—	Cohen and Omenn 1972a, Shows and Ruddle 1968, Baker and Mintz 1969

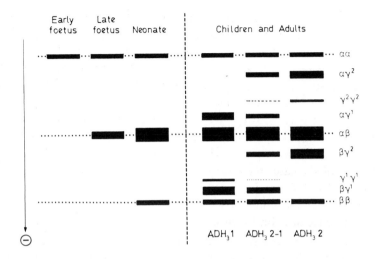

Fig. 2.24. Sequential changes of alcohol dehydrogenase isozyme patterns in liver during development. The ADH_1 locus determines the α-polypeptides, the ADH_2 locus determines the β-polypeptides and the ADH_3 locus determines the γ-polypeptides. There are two common alleles at the ADH_3 locus which determine respectively the polypeptides γ^1 and γ^2. The ADH_3 locus is not active in foetal liver, but in children and adults three different phenotypes occur, ADH_3 1, ADH_3 2-1 and ADH_3 2, and the corresponding isozyme patterns are indicated. There is also polymorphism at the ADH_2 locus (see p. 299) but its effects on the isozyme patterns are not shown here. For full details see Smith et al. (1971, 1972 and 1973).

($\alpha\alpha$, $\beta\beta$ and $\gamma\gamma$) and three heterodimeric ($\alpha\beta$, $\alpha\gamma$ and $\beta\gamma$). In fact the situation is even more complex because there are two common alleles at the ADH_3 locus which determine the structure of alternative forms of the γ polypeptide (γ^1 and γ^2) and the isozyme pattern in adult liver in different individuals varies according to which alleles are present. Allelic differences have also been shown to occur at the ADH_2 locus and this further complicates the picture. It is of interest to note that the same changes in development are not observed in lung or gastro-intestinal mucosa, two other tissue sites at which appreciable amounts of alcohol dehydrogenase activity can be detected. In lung the ADH_2 locus and in the gastrointestinal mucosa the ADH_3 locus predominate throughout from early foetal life.

Certain enzymes occur in distinct molecular forms which are present in the same cell but are separated within the cell, one form being localised in mitochondria and the other occurring in the cytosol. These so-called mito-chondrial and soluble forms of the enzyme (e.g. glutamate-oxalate transaminase, isocitrate dehydrogenase, malate dehydrogenase and malic enzyme table 2.5), are coded by separate gene loci. They often appear to be very similar to one another both in molecular size and in subunit structure. However in no case does one find hybrid isozymes containing polypeptides determined by the separate loci, even though the isozymes are usually multimeric and are both formed in the same cell. The reason for this is not known. It may be that there is in these cases some special structural feature of the polypeptides formed by the two loci which prevents their stable association together in the same hybrid molecule. It is of interest that in several of the cases the mitochondrial and soluble forms of the enzyme are markedly different in molecular charge, and this may reflect certain critical structural features of the molecules which are perhaps the reason for the absence of hybrid forms.

Where a set of isozymes are determined by two or three separate gene loci as occurs in each of the cases listed in table 2.5, it is of interest to know whether the separate loci are closely linked on the same chromosome or whether they are well separated or located on different chromosomes. Information on this point is not yet available for most of the examples given in table 2.5. But among those where it has been possible to carry out relevant investigations, evidence for linkage has been obtained only in the case of salivary and pancreatic amylase. Here the loci Amy_S and Amy_P have been shown to be closely linked (Merritt et al. 1972) and located on chromosome No. 1 (Merritt et al. 1973). In several cases however it has been found that the loci determining a set of isozymes are not closely linked and are often indeed located on separate chromosomes. Examples are the lactate dehydrogenase loci A and B (p. 60); phosphoglucomutase loci PGM_1, PGM_2 and PGM_3 (p. 69); the malate dehydrogenase loci MOR_S and MOR_M (Van Heyningen et al. 1973) and the superoxide dismutase loci SOD_A and SOD_B (Ruddle 1973).

2.3.2. *Multiple allelism at a single locus*

At any given gene locus a number of different alleles may occur in a population of individuals, each allele coding for a structurally distinct version of the particular polypeptide chain, so that the primary structure of the enzyme or protein involved will differ from one individual to another, according to the alleles that they happen to carry at the locus in question. Heterozygotes, since they carry two different alleles, will in general be expected to show a more complex pattern of isozymes than homozygotes. In some cases, the pattern of isozymes in a heterozygote will represent a simple mixture of those occurring in the two corresponding homozygous types. But in other cases, where the enzyme is polymeric, additional 'hybrid' isozymes not present in either type of homozygote occur (see pp. 32–37).

The extent to which the several polypeptide products of different gene loci which are concerned in determining a set of isozymes differ in their primary aminoacid sequences no doubt varies very widely from case to case. But it is perhaps not unusual for quite extensive differences to be present. By contrast the different polypeptide products due to allelic differences at a single locus probably differ from each other in most cases by only single aminoacid substitutions. Thus, in general, isozyme differences arising because of the occurrence of multiple gene loci, are likely as a rule to involve a greater degree of molecular difference than isozymes arising because of multiple allelism at a single locus.

However, the critical point of distinction between these separate genetical causes of isozyme formation is that multiple allelism results in differences between individual members of a population in the pattern of isozymes they form, whereas multiple loci will in general be common to all members of the species, and thus define the overall pattern of isozyme formation that occurs.

2.3.3. *Secondary isozymes*

Multiple gene loci and multiple allelism at single loci provide, as it were, the basic genetical framework which defines the main characteristics of the various molecular forms of functionally similar enzymes or proteins which occur in individual members of a species. But the complexity of many isozyme systems which have been studied cannot be fully explained in these terms alone. It seems that secondary modifications of protein structures subsequent to the primary synthesis of their polypeptide chains on the mRNA templates in the ribosomes, is also an important and general cause of the multiplicity of separable components which are often observed when particular enzymes or proteins are studied by such techniques as electrophoresis.

Nucleoside phosphorylase: A particularly striking example of the generation of these so-called secondary isozymes is provided by the enzyme purine-nucleoside phosphorylase, which catalyses the cleavage of inosine to hypoxanthine and ribose-1-phosphate.

When red cell haemolysates are examined electrophoretically a series of' seven or eight nucleoside phosphorylase isozymes are seen (fig. 2.25). Red cells of course have a relatively long life span (> 100 days) and lose their nuclei and cease enzyme and protein synthesis at an early stage. A sample of red cells taken from a normal individual therefore consists of cells of a wide range of differing ages. Because of their differences in specific gravity it is possible by centrifugation in a density gradient to separate the cells into fractions of differing age. At one end of the fraction there will be pre-

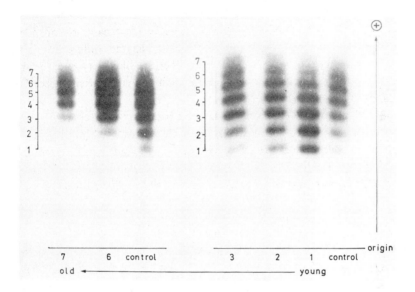

Fig. 2.25. Isozyme components of purine nucleoside phosphorylase (NP) in red cells fractionated by density centrifugation (Edwards et al. 1971, Turner et al. 1971). The controls shown represent a complete haemolysate from a normal individual. Fractions 1, 2 and 3 from the density centrifugation represent, in order, the youngest cell fractions. Fractions 6 and 7 represent the oldest cell fractions. Fractions 4 and 5 representing cells of intermediate age are not shown.

dominantly young cells (with relatively low specific gravities) and at the other predominantly older cells (with relatively high specific gravities). In all fractions the cells lack nuclei but the very youngest cell fractions will contain a proportion of reticulocytes. Thus it is possible to compare the isozyme patterns in red cells of differing average ages. When nucleoside phosphorylase is examined in this way, it is found that the faster moving (more anodal) isozymes are relatively more prominent in the older cell fractions, and the slower moving (i.e. less anodal) isozymes are relatively more prominent in the younger cell fractions, the least anodal isozymes being the most active in the youngest cells (Edwards, Hopkinson and Harris 1971, Turner, Fisher and Harris 1971). There is in fact a progressive change in the isozyme patterns as the red cell ages (fig. 2.25).

Similar effects are observed when comparisons of whole red cell samples from normal individuals and from individuals with various kinds of hae-molytic disease are made. The haemolytic disease samples, which in general have a greater proportion of younger cells and reticulocytes, show nucleoside phosphorylase patterns in which the slower moving isozymes are relatively much more active than in the normal individual, while the faster moving isozymes are relatively weaker. In other tissues such as liver and kidney, isozymes with similar mobilities to those seen in red cells are found, but the slower moving ones predominate, the slowest being the most active and the others diminishing in activity with increasing electrophoretic mobility. The cells in such tissues though they no doubt vary in age are all nucleated and capable of continuing enzyme and protein synthesis. Fibroblasts and lymphocytoid cells grown in tissue culture divide relatively rapidly and are actively synthesising new enzyme protein, and in such cells most of the activity is attributable to a single isozyme corresponding to the least anodal isozyme of the red cell pattern, though much weaker isozymes with mobilities corresponding to the next one or two series may also be observed.

Taken as a whole these findings suggest that the slowest moving (i.e. least anodal) nucleoside phosphorylase isozyme is the primary form of the enzyme synthesised in all cells, and that the multiple series of discrete, more rapidly moving, isozymes which are observed in varying amounts in different tissues, represent secondary forms of the enzyme protein which are generated in a stepwise manner from the primary form and from one another, in the course of the life span of the cell. The various isozymes all appear to be of the same molecular size (Edwards et al. 1971). However they differ to some extent in their kinetics (Turner et al. 1971).

Nucleoside phosphorylase is a trimer (p. 87). A number of rare electro-phoretic variants have been found (Edwards et al. 1971) and in each case when the isozymes were examined in fibroblasts in tissue culture, typical four banded patterns expected in heterozygotes for trimeric enzyme proteins (p. 35) were obtained. In red cells however the isozyme patterns in these heterozygous variants were very complex indeed. They appeared to be made up of the four primary isozymes seen in the fibroblasts and also of four separate series of secondary isozymes derived from each of them (fig. 2.26).

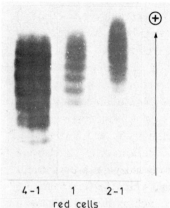

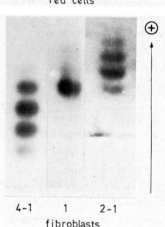

Fig. 2.26. Purine nucleoside phosphorylase (NP) isozyme patterns in red cells and in fibroblasts growing in tissue culture from individuals of different genotypes. Type 1 shows the pattern observed in the usual phenotype (genotype NP^1 NP^1). Types 2-1 and 4-1 show the patterns in individuals heterozygous for the rare alleles NP^2 and NP^4, i.e. genotypes NP^2 NP^1 and NP^4 NP^1 (Edwards et al. 1971).

Possible structural changes: Numerous other examples of secondary isozymes have been identified and the phenomenon appears to be quite common. In most cases the generation of such secondary isozymes from the primary form appears to involve some structural change in the molecule. However the precise nature of such structural modifications are in most cases not known. A variety of possibilities can be envisaged. They may involve for example the removal of amide groups from one or more glutamine or asparagine residues in the polypeptide sequences. This would increase the net negative charge and so increase the electrophoretic mobility towards the anode. In fact a change in electrophoretic mobility in this direction is a common feature. It is seen for instance in the secondary isozymes of phosphoglucomutase (p. 69) and nucleoside phosphorylase, and other examples are the secondary isozymes of adenosine deaminase (Hopkinson and Harris

1968), adenylate kinase (Fildes and Harris 1966), peptidase B (Lewis and Harris 1967), phosphohexoseisomerase (Detter et al. 1968), the soluble and mitochondrial forms of malate dehydrogenase (Davidson and Cortner 1967a, b), esterase D (Hopkinson et al. 1973). Evidence for deamidation as a cause of secondary isozyme formation has in fact been obtained in the cases of carbonic anhydrase (Funakoshi et al. 1969), aldolase (Midelfort and Mahler 1972) and salivary amylase (Keller et al. 1971, Karn et al. 1973).

Other possible structural modifications which may occur in particular cases could involve acetylations, the oxidation of sulphydryls, the addition of phosphate groups, the addition of carbohydrate groupings with perhaps differing numbers of sialic acid residues to the molecules, the cleavage of a portion of a polypeptide chain by proteolytic enzymes with the loss of part of the amino acid sequence, and so on.

If such 'secondary' modifications affected some, but not all, of the enzyme protein molecules or affected them in different degrees, or represented a series of steps in a process through which all of them pass, then a characteristic set of secondary isozymes will be consistently found.

Conformational isomerism: It has been suggested than in some cases the several members of a set of isozyme proteins may have the same primary structure but differ in their three-dimensional, tertiary or quaternary structures, because the polypeptide chain or chains of which they are composed may be able to assume two or more relatively stable configurations (Kitto et al. 1966, Kaplan et al. 1968). Thus the isozymes would represent a set of conformational isomers, which because of the manner in which particular groups are exposed or concealed in the three-dimensional structures would differ from one another in particular properties such as electrophoretic mobility.

If this were so one would expect that the several isozymes of the set would be interconvertible. That is to say, if any one of the isozymes were isolated, it should be possible to generate from it the complete set. If the several forms were particularly stable, such interconversion might be achieved only by using a technique involving reversible denaturation (Epstein and Schechter 1968a). It was originally suggested that the isozymes of mitochondrial malate dehydrogenase might represent such a set of conformational isomers (Kitto et al. 1966), but subsequent studies involving reversible denaturation failed to demonstrate the expected interconversions (Schechter and Epstein 1968, Kitto et al. 1970). One possible example of conformational isomerism (Fisher and Harris 1971) may be the case of the two isozymes of red cell acid phosphatase which are consistently found in homozygous individuals, since here ready interconversion of the two forms appears to occur (see pp. 182–183).

In the case of certain dehydrogenases which require the coenzymes NAD or NADP, different electrophoretic patterns are observed according to the degree to which the enzyme is saturated with the coenzyme. When unsaturated, various bands are seen which apparently represent enzyme molecules to which 0, 1, 2 or more coenzyme molecules are bound. In order to obtain consistent results when working with tissue extracts it is necessary to keep the enzyme saturated with the coenzyme throughout the electrophoretic separation, and this is achieved by including the coenzyme as a constituent of the buffers used in the electrophoretic system. An example is the case of alcohol dehydrogenase (Smith et al. 1971), where if this precaution is not taken additional isozyme bands may be observed.

Aggregation and polymerisation: In certain cases isozyme molecules may aggregate or polymerise, so that molecular species of higher molecular weight are formed, and since they may be separated by electrophoresis in such media as starch gel or acrylamide, they will appear as extra isozymes. This phenomenon is well illustrated by the serum protein haptoglobin, types Hp 2-1 and Hp 2-2 (see p. 93).

Other possible causes: In the analysis of isozyme patterns it is always necessary to keep in mind the possibility that the process of extraction of the enzyme from the particular tissues, or the conditions under which the extract is kept prior to electrophoresis, may be responsible for producing some of the electrophoretically separable forms which are observed. For example some of the enzyme molecules may remain bound to other types of molecule (large or small) in the intracellular millieu and these may not be separated from the enzyme by the electrophoretic process. Another possible source of confusion may be the action of proteolytic or other enzymes in the tissue extract on the particular enzyme protein being studied, if proper conditions of storage are not used.

A further possibility is the occurrence of chemical reactions with other constituents of the tissue extract. A well recognised example of this is the reaction of certain enzymes with the oxidised glutathione which accumulates in haemolysates on storage. Reduced glutathione is present in relatively high concentrations in the intact red cell, and is rapidly converted to oxidised glutathione after haemolysis. Oxidised glutathione can undergo disulphide bond interchange with certain cysteine residues in particular enzyme proteins and the addition of one or more glutathione moieties to the enzyme protein increases its net charge and hence its electrophoretic mobility. This has been shown to be the cause of anomalous isozyme patterns observed in a number of red cell enzymes after storage of the haemolysates. The effect can be easily prevented by the addition of excess mercaptoethanol or dithioerythritol to the fresh haemolysate which maintains the reactive cysteine residues in the enzyme protein in the reduced state. Enzymes which have been found to be particularly susceptible to this type of effect are adenosine deaminase (Spencer et al. 1968, Hopkinson and Harris 1969), red cell acid phosphatase (Bottini et al. 1964, Fisher and Harris 1969), peptidase C and D (Lewis and Harris 1967, 1969b), phosphoglucomutase PGM_3 (Fisher and Harris 1972) and also certain mutants of peptidase A (Sinha and Hopkinson 1969) and phosphohexoseisomerase (Hopkinson 1970).

2.4. *Subunit structures of enzymes*

It has been seen how quite simple electrophoretic separations using small quantities of haemolysates, extracts of other tissues or of cells grown in tissue culture, from heterozygous individuals will often give basic information about the subunit structure of enzyme molecules (p. 35). If hybrid molecules not present in either type of homozygote are demonstrable in heterozygotes then one can infer that the separate enzyme proteins are each made up of two or more polypeptide subunits. Furthermore from the number of such hybrid forms observed in the electrophoretic pattern one can obtain an idea of the number of subunits involved. The electrophoretic pattern in heterozygotes for monomeric enzymes will represent a simple mixture of the molecular forms seen in the respective homozygotes, with no hybrid forms. But as has already been noted (p. 36) the failure to detect hybrid isozymes in a heterozygote does not conclusively exclude the possibility of a multimeric structure.

Information about subunit structure is also of course provided by the electrophoretic patterns of isozymes observed in cases where two or more gene loci are concerned in determining the enzyme structure, provided that at least two of them are active in the same cells. Typical examples are lactate dehydrogenase (p. 53) and alcohol dehydrogenase (p. 72). It may be noted that in such cases material from homozygous individuals provides the relevant information.

Studies on somatic cell hybrids obtained by fusing tissue cultured cells from two different species (see p. 59) represent yet another way in which

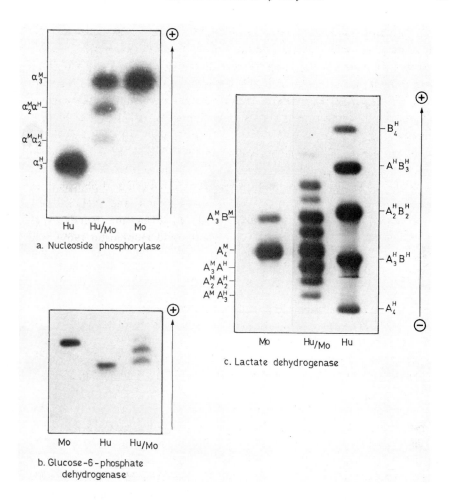

a. Nucleoside phosphorylase

b. Glucose-6-phosphate dehydrogenase

c. Lactate dehydrogenase

Fig. 2.27. Hybrid isozyme formation in human-mouse somatic cell hybrids. a) Purine nucleoside phosphorylase (NP) which is trimeric, b) Glucose-6-phosphate dehydrogenase (G-6-PD) which is dimeric, and c) Lactate dehydrogenase (LDH) which is tetrameric. In each case the isozymes of the human (Hu) and mouse (Mo) parental lines, and also of the human-mouse somatic cell hybrid line (Hu/Mo), are shown. For NP and LDH the subunit composition of some of the isozymes are indicated. In the case of LDH there are two separate loci determining distinct polypeptides in both the human and the mouse cells (A^H and B^H in man, and A^M and B^M in mouse). A^H has a greater positive charge than A^M. B^H and B^M carry the same charge. There is relatively more B^H than B^M in these cell lines.

the subunit structure of an enzyme may be inferred from the isozyme pattern. In such cases provided the enzyme has a different electrophoretic mobility in the two species and the enzyme is multimeric, hybrid isozyme formation is readily observed in the somatic cell hybrids (fig. 2.27). Numerous examples of this effect have now been found in studies on human–mouse and human–chinese hamster somatic cell hybrids (Ruddle 1972). As a general rule the electrophoretic patterns are in essentials similar to those observed in heterozygotes in either of the species alone, and the findings complement one another very exactly. Exceptions occur however in the case of enzymes determined by genes located on the X-chromosome. Thus when somatic cell tissue extracts are examined for the enzyme glucose-6-phosphate dehydrogenase in heterozygous females, no hybrid form is observed because only one or other of the two alleles present is on the so-called 'active' X-chromosome in any single cell (p. 174). However somatic cell hybrids between two different species can contain active X-chromosomes from both species. So both types of polypeptide are produced and if the enzyme is multimeric hybrid isozymes are observed. In the case of glucose-6-phosphate dehydrogenase a triple banded electrophoretic pattern typical of a dimeric enzyme is seen (fig. 2.27).

Further evidence supporting the subunit structure inferred by these methods can often also be obtained by *in vitro* hybridisation experiments, again using electrophoresis to examine isozyme products. In such experiments the aim is to find conditions under which a multimeric enzyme can be dissociated into its subunits which can then be allowed to recombine to give enzymically active products. If a mixture of homologous enzymes from two different species, or from two different loci in the same species, or from two different alleles at a single locus in the same species, are then subjected to this procedure the appropriate hybrid isozymes may be generated. Such *in vitro* hybridisation has now been accomplished in a number of cases (e.g. lactate dehydrogenase (p. 54), peptidase A (p. 34), alcohol dehydrogenase (Smith, Hopkinson and Harris 1973), aldolase (Penhoet et al. 1967) and nucleoside phosphorylase (Edwards et al. 1971)). In each case the resulting isozyme pattern obtained has confirmed the subunit structure expected from isozyme studies in the various *in vivo* situations discussed above. Similarly, from an isolated 'hybrid' isozyme the expected homomeric isozymes can be formed by applying the appropriate procedure to produce dissociation and recombination of the subunits. A typical example is shown in fig. 2.28.

Proof of a particular subunit structure can of course best be obtained by comparing the molecular weights of the native enzyme and of its products after complete dissociation by for example denaturation in the presence of high concentrations of urea, guanidine or sodium dodecyl sulphate. This is often difficult because it involves the use of highly purified enzyme preparations whereas most of the methods discussed above can be carried out on crude tissue extracts or on only partially purified preparations.

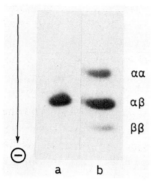

Fig. 2.28. Dissociation and recombination of subunits in the heterodimeric isozyme $\alpha\beta$ of human alcohol dehydrogenase (Smith, Hopkinson & Harris 1973). The α-subunit is the product of the ADH_1 locus and the β-subunit is the product of the ADH_2 locus (see p. 72). Key: a) $\alpha\beta$ isozyme isolated from foetal liver by ion-exchange chromatography, b) $\alpha\alpha$, $\alpha\beta$ and $\beta\beta$ isozymes formed by dissociation of the $\alpha\beta$ isozyme into its subunits and their subsequent recombination (for details see Smith et al. 1973).

Purine nucleoside phosphorylase (p. 78) provides an interesting example of a case where all these different methods have been applied. The original suggestion that this enzyme is a trimer came from electrophoretic studies on heterozygotes (Edwards, Hopkinson and Harris 1971). This was supported by the findings in human–mouse somatic cell hybrids grown in tissue culture and by *in vitro* dissociation and recombination experiments using the human enzyme with the mouse enzyme and also the human enzyme with the bovine enzyme. In all cases a four-banded isozyme pattern expected from a trimer was obtained. Further supporting evidence came from the observation that three moles of hypoxanthine appeared to bind to one mole of the enzyme (Agarwal and Parks 1969). The situation appeared however to be very unusual since at that time no certain example of an enzyme with a trimeric structure had been identified. Eventually however molecular size determinations by ultracentrifugation using the purified bovine enzyme was found to give a value of 84,000 which was in agreement with estimates obtained by gel filtration on non-purified human material; and molecular size determina-

One gene–one polypeptide chain

TABLE 2.6

A series of enzymes classified by subunit structure. The enzymes represent the products of fifty different loci. The classification is based on the electrophoretic patterns observed in heterozygotes, somatic cell hybrids, or *in vitro* dissociation-recombination experiments.

Monomers		Dimers		Trimers		Tetramers	
1) phosphoglucomutase	PGM_1	1) alcohol dehydrogenase	ADH_1	1) purine nucleoside phosphorylase	NP	1) lactate dehydrogenase	LDH_A
2) phosphoglucomutase	PGM_2	2) alcohol dehydrogenase	ADH_2			2) lactate dehydrogenase	LDH_B
3) phosphoglucomutase	PGM_3	3) alcohol dehydrogenase	ADH_3			3) lactate dehydrogenase	LDH_C
4) carbonic anhydrase	$CA\ I$	4) glycerol-3-phosphate dehydrogenase	GPD_1			4) aldolase	ALD_A
5) carbonic anhydrase	$CA\ II$	5) glycerol-3-phosphate dehydrogenase	GPD_2			5) aldolase	ALD_B
6) red cell acid phosphatase	ACP_1	6) acid phosphatase	ACP_2			6) aldolase	ALD_C
7) adenylate kinase	AK	7) acid phosphatase	ACP_3			7) pyruvate kinase	PK_3
8) adenosine deaminase	ADA	8) glutamate oxalate transaminase	GOT_S			8) superoxide dimutase	SOD_B
9) NADH diaphorase	DIA	9) glutamate oxalate transaminase	GOT_M			9) serum cholinesterase	E_1
10) mannosephosphate isomerase	MPI	10) malate dehydrogenase	MOR_S			10) malic enzyme	MOD_S
11) peptidase B	$PEP\text{-}B$	11) malate dehydrogenase	MOR_M				
12) peptidase C	$PEP\text{-}C$	12) adenine phosphoribosyl transferase	$APRT$				
13) phosphoglycerate kinase	PGK	13) phosphohexose isomerase	PHI				
		14) phosphogluconate dehydrogenase	PGD				

15) peptidase A — *PEP-A*
16) peptidase D — *PEP-D*
17) pyrophosphatase — *PP*
18) glucose-6-phosphate dehydrogenase — *Gd*
19) superoxide dismutase — *SOD*ₐ
20) enolase (phospho-pyruvate hydratase) — *PPH*
21) esterase D — *ES-D*
22) isocitrate dehydrogenase — *ICD*ₛ
23) glutamate-pyruvate transaminase — *GPT*
24) 2,3 diphosphoglycer-ate mutase — *DPGM*
25) inosine triphosphat-ase — *ITP*
26) placental alkaline phosphatase — *PL*

References: PGM_1, PGM_2, PGM_3 (pp. 60–69); *CA I, CA II* (Table 2.5, p. 74); ACP_1 (p.180); *AK* (Fildes & Harris 1966); *ADA* (Spencer et al 1968); *DIA* (Hopkinson et al 1970); *MPI* (Nichols et al 1973); *PEP B* (Lewis & Harris 1967); Pep C (Povey et al 1972); *PGK* (Meera Khan et al 1971); ADH_1, ADH_2, ADH_3 (Smith et al 1973); GPD_1, GPD_2 (Hopkinson et al 1974); ACP_2, ACP_3 (Swallow & Harris 1972); GOT_S, GOT_M, MOR_S, MOR_M (Table 2.5 p. 75); *APRT* (Mowbray et al 1972); *PHI* (Detter et al 1968); *PGD* (Parr 1966); *PEP A* (p. 33); *PEP D* (Lewis & Harris 1969b); *PP* (Fisher et al 1974); *Gd* (p. 86); SOD_A (Brewer 1967, Beckman 1973); *PPH* (Cohen et al 1973); *ES-D* (Hopkinson et al (1973); ICD_S (Chen et al 1972); *GPT* (Chen & Giblett 1971a); *DPGM* (Chen et al 1971b); *ITP* (Hopkinson, D.A. personal communication); *PL* (p. 39); *NP* (p. 87); LDH_A, LDH_B, LDH_C (pp. 52–59); ALD_A, ALD_B, ALD_C (p. 199); PK_3 (p. 203); SOD_B (Ruddle 1973); E_1 (Scott & Powers 1972); MOD_S (Table 2.5 p. 75).

tions on the bovine enzyme dissociated into its subunits by treatment with sodium dodecyl sulphate gave a value of 28,000, just one third of the molecular size of the native enzyme (Edwards, Edwards and Hopkinson 1973).

Table 2.6 lists a series of enzymes which on the basis of electrophoretic studies have been classified as monomers, dimers, trimers or tetramers. More than two thirds are multimeric. Since the enzymes in the table were selected only in that sufficiently precise electrophoretic techniques were available for the study of the isozymes formed by heterozygotes, somatic cell hybrids, or by multiple homozygous loci it is reasonable to regard them as representative of enzymes in general. If so one can conclude that a dimeric structure is the most common form (52%). A tetrameric structure is not uncommon (20%) but a trimeric structure is rather unusual, since only one example was found in the 50 cases listed. In 26% of the cases no hybrid isozymes were detected and these are probably almost all monomers.

CHAPTER 3

Duplications, deletions, unequal crossovers, chain elongations and other rearrangements

3.1. The haptoglobin variants

A gene mutation which results in a single aminoacid substitution in a protein generally represents, as we have seen, a single base change in the DNA sequence of the gene. But other kinds of mutational events, which may involve changes such as duplications or deletions of particular stretches of the DNA sequence can also occur. Like the single base changes, these alterations in the sequence will be perpetuated by the ordinary processes of DNA replication. They will also be reflected by corresponding alterations in the structure of the polypeptide or polypeptides coded by the gene or genes affected by the change. An example of an alteration in protein structure that has apparently been brought about by a mutational event of this sort is provided by one of the common variant forms of the serum protein, haptoglobin.

Blood serum contains a complex mixture of many different proteins which originate in various tissues. Haptoglobin constitutes part of the so-called α_2 serum globulin fraction, and is originally synthesised in the liver. It has the very distinctive property of binding free haemoglobin tightly and specifically in much the way that antibodies bind specific antigens, though the haptoglobin–haemoglobin complex so formed remains soluble and does not precipitate.

The precise significance of haptoglobin in normal function is not known. The quantity of haemoglobin usually required to saturate the binding capacity of the haptoglobin present in normal serum is only about 50–150 mg/ 100 ml (Nyman 1959), and any excess haemoglobin remains in the free state. Haptoglobin–haemoglobin complexes formed as a result of red cell breakdown, are rapidly removed from the serum and are degraded in the tissues, but free haemoglobin is mainly excreted in the urine. So it is possible that one function of haptoglobin is to minimise the loss of iron by the body. Haptoglobin may also be significant in bile pigment formation because it has been shown that while the haem in native haemoglobin is resistant to the attack of a liver enzyme, known as α-methenyl oxygenase, which converts haem to a precursor of biliverdin, haem in the haptoglobin–haemoglobin complex is readily attacked by this enzyme (Nakajima et al. 1963).

Smithies (1955) discovered that when serum from different individuals is examined by starch gel electrophoresis several distinct and quite characteristic haptoglobin types can be recognised (fig. 3.1). The most commonly

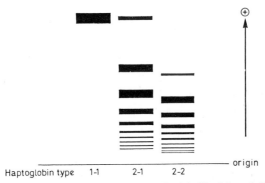

Fig. 3.1. The three common haptoglobin types: Hp 1-1, Hp 2-1 and Hp 2-2. Diagram of patterns of components after electrophoresis in starch gel (pH 8.6).

occurring types are called Hp 1-1, Hp 2-1 and Hp 2-2, and they differ from each other both in the electrophoretic mobilities and also in the number of protein components present. In type 1-1 only a single haptoglobin component is seen. A component with the same mobility can be detected in type 2-1 but not in type 2-2. However both the 2-1 and 2-2 types show a whole series of other haptoglobin components migrating more slowly than the 1-1 component, and each of these components in type 2-1 has a different mobility from those in type 2-2. Thus the pattern of components is qualitatively distinct in the three types, and simple mixing for example of 1-1 and 2-2 sera will not produce a 2-1 pattern.

In European populations virtually all individuals can be classified into one or other of these three types; about 16% are type 1-1, about 48% are type 2-1 and about 36% are type 2-2. The types are genetically determined and the initial family studies led to the suggestion that a pair of alleles at an auto-somal locus were involved (Smithies and Walker 1955, 1956). These were called Hp^1 and Hp^2, and it was supposed that Hp 1-1 individuals were homozygous for Hp^1, Hp 2-2 individuals homozygous for Hp^2 and Hp 2-1 individuals were heterozygous Hp^1Hp^2. It appeared therefore that the homozygotes differed from one another both in the mobility and the number of protein components they formed, and also that the heterozygotes formed a series of protein components qualitatively different from those present in either type of homozygote.

Structural studies on the purified haptoglobins from sera of the different types carried the matter much further. It turned out that the multiple components characteristic of types 2-1 and 2-2 represent a series of polymers of increasing molecular weight (Smithies and Connell 1959). Both they and the single component in type 1-1 each contain two sorts of non-identical polypeptide chains, which are represented two or more times in a haptoglobin molecule according to its molecular size. The two sorts of polypeptide are called the α- and β-chains, and the differences between the three common haptoglobin types depend on structural differences in the α-chain (Smithies et al. 1966). The β-chains are evidently the same in each type (Cleve et al. 1967). In the haptoglobin molecule the chains are apparently cross-linked by disulphide bonds.

Electrophoretic studies on the separated α-chains from the different types revealed further genetically determined heterogeneity (Connell et al. 1962, Smithies et al. 1962a, 1966). Two sorts of α-chain can be obtained from preparations of type 1-1 haptoglobin. These are called hp1Fα and hp1Sα, the former having a somewhat faster electrophoretic mobility than the latter

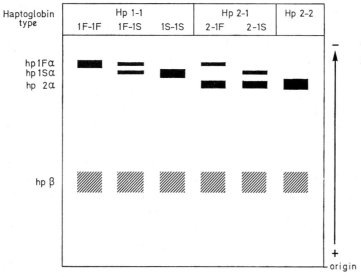

Fig. 3.2. Electrophoretic separation of hp α-chains in different haptoglobin types. Starch gel electrophoresis in formate buffer (*p*H 4.0) and 8.0 M urea (Smithies et al. 1962a, 1966).

under the particular conditions used for their separation (fig. 3.2). Type 1-1 haptoglobin from some individuals contains only hp1Fα, that from other 1-1 individuals contains only hp1Sα, and there is a third group of type 1-1 individuals in which both hp1Fα and hp1Sα are present. These 'concealed' differences within the Hp 1-1 type are genetically determined and their manner of segregation within families shows that two sorts of Hp^1 allele must exist. These are designated Hp^{1F} and Hp^{1S}. Homozygotes for one of these alleles form a type 1-1 haptoglobin containing only one kind of α-chain. Heterozygotes ($Hp^{1F}Hp^{1S}$) form type 1-1 haptoglobin which contains molecules with both kinds of α-chain.

The α-chain obtained in most cases from type 2-2 haptoglobin migrates electrophoretically as a single slower-moving zone. It is referred to as hp2α. As would be expected two subtypes of type 2-1 can be distinguished, one corresponding to the genotype $Hp^{1F}Hp^2$ contains the α-chains hp1Fα and hp2α, the other from the genotype $Hp^{1S}Hp^2$ contains hp1Sα and hp2α. It is of interest that the three Hp 1-1 types cannot be clearly distinguished from one another by routine electrophoresis of the native haptoglobins, nor can the two different types of Hp 2-1.

Studies on the structure of the isolated α polypeptides revealed remarkable and at the time quite unexpected differences between hp1Fα and hp1Sα on the one hand, and hp2α on the other (Smithies et al. 1962b, Connell et al. 1966). It was found that hp1Fα and hp1Sα polypeptides each have a molecular weight of about 9,000, whereas the hp2α polypeptide is nearly twice as big, having a molecular weight of about 16,000. The hp1Fα and hp1Sα polypeptides each contain 83 aminoacids, while the hp2α polypeptide has 142 aminoacids in its sequence (Black and Dixon 1968). Thus the Hp^2 allele evidently determines a polypeptide chain nearly twice the size of the polypeptide chains determined by the Hp^{1F} and Hp^{1S} alleles.

Peptide 'fingerprint' patterns obtained from chymotryptic digests of the hp1Fα and hp1Sα polypeptides showed that these two chains only differed in a single peptide. The 'fingerprint' pattern from hp1Fα contained one peptide 'F' not present in the hp1Sα pattern, and the hp1Sα pattern contained one peptide 'S' not present in the hp1Fα pattern (Smithies et al. 1962b). Aminoacid analysis revealed that a lysine residue in 'F' is replaced by a glutamic acid residue in 'S', and the difference between hp1Fα and hp1Sα was subsequently shown by full sequence studies (Black and Dixon 1968) to involve only this single aminoacid substitution. Thus it may be concluded that the alleles Hp^{1F} and Hp^{1S} have arisen one from the other by a mutation of the standard kind, involving only a single base change in the DNA.

```
      1                                                    10                              20
NH₂ - Val - Asn - Asp - Ser - Gly - Asn - Asp - Val - Thr - Asp - Ile - Ala - Asp - Asp - Gly - Gln - Pro - Pro - Lys -
                                                           30                              40
    - Cys - Ile - Ala - His - Gly - Tyr - Val - Glu - His - Ser - Val - Arg - Tyr - Gln - Cys - Lys - Asn - Tyr - Lys -
                                                           50                     Lys      60
    - Leu - Arg - Thr - Gln - Gly - Asp - Gly - Val - Tyr - Thr - Leu - Asn - Asn - Glu - Lys - Gln - Trp - Ile - Asn - Lys -
                                                           70                              80
    - Ala - Val - Gly - Asp - Lys - Leu - Pro - Glu - Cys - Glu - Ala - Val - Gly - Lys - Pro - Lys - Asn - Pro - Ala - Asn -
              83
    - Pro - Val - Gln - COOH
```

(a)

```
      1                                                    10                              20
NH₂ - Val - Asn - Asp - Ser - Gly - Asn - Asp - Val - Thr - Asp - Ile - Ala - Asp - Asp - Gly - Gln - Pro - Pro - Lys -
                                                           30                              40
    - Cys - Ile - Ala - His - Gly - Tyr - Val - Glu - His - Ser - Val - Arg - Tyr - Gln - Cys - Lys - Asn - Tyr - Lys -
                                                           50                     Lys      60
    - Leu - Arg - Thr - Gln - Gly - Asp - Gly - Val - Tyr - Thr - Leu - Asn - Asn - Glu - Lys - Gln - Trp - Ile - Asn - Lys -
                                                           70                              80
    - Ala - Val - Gly - Asp - Lys - Leu - Pro - Glu - Cys - Glu - Ala - Asp - Asp - Gly - Gln - Pro - Pro - Lys - Cys -
                                                           90                              100
    - Ile - Ala - His - Gly - Tyr - Val - Glu - His - Ser - Val - Arg - Tyr - Gln - Cys - Lys - Asn - Tyr - Tyr - Lys - Leu -
                                                           110                     Lys     120
    - Arg - Thr - Gln - Gly - Asp - Gly - Val - Tyr - Thr - Leu - Asn - Asn - Glu - Lys - Gln - Trp - Ile - Asn - Lys - Ala -
                                                           130                             140
    - Val - Gly - Asp - Lys - Leu - Pro - Glu - Cys - Glu - Ala - Val - Gly - Lys - Pro - Lys - Asn - Pro - Ala - Asn - Pro -
             142
    - Val - Gln - COOH
```

(b)

Fig. 3.3. The aminoacid sequences of the different haptoglobin α-chains hp1Fα, hp1Sα and hp2α (Black and Dixon 1968). (a) hp1Fα: Lys at position 54; hp1Sα: Glu at position 54. (b) hp2α: because the order of the F/S substitutions is not known both Lys and Glu are shown at positions 54 and 113. (Aminoacid abbreviations as in caption to fig. 1.5, p. 14.)

The peptide 'fingerprint' pattern obtained from the hp2α chain revealed a quite unusual situation. It was found to contain all the peptides present in both hp1Fα and hp1Sα including both of the specific peptides 'F' and 'S'. It also had an additional peptide, 'J', not present in either the hp1Fα or hp1Sα patterns (Smithies et al. 1962b).

From studies on the separated peptides it was possible to construct the full aminoacid sequences of the three polypeptide chains hp1Fα, hp1Sα and Hp2α determined by the three common alleles Hp^{1F}, Hp^{1S} and Hp^2 (Black and Dixon 1968). These sequences are shown in fig. 3.3. The hp1Fα and hp1Sα sequences each of which has 83 aminoacids, differ only in position 54, where lysine in hp1Fα is replaced by glutamic acid in hp1Sα. The remarkable thing about the hp2α sequence which contains 142 aminoacids, is that the sequence of the first 71 aminoacids in this chain is the same as the sequence of the first 71 aminoacids in the hp1α chains, whereas the sequence from position 71 to the carboxyl terminal end of the hp2α chain is the same as the sequence in the hp1α chains from position 12 to their carboxyl terminal ends. The sites of the lysine–glutamic acid substitution in the hp1Fα and hp1Sα chains correspond to positions 54 and 113 in the hp2α chain. One of these positions in hp2α is occupied by lysine and the other by glutamic acid, but it is difficult to determine which is which. The 'J' peptide characteristic of the 'fingerprint' pattern of hp2α represents an aminoacid sequence in the middle of the chain, and its relation to the amino-terminal sequences and the carboxyl-terminal sequences in the hp1α chains is illustrated in fig. 3.4.

These findings suggest that the hp2α chain can be considered to represent an end to end fusion of an hp1Fα polypeptide chain and an hp1Sα chain, but with the loss at the site of fusion of 12 (or 13) residues from the carboxyl terminal end of one of the hp1α polypeptides, and of 12 (or 11) residues from the amino terminal end of the other hp1α polypeptide. This conclusion implies that the Hp^2 allele represents an almost complete duplication of an Hp^1 allele. The base pair sequence must be almost twice as long as that in either the Hp^{1F} or Hp^{1S} alleles, and the second half of the sequence is presumably an exact replica of the first half of the sequence except for the loss of some 24×3 base pairs at the site of junction, and the substitution of one base pair in the first half of the sequence by another in the second half corresponding to the aminoacid substitutional difference between hp1Fα and hp1Sα.

How did such an allele originate? The likely explanation is that it first arose in a heterozygous individual $Hp^{1F}Hp^{1S}$ as a result of a chromosomal rearrangement which caused virtually the whole of the base sequences of the two alleles Hp^{1F} and Hp^{1S} to be brought together end to end on a single chromosome. One way in which this might have occurred (Smithies 1964) is as a consequence of an event in which breaks happened to occur more or less simultaneously in each of the two homologous chromosomes (or chromatids) and this was followed by cross reunion of the broken ends (fig. 3.5). We may suppose that one of these breaks occurred in the region of the distal end of the haptoglobin locus on one of the chromosomes, and the other at the proximal end of the haptoglobin locus in the homologous chromosome.

(a)		(b)		(c)
			58	Ile		58	Ile
				Asn			Asn
1	Val		60	Lys		60	Lys
	Asn			Ala			Ala
	Asp			Val			Val
	Ser			Gly			Gly
	Gly			Asp			Asp
	Asn			Lys			Lys
	Asp			Leu			Leu
	Val			Pro			Pro
	Thr			Glu			Glu
10	Asp			Cys			Cys
	Ile		70	Glu		70	Glu
	Ala		----	Ala ----			Ala
	Asp			Asp			Val
	Asp			Asp			Gly
	Gly			Gly			Lys
	Gln			Gln			Pro
	Pro			Pro			Lys
	Pro			Pro			Asn
	Pro			Pro			Pro
20	Lys			Lys			Ala
	Cys		80	Cys		80	Asn
	Ile			Ile			Pro
	Ala			Ala			Val
	His			His		83	Gln
	Gly			Gly			
	Tyr		85	Tyr			

Fig. 3.4. Relationship of sequence of 'J' (junction) peptide found in chymotryptic digest of hp2α polypeptide chain, to the amino and carboxyl terminal sequences of the hp1α chains. (a) Amino-terminal sequence of hp1α (residues 1-26). (b) 'Junction' peptide obtained from chymotryptic digest of hp2α (residues 58-85). (c) Carboxyl-terminal sequence of hp1α (residues 58-83).

This was followed by cross-reunion so that one chromosome came to contain the major portions of the two alleles aligned end to end, while the other contained only a fragment of each. Thus two new chromosomes would

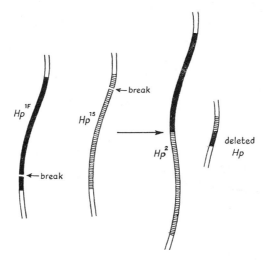

Fig. 3.5. Diagram illustrating postulated formation of Hp^2 allele as a result of breaks in Hp^{1F} and Hp^{1S} alleles followed by aberrant reunion.

appear. One would contain an almost complete duplication of the Hpα locus. This would be the Hp^2 allele. In the other new chromosome the locus would be effectively deleted, except for a short base pair sequence which would probably not define a viable polypeptide chain. Formally the event could be regarded as a reciprocal translocation occurring between two homologous chromosomes.

Translocation of segments between different chromosomes presumably resulting from random breaks followed by aberrant reunions are known to occur, and in population surveys carried out using standard cytogenetical techniques, such chromosomal translocations have been shown to exist in a small but appreciable fraction of the general population (Court Brown and Smith 1969). For the most part these have only been identified when the chromosomal changes produced are sufficiently gross that they can be recognised microscopically either because of an alteration in the dimensions of particular chromosomes observed in mitosis, or because of aberrations of pairing observed in meiosis. One may presume that similar changes not detectable under the microscope are at least as common if not more frequent. Thus the idea that the Hp^2 gene originally arose in this way is a plausible one.

Another way in which it has been suggested that the Hp^2 allele might have arisen, is as a consequence of what is known as unequal crossing over (Smithies et al. 1962b, Black and Dixon 1968). If in the course of meiosis in a heterozygous individual $Hp^{1F}Hp^{1S}$ there were aberrant pairing of the two homologous chromosomes so that the Hp^{1F} and the Hp^{1S} alleles were misaligned, then should crossing-over occur between the distal end of one allele and the proximal end of the other, the new Hp^2 allele could have been produced on one of the chromosomes (fig. 3.6). The phenomenon of unequal crossing-over (see pp. 103–112) is known to happen but is thought to require close

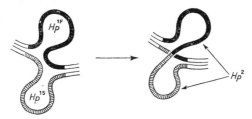

Fig. 3.6. Diagram illustrating postulated formation of Hp^2 allele as a result of unequal crossing over between Hp^{1F} and Hp^{1S} alleles.

homology between the segments of DNA in the two regions of the chromosomes where the crossing-over occurs, so that aberrant pairing is possible. Black and Dixon (1968) arguing from the aminoacid sequences have suggested that there may indeed be sufficient homology between the proximal and distal regions of the DNA coding for hp1Fα and hp1Sα, to make this hypothesis of the origin of Hp^2 reasonable. But although the matter is still unresolved, the hypothesis of random breaks and reunions seems perhaps the most plausible one at the present. An essential difference between the two hypotheses is that the unequal crossover hypothesis requires that the aberrant event took place during pairing at meiosis, while the other hypothesis does not necessarily require this.

Whether or not the original appearance of the Hp^2 allele was a consequence of unequal crossing-over, it seems that this general process is significant in the formation of other haptoglobin variants derived from hp2α. This point is discussed later (pp. 109–112).

3.2. *Gene duplication and protein evolution*

One apparent consequence of the extended length of the polypeptide Hp2α is that haptoglobin proteins containing it tend to form a series of stable polymers of increasing molecular size. This is probably because the additional cysteine residues on the extended polypeptides can form disulphide bridges between the chains and thus allow the generation of stable haptoglobin molecules with larger numbers of polypeptide subunits. It is of some interest that the multiple polymers characteristic of haptoglobin types 2-2 and 2-1 in man have not been found in comparative studies on a variety of other animal species including a number of primates. In these other species the haptoglobin formed appears to be homologous with human Hp 1-1 (for review see Sutton 1970). These findings have led to the suggestion that the chromosomal rearrangement giving rise to the duplicated polypeptide structure was probably a single event which occurred some time after the evolutionary separation of the human line. Studies of the world-wide distribution of haptoglobin types show that Hp^2 has remarkably high frequencies in many different populations (Shim and Bearn 1964). In Europe for example about 60% of all the Hp alleles are Hp^2, and even higher frequencies have been found in parts of India and Asia. It seems then that the Hp^2 allele has, despite its relatively recent origin, spread widely through the species (see fig. 8.2, p. 292). Thus one may in fact be witnessing an evolutionary change in the gross structure of a particular protein (Dixon 1966).

In most proteins a polypeptide chain of increased length resulting from such a partial gene duplication is not likely in the great majority of cases to give rise to a stable conformation. The three-dimensional arrangement would generally be profoundly altered, and particularly when the polypeptide is required to fit together with other polypeptides in the protein it seems improbable that a viable and functional protein product would usually emerge. Thus one would expect that only very rarely would what appears to be the essentially random process of breaks and reunions of chromosomes give rise to a viable protein with a grossly altered structure.

Quite apart from this possibility however, chromosomal rearrangements have probably played an important role in protein evolution in a somewhat different way. If the breaks were to occur not in the middle of the DNA sequence defining a particular polypeptide chain, but at the very end of it or at its beginning, and if following two such breaks at appropriate sites in homologous chromosomes (or in a pair of chromatids) cross-reunion should take place, then complete gene duplication could result. In other words one chromosome would come to contain two similar genes aligned sequentially, each capable of determining separately a polypeptide chain of the same length and with unaltered structure. This sort of duplication could of course involve a longer segment of chromosome containing a number of genes. A variety of other kinds of chromosomal rearrangement due to breaks and aberrant reunions and resulting in gene duplication may also be envisaged (Watts and Watts 1968). They may result in the duplicates being aligned sequentially in the same chromosome as in the haptoglobin case or separated by a non-duplicated region, and the duplicated sequences may run in the same direction or be reversed relative to one another. Furthermore under certain circumstances the duplicates could come to lie in two quite different chromosomes. For instance if two breaks occurred in one chromosome (or its chromatid) and a single break in another the free segment from the one could become inserted between the broken ends of the other with subsequent reunion.

Complete gene duplication, however it comes about, is potentially of considerable evolutionary significance because although initially no structural alteration in the corresponding enzyme or protein need be present, mutations of the single base change type could subsequently occur in one or other of the duplicates and if favoured by natural selection might eventually lead to divergence in the structure and function of the polypeptide products of the two loci.

The occurrence of duplications of genetic material in chromosomes has been recognised for many years, and their likely significance of evolution was appreciated long before their immediate relevance to protein structure was appreciated. Writing in 1951, E.B. Lewis summarised the argument succinctly. 'A gene which mutates to a new function should in general lose its ability to produce its former immediate product, or suffer an impairment in that ability. Since it is unlikely that this old function will usually be an entirely dispensable one from the standpoint of the evolutionary survival of the organism, it follows that the new gene will tend to be lost before it can be tried out, unless, as a result of establishment of a duplication the old gene has been retained to carry out the old function'.

The concept can effectively be illustrated by considering the distinct though structurally similar polypeptide chains, α, β, γ and δ which occur in the different forms of normal haemoglobin; A, A_2 and F. There are considerable homologies in the aminoacid sequences of these four polypeptides, and these can be most simply accounted for by assuming that the genes which determine them were originally derived from a common ancestral gene. The evolutionary process can be envisaged as having involved a successive series of gene duplications followed in each instance by the divergent evolution of the products formed as a result of point mutations causing different amino-acid substitutions. A simple scheme proposed by Ingram (1961) is illustrated in fig. 3.7. The order in which it is assumed that the duplications occurred to

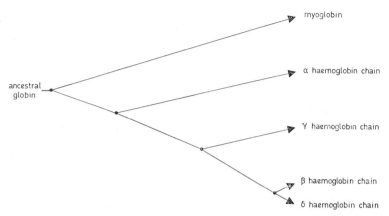

Fig. 3.7. Scheme for the evolution of the α, β, γ and δ haemoglobin polypeptide chains, and also the myoglobin chain from a common ancestral form of globin (after Ingram 1961). Each ● indicates the occurrence of a gene duplication.

give rise to the ancestral forms of the different loci is consistent with the degree of homology of the aminoacid sequences in the different polypeptide chains as they occur in the human species today. Thus the β- and δ-chains differ in only ten out of 146 positions in their aminoacid sequences and can therefore be regarded as having appeared as separate entities fairly recently in evolutionary history. It is significant that these two loci have been shown by linkage studies to lie very close together on the same chromosome (pp. 49–50), and this of course fits in very well with the idea that they might originally have arisen from a gene duplication. The discovery that there are two distinct γ-chains, $^{G}\gamma$ and $^{A}\gamma$, differing in a single aminoacid and evidently determined by separate loci (p. 45), suggests a further very recent duplication.

The α polypeptide chain has only 141 aminoacid residues whereas the β-, γ- and δ-chains each have 146 and the correspondences in sequence between the α-chain and the other chains, though striking, are much fewer. This suggests therefore an earlier divergence of the ancestral α-chain from the others. The difference in number of residues can be visualised as having arisen either from small deletions or from accretions of base sequences resulting from chromosome breaks and reunions subsequent to divergence. The finding that the α-locus is not closely linked to the β- or δ-loci (p. 48) presumably implies that the original duplication was associated with separation of the loci or that a translocation of a chromosomal segment causing the separation of these loci occurred at some subsequent stage in their evolutionary history. If, as now seems probable, two α-loci occur (p. 46) then this may be presumed to be the consequence of a further duplication.

Ingram also suggested that the gene determining myoglobin was also originally evolved from the ancestral gene which was the progenitor of the haemoglobin genes. Myoglobin is a haem protein concerned with oxygen transfer in muscle. Unlike haemoglobin it consists of only a single polypeptide chain, which is however in size (153 aminoacid residues) not very different from the polypeptide chains found in haemoglobin. The three-dimensional structure of myoglobin has been shown to have a remarkable similarity to each of the haemoglobin polypeptides, and it also exhibits some degree of homology with their aminoacid sequences. One may imagine that the evolution of haemoglobin chains led at some stage to the possibility of their combining together to form dimers and later tetramers while this did not happen with the myoglobin chain. The tetrameric form allows the possibility of so called haem–haem interaction, and the ultimate development of the characteristic haemoglobin molecule with its sigmoid oxygen dissociation curve specially adapted for the transfer of oxygen from the lungs to the tissues. Myoglobin with its single polypeptide chain and its single haem group gives a hyperbolic oxygen dissociation curve. This is appropriate to the role it plays in providing a kind of oxygen store in muscle.

Although the details of such evolutionary changes are still very far from clear, the underlying idea that structurally different polypeptides which occur in the same or in similar proteins may have originated by duplications from a single ancestral gene, is an important one. It provides a new insight into the structural complexities of many enzymes and other proteins where the occurrence of more than one structurally distinct but nevertheless very similar polypeptide chain is proving to be a not uncommon feature.

3.3. Unequal crossing-over

Crossing-over is a normal phenomenon which takes place between the chromatids of homologous chromosomes when they come together and pair at meiosis. The mechanism by which pairing or synapsis takes place in meiotic cell division is not understood. It is usually a very exact process, and it presumably depends in some way on a precise matching gene for gene along the whole length or at least long stretches of the synapsing chromosomes. This in turn is presumably a reflexion of the close correspondence of the nucleotide sequences in the separate pairs of alleles. As a consequence of this close matching of the pairs of homologous chromosomes, crossing-over, when it takes place at some particular site along the chromatids, does not normally alter or disrupt the arrangement of the DNA sequences, as might be expected to occur if they were misaligned.

But misalignment or mispairing at synapsis is more likely to happen in regions where a duplication of the DNA sequence exists, and if a cross-over should take place in such a misaligned region, then rearrangements of the DNA sequences and hence alterations in the structures of the polypeptides coded by these sequences may be brought about. This phenomenon is usually referred to as 'unequal crossing-over'. It provides an elegant explanation for the characteristic structural abnormalities found in a particular group of rare haemoglobin variants known as the 'Lepore' haemoglobins (Baglioni 1962a).

3.3.1. The Lepore haemoglobins

The Lepore haemoglobins, so-called from the name of the family in which the defect was first recognised (Gerald and Diamond 1958), contain normal haemoglobin α-chains, but the non-α chains have a quite unusual structure. Like the normal β- and δ-haemoglobin chains, the Lepore non-α chains have 146 aminoacids. But while the first part of the sequence is the same as that of the first part of the sequence of the δ-chain, the remainder of the sequence corresponds to the latter part of the sequence of the normal β-chain.

Three sorts of Lepore haemoglobin have been identified (fig. 3.9, p. 106). In one of these, known as Hb-Lepore Washington or Boston, the change over from the δ-like sequence to the β-like sequence occurs between positions 87 and 116 (Baglioni 1962, Labie et al. 1966). In another, known as Hb-Lepore Baltimore, the change over occurs between positions 50 and 86 (Ostertag and Smith 1969). In a third, known as Hb-Lepore Hollandia, the change over occurs between positions 22 and 50 (Barnabas and Miller 1962, Curtain 1964).

These Lepore haemoglobins have been mainly found in heterozygotes, where the variant haemoglobin constitutes about 5–15% of the total haemoglobin present and Hb A most of the rest. Some homozygotes have however also been identified. Characteristically they have no Hb A or Hb A$_2$, and they generally show a moderate to severe degree of chronic haemolytic anaemia (Neeb et al. 1961, Quattrin et al. 1967, Duma et al. 1968).

The sequences of normal β and δ polypeptide chains differ from one another in 10 residues, at positions 9, 12, 22, 50, 86, 87, 116, 117, 125 and 126 (fig. 2.8, p. 44). Thus the nucleotide sequences in the corresponding genes are likely to be very similar over most of their lengths. Furthermore it is evident from linkage studies that the two loci are on the same chromosome and probably lie very close to one another, if not in immediate juxtaposition (p. 50). Thus there exists a possibility that occasionally at meiosis the β-locus on one chromosome will pair aberrantly with the δ-locus on its partner. If following such a displaced synapsis crossing over were to occur in this region, then a new allele could arise which would specify a polypeptide chain with the δ sequence at one end and the β sequence at the other, and this is precisely the structure found in the Lepore haemoglobins.

Fig. 3.8 illustrates how such unequal crossing-over between the β- and δ-loci could give rise to two abnormal chromosomal products. In one chromosome the normal β- and δ-alleles would be absent, but a new gene

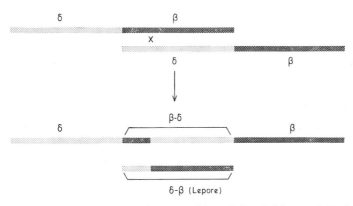

Fig. 3.8. Diagram illustrating how aberrant pairing of β and δ haemoglobin loci and crossing over may give rise to a 'Lepore' gene. The $\times$ indicates the site of crossing over.

(designated δ-β) would occur. In the other chromosome the normal β- and δ-alleles would be present and between them there would be a new gene (designated β-δ). Since the different types of Lepore haemoglobins are associated with deficiencies of both normal Hb A ($\alpha_2\beta_2$) and normal Hb A$_2$ ($\alpha_2\delta_2$), it seems probable they are derived from the type of cross-over product in which the normal β- and δ-genes are lost.

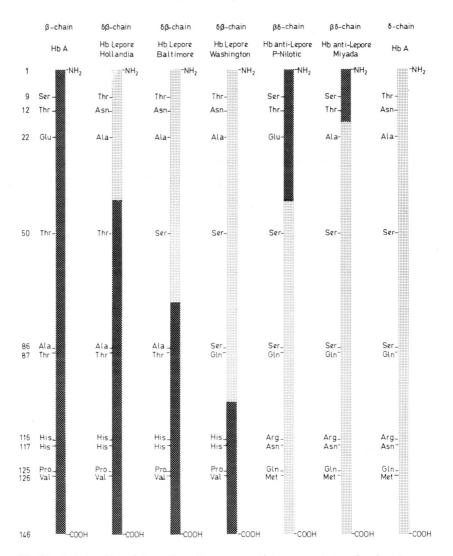

Fig. 3.9. Relationships of the aminoacid sequences of the non-α chains of various Lepore ($\delta\beta$) and anti-Lepore ($\beta\delta$) haemoglobins to those of the normal β- and δ- haemoglobin chains. Only the aminoacids which differ in the normal β- and δ-polypeptide sequences are shown.

It would be expected that the other type of cross-over product would result in the synthesis of both the normal β- and δ-chains as well as the abnormal βδ-chain. In such cases the abnormal chain would be expected to be the same as the β-chain in the first part of its sequence and the same as the δ-chain subsequently. Two abnormal haemoglobins where the non-α-chains have this structure have been found (fig. 3.9). In one, known as the Hb Miyada, the changeover from the β-like sequence to the δ-like sequence occurs between positions 12 and 22 (Ohta et al. 1971). In the other, known as Hb P-Nilotic, the changeover occurs between positions 22 and 50 (Badr et al. 1973). These haemoglobins are conveniently referred to as anti-Lepore haemoglobins.

The structures of the various Lepore and anti-Lepore haemoglobins are shown diagrammatically in fig. 3.9. They illustrate an important general characteristic of unequal crossing-over. This is that the actual position of the cross-over may vary quite widely and yet still yield the same products. For instance, in Hb-Lepore Baltimore the aminoacid residues up to and including the serine at position 50 are the same as in the normal δ-chain, whereas the aminoacid residues from the alanine at position 86 onwards are the same as in the normal β-chain. Since the amino acid sequence between positions 50 and 86 is identical in the δ- and β-chains, it is very probable that the base sequences of the genes which code them are very similar if not identical in this region. If so the same structurally abnormal polypeptide chain could result from cross-overs occurring at many different points in the DNA sequence between the base triplets coding for the amino acids at positions 50 and 86.

It will be noted that the results of structural studies on the Lepore and anti-Lepore haemoglobins strengthen the conclusion derived from family studies (see pp. 49–50) that the δ- and β-loci are closely linked and quite possibly lie immediately adjacent to one another on the same chromosome. They also provide some information about their probable order in the chromosome. This can be represented conventionally as δβ rather than βδ, because the amino-terminal end of the Lepore haemoglobins is evidently determined by the first part of the δ-locus, while the carboxyl terminal end is determined by the second part of the β-locus.

3.3.2. *Hb Kenya: a γ-β fusion product*

Hb Kenya is a rare haemoglobin variant in which the α-chains are normal, but the non-α-chains appear to represent a Lepore-like fusion between a γ- and a β-chain (Huisman et al. 1972, Kendall et al. 1973). The first part of the abnormal chain from the amino-terminal end is the same as the sequence in the first part of the normal γ-chain, while the second part of the amino acid sequence corresponds to that of the β-chain. The normal γ-chain like the β-chain has 146 aminoacids and the two chains have identical aminoacids at 107 positions, but different ones at 37 positions (see fig. 2.8, p. 44). The changeover from the γ-like sequence to the β-like sequence in the non-α-chain of Hb Kenya apparently occurs between positions 81 and 86 inclusive (Kendall et al. 1973).

This abnormal $\gamma\beta$-chain by analogy with the $\delta\beta$-chains of the Lepore haemoglobins is thought to have arisen from an event involving mispairing of a γ-gene with a β-gene followed by unequal crossing over. If so this would imply that the γ-gene locus and the β-gene locus are on the same chromosome and closely linked. In fact at least two different γ-gene loci are known to occur, one determining the so-called $^G\gamma$-chain and the other the $^A\gamma$-chain (p. 45), and the data indicates that it is the $^A\gamma$-gene locus which has undergone crossing-over with the β-gene locus.

As in the case of the Lepore haemoglobins (fig. 3.8) two chromosomal products would be expected from such an unequal crossing-over event, one in which other closely linked gene loci would be lost or deleted and the other in which they are retained. Analysis of the other haemoglobins formed in individuals heterozygous for the γ-β gene determining the Hb Kenya indicate that the gene occurs on a chromosome on which the normal δ- and

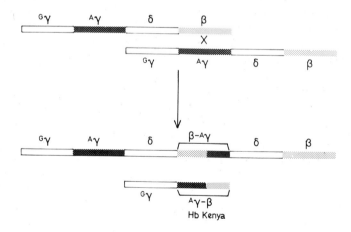

Fig. 3.10. Postulated unequal crossing over event giving rise to the fused γ-β gene determining the non-α chain in Hb Kenya.

the normal β-loci are not present. On the other hand the $^{G}\gamma$ gene which may be plausibly supposed to be closely linked to the $^{A}\gamma$ gene, is evidently retained (Kendall et al. 1973; Smith et al. 1973).

The data assembled so far suggests the order of these gene loci on the chromosome in the normal individual is probably $^{G}\gamma:^{A}\gamma:\delta:\beta$. The postulated unequal crossing-over event giving rise to fused γ-β gene determining Hb Kenya is illustrated in fig. 3.10.

It is of interest that the chromosome carrying the γ-β fusion gene appears to lack normal δ- and β-genes, in the same way that the chromosome carrying the δ-β Lepore gene lacks the normal δ- and β-genes. But the effects in heterozygotes on the synthesis of haemoglobin are very different. The Lepore chromosome gives rise to a thalassaemia-like trait whereas the chromosome carrying the γ-β fusion gene is associated with hereditary persistence of foetal haemoglobin. This point is considered later (p. 139).

3.3.3. *Unequal crossing-over as a cause of further haptoglobin variants*

The duplicated sequence of the Hp^2 allele would also be expected to pre-dispose to misalignment at synapsis and hence to unequal crossing over. This may generate further alleles which determine haptoglobin α-chains with distinctive structures (Smithies et al. 1962b, Nance and Smithies 1963).

For example at meiosis in a heterozygote of genotype $Hp^{1F}Hp^2$, the Hp^{1F} allele in one chromosome might pair with the 'S' segment of the Hp^2 allele because of the close similarity of the sequences. If crossing-over should then happen to occur at an appropriate site in the paired chromatids then an Hp^{1S} allele could be formed as one cross-over product and a new version of the Hp^2 allele as the other (fig. 3.11). This new Hp^2 allele would be in effect

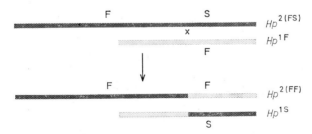

Fig. 3.11. Diagram illustrating how crossing-over in a heterozygote for the haptoglobin alleles $Hp^{2(FS)}$ and Hp^{1F} may result in the formation of the alleles $Hp^{2(FF)}$ and Hp^{1S}. F and S indicate the sites of the base differences in the DNA sequence of the haptoglobin genes which give rise to the lysine/glutamic acid substitutions in the polypeptides shown in fig. 3.3. The $\times$ indicates the site of crossing over.

an almost complete duplication of Hp^{1F}, and if we write the standard form of Hp^2 as $Hp^{2(FS)}$ then the new allele could be written $Hp^{2(FF)}$. In the same way another possible form of the Hp^2 allele, namely $Hp^{2(SS)}$ could arise from crossing-over in a heterozygote $Hp^{2(FS)}Hp^{1S}$. Variants of the polypeptide hp2α with the electrophoretic properties expected of polypeptides with structures hpFFα and hpSSα have indeed been detected in certain individuals and it seems likely that they are the products of the alleles $Hp^{2(FF)}$ and $Hp^{2(SS)}$ which have arisen in just this way (Nance and Smithies 1963).

It will be seen that if this is so, four different versions of the Hp^2 allele may be expected to exist, and the hp2α polypeptide chains which they determine will differ according to whether lysine or glutamic acid is present at positions 54 or 113 in the aminoacid sequence. The four different alleles and the polypeptide chains they determine may be written as follows:

$Hp^{2(FS)}$ → hp 2(FS)α   i.e. 54 Lys, 113 Glu
$Hp^{2(SF)}$ → hp 2(SF)α   i.e. 54 Glu, 113 Lys
$Hp^{2(FF)}$ → hp 2(FF)α   i.e. 54 Lys, 113 Lys
$Hp^{2(SS)}$ → hp 2(SS)α   i.e. 54 Glu, 113 Glu

It follows that ten different genotypic combinations may be included in the so-called Hp 2-2 phenotype (table 3.1), although only some of these may be distinguishable electrophoretically even when the separated hp 2α chains are examined. Thus while hp 2(FF)α and hp 2(SS)α are probably distinguishable

TABLE 3.1

Possible Hp 2-2 and Hp 2-1 types.

	Haptoglobin 2-2		Haptoglobin 2-1
Genotype	α-polypeptides	Genotype	α-polypeptides
$Hp^{2(FS)}$ $Hp^{2(FS)}$	hp 2(FS)α	Hp^{1F} $Hp^{2(FS)}$	hp 1Fα+hp 2(FS)α
$Hp^{2(FS)}$ $Hp^{2(SF)}$	hp 2(FS)α+hp 2(SF)α	Hp^{1F} $Hp^{2(SF)}$	hp 1Fα+hp 2(SF)α
$Hp^{2(FS)}$ $Hp^{2(FF)}$	hp 2(FS)α+hp 2(FF)α	Hp^{1F} $Hp^{2(FF)}$	hp 1Fα+hp 2(FF)α
$Hp^{2(FS)}$ $Hp^{2(SS)}$	hp 2(FS)α+hp 2(SS)α	Hp^{1F} $Hp^{2(SS)}$	hp 1Fα+hp 2(SS)α
$Hp^{2(SF)}$ $Hp^{2(SF)}$	hp 2(SF)α		
$Hp^{2(SF)}$ $Hp^{2(FF)}$	hp 2(SF)α+hp 2(FF)α		
$Hp^{2(SF)}$ $Hp^{2(SS)}$	hp 2(SF)α+hp 2(SS)α	Hp^{1S} $Hp^{2(FS)}$	hp 1Sα+hp 2(FS)α
$Hp^{2(FF)}$ $Hp^{2(FF)}$	hp 2(FF)α	Hp^{1S} $Hp^{2(SF)}$	hp 1Sα+hp 2(SF)α
$Hp^{2(FF)}$ $Hp^{2(SS)}$	hp 2(FF)α+hp 2(SS)α	Hp^{1S} $Hp^{2(FF)}$	hp 1Sα+hp 2(FF)α
$Hp^{2(SS)}$ $Hp^{2(SS)}$	hp 2(SS)α	Hp^{1S} $Hp^{2(SS)}$	hp 1Sα+hp 2(SS)α

from one another and also from hp 2(FS)α and hp 2(SF)α with appropriate electrophoretic techniques, hp 2(FS)α and hp 2(SF)α are evidently indistinguishable (Nance and Smithies 1963). Similarly eight different genotypic combinations causing the Hp 2-1 phenotype may occur (table 3.1) though again not all of these may be distinguishable by electrophoresis. The differentiation of these various types is of course likely to be even more complex if as one would expect there also occasionally occurred other haptoglobin variants due to mutations of the single base change type and resulting in an aminoacid substitution in one or other of the α-chains with an alteration in charge. In general it seems that the alleles $Hp^{2(FF)}$ and $Hp^{2(SS)}$ are much less frequent than the other Hp^2 alleles ($Hp^{2(FS)}$ or $Hp^{2(SF)}$). Thus in one extensive population study (Nance 1967) it was found that although the total Hp^2 frequency was close to 0.50, the separate frequencies of $Hp^{2(SS)}$ and $Hp^{2(FF)}$ were each only be about 0.01.

In homozygotes Hp^2Hp^2, pairing at meiosis will presumably usually be of the ordinary type expected between any pair of alleles. However because of the close homology between the first and the second halves of the sequence of Hp^2, there is likely to be a significant tendency for occasional misalignment or displaced synapsis; that is the pairing of the first half of one allele with the second half of the other (fig. 3.12). If crossing over were then to occur

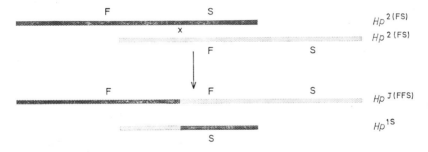

Fig. 3.12. Diagram illustrating how unequal crossing over between the haptoglobin alleles in a homozygote $Hp^{2(FS)}Hp^{2(FS)}$ may result in the appearance of a 'triplicated' allele $Hp^{J(FFS)}$ and a normal Hp^{1S} allele. The × indicates the site of crossing over.

in the paired segment, this could give rise to a new 'triplicated' allele, as well as an ordinary Hp^1 allele (Smithies 1964). Such a 'triplicated' allele would be expected to result in the formation of an hpα chain nearly three times the size of Hp1Fα or Hp1Sα. What appears to be a polypeptide chain of this size (Dixon 1966) has in fact been separated from a very unusual haptoglobin type (the so-called 'Johnson' phenotype — Giblett 1964, 1968, Cleve and Herzog 1969). Although the detailed structure of this extended polypeptide has not yet been determined, it seems quite likely that it represents the product of a 'triplicated' haptoglobin allele of the sort which may be expected to arise occasionally as a result of unequal crossing-over in the genotype Hp^2Hp^2.

3.4. *Deletions*

Certain types of mutational event can apparently result in the loss or deletion of part of a chromosome or chromatid, and the abnormality may then be transmitted to subsequent generations by the ordinary process of chromosome replication. Such mutations may be the consequence of two or more simultaneous chromosome or chromatid breaks with aberrant reunions, or they may arise because of unequal crossing over or from some other cause. Often they may result in the deletion of the DNA representing a whole series of genes, but in some cases the deletion may be much less extensive and involve only a sequence of base pairs within the confines of a single gene. For example the mutational event which originally gave rise to the Hp^2 allele (see pp. 96–99) at the α-haptoglobin locus, probably produced as its other product a partially deleted haptoglobin allele in which a sequence of 72 base pairs (24×3) occurred instead of the 249 (83×3) coding for the eighty-three aminoacids in the original Hp^{1F} and Hp^{1S} alleles.

A mutant allele with a deletion of part of its DNA sequence may be expected in most cases to code for an abnormal polypeptide with a very unusual structure. The nature of the abnormality will depend not only on the total length of the DNA sequence which is missing but also on the exact number of base pairs involved in the deletion. If the number of base pairs deleted is three or a multiple of three, then the polypeptide chain determined by the mutant allele can be expected simply to lack the series of aminoacids normally coded by the missing sequence of base pairs. That is to say, the sequence of aminoacids proximal to the deleted sequence will be the same as in the corresponding normal polypeptide chain and this will be immediately followed by the sequence of aminoacids which normally occur distal to the deleted segment. However if the number of base pairs deleted is not exactly three or a multiple of three, then an even more drastic alteration in the structure of the polypeptide for which it codes will be expected. This is because in the translation of the base pair sequence of the nucleic acid into the corresponding sequence of aminoacids in a polypeptide chain, the bases are read off three at a time consecutively, and if a number of bases which is not a multiple of three are missing all the subsequent bases in the sequence will be misread and will in general now designate an entirely different aminoacid sequence. So such a mutation will be expected to result in an abnormal allele coding for polypeptide which only resembles its normal counterpart in the sequence of aminoacids proximal to the beginning of the deleted segment. Such mutations are said to involve 'frame shifts'.

Although a very large number of different alleles with partially deleted DNA sequences may be generated by different mutational events, the polypeptide structures that they define will in most cases be very unlikely to result in a viable protein which is capable of being detected. If the protein is formed at all, it will usually be extremely unstable and generally functionally inactive. Only in what must be a very small proportion of all the mutants of this sort that appear, can one expect to detect the abnormal protein product. Nevertheless certain haemoglobin variants have in fact been found in which the structural abnormality implies that they are determined by such partially deleted mutant alleles.

One example is Hb Freiburg, in which a single aminoacid residue has been shown to be missing (Jones et al. 1966). This is a valine normally present at position 23 in the β polypeptide chain. The rest of the aminoacid sequence of the β-chains, and also the complete sequence of the α-chain are apparently normal. The abnormal haemoglobin was discovered in a woman with a mild haemolytic anaemia, and also in two of her children, but it was not present in either of her parents or in her three brothers. In the affected individuals it occurred along with normal Hb A. So one may infer that the mother and those of her children who showed the abnormal haemoglobin were heterozygous for a rare allele at the β-gene locus which had probably arisen as a new mutation in the germ cells of one of the mother's parents. The mutation had presumably resulted in the deletion of the three consecutive base pairs which normally constitute the base triplet coding for valine at position 23 in the β polypeptide chain.

Another example is Hb Gun Hill (Bradley et al. 1967, Rieder and Bradley 1968) in which the abnormality also occurs in the β-chain. Here five aminoacid residues were found to be missing and although it could be concluded that they represented the sequence which normally occurs either in positions 91-95, or 92-96, or 93-97 in the β-chain, the exact position of the deletion could not be determined from the analytical data. The abnormal haemoglobin was found as well as normal Hb A in an adult with a long history of chronic haemolytic disease, and in his daughter. They were presumably heterozygous for a rare allele at the β-locus, in which a sequence of fifteen consecutive base pairs had been lost.

Both Hb Freiburg and Hb Gun Hill belong to the class of 'unstable' haemoglobins (pp. 24–26), the instability presumably being a consequence of a distortion of the three dimensional configuration of the protein due to the missing aminoacids. The distortion of the three-dimensional structure was also indicated in other ways. Thus in the case of Hb Freiburg it was found that the spectral absorption curve of the methaemoglobin derivation was significantly different from that found with the met-forms of either Hb A or any of the Hb 'M's (p. 24) which have so far been identified. Although the site of the missing aminoacid (β 23) in the three-dimensional arrangement of the polypeptide chain is not apparently very close to the haem group, nevertheless the deletion evidently distorts the molecular structure in such a way as to modify the haem–globin interrelationships. It was also found that the heterozygous individuals with this abnormality showed a consistent though mild degree of cyanosis due to methaemoglobinaemia, presumably because the iron in the haems attached to the abnormal β-chains when oxidised to the ferric form is less readily reduced than in normal haemoglobin. In the case of Hb Gun Hill, both the haems normally attached to the β polypeptide chains were found to be missing, so that the protein was functionally quite abnormal. The five aminoacids which are deleted in this haemoglobin normally occur in a region (β 90-97) closely associated with the haem group, and presumably their absence creates a situation in which a haem cannot be held in a stable association with the polypeptide chain.

TABLE 3.2

Haemoglobin variants with deletions.

Abnormal Hb	Number of deleted aminoacids	Site of deletion	References
Freiberg	1	β 23	Jones et al. (1966)
Gun Hill	5	β 91–95 or 92–96 or 93–97	Bradley et al. (1967), Rieder and Bradley (1968)
Leiden	1	β 6 or 7	De Jong et al. (1968)
Tochigi	4	β 56–59	Shibata et al. (1970)
Tours	1	β 87	Wajcman et al. (1973)
St. Antoine	2	β 74–75	Wajcman et al. (1973)

Other examples of haemoglobin variants which have been shown to have deletions of one or more aminoacids are listed in table 3.2. They all involve the β-chain and the actual number of aminoacids missing varies from one to five in the different cases. The abnormal haemoglobin is to some degree unstable in each case, but the degree of this and hence the severity of the functional disturbance and haemolytic anaemia which results varies widely, being quite mild for example in the cases of Hb Leiden and Hb St. Antoine. The abnormality in Hb Tours resembles that found in Hb Gun Hill in that it leads to a loss of the haem groups from the abnormal chain with considerable functional impairment.

The severity of the haemolytic disorder produced in the heterozygous state clearly depends both on the actual number of consecutive aminoacids which have been lost from the sequence, and on the site where this occurs. Disruption of the three-dimensional structure will in general be progressively increased as the number of aminoacids lost increases, and it is perhaps not surprising that abnormal haemoglobins with deletions of more than five aminoacids have not yet been found. There is no reason to think that mutations resulting in the deletion of longer sections of the sequence occur with any lower frequency. However the likelihood of detecting abnormal haemoglobins with severely disrupted structure is presumably much decreased.

3.5. *Chain elongations*

Haemoglobin Constant Spring (Hb CS) is a variant with an abnormal α-chain which has an additional 31 aminoacid residues extending beyond the carboxyl terminal arginine of the normal α-chain. Thus it has a sequence 172 aminoacids long and the first 141 aminoacids are identical in sequence with those of the normal α-chain (Milner et al. 1971, Clegg et al. 1971). The additional aminoacid sequence shows no obvious relationship to other parts of the α-chain or to any other human globin chain.

Only about 1 % of the total haemoglobin in heterozygotes for the allele determining this abnormal α-chain and its normal counterpart, is Hb CS and the remainder is almost entirely normal Hb A. In the homozygote Hb CS accounts for about 7% of the haemoglobin, and the rest is mainly Hb A (Lie-Injo et al. 1973). This and other evidence (p. 46 and p. 144) supports the idea that there are two α-chain loci, with the mutant allele determining Hb CS occurring at only one of them. The rate of synthesis of the Hb CS α-chain is evidently much reduced compared with normal α-chains (Clegg et al. 1971, Clegg and Weatherall 1974), and the mutant allele concerned is essentially similar in its effect and indeed probably identical with one of the alleles previously postulated as the cause of certain forms of α-thalassaemia (p. 144). It appears to be quite widely distributed, particularly among people of Oriental origin, and in certain populations (e.g. in Thailand) heterozygous individuals may represent 4–5% of the total population (Clegg and Weatherall 1974).

The simplest explanation of the abnormal elongation of the α-chain in Hb CS, is that it is the consequence of a mutation in the normal chain terminating codon of the α-gene, which involved the replacement of one base by another in the base triplet. Thus either a UAA or a UAG triplet in the messenger RNA (fig. 1.5, p. 14) could have been changed to CAA or CAG and this would then code for glutamine which is the first amino acid added to the normal α-chain sequence in Hb CS. If this occurred then presumably the α-chain messenger RNA would be read until the next termination codon is reached at a position immediately following the codon for aminoacid number 172 in the whole sequence (Clegg et al. 1971).

This hypothesis assumes of course that the normal α-chain genes have additional genetic material of the order of 100 bases beyond the normal chain terminating codon. There is indeed direct evidence that this is the case in certain animal species, since isolated haemoglobin messenger RNA has been found to consist of strands about 650 bases in length (Pemberton et al. 1972, Labrie 1969, Gaskill and Kabat 1971) of which only about 450 are needed to code for the aminoacid sequences of α- or β-chains.

Residue number		138	139	140	141	142	143	144	145	146	147	148·········172
normal α chain		Ser	Lys	Tyr	Arg	Term						
α^{cs} chain		Ser	Lys	Tyr	Arg	Gln	Ala	Gly	Ala	Ser	Val	Ala·········Glu
Normal RNA codons			$AA{A \atop G}$	UAC	CGU	UAA	GCU	GGA	GCC	UCG	GUA	$GC{U,C \atop A,G}$
new α^{Wayne} codons			AA	UAC	CGU	UAA	GCU	GGA	GCC	UCG	GUA	G
α^{Wayne} chain		Ser	Asn	Thr	Val	Lys	Leu	Glu	Pro	Arg	Term	

Fig. 3.13. Postulated messenger RNA sequences coding for the end of the normal α-chain, the α^{cs}-chain and the α^{Wayne}-chain, showing how the additional aminoacids in the α^{Wayne}-chain could have resulted from a deletion of a single base in the normal α-sequence and a consequent frame-shift. The deleted base would be the third in the codon normally determining lysine at position α-139. It is supposed that the mutation determining Hb Constant Spring involved a change in the normal α-chain terminating codon UAA to CAA coding for glutamine. (D. J. Weatherall and J. B. Clegg, personal communication.)

These messenger RNAs contain variable amounts of poly A sequencies amounting up to about 100 bases (Pemberton and Baglioni 1972), thus leaving approximately 100 bases unaccounted for.

Further evidence in favour of the hypothesis has come from the discovery of another haemoglobin variant (Hb Wayne) with an elongated α-chain (Seid Akhavan et al. 1972). This has 146 aminoacid residues instead of the normal 141, and it differs in sequence from the normal α-chain from position 139 onwards. If hypothetical messenger RNA sequences for the end of the normal α-chain and for $α^{CS}$ and $α^{Wayne}$ are written out (fig. 3.13), it can be shown that the latter can be obtained by the deletion of one base in the codon for the lysine at position 139 (Clegg and Weatherall 1974). Thus the abnormal elongated α-chain of Hb Wayne can be regarded as the product of a mutation resulting in the deletion of a single base and a consequent frame shift (see p. 113). Its existence confirms the presence of an extended sequence at the end of human α-messenger RNA. It also suggests that the normal termination codon for α-mRNA is UAA, but that the alternative termination codon UAG (fig. 1.5, p. 14) can also be read in the haemoglobin synthesising system. The proposed messenger RNA sequences shown in fig. 3.13 nicely account for the fact that the abnormal α-chain in Hb Wayne is only elongated by five aminoacids.

Elongation of the β-chain has been found to occur in haemoglobin Tak (Flatz et al. 1971). This appears to have eight aminoacid residues added to the end of the normal β-chain sequence. The residue at position 147 is threonine, and this could not arise from a single base change mutation in any of the three chain-terminating codons (fig. 1.5, p. 14). At present the nature of the mutation involved in this case is uncertain.

3.6. The immunoglobulins: an exception to the rules

The biosynthesis of immunoglobin proteins is an exception to the general rule that the aminoacid sequence of a single polypeptide chain is coded by the sequence of bases in a single gene. Here the formula appears to be two genes – one polypeptide chain. The production of immunoglobulins is also unusual in a number of other important respects.

Immunoglobulins function as antibodies. Every normal individual synthesises an enormous variety of structurally distinct immunoglobulin molecules, each capable of binding specifically to a particular antigen. An individual is exposed in his lifetime to a very great variety of antigenic foreign substances (derived from bacteria, viruses, plant products, artificial chemicals, etc.), against which he is capable of producing quantities of specific immunoglobulins (antibodies) to protect him from their effects. Thus the system which generates the great diversity of structurally distinct immunoglobulin molecules in the individual, represents in effect a defence mechanism to cope with the intrusion of a wide variety of foreign organisms and substances from the environment. This requirement for an enormous degree of molecular diversity in a single individual distinguishes the immunoglobulins from virtually all other types of protein. It is therefore perhaps not surprising that the genetic basis of the biosynthesis of the immunoglobulins should turn out to be rather special, and to differ in a number of important respects from that which underlies the synthesis of other enzymes and proteins.

The immunoglobulins are found in blood serum and other body fluids, but they are originally synthesised in lymphoid cells. In general each cell actively producing immunoglobulin appears to produce only a single type of immunoglobulin molecule or antibody. The immunoglobulins present in normal serum however are derived from a very large number of different lymphoid cells, so that they are highly heterogenous. This makes the direct study of their structures very difficult.

In fact the elucidation of the molecular structures of individual immunoglobulins has been largely possible because in the rare malignant disorder known as multiple myelomatosis the patient's serum contains very large amounts of a single immunoglobulin in an essentially pure form. Such immunoglobulins are usually referred to as myeloma proteins. Myelomatosis is a disease resulting from a neoplastic proliferation of cells of the bone marrow. In a particular patient these cancerous cells are all apparently committed to the synthesis and secretion of large amounts of a single immunoglobulin. These myeloma proteins are not abnormal proteins in the sense that their structures differ intrinsically from that of the immunoglobulins of a normal individual. Instead they result from the enormous magnification in the rate of synthesis of just one member of the population of immunoglobulin molecules made in the patient prior to the development of the malignancy. The myeloma protein produced in excess in any one patient with myelomatosis, differs in its structure from that produced by any other patient with the condition. However a series of such myeloma proteins can be regarded as an essentially random sample of all the many different immunoglobulins present in the serum of the normal individual. Since such myeloma proteins can be obtained from individual patients in large amounts and subjected to structural studies and aminoacid sequence analysis, they provide a direct insight into the character of the molecular diversity of the immunoglobulins of the normal individual.

In addition many patients with myelomatosis produce large amounts of another protein – called a Bence-Jones protein – which because of its relatively small size passes from serum through the renal glomeruli and appears in the urine. The Bence-Jones protein formed by an individual' patient generally represents one of the polypeptide chains (the so-called light-chain) of the particular myeloma protein formed by that patient. Bence-Jones proteins can be isolated from urine and their aminoacid sequences determined. The sequence of a Bence-Jones protein from any one patient differs from that of any other, so that the aminoacid sequences of a series of Bence-Jones proteins from different patients furnish an insight into the variations in structure of the light-chains of the immunoglobulins produced by a normal individual.

There is a very extensive literature dealing with structural, immunological and genetical studies on immunoglobulins in man and in other species. From the considerable work already carried out a general picture of the genetical basis of immunoglobulin diversity has emerged, though some basic problems still remain to be resolved. Only a brief summary of the main conclusions can be given here. For detailed accounts and full references the reader is referred to the extensive reviews by Gally and Edelman (1972), Pink et al. (1971), Fudenberg and Warner (1970) and Grubb (1970).

3.6.1. Immunoglobulin structure

Each immunoglobulin molecule contains two sorts of polypeptide chain which differ in size. These are known as the L- or light-chains which are about 220 aminoacids long, and the H- or heavy-chains which have sequences of 400 aminoacids or more.

The crucial feature of these polypeptide chains is that their aminoacid sequences are made up of two regions, a so-called variable region (V) starting at the amino-terminal end and a so-called constant region (C) which is the rest of the sequence to the carboxyl-terminal end. In the L-chains the V-region (V_L) is about half the complete sequence (approximately 110 aminoacids). In the H-chains, the variable region (V_H) represents only about one quarter of the whole sequence. Thus the length of the aminoacid sequence in the variable regions of both the light and heavy chains is about the same. The V-region of the L-chain in any given immunoglobulin differs in the details of its aminoacid sequence from that of all other immuno-globulin chains. Similarly the V-region of the H-chains in any given immunoglobulin is effectively unique. Thus most of the diversity in immunoglobulin structure derives from the variations in the V-region sequences.

Fig. 3.14 shows diagrammatically the general structure of one class of immunoglobulins, known as IgG. There are two identical L-chains and two

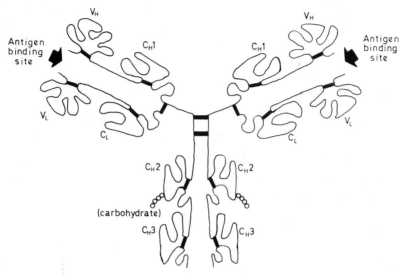

Fig. 3.14. Diagram of the structure of immunoglobulin molecules of class IgG (from Galley and Edelman 1972). Key: a) V_L and C_L are the variable and constant regions of the light chains, b) V_H is the variable region of the heavy chain. C_H1, C_H2 and C_H3 make up the constant region of the heavy chain, and are thought to represent homologous segments of the chain, c) A single inter-chain disulphide bond links C_L and C_H, d) Inter-chain disulphide bonds link the two heavy chains in the molecule, e) Intra-chain disulphide bonds occur in C_H1, C_H2 and C_H3; and also in V_H and V_L.

identical H-chains which in this class of immunoglobulins are called γ-chains. The chains are linked together by disulphide bridges, and there are also several intra-chain disulphide bridges within each of the chains. The unique configuration of the variable regions of a pair of light- and heavy-chains provides, it is thought, the specific antigen-binding site. Thus two such binding sites are present in the molecule. There are similarities in the aminoacid sequence of consecutive parts of the constant region of the heavy-chain which suggest that there are at least three homologous regions which probably arose in evolution by partial gene duplication. These are indicated in the diagram as C_H1, C_H2 and C_H3.

Immunoglobulin classes

Besides IgG a number of other classes of immunoglobulin also occur (table 3.3). They are known as IgA, IgM, IgD and IgE, and are differentiated by the structures of the C-regions of their heavy-chains which are referred to as α, μ, δ and ε respectively. There are in each class of immunoglobulins equal numbers of L- and H-chains, but the number of light–heavy pairs can differ in different classes. In any one molecule each of the light–heavy pairs is identical in structure.

TABLE 3.3

Immunoglobulin classes.

Immunoglobulin class	IgG	IgA	IgM	IgD	IgE
Heavy-chain	γ	α	μ	δ	ε
Light-chain	κ or λ	κ or λ	κ or λ	κ or λ	κ or λ
Molecular formula	$\kappa_2 \gamma_2$	$\kappa_2 \alpha_2$	$(\kappa_2 \mu_2)_5$	$\kappa_2 \delta_2$	$\kappa_2 \varepsilon_2$
	or $\lambda_2 \gamma_2$	or $\lambda_2 \alpha_2$	or $(\lambda_2 \mu_2)_5$	or $\lambda_2 \delta_2$	or $\lambda_2 \varepsilon_2$
Carbohydrate content	2.5%	5–10%	5–10%	10%	12%

Note: IgA can have additional unrelated chains called SC and J;
      IgM can also have J chains.

Immunoglobulins are glycoproteins and have polysaccharide groupings attached to their heavy-chains. However the amount of carbohydrate varies from one class of immunoglobulin to another (table 3.3).

In addition to the light- and heavy-chains certain classes of immunoglobulin have been found to contain additional polypeptide chains. For example IgA molecules present in external secretions may contain a so-called secretory component (SC) and also another polypeptide known as the 'J' component or 'J' piece.

Two sorts of light-chain, known as κ- and λ-chains are formed in normal individuals, their C-regions having distinctive aminoacid sequences. They may each occur in any of the immunoglobulin classes; thus any H-chain can be associated with either a κ or a λ light-chain in a given molecule (table 3.3).

C-region subclasses: The main immunoglobulin classes (table 3.3) can be further subdivided into discrete sub-classes according to the structures of the C-regions of their H-chains. Thus four sub-classes of IgG with distinctive γ-chains (γ_1, γ_2, γ_3 and γ_4) have been recognised, two of IgA with distinctive α-chains and two of IgM with distinctive μ-chains. Similarly three subtypes of λ light-chains have been identified which differ in certain aminoacids in their C-regions. All these various sub-classes of immunoglobulins evidently occur together in all normal individuals so it is inferred that they are each coded by a specific gene locus (see fig. 3.15).

Allotypes: There also exist differences between normal individuals which are apparently due to the occurrence of different alleles at certain of these loci. These individual differences are usually identified by examining the heterogeneous immunoglobulins of serum by specific serological techniques (Grubb 1970). One set of individual differences is known as the Gm system and evidently reflects structural differences in γ H-chains. Another set of individual differences known as the InV system reflects differences in the κ light-chains. A third set known as the Am system involves differences in α H-chains. Individuals can by appropriate serological tests using specific anti-sera, be classified into a series of types (usually referred to as allotypes) of the Gm system, and also of the InV system and the Am system. The Gm allotypes behave in family studies as if they are determined by a series of alleles at a single locus, or at a set of very closely linked loci, on an autosomal chromosome. Similarly the InV allotypes behave as an autosomal allelic series. At least 21 different allotypic variants have been identified in the Gm system, 3 in the InV system and 2 in the Am system. Family studies show no demonstrable linkage between the 'alleles' of the Gm system and those of the InV system. The loci are presumably on different chromosomes or well separated on the same chromosome. However the 'alleles' of the Gm system appear to be very closely linked to those of the Am system.

In a few cases specific aminoacid substitutions in particular immunoglobulin polypeptide chains have been identified as occurring in association with certain allotypes. For example, two of the InV allotypes differ in the κ light-chain by a single aminoacid at position 191 (valine or leucine). This difference is in the C-region of the chain as are other aminoacid substitutions which have been identified in various γ heavy-chains associated with particular Gm allotypes.

V-region classes and subclasses: Despite the great diversity in the V-regions of the immunoglobulin chains, it is nevertheless possible to classify them in terms of the degree of homology between their aminoacid sequences. There appear to be three main classes corresponding to the V-regions of the κ- and λ- and H-chains. These may be further divided into subclasses. Thus there are at least 3 sub-classes of the V-regions of κ-chains, 5 sub-classes of the V-regions of λ-chains, and 4 subclasses of the V-regions of the H-chains (fig. 3.15). Within any subclass there still remains much variation in amino-acid sequence, but the individual members of a subclass resemble each other to a much greater degree than they resemble members of other sub-classes.

3.6.2. *Immunoglobulin genes*

To account for the now very extensive structural, immunological and genetical data which has been assembled it seems necessary to postulate that there are three distinct genetic systems which determine respectively the κ light-chains, the λ light-chains and the various classes of heavy-chains (fig. 3.15). The three systems quite possibly occur on different autosomal chromosomes. Each system must include a series of separate gene loci.

Furthermore the V-region and the C-region of any given immunoglobulin polypeptide chain must evidently be coded by genes at two different loci,

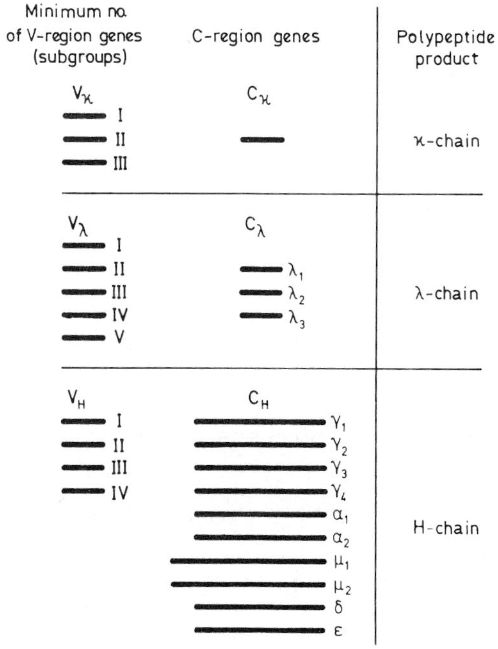

Fig. 3.15. Minimum number of genetic systems which it is necessary to postulate for the human immunoglobulin chains. Each bar represents at least one gene, but there may be many more (from Pink et al. 1971).

though in the same system. In other words two different gene loci define the aminoacid sequence of a single polypeptide chain.

There is evidence which indicates that a single polypeptide chain of an immunoglobulin is transcribed from a single strand of messenger RNA. This and other evidence suggests that the fusion of the V- and C-regions of the chain may be brought about at the DNA level and not for example by the joining together of separate V- and C-polypeptides determined by the two loci. Consequently it has been postulated that in some way a V-gene in a given cell is translocated into immediate juxtaposition with a C-gene, and a single strand of messenger RNA coding for the full aminoacid sequence of the particular immunoglobulin chain is transcribed from the fused VC-gene. Exactly how this comes about is not clear.

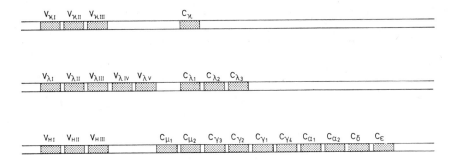

Fig. 3.16. Diagram showing the three postulated linked sets of genes (translocons) for the human immunoglobulins. The exact number and arrangement of genes within each translocon is not known (from Gally and Edelman 1972).

It is further suggested that the three genetic systems, called 'translocons' by Gally and Edelman (1972), each contain two sets of closely linked loci; one set of loci coding for C-regions, and another set of loci coding for V-regions. A possible arrangement is shown in fig. 3.16. The number of C-region gene loci in each translocon corresponds to the number of distinct classes and subclasses of the various immunoglobulin chains. Thus in the case of the H-chains, ten different C-regions have been defined and each is presumably coded by a separate locus. Similarly in the case of the λ light-chains, three distinct C-regions have been defined.

However the number of V-region genes in each of the systems is another matter, because two quite different hypotheses have been put forward to account for the enormous diversity in the amino acid sequences of the V-regions. One hypothesis suggests that only a limited number of different loci are involved, one perhaps for each V-region subgroup, and that the diversity in V-region sequences is largely the consequence of somatic mutation (Jerne 1971) or somatic recombination (Smithies 1967, Edelman and Gally 1967). The other hypothesis is that each of the many different V-regions is coded by a specific gene locus (Dreyer and Bennett 1965, Hood and Talmadge 1970). If so there might be a thousand or more different V-region loci for each of the V-region subgroups. The essential difference between the hypotheses is that the first hypothesis suggests that most of the diversity is generated somatically from a relatively small number of loci, while the second suggests that the diversity is already present in the germ line and involves a large number of different loci.

Allelic exclusion: There is a further feature of the genetics of immunoglobulin synthesis which is exceptional. In general any cell actively producing immunoglobulin, appears to produce only a single molecular type. Thus the C-region of the H-chain is the product of just a single allele at one of the C-region loci. The other allele at the same locus is not expressed, nor are the alleles at the other heavy-chain C-region loci. Similarly for the C-region of the light-chain.

Individuals who are heterozygous for allotypic markers are capable of synthesising the products of both alleles, and this can be demonstrated by serological tests on the heterogeneous collection of immunoglobulins in whole serum. However only one allele is evidently expressed by a single immunoglobulin producing cell, not both. This is illustrated by the results of allotype tests on myeloma or Bence-Jones proteins. Only a single allotype specificity for each chain is detected even though the individual may be heterozygous.

The phenomenon is known as allelic exclusion. It is similar in certain respects to the inactivation of the genes on one of two X-chromosomes in females (p. 174). But the immunoglobulin loci are located on autosomal chromosomes and are very unusual in this respect because where it has been possible to test the matter both alleles at all other autosomal loci appear to be expressed in the same cell. There are, for example, many autosomal loci which code for the polypeptide chain of a multimeric enzyme, and in these cases it is a general rule that in heterozygotes 'hybrid' enzymes containing polypeptide chains derived from each of the two alleles are observed (p. 32). This implies that the polypeptide products of the two alleles are both being produced in the same cell.

CHAPTER 4

Gene mutations affecting rates of protein synthesis

4.1. Genetic regulation of enzyme and protein synthesis

The genetical constitution of an individual not only defines the primary structures of all the many different enzymes and proteins which he is capable of synthesising, but it is also concerned with the regulation of their rates of synthesis. However the detailed manner in which such regulatory functions are exercised in higher organisms is in most respects still very obscure.

Some of the problems can be illustrated by considering the synthesis of haemoglobin in the normal individual. This protein is formed in quite considerable amounts by certain erythropoietic cells, but is not synthesised appreciably by the many other types of cells in the organism. Yet one supposes that all these cells contain the same complement of genes. So we must ask why in some cells the haemoglobin genes are effective and in others their activity is repressed. Even in cells which are capable of synthesising haemoglobin, quite marked disparities are observed. They all presumably contain the β-, γ- and δ-loci. Yet in the foetus the rate of synthesis of γ-chains is considerably greater than that of β-chains, while in the adult β-chain synthesis predominates and γ-chain synthesis is barely detectable. One also finds that δ-chain synthesis which occurs in the adult, goes on at about one fortieth the rate of β-chain synthesis. These relationships are illustrated in fig. 4.1.

The same kind of problem is posed by the distribution of enzyme proteins. Certain enzymes are found in most tissues of the organism and the particular proteins concerned appear to be the same in all cells. Others are much more localised. In some cases the specific enzyme activity appears to be confined to only one or a very limited number of tissues, and there are of course many examples where a particular enzyme though widely distributed appears to be formed at markedly different rates in different tissues. Furthermore what is apparently the same metabolic function may be performed by structurally distinct forms of an enzyme (isozymes) in different tissues. Liver phosphory-

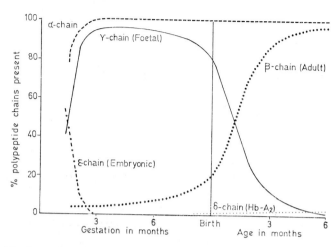

Fig. 4.1. The formation of the a, β, γ, δ and ε haemoglobin polypeptide chains in early life (Huehns et al. 1964).

lase for example differs in its structure from muscle phosphorylase and a gene mutation may affect one and not the other (see p. 214). Yet they each seem to be involved in the same way in glycogen degradation. Similarly enzymes such as aldolase and pyruvate kinase exist as structurally distinct proteins in different tissues, and a single mutation may affect one form and not another (see pp. 198–204). The variation of lactate dehydrogenase and other enzymes from one tissue to another (see pp. 71–77) illustrates the same general point.

All this is perhaps not very surprising because it is in effect simply another way of describing tissue differentiation as it occurs in multicellular organisms. Individual cells each exhibit a characteristic pattern of enzyme and protein synthesis, and they do not make the full range of proteins and enzymes of which the organism as a whole is capable. But since in general all nucleated cells in a single individual with the exception of the gametes are thought to have the same complement of genes, one must suppose that in any one cell the activity of a substantial fraction of these is continuously repressed.

Thus the problem of the genetic regulation of the rates of protein synthesis must be viewed as part of the wider problem of embryological development and of tissue differentiation. As yet no satisfactory general theory which adequately accounts for these phenomena has been advanced.

4.2. 'Structural' genes and 'regulator' genes

One hypothesis which has been widely discussed in this connexion was originally developed by Jacob and Monod (1961) from studies of mutants affecting the synthesis of certain inducible enzymes in *E. coli*. These authors proposed that in general genes may be divided into two main classes according to the different functions they fulfil in determining protein synthesis; 'structural' genes which determine the primary aminoacid sequence of individual proteins, and 'regulator' genes which govern the rates at which particular proteins are synthesised in a given intracellular environment. It was also suggested that frequently several structural genes which determine the aminoacid sequences of proteins with related metabolic functions may occur on a chromosome in a closely linked group, and that whether they are functionally active (i.e. form messenger RNA) or are repressed, depends on the state of a segment of DNA, called the 'operator', located at one end of the set of linked 'structural' genes. Such a group of adjacent 'structural' genes under the control of a single operator is referred to as an 'operon', and is thought to behave as a unit in the formation of messenger RNA (i.e. in transcription). The hypothesis suggests that the state of the operator and hence the activity of the associated structural genes is controlled by a repressor substance which is the product of a 'regulator' gene.

Originally there was some doubt about the chemical nature of the postulated repressor, but it was subsequently shown in more than one case that it is in fact a protein (Gilbert and Müller Hill 1966, Ptashne 1967). So one may presume that both 'regulator' and 'structural' genes code for the aminoacid sequences of specific proteins, and that the distinction between them essentially depends on the functional role of the proteins they determine.

The suggested interrelations between these different genetic units are illustrated by the model scheme shown in fig. 4.2. S_1, S_2 and S_3 are structural loci whose DNA base sequences define the aminoacid sequences in the polypeptide chains P_1, P_2 and P_3. M_1, M_2 and M_3 are the intermediary messenger RNA's, probably formed as a single continuous strand. O is the operator and lies immediately adjacent to S_1. O, S_1, S_2 and S_3 together constitute an 'operon'. RG is a regulator gene which forms the repressor protein R. R is capable of binding specifically with the operator O and when this occurs structural gene activity along the whole length of the operon is repressed. R however is also capable of complexing with certain small molecules (inducers) which may be present in the intracellular milieu. When this happens its configuration is altered in such a way that it no longer acts on the

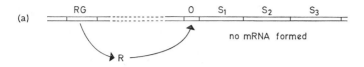

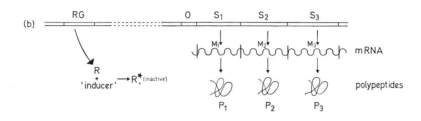

Fig. 4.2. Diagram illustrating the regulation of 'structural' gene activity according to the scheme of Jacob and Monod (1961), as described in the text. (a) Repression of 'structural' gene activity. (b) Derepression of 'structural' gene activity.

operator, thus allowing structural gene activity to proceed. So the presence of the particular metabolites which complex in a specific way with R, will affect the synthesis of all the polypeptides (P_1, P_2 and P_3) coded by the 'structural' genes in the operon. If such a metabolite is absent the structural genes will be repressed. If it is present they will show more or less activity depending on its concentration and its affinity for R.

Mutation at one of the structural loci (e.g. S_2) will usually affect only the synthesis of the corresponding polypeptide chain (P_2), altering its structure by causing, for example, a single aminoacid substitution. Mutation at the regulator locus on the other hand will be expected to affect the rates of synthesis of all the polypeptides determined by the structural genes in the operon, but will not alter their structures. Thus it might result in a modification of the repressor substance R such that it can no longer act on the operator. If so there would be continuous and uncontrolled synthesis of P_1, P_2 and P_3 irrespective of changes in the intracellular milieu. Alternatively it might modify R so that it no longer complexes with the inducer molecules but can still act on the operator. If so continuous repression of synthesis of the polypeptides would occur. Mutations of the operator may also be envisaged. According to their nature they might result in continuous repression or continuous activation of the structural genes, irrespective of whether the repressor R was present in the active state or not. An important feature of the scheme is that in heterozygotes a mutant of an operator would be expected to affect only the activity of the closely linked structural loci on the same chromosome, and not those on the homologous chromosome; whereas a mutant regulator allele in a heterozygote would be expected to affect the activity of the structural genes on both chromosomes equally.

Since this general scheme for enzyme and protein regulation was first put forward by Jacob and Monod, a considerable amount of work in micro-organisms involving a variety of enzyme systems, has been carried out and the results in their main essentials have substantiated the key features of the hypothesis, although the details vary from system to system and in some cases 'positive' as opposed to 'negative' control systems appear to operate.

However the regulatory systems of multicellular differentiated organisms are for various reasons likely to be more complex than those of unicellular organisms. Indeed it seems quite probable that the evolutionary process has involved the addition of novel patterns of regulation as well as the reorganisation of pre-existing patterns, and this has perhaps been a more important general phenomenon in the evolution of higher organisms than the emergence of new 'structural' genes (Britten and Davidson 1971). In this connexion the discovery that in higher organisms a considerable amount of the DNA in chromosomes occurs in the form of repetitive base-pair sequences (Britten and Kohne 1968) which evidently for the most part do not code for proteins, is of considerable interest. A possible function for at least some of this reiterated DNA lies in the regulation of the activity of genes which do code for the structures of polypeptide chains and also for the structures of transfer RNAs. Britten and Davidson (1969, 1971), for example, by developing this idea have postulated a complex regulatory network depending on a series of functionally different DNA elements whose integrated effect is to control the production of messenger RNA from what they term 'producer' **genes** (analogous to the 'structural' genes of the Jacob–Monod model).

But there is yet very little direct evidence one way or the other for these various kinds of regulatory systems. In particular there is virtually no critical data bearing on the question as to whether 'regulator' or 'operator' loci with properties of the kind postulated by Jacob and Monod occur in mammals such as man. Nor is there direct evidence for the more complex regulatory system postulated by Britten and Davidson. Nevertheless it seems necessary to assume that some sort of rather precise genetical control of the rates of synthesis of enzymes and proteins in individual cells of higher organisms must exist.

A considerable variety of inherited disorders involving specific deficiencies of particular enzymes and proteins are known in man, and it has been suggested, at various times, that a number of these may be due to mutations of regulator genes or of operators. In some instances where this has been postulated, subsequent work has shown that in fact a structural abnormality of the particular protein occurred and so the mutation must have affected the so called 'structural' gene for the protein. In the other cases the available data is generally inadequate to enable a critical distinction between the different possibilities to be made.

4.3. Structure-rate relationships

Another possibility which must be considered in connexion with the general problem of the genetic control of rates of synthesis of specific proteins is that the nucleotide sequence of a given 'structural' gene besides coding for the aminoacid sequence of a polypeptide chain may also in some degree determine the rate at which it is synthesised (Itano 1957, 1965, 1966; Boyer et al. 1964).

In the synthesis of a specific polypeptide chain the DNA chain coding for its aminoacid sequence is first transcribed into a corresponding mRNA chain which then becomes attached to ribosomes where it serves as a template for the formation of the polypeptide chain by the sequential binding of tRNA molecules with their attached aminoacids. The overall rate of synthesis of a specific protein may be limited by the particular rates at which these different stages in the process take place, and also of course by the rates of subsequent stages required for the final completion of the protein molecule. Thus the overall rate of synthesis of the protein may be limited by the rate of mRNA formation, by its stability, by the rate of its attachment to and release from the ribosomes, by the rate of initiation of tRNA binding and the rates at which successive tRNA molecules with their appropriate aminoacids bind successively to the mRNA base triplets, by the rate of release of the finished polypeptide chain from the ribosomes, and by the rate at which the folded polypeptides may combine with other polypeptides in the case of polymeric proteins. The rate of formation of mRNA, its stability and its binding to and release from ribosomes could well be affected by its precise nucleotide sequence, which in turn is determined by the nucleotide sequence of the particular 'structural' gene concerned. Furthermore the rate of polypeptide chain formation on the ribosomes could be affected by the particular mRNA sequence, or the availability of the different tRNA molecules needed to assemble the polypeptide chain. Also the particular sequential arrangement of the aminoacids in the protein could affect the rate at which it folds into an appropriate configuration as the chain is assembled, and also perhaps the rate at which it is released from the ribosomes. Finally the precise polypeptide structure may affect the rate at which it is capable of combining with other polypeptides or prosthetic groups to give the final protein. Thus the specific base sequences of the DNA of any particular 'structural' gene or genes could in a variety of ways and at several different levels influence the rate of synthesis of a particular protein.

Heterozygotes for 'structural' gene mutations offer one approach to investigating whether this kind of effect actually occurs. It has been suggested for example that differences in the amounts of the two polypeptide products observed in certain heterozygotes, may in some cases be due to differences in their rates of synthesis imposed by the difference in base sequences of the two alleles. Heterozygotes with the sickle-cell trait and also those with the haemoglobin C trait illustrate the general argument. In the sickle-cell trait it is consistently found that there is more Hb A in the red cells than Hb S. On average about 60–65% of the haemoglobin present is Hb A ($\alpha_2\beta_2$) and about 35%–40% is Hb S ($\alpha_2\beta_2^{6\ \text{Glu}\to\text{Val}}$) (Wells and Itano 1951, Wrightstone and Huisman 1968). A similar disproportion is observed in haemoglobin C trait, where again the amount of Hb A present is significantly more than the amount of Hb C. The total amount of haemoglobin formed in these heterozygous states is not on average significantly different from that in normal homozygotes. Furthermore although there may be minor differences in stability between the variant haemoglobins and Hb A, they would appear to be too slight to account in terms of preferential breakdown of the variant haemoglobin for the disparity observed. It seems then that the characteristic disproportion in the amounts of the two haemoglobins in these heterozygotes is quite possibly a consequence of differences in rates of synthesis (Itano 1965, Boyer et al. 1964). This implies that the rate of synthesis of for example the $\beta^{6\ \text{Glu}\to\text{Val}}$ polypeptide chain in the sickle-cell trait, or the $\beta^{6\ \text{Glu}\to\text{Lys}}$ polypeptide chain in the haemoglobin C trait, is significantly less than the rate of synthesis of normal β polypeptide chains in these heterozygotes. The mutation in each case is thought to alter by a single base change the nucleotide sequence of the messenger RNA, and to alter by a single aminoacid substitution the aminoacid sequence of the corresponding polypeptide. Either of these changes could in theory limit the overall rate of synthesis, as compared with that of the normal polypeptide. But the exact nature of the proposed structure–rate relationship in these cases remains obscure.

Such structure–rate relationships could well be relevant not only to the synthesis of variant polypeptides determined by mutant or abnormal genes, but also to so-called normal alleles. Indeed it is possible that in any gene there may be one or just a few triplets in the nucleotide sequence which by the restrictions they impose at some step in the process of polypeptide synthesis, are rate limiting. So they will effectively determine the maximal overall rate of synthesis of the particular polypeptide. Relevant to this is the fact that any one aminoacid may be coded by several distinct mRNA triplets which may require different tRNA molecules. So the triplet which actually codes for a particular aminoacid in a sequence, and the availability of the corresponding tRNA in the cell, could be important in deciding the overall rate of synthesis of the polypeptide.

An interesting case where marked disparities in the rates of assembly of two distinct polypeptide chains determined by separate loci, have been demonstrated directly, is provided by the normal β- and δ-chains of haemoglobin (Winslow and Ingram 1966). These chains differ in only ten aminoacids so the genes coding for them must be very similar. Yet under normal circumstances the rate of assembly of the β-chain is much greater than that of the δ-chain, though it is not known exactly how this comes about.

But whatever the detailed mechanisms involved, the idea that a gene by the very nature of its nucleotide sequence may provide an inbuilt control on the rate of synthesis of the polypeptide it determines may well prove an important one. It could in principle account for many of the characteristic differences that are observed between the amounts of different enzymes or proteins in a single cell.

4.4. Inherited defects in rates of protein synthesis - the thalassaemias

Quite a number of inherited abnormalities are known in which the central defect is the deficiency of a particular protein, and many of these are probably due to mutations which result in a gross but specific reduction in the rate of synthesis of one or more polypeptide chains. The most extensively studied of such conditions are those which involve defects in the synthesis of haemoglobin. Among them are a series of chronic haemolytic anaemias collectively known as the thalassaemias. They appear to be determined by a series of distinct abnormal genes, in different heterozygous and homozygous combinations. But it has not yet proved possible to define the precise nature of the mutations responsible, or the manner in which these result in the observed defect in haemoglobin synthesis. However since the conditions are probably representative of a whole class of similar disorders in which specific defects in rate of synthesis of a variety of other proteins and enzymes occur, the problems and difficulties posed by the thalassaemias are of some general relevance.

It is useful to classify the various thalassaemias according to the poly-peptide chain primarily concerned in causing the haemoglobin deficit (Ingram and Stretton 1959). Thus in the so-called β-thalassaemias, the main defect appears to involve β-chain synthesis, whereas in α-thalassaemias α-chain synthesis is primarily affected.

The very extensive literature of the various thalassaemic conditions has been reviewed in detail by Weatherall and Clegg (1972).

4.4.1. β-Thalassaemia

β-Thalassaemia occurs relatively commonly in certain populations in the Mediterranean countries (e.g. in Southern Italy and Greece) and it is also not uncommon among populations living in India and the Far East. It is prob-able that a number of different mutant genes can cause this type of abnor-mality and although they each result in depression of β-chain synthesis, the degree to which this occurs appears to vary considerably from one mutant to another. In some cases β-chain synthesis may be completely absent. In others synthesis of β-chains does occur but at a much reduced rate.

Homozygotes for such genes show a severe form of anaemia classically known as thalassaemia major (or Cooley's anaemia). The disease generally becomes manifest shortly after birth, at the time when γ-chain synthesis is normally giving place to β-chain synthesis. The severe anaemia results from a deficiency of haemoglobin A ($\alpha_2\beta_2$). This produces a stress situation in the erythroid marrow which by a mechanism not yet understood, leads to the continued production of cells still capable of forming haemoglobin F ($\alpha_2\gamma_2$). Consequently most of the haemoglobin present is Hb F. The remainder is mainly Hb A but in the most severe cases this may be completely or almost completely absent. Normal or slightly increased levels of Hb A_2 are also usually found. The continued production of γ-chains is generally quite insufficient to compensate for the deficiency of β-chain synthesis. So the mean red cell haemoglobin concentration is markedly reduced. Also the total number of red cells in the circulation is much less than normal, and the red cells themselves show very marked variations in shape and size. They also vary in the amounts of Hb F they contain, those red cells with higher Hb F surviving longer than others.

Studies on the kinetics of globin synthesis in red cells in β-thalassaemia (Weatherall et al. 1965, Bank and Marks 1966, Huehns and Modell 1967, Bargellesi et al. 1967) have demonstrated the grossly reduced rate of β-chain formation and have also shown that α-chain production, which appears to go on at a normal rate, greatly exceeds β- and γ-chain formation (Bank et al. 1968). Thus a great excess of α-chains is produced. These are evidently very unstable and tend to precipitate and become associated with the red cell membrane. This in turn renders the red cells particularly susceptible to premature destruction. Consequently the severe anaemia characteristic of β-thalassaemia results not only from an overall deficit in haemoglobin synthesis, but also from an excessive rate of red cell destruction.

Heterozygotes often show a mild anaemia (thalassaemia minor), which is however very variable in degree and indeed may often not be clinically apparent. There are usually typical abnormalities of red cell morphology (microcytosis, anisocytosis, and target cells). Most of the haemoglobin present is haemoglobin A, but characteristically the proportion of haemoglobin A$_2$ ($\alpha_2\delta_2$) is increased, assuming values of perhaps 4 to 7% of the total instead of the normal 2 to 3% (Kunkel et al. 1957). Thus δ-chain synthesis appears to be somewhat augmented. Hb F may also occur in slightly increased amounts (0.5–4% of the total haemoglobin), and it is unevenly distributed between different red cells as in the homozygotes.

Heterozygotes for a β-thalassaemia gene may be heterozygous as well for one or another of the genes which determine variant haemoglobins with abnormal β-chains such as Hb S, Hb C or Hb E. The anaemias that result are referred to as thalassaemia–sickle-cell disease, thalassaemia–haemoglobin C disease etc. In these conditions the variant haemoglobin predominates and accounts for 70% or more of the total haemoglobin present. It appears that in such cases synthesis of the variant β-chain (i.e. β^S, β^C or β^E) is not depressed, so that the proportion of Hb A found gives some indication of the degree of depression of synthesis of the normal β-chain caused by the particular β-thalassaemia gene present. In some cases normal β-chains may be virtually absent.

Two types of hypotheses have been advanced about the nature of the mutations that result in the defective β-chain synthesis characteristic of β-thalassaemia. One is that the mutations occur at the structural locus which determines the aminoacid sequence of the β-chain. In this case the β-thalassaemia genes would be allelic to the genes which determine β-chain variants such as Hb S or Hb C. The other hypothesis is that the mutations affect a different locus, which however is specifically involved with the regulation of the rate of β-chain synthesis in the normal individual.

To some extent information about this question can be obtained from studies of families in which children of doubly heterozygous individuals (e.g. individuals with either thalassaemia–sickle-cell disease or thalassaemia–haemoglobin C disease) occur. A considerable number of such informative families (Ceppellini 1959, Motulsky 1964, Weatherall 1972) have now been investigated and it is clear that with only a few possible exceptions the great majority of the children have received from their doubly heterozygous parent either the β-thalassaemia gene or the gene for the β-chain variant, but not both or neither. The pedigrees therefore suggest that the genes are either at the same locus (i.e. are allelic), or if they do occur at separate loci then these must be relatively closely linked, if not immediately adjacent to one another on the same chromosome. Since the doubly heterozygous parent in these different families appeared to transmit one or other of the abnormal genes but not both to any particular child, then we must conclude that if two loci are indeed involved then the doubly heterozygous parent in every case carried the two abnormal genes on the separate members of the particular pair of homologous chromosomes (i.e. in repulsion).

But it is not possible from the available data to distinguish critically between allelism and the alternative possibility that two separate but closely linked loci are concerned. It is important to note that studies of this kind are beset with considerable difficulties. If linkage is close a very large number of informative families need to be studied to identify a cross-over, and confusion because of undetected illegitimacy can easily occur. There are also special problems because of the variability in manifestation of the β-thalassaemia genes, and the absence of any unequivocal criterion of their presence or absence. Thus it may prove difficult to distinguish with certainty between a milder manifesting thalassaemia heterozygote and a normal homozygote. It could also prove difficult, and in certain cases perhaps impossible, to distinguish phenotypically between an individual who is a simple heterozygote for a β-thalassaemia gene which results in severe or perhaps complete suppression of β-chain synthesis, and one who is doubly heterozygous and carries this gene and also a gene determining a β-chain variant on the same chromosome (i.e. in coupling). Such a doubly heterozygous combination might be expected to result in a complete or nearly complete suppression synthesis of the β-chain variant and so the individual could resemble an ordinary β-thalassaemia heterozygote (Motulsky 1964). These situations are of course those precisely involved in the detection of presumptive cross-overs.

If the β-thalassaemia gene does indeed occur at a locus separate from the

β-chain structural locus but closely linked to it, then the findings in double heterozygotes such as β-thalassaemia–sickle-cell disease or β-thalassaemia–haemoglobin C disease indicate that the thalassaemia gene can only repress activity of the β-structural locus on its own chromosome. This is so because in these conditions the rate of synthesis of the β-chain variant (e.g. Hb S or Hb C) is apparently not affected. Thus on this hypothesis the β-thalassaemia locus might correspond in the Jacob–Monod scheme to an operator which was specifically concerned with controlling the activity of the adjacent β-chain structural gene. It may be noted that since δ-chain synthesis is not repressed in β-thalassaemia it would be necessary to suppose that the closely linked β- and δ-structural loci are not under the control of the same operator, and thus do not constitute an operon.

If, on the other hand, the β-thalassaemia genes represent mutations at the locus which determines the structure of the β-chain, the question arises as to exactly how such mutants result in the marked depression in β polypeptide synthesis which is observed. In theory a change in the nucleotide sequence of the β 'structural' gene might produce such an effect by imposing a severe rate limiting restriction at one of the several different stages in the normal processes of transcription and translation which lead to the final synthesis of the polypeptide and its incorporation into haemoglobin (see discussion of structure-rate relationships pp. 131–133). But the identification of such abnormalities poses exceedingly difficult technical problems which still in large part remain to be resolved, although some important results have been obtained. For example it has been shown that the actual quantity of β-chain messenger RNA in reticulocytes from patients with β-thalassaemia is very much less than α-chain messenger RNA, whereas in normal reticulocytes the amounts of α and β mRNAs are about equal (Kacian et al. 1973, Housman et al. 1973, Forget et al. 1974). These results appear to exclude the possibility that the defect is primarily at the level of translation. However they leave open the question as to whether the messenger RNA is actually formed at a reduced rate, or whether the decreased amounts which are observed occur because it is degraded more rapidly than is normally the case.

Only a limited number of cases of β-thalassaemia have been studied in detail from this point of view, and in view of the evident heterogeneity of the condition, it is quite possible that different mutants could affect the synthetic process in different ways and yet produce a similar end-result, that is a severe limitation in β-chain production or its complete failure.

4.4.2. Hereditary persistence of foetal haemoglobin (African type)

It is of interest to contrast the effects of the β-thalassaemia genes with those of another mutant which results in complete suppression of β-chain synthesis. This mutant is usually referred to as the 'high Hb F' gene and it gives rise to a condition known as 'hereditary persistence of foetal haemoglobin $(\alpha_2\gamma_2)$'. It appears from family studies that the gene occurs at a locus which, like the β-thalassaemia locus, is either closely linked to the β-chain 'structural' locus or is in fact identical with it. It has mainly been observed in Africans or in people of African descent.

In adults heterozygous for this 'high F' gene and its normal allele (Conley et al. 1963), about 25% of the total haemoglobin is Hb F $(\alpha_2\gamma_2)$. The rest is. mainly Hb A $(\alpha_2\beta_2)$ and the total amount of haemoglobin present is within the normal range. Some Hb A_2 $(\alpha_2\delta_2)$ also occurs, but the proportion is somewhat less than is normally found. Thus δ-chain synthesis appears to be depressed, a situation which is different from that in heterozygotes for β-thalassaemia genes where the proportion of Hb A_2 is usually increased. Another difference is that the foetal haemoglobin apart from being much greater in amount than in β-thalassaemia heterozygotes is distributed evenly among the red cells, whereas in β-thalassaemia heterozygotes it is very unevenly distributed.

The homozygote for this 'high F' gene is evidently only capable of forming Hb F (Wheeler and Krevans 1961, Baglioni 1963, Charache and Conley 1969, Seigel et al. 1970). No Hb A or Hb A_2 are present, but compensation by Hb F synthesis is evidently quite adequate so that severe anaemia does not result.

Individuals heterozygous for the 'high F' gene and heterozygous as well for one of the genes determining a β-chain variant such as β^S or β^C have also been identified (Edington and Lehmann 1955, MacIver et al. 1961, Conley et al. 1963). Here the only haemoglobins present are the variant haemoglobin, and Hb F. There is no Hb A or Hb A_2.

Thus the abnormality produced by this 'high F' gene appears to represent a complete suppression of both β- and δ-chain synthesis, associated with a continuing synthesis of γ-chains which more or less completely compensates for the deficit of β. This contrasts strikingly with the situation with the β-thalassaemia genes where usually only β-chain synthesis is suppressed and not δ-chain synthesis, and where compensatory synthesis of γ-chains although it occurs is very much less efficient so that severe anaemia results.

In the normal individual two sorts of γ-chain are synthesised, $^G\gamma$ and $^A\gamma$, which differ according to whether glycine or alanine occupies position 136 in the aminoacid sequence (p. 45). They are evidently determined by separate loci. Studies on the foetal haemoglobin in heterozygotes for the 'high Hb F' gene in Blacks have shown that in most cases both $^G\gamma$ and $^A\gamma$ are increased (Huisman et al. 1969). In a few cases however only $^G\gamma$ is increased and in these individuals the proportion of the total haemoglobin which is Hb F is somewhat less (10–20%) than in heterozygotes where both $^A\gamma$ and $^G\gamma$ are present in excess (Hb F 20–35%). There are also apparently a few cases where although both $^A\gamma$ and $^G\gamma$ are produced, $^A\gamma$ is in considerable excess over $^G\gamma$ (Huisman et al. 1970). These findings indicate that the condition is heterogeneous and two or more different mutants may produce it.

The nature of the mutations which give rise to hereditary persistence of foetal haemoglobin are not known with certainty. However a plausible explanation of at least some types of this condition is that they represent actual deletions of the β- and δ-loci on the affected chromosome. As a result β- and δ-chains are not produced and γ-chain synthesis persists at a high level into adult life. In those cases where only $^G\gamma$-chains are found in the increased foetal haemoglobin, then one could infer that the $^A\gamma$ locus as well as the β- and δ-loci are deleted.

This hypothesis is supported by the findings in the case of Hb Kenya which is an abnormal haemoglobin in which the non-α-chain represents a $\gamma\beta$ fusion polypeptide. This is determined by an abnormal gene which is presumed to have arisen by unequal crossing-over between a mispaired β-gene and a $^A\gamma$-gene (see p. 107). Of the total haemoglobin present in adult heterozygotes, Hb Kenya represents about 10%, Hb F about 6%, Hb A_2 about 1.5% and the rest is Hb A (Clegg and Weatherall 1973). Thus there is, besides the presence of Hb Kenya, a marked increase in Hb F and a decrease in Hb A_2. The excess Hb F contains only $^G\gamma$-chains and this would be consistent with a deletion of the β-, δ- and $^A\gamma$-loci occurring as a result of an unequal cross-over (see fig. 3.10, p. 108) which would leave the $^G\gamma$-locus and $\gamma\beta$ fused locus on the chromosome (Clegg et al. 1973, Kendall et al. 1973). The abnormality in Hb Kenya can be regarded as representing one type of hereditary persistence of foetal haemoglobin. The common African type in which both $^G\gamma$- and $^A\gamma$-chains are synthesised in excess and there is no β- or δ-chain synthesis may well have arisen by a similar unequal crossing-over event which led to the loss of the β- or δ-locus but allowed the retention of both the $^G\gamma$- and $^A\gamma$-loci.

4.4.3. Other abnormal genes affecting β- and δ- and γ-chain synthesis

A number of other apparently specific hereditary abnormalities involving
defective synthesis of the β- and δ-haemoglobin chains have also been
identified. One of these is known as δβ-thalassaemia (Brancati and Baglioni
1966, Comings and Motulsky 1966) and has also been referred to as F-
thalassaemia (Motulsky 1964). It resembles β-thalassaemia clinically but
tends to be less severe. However, both β- and δ-chain synthesis is depressed.
Compensatory γ-chain synthesis occurs and is significantly more effective
than in β-thalassaemia, though it does not completely compensate for the
deficit of β-chains. In the homozygote where there is marked anaemia, no
Hb A or Hb A$_2$ can be detected and Hb F appears to be the only haemoglobin
present.

The mutants which determine the Lepore haemoglobins (p. 104) can also
be regarded as determining a form of δβ-thalassaemia. In the Lepore haemoglobins the non-α-chain represents a δβ fusion product, and the mutants
evidently arise by unequal crossing-over between misaligned δ- and β-loci
(fig. 3.8, p. 105). The unequal cross-over results not only in the formation
of the abnormal δβ fusion gene, but also in the deletion of the normal β- and
δ-gene from the same chromosome. In a heterozygote about 10–15% of
total haemoglobin is Hb Lepore and the rest is mainly Hb A, though some
increase in Hb F also occurs. There is usually no overt anaemia but the
haematological picture resembles that seen in thalassaemia minor. In homozygotes (Neeb et al. 1961, Quattrin et al. 1967, Duma et al. 1968) there is a
severe anaemia similar to thalassaemia major. No Hb A or Hb A$_2$ is formed
and the haemoglobin present is mainly Hb F (70–80%) the rest being Hb
Lepore.

A further distinct condition in which suppression of both β- and δ-chain
formation is found has been regarded as another and somewhat less marked
form of 'hereditary persistence of foetal haemoglobin' (Fessas and Stamatoyannopoulos 1964). It is not uncommon in Greece and has been referred to
as the 'Greek type', in contradistinction to the more marked form known as
the 'African type'. The heterozygotes differ from those with the 'African'
type of hereditary persistence of foetal haemoglobin in that the average level
of Hb F is only on average about 14%, compared with an average of about
25% in the 'African' type. Also the excess Hb F appears to contain only
^{A}γ-chains (Huisman et al. 1970). Furthermore the mutation does not appear
to abolish β- and δ-chain synthesis completely as it does in the 'African'
type, though the rate of synthesis of these chains is evidently much reduced.

TABLE 4.1

Effects of different mutations of the Hb δ-β complex on the formation of HbF ($\alpha_2\gamma_2$) in heterozygotes (Motulsky 1964).

Type of mutation		Effect of mutation on		HbF ($\alpha_2\gamma_2$) in heterozygotes		Mean corpuscular haemoglobin in heterozygotes
		β-chain synthesis	δ-chain synthesis	% of total Hb	Distribution of HbF among erythrocytes	
'Thalassaemia'	β-thalassaemia	Mild to severe reduction	Increased	Small increase	Uneven	Low
	$\delta\beta$-thalassaemia	Mild to severe reduction	Normal or slight reduction	5–20%	Uneven	Low
'Hereditary persistence of foetal haemoglobin'	'Greek' type	Moderate reduction	Moderate reduction	10–20%	Even	Normal
	'African' type	Complete suppression	Complete suppression	15–35%	Even	Normal

An important difference between this abnormality and the $\delta\beta$-thalassaemia referred to above is that in heterozygotes, although Hb F may comprise on average much the same proportion of the total haemoglobin present, it is evenly distributed between the circulating red cells, whereas in $\delta\beta$-thalassaemia Hb F is very unevenly distributed between different red cells.

A summary of the findings in heterozygotes for these different mutants is given in table 4.1.

Various hypotheses similar to those advanced to explain β-thalassaemia and the so-called 'African type' of 'hereditary persistence of foetal haemoglobin' have been put forward to account for these other different though apparently related conditions. But at present there is no critical way of deciding between the various theoretical possibilities. It is clear however that there must occur a variety of distinct mutants affecting in different ways the synthesis of β-, δ- and γ-chains, and giving rise to these different abnormalities. The precise way in which they produce their effects, however, remains obscure. One may anticipate that similar complexity and heterogeneity will be encountered in connexion with defects in rates of synthesis of other proteins and enzymes.

4.4.4. α-Thalassaemia

Since α-chains occur in both adult and foetal haemoglobins a mutation causing severe depression in α-chain synthesis would be expected to become manifest in foetal life. In fact a condition of this sort has been recognised. The affected foetus is grossly oedematous, a condition known as hydrops foetalis, and there is enlargement of the liver and other abnormalities (Lie-Injo et al. 1962, Pootrakul et al. 1967a). Death occurs *in utero* with abortion in the latter part of the pregnancy. All the haemoglobin present in the red cells has an abnormal structure. It mainly consists a tetramer made up of normal γ-chains and its structure may be written as γ_4 (Weatherall et al. 1970, Todd et al. 1970). It is usually referred to as Hb Bart's (Ager and Lehmann 1958, Hunt and Lehmann 1959). Evidently there is a complete absence of α-chain synthesis, but γ-chain synthesis proceeds normally and in the absence of α-chains the γ-tetramer is formed. The severe pathological consequences are due both to the haemoglobin deficit and also to the abnormal oxygen dissociation curve of Hb Bart's (γ_4) which tends to make oxygen less readily available to the tissues.

Prior to the idea that there might be two separate loci coding for α-chains (p. 46), this condition was attributed to a mutant either at the single locus thought to code for the α-chain or at another locus (probably closely linked) which in some way controlled its synthesis. This mutant was called α-*Thal*$_1$. The mutant evidently prevented α-chain synthesis, since none could be detected in homozygotes. However heterozygotes for the mutant and its normal allele appear to be quite healthy and to have at most only a minimal anaemia (Pootrakul et· al. 1967a). In the newborn significant amounts of Hb γ_4 are present, but this tends to disappear along with Hb F ($\alpha_2\gamma_2$) during the following few months (Wasi et al. 1969).

Another condition, known as Hb H disease, is also caused by defective α-chain synthesis (for review see Weatherall and Clegg 1972). The disease is a variable though often moderately severe anaemia in which some 5–20% of the haemoglobin present is an unusual form known as Hb H, while the rest is mainly Hb A. Hb H is a tetramer made up of normal β-chains, i.e. β_4 (Hunt et al. 1959). It is very unstable and gives rise to intraerythrocytic inclusion bodies which are a characteristic feature of the anaemia. Where it has been possible to study the cord blood of infants who subsequently develop Hb H disease (Pootrakul et al. 1967b), about 25% of the haemoglobin was found to be Hb γ_4, and this subsequently decreased progressively with a reciprocal increase in Hb β_4. Thus in Hb H disease there appears to be a partial deficiency of α-chains which results in a relative excess of γ-chains in foetal life and of β-chains later on.

Hb H disease occurs quite frequently in S.E. Asia, particularly in Thailand. Extensive family and population studies showed that individuals with this condition are heterozygous. Assuming a single α-chain locus, one of the alleles present appeared to be the α-*Thal*$_1$ gene which in homozygotes results in a complete failure of α-chain synthesis and hydrops foetalis. The other appeared to be an allele which also depressed α-chain synthesis but to a lesser degree. This was called α-*Thal*$_2$ (Wasi et al. 1969, Na-Nakorn and Wasi 1970).

Studies of haemoglobin synthesis in Hb H disease showed a much reduced rate of production of α-chains relative to β-chains (Clegg and Weatherall 1967). Since homozygosity for the α-*Thal*$_1$ gene evidently resulted in no α-chain synthesis at all, any α-chain synthesis in Hb H disease (presumed genotype α-*Thal*$_1$ α-*Thal*$_2$), was attributed to the α-*Thal*$_2$ gene. The heterozygous states for the presumed α-*Thal*$_1$ and α-*Thal*$_2$ genes and the normal allele were found to be quite benign clinically. Some haematological changes were detectable in α-*Thal*$_1$ heterozygotes, but barely so in α-*Thal*$_2$ heterozygotes. However in both these states imbalance of α- and β-chain production has been demonstrated (Schwartz et al. 1969), reduced α-chain production relative to β-chain production in α-*Thal*$_1$ heterozygotes being somewhat more marked than in the α-$Thal_2$ heterozygotes.

An important development in the subject came with the discovery of Hb Constant Spring (Hb CS) a haemoglobin variant in which the α-chain is elongated at the carboxy-terminal end by an additional sequence of 31 aminoacids (p. 115). It is thought to be due to a single base change mutation in the α-chain terminating codon. Hb CS was found to occur not infrequently in individuals who had typical Hb H disease. In some parts of S.E. Asia, for example, as many as 50% of the Hb H patients appear to have this abnormal haemoglobin, though in other populations its occurrence in Hb H disease appears to be less common (Clegg and Weatherall 1974). The amount of Hb CS present is always very small. Where it occurs in Hb H disease it amounts to only about 2–3% of the total haemoglobin, most of the rest being Hb A (i.e. $\alpha_2\beta_2$) and Hb H (i.e. β_4). In a family study Milner et al. (1971) found that of the two parents of cases of Hb H disease with Hb CS, one appeared to be heterozygous for the α-$Thal_1$ gene and its normal allele, and the other heterozygous for the gene determining Hb CS; but the amount of Hb CS present was only about 1% of the total haemoglobin, the rest being mainly Hb A. Further family studies and other work indicated that the Hb CS mutant had to be considered as a type of α-$Thal_2$ gene (Milner et al. 1971, Clegg and Weatherall 1974). But if so it would be necessary to account for the occurrence of normal α-chains in Hb H disease where Hb CS occurs, since the other allele α-$Thal_1$ is from other evidence incapable of leading to α-chain synthesis. This apparent anomaly is most easily explained by the hypothesis that there are two loci for α-chain production. A variety of other findings in α-thalassaemia are also probably most easily accounted for by a hypothesis involving two loci each coding for α-chains (Lehmann 1970). There are also other lines of evidence which lend strong support to such a hypothesis (see p. 46).

One scheme embodying this concept is shown diagrammatically in fig. 4.3. It is assumed that the two loci are closely linked and that mutations at either or both loci may be involved in determining the various forms of α-thalassaemia. If mutations resulting in a complete failure of α-chain production were present at both loci on the same chromosome this would correspond to the α-$Thal_1$ gene in the single locus hypothesis. If one of the loci carried such a mutant, but the other one on the same chromosome had normal function, this would correspond to the previously postulated α-$Thal_2$ gene. On this basis the form of α-thalassaemia with hydrops foetalis would have four mutant genes, two at each locus. In Hb H disease there would be three mutant genes, so that α-chain synthesis would be directed by the single

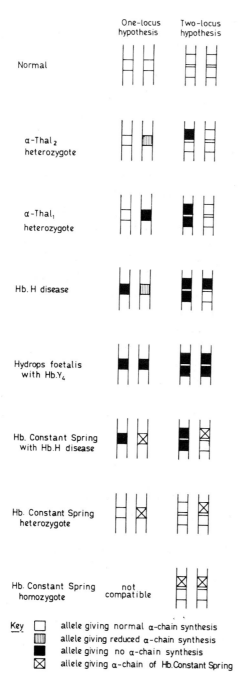

Fig. 4.3. Diagram illustrating the genetic determination of the α-thalassaemias according to the one-locus or two-locus hypotheses for α-chain formation. For the two-locus hypothesis it is assumed in this diagram that the two loci are closely linked. (The diagram is based on those of Lehmann (1970) and Weatherall and Clegg (1972)).

remaining normal gene. In the case of heterozygotes previously considered to carry either the α-*Thal*$_1$ gene or the α-*Thal*$_2$ gene in heterozygous combination with a normal allele, there would according to the new scheme be two and three normal genes respectively and the restriction on α-chain synthesis which results would only lead to minimal haematological and clinical consequences.

On this scheme the mutant determining Hb CS would be presumed to occur at one locus, and the other locus on the same chromosome would have normal function. Since the rate of synthesis of Hb CS is apparently reduced to only a small fraction (perhaps 3–5%) of the rate of synthesis of α-chains by the normal allele, the combination would in effect be very similar to that postulated for the α-*Thal*$_2$ gene.

The finding of more than 50% normal Hb A in individuals with two different α-chain variant haemoglobins (Hollan et al. 1972), and the presence of normal Hb A in individuals apparently homozygous for the mutant determining Hb CS (Lie-Injo et al. 1974) are clearly consistent with the two locus hypothesis, and cannot indeed be accounted for on the basis of only a single α-chain locus. An observation however which does not fit the hypothesis very easily is the absence of Hb A in individuals homozygous for an α-chain mutant which determines the variant Hb Tongariki, and which has an appreciable incidence in Melanesia (Abramson et al. 1970, Beaven et al. 1970). To account for this it would be necessary to suppose either that the other locus on the same chromosome was completely defective in α-chain production for which there is no good evidence; or that certain populations are polymorphic for chromosomes bearing duplicated or single α-chain loci.

CHAPTER 5

Quantitative and qualitative variation of enzymes

5.1. Quantitative variation in enzyme activity

Many gene mutations are expressed by a characteristic alteration in the activity of a specific enzyme. In some cases individuals homozygous for the mutant gene appear to have a complete or nearly complete deficiency of the particular enzyme activity and this may result in a major metabolic disturbance (see Ch. 6). In other cases the reduction in activity produced by the mutant is less profound, and very occasionally increased activity has been noted. In fact it appears that almost any degree of alteration of an enzyme activity may be produced by different mutations.

The assessment of such changes in different individuals is in general based on quantitative measurements of the level of the enzyme activity in extracts of cells from a particular tissue or tissues (e.g. red cells, white cells, liver etc.). In considering the significance of these observations it is important to recognise that the measured level of activity of an enzyme in the cells of a tissue is a complex parameter. In the first place it depends on the specific catalytic properties of the enzyme, and these are determined by the detailed molecular structure of the enzyme protein. Secondly, it depends on the actual quantity of enzyme protein present in the tissue. This represents the resultant of the rates of two quite distinct and opposite processes; the rate of synthesis of the protein, and the rate at which it is broken down or decays. Finally extraneous factors such as the presence of specific inhibitors, cofactors, metal ions and so on may affect the observed activity, as well of course as the particular experimental conditions used in the assay procedure (e.g. substrate concentrations, pH, etc.). Other complicating factors are the localisation or compartmentalisation of the enzyme within the cell, and the particular methods used to disintegrate the cell for enzyme analysis.

It is apparent, then, that a particular mutant gene might cause an observed alteration in the level of enzyme activity in a variety of different ways. The main possibilities can be conveniently summarised as follows:

147

(a) The mutational change may have led to the synthesis of a structurally altered protein with defective or modified catalytic properties. A single aminoacid substitution, for example, if it affected the active centre of the enzyme might, by altering the facility with which substrate or coenzyme is bound to the protein during catalysis, result in a marked modification in the kinetics of the catalytic process, and hence in the apparent enzyme activity. In such circumstances the actual quantity of enzyme protein present could be unaffected, even though the enzyme activity is reduced or perhaps completely absent.

(b) The mutant gene may lead to the synthesis of a structurally altered protein whose catalytic activity is not significantly affected, but whose inherent stability is less than that of its normal counterpart. The enzyme protein would tend to be more rapidly denatured *in vivo*, and its half life consequently reduced. The increased rate of breakdown would mean that the actual quantity of functionally active enzyme protein present at any one time, and hence the level of activity, would be correspondingly diminished. It might indeed be so low as to be undetectable. Occasionally the structural alteration might enhance the stability of the enzyme protein and hence result in an increase in the observed activity.

(c) The rate of synthesis of the enzyme protein may be altered. This could occur following mutations in the so-called 'structural' gene (or genes) coding the aminoacid sequences of the protein, or it could result from a mutation in another gene, normally concerned in regulating the activity of such a 'structural' gene. The enzyme protein may simply not be formed, or may be formed at a reduced rate so that the quantity present at any given time is less than normal. Occasionally an increased rate of synthesis might result, and lead to an apparent enhancement of the enzyme activity.

(d) The particular enzyme may be affected only indirectly. A mutation occurring in a gene not normally involved in controlling the synthesis of the enzyme itself, may nevertheless influence its activity by for example causing an alteration in the intracellular concentration of some activator or inhibitor.

Some of these possibilities are of course not mutually exclusive and so in certain cases a mutant might bring about an altered enzyme activity in more than one way. For example, a single aminoacid substitution or other kind of structural change in an enzyme protein could lead to an alteration in its kinetic properties and also in its inherent stability.

These different possibilities, as well as a variety of technical problems make the analysis of the genetical and biochemical nature of inherited quantitative variation in enzyme activity peculiarly difficult. The technical problem largely arises because enzyme proteins, unlike a protein such as haemoglobin, generally occur only in trace amounts, so they cannot easily be isolated and characterised from single individuals. Also they are usually only readily detected and specified by virtue of their catalytic activity, and if this is considerably reduced or even absent the elucidation of the underlying abnormality is not easy to investigate.

In some cases it has been found possible to prepare specific antibodies to the enzyme protein. If by immunoprecipitation reactions one can then demonstrate the presence of cross-reacting material in a tissue extract from an individual in whom the normal enzyme activity is lacking, this suggests that a structurally abnormal enzyme protein with defective catalytic activity is being synthesised (see, for example, Tedesco 1972 re. galactose-1-phosphate uridyl transferase, Rubin et al. 1971 and Arnold et al. 1971 re. hypoxanthine-guanine phosphoribosyl transferase); on the other hand if the amount of enzyme protein as estimated immunochemically is reduced to the same extent as the amount of activity as estimated enzymatically, this indicates that there is a true reduction in enzyme protein due either to its instability or its reduced rate of synthesis (see, for example, Aebi and Suter 1971, re. catalase).

There is a further general problem which commonly arises in the study of situations where inherited differences between individuals in the level of activity of a particular enzyme have been identified. Although occasionally it is possible to classify people into sharply distinct groups in terms of enzyme activity, for example those with relatively high and those with relatively low activity (or perhaps no detectable activity at all), more often the variation in activity levels is continuously graded. Although the variation may be wide, and any particular individual may consistently show a level of activity characteristic of one region of the distribution rather than another, it is often not possible to classify individuals on a non-arbitrary basis into distinct groups because of the absence of any well defined discontinuities in the distribution of activities. Clearly defined bimodal or trimodal distributions of activities appear to be the exception rather than the rule, even when much of the variation is due to only a small number of genes. This makes precise genetical analysis difficult, unless some other more clear-cut criteria can be found to characterize the enzyme in individuals with different levels of activity.

The examples discussed below illustrate the manner in which some of these general problems have been, at least in part, resolved in the case of certain enzymes. They show how single gene differences may, in a variety of ways, lead to characteristic differences between individuals in enzyme activity. They also illustrate how the discrete and quite specific effects of a small number of different genes may underlie apparently continuously graded variation in activity levels.

5.2. Serum cholinesterase

This enzyme is normally found in quite high activity in human blood serum. It is probably mainly synthesised in the liver. A variety of choline esters such as acetylcholine, butyrylcholine and benzoylcholine are readily hydrolysed as are a number of non-choline esters such as acetylsalicylic acid. Thus there is a fairly wide substrate specificity, and it is not known exactly what ester may serve as its natural substrate under normal conditions. The enzyme (sometimes also called pseudocholinesterase) must be distinguished from acetylcholinesterase (so-called 'true cholinesterase') which occurs in nervous tissue and also in red cells and which has a much more restricted substrate specificity, particularly directed to acetylcholine.

The inherited variations of serum cholinesterase first came to light following the introduction and widespread use of the drug suxamethonium (succinyl dicholine) as a muscle relaxant in surgery and electroconvulsion therapy (fig. 5.1). Normally the effects of this drug are quite short because it is rapidly hydrolysed into inactive products by serum cholinesterase. How-

$$CH_2.CO.OCH_2.CH_2.\overset{+}{N}(CH_3)_3$$
$$CH_2.CO.OCH_2.CH_2.\overset{+}{N}(CH_3)_3$$

Fig. 5.1. Suxamethonium (succinyl dicholine).

ever occasional individuals (about 1 in 2000 in European populations) are unusually sensitive to its effects. Following a normal dose of the drug such people develop an extremely prolonged muscular paralysis and respiratory apnoea often lasting two hours or more, instead of the usual period of just a few minutes. It was found that in these individuals the level of serum cholinesterase was consistently low, and it seemed reasonable therefore to suppose that this was the cause of the sensitivity (Bourne et al. 1952, Evans et al. 1952). Furthermore a significant number of the immediate relatives of suxamethonium sensitive individuals were also found to have in various degrees reduced levels of serum cholinesterase activity (Lehmann and Ryan 1956). This indicated that the peculiarity might be genetically determined, and led to the suggestion that suxamethonium sensitive individuals with the low levels of serum cholinesterase are homozygous for an abnormal gene which in heterozygotes results in a moderate reduction in enzyme level. The three postulated genotypes could not however be clearly distinguished one from another on the basis of the levels of serum cholinesterase activity. In fact the distribution of serum cholinesterase levels in the population as a whole is effectively continuous, although the spread of values is very wide, and the suxamethonium sensitive individuals themselves show values at the lower end of the distribution.

5.2.1. 'Atypical' serum cholinesterase

An important advance was made when it was discovered that the serum cholinesterase present in suxamethonium sensitive individuals is atypical in certain of its properties (Kalow and Genest 1957, Kalow and Davies 1958, Davies et al. 1960). It was found for example, that when Michaelis constants are determined on the enzyme obtained from suxamethonium sensitive people, using a number of different choline esters as substrates, the values obtained are for each substrate significantly greater than the corresponding values determined in unselected controls (Davies et al. 1960). Furthermore the magnitude of the difference varies from substrate to substrate (table 5.1).

TABLE 5.1

Michaelis constants for various choline esters with the 'usual' and 'atypical' serum cholinesterase (Davies et al. 1960).

	Michaelis constant (K_m) (mmole/l)		Ratio of K_m
	'Usual' enzyme	'Atypical' enzyme	'atypical' : 'usual'
Acetylcholine	1.40 $\pm$0.04	9.0 $\pm$0.10	6.4:1
Propionylcholine	0.41 $\pm$0.04	2.3 $\pm$0.35	5.6:1
Butyrylcholine	0.29 $\pm$0.03	1.2 $\pm$0.08	4.1:1
Pentanoylcholine	0.72 $\pm$0.04	1.5 $\pm$0.17	2.1:1
Hexanoylcholine	0.57 $\pm$0.09	0.82 $\pm$0.06	1.4:1
Heptanoylcholine	0.38 $\pm$0.22	1.11 $\pm$0.14	2.9:1
Benzoylcholine	0.004$\pm$0.0003	0.022$\pm$0.003	5.5:1

Significant differences are also found with serum cholinesterase inhibitors of the type which contain a positively charged nitrogen atom either as a quaternary ammonium group (as in choline), or a substituted amino group (Kalow and Davies 1958). The enzyme present in suxamethonium individuals is less readily inhibited by each of these substances than the normal enzyme, and again there are differences in the magnitude of the effect from inhibitor to inhibitor (table 5.2). From these and other results (e.g. Clark et al. 1968) it became clear that in suxamethonium sensitive people the serum cholinesterase enzyme protein must be qualitatively different in structure from that which occurs in most other people. However it should be noted that in a number of properties, for example thermostability and electrophoretic mobility the two forms of the enzyme do not apparently differ.

The kinetic findings can be largely accounted for in terms of less efficient binding of the substrate or of the inhibitor by the atypical enzyme. Possibly

TABLE 5.2

Action of various inhibitors on 'usual' and 'atypical' serum cholinesterase (Kalow and Davies 1958).

Inhibitor	Log_{10} molar concentration giving 50% inhibition (pI_{50})		Ratio of concentrations giving 50% inhibition
	'Usual' enzyme	'Atypical' enzyme	I_{50} 'atypical' : I_{50} 'usual' (non-logarithmic)
Procaine	−4.36	−3.23	14:1
Decamethonium	−4.88	−2.74	141:1
Chlorpromazine	−5.36	−4.13	17:1
Dibucaine	−5.57	−4.27	20:1
Neostigmine	−6.89	−5.48	25:1
Physostigmine	−7.84	−6.57	18:1

the peculiarity lies in an altered configuration at the so-called 'anionic site' on the enzyme surface where the charged choline or choline-like grouping of the substrate or inhibitor is thought to become attached. However, whatever the structural nature of the peculiarity, the results from kinetic experiments indicate that the low levels of activity of the enzyme observed in suxamethonium sensitive individuals can be largely, if not entirely, accounted for in terms of its altered catalytic properties. The actual amount of the enzyme protein present in the serum of such people is probably not on average very different from that present in other individuals.

The inhibition studies led to the development of a rather simple test for detecting the atypical enzyme in routine studies. It involves the determination of the degree of inhibition of serum cholinesterase activity by the inhibitor dibucaine (fig. 5.2) under certain standard conditions (Kalow and Genest

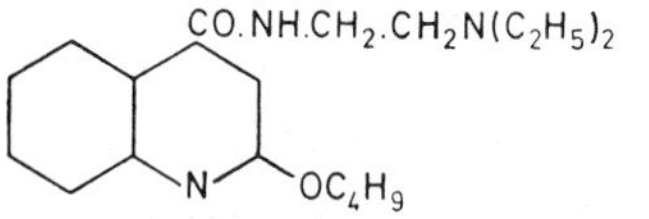

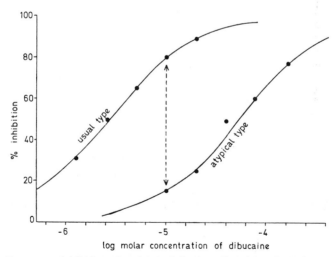

$CO.NH.CH_2.CH_2N(C_2H_5)_2$

OC_4H_9

Fig. 5.2. Dibucaine (nupercaine, percaine).

1957). The percentage inhibition obtained is called the dibucaine number (DN), and has a value of about 20 ± 4 for the 'atypical' enzyme and about 80 ± 2 for the normal or 'usual' type enzyme (fig. 5.3). All individuals with dibucaine numbers around 20 are extremely sensitive to suxamethonium.

When dibucaine number determinations are made on a random sample of the general population, a third group of people can also be clearly identified. They have dibucaine numbers of about 62 ± 4, and they are said to have the

Fig. 5.3. Percentage inhibition of activity of the 'usual' and 'atypical' forms of serum cholinesterase by different concentrations of dibucaine (Kalow and Genest 1957). Optimal discrimination between the two forms of the enzyme is obtained with 10^{-5} M dibucaine. Activity is determined spectrophotometrically with benzoylcholine as substrate.

'intermediate' phenotype. They occur with a frequency of about 1 in 25 in most European populations and they are not found in general to be particularly sensitive to suxamethonium. Such individuals appear to synthesise both the 'usual' and the 'atypical' forms of serum cholinesterase and do so in roughly equal amounts.

With the advent of dibucaine number determination detailed genetical studies became possible, and large numbers of families selected either through suxamethonium sensitive individuals with the 'atypical' enzyme or through individuals of the 'intermediate' type identified in population studies, were investigated (Kalow and Staron 1957, Harris et al. 1960, Harris and Whittaker 1962). Most of the results can be simply explained in terms of the segregation of two allelic genes, E_1^u which determines the 'usual' type enzyme, and E_1^a which determines the 'atypical' enzyme. Homozygotes $E_1^u E_1^u$ and $E_1^a E_1^a$ form only the corresponding enzyme. Heterozygotes $E_1^u E_1^a$ synthesise both types of enzyme and consequently give 'intermediate' dibucaine numbers.

It is of interest to compare the discrimination of three phenotypes 'usual', 'intermediate' and 'atypical' obtained by dibucaine number determination with that obtained by simple measurement of levels of activity as determined by standard methods (fig. 5.4). On the average, individuals with the 'atypical' phenotype as defined by dibucaine number have lower levels of activity than individuals with the 'usual' phenotype, and 'intermediate' individuals by dibucaine number have intermediate levels of activity (Kalow and Staron 1957, Harris et al. 1960). However, while dibucaine number determinations provide a sharp distinction between the three types, activity determinations do not, since there is considerable overlap of the three distributions. This is perhaps not surprising because, while the dibucaine number is a reflection of the qualitative characteristics of the enzyme and is independent of the quantity of enzyme present, the level of activity depends both on the qualitative characteristics of the enzyme and also on the amount present. Since there are evidently many extraneous factors which can influence the amount of enzyme protein present in plasma at any given time, one would expect the activity determination to be a much more variable and less discriminating characteristic. In practice, one can find occasional individuals with the 'usual' type enzyme with levels of activity as low as those commonly found in the 'atypical' type. However, such individuals are very much less sensitive to suxamethonium and do not generally have a particularly marked apnoea.

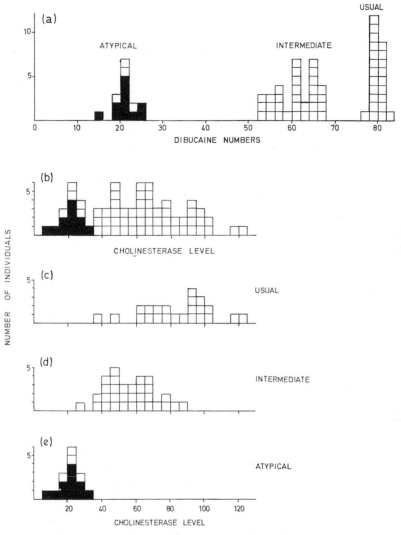

Fig. 5.4. Distributions of dibucaine numbers and of levels of cholinesterase activity in sera from 11 individuals found to be excessively sensitive to suxamethonium, and 58 of their relatives (Harris et al. 1960). Each square represents one individual, the suxamethonium sensitive propositi being marked in black. (a) Distribution of dibucaine numbers showing clear discrimination of the three phenotypes 'usual', 'intermediate' and 'atypical' (see text). (b) Distribution of levels of serum cholinesterase activity in the same serum samples as in (a). The activity levels were determined manometrically with acetylcholine as substrate. (c) Distribution of activity levels of those individuals classified as having the 'usual' phenotype in (a). (d) Distribution of activity levels of those individuals classified as having the 'intermediate' phenotype in (a). (e) Distribution of activity levels in those individuals classified as having the 'atypical' phenotype in (a). *Note:* Distribution (b) represents the sum of distributions (c), (d) and (e).

5.2.2. *The 'silent' allele*

In the course of dibucaine number studies on families in which individuals of the 'atypical' type occurred, it became apparent that there are occasional exceptions to the rules of inheritance predicted by the simple two-allele hypothesis. If all individuals with only the 'atypical' enzyme are homozygous $E_1^a E_1^a$ then both their parents and all their children should also carry the gene E_1^a, and hence show either of the 'intermediate' or 'atypical' dibucaine number phenotypes. However, although this is usually the case, there are certain families in which one of the parents or children of an 'atypical' individual appear to have only the 'usual' type enzyme (Kalow and Staron 1957, Harris et al. 1960, Liddell et al. 1962, Simpson and Kalow 1964). Furthermore, non-biological parenthood as a general explanation for this phenomenon can be excluded.

The findings are in fact most simply accounted for by postulating the segregation in these exceptional families of a rare third allele, E_1^s, which is completely or almost completely ineffective in producing functionally active enzyme. This has been referred to as the 'silent' allele. If such a gene occurred then one would expect that heterozygotes $E_1^a E_1^s$ would form only the 'atypical' enzyme, and heterozygotes $E_1^u E_1^s$ only the 'usual' enzyme (table 5.3).

TABLE 5.3

Serum cholinesterases formed by individuals with different combinations of alleles E_1^u, E_1^a and E_1^s

Genotype	Enzymes formed
$E_1^u E_1^u$ $E_1^u E_1^s$	'usual'
$E_1^u E_1^a$	'usual' + 'atypical'
$E_1^a E_1^a$ $E_1^a E_1^s$	'atypical'
$E_1^s E_1^s$	none

Also one would expect the occurrence of a further genotype $E_1^s E_1^s$ in which there is a complete or nearly complete absence of serum cholinesterase activity. A number of individuals with this rare condition have indeed been identified (Liddell et al. 1962, Doenicke et al. 1963, Hodgkin et al. 1965, Goedde et al. 1965, also see p. 159). They are, as would be expected, extremely sensitive to suxamethonium.

One might also anticipate that $E_1^u E_1^s$ individuals would on the average have lower levels of activity than $E_1^u E_1^u$ individuals, and similarly $E_1^a E_1^s$ individuals would have lower levels of activity than $E_1^a E_1^a$ individuals. This has been found to be the case, but because there is considerable variation in serum cholinesterase level within each of these types it is not usually possible to identify them unequivocally through activity determinations alone.

Apart from excessive sensitivity to suxamethonium, $E_1^s E_1^s$ individuals who are virtually devoid of serum cholinesterase, are usually quite healthy.

5.2.3. *'Fluoride resistant' serum cholinesterase*

A further variant of the enzyme was discovered in a rather unexpected way (Harris and Whittaker 1961). Serum cholinesterase is inhibited by low concentrations of sodium fluoride, and this is chemically so different from dibucaine and related substances which differentiate apparently very specifically between the 'usual' and 'atypical' forms of the enzyme, that it must be presumed to act in a quite different manner. It was therefore surprising to find that the two forms of the enzyme, 'usual' and 'atypical' can also be distinguished using fluoride as the inhibitor. The 'atypical' enzyme is much less readily inhibited than the 'usual' enzyme, and when a large number of different sera previously classified into 'usual', 'intermediate' and 'atypical' types by dibucaine number determination were examined using fluoride as inhibitor, the discrimination into the three phenotypes was, with a few exceptions, found to be virtually the same.

The occasional exceptions, however, turned out to be of particular interest. Family studies showed that they represented new serum cholinesterase phenotypes which could be correctly identified only by a combination of dibucaine and fluoride inhibitor tests (Harris and Whittaker 1962, Liddell et al. 1963, Whittaker 1967). It emerged that there is a further allele (E_1^f) which determines a third form of serum cholinesterase with properties different from the 'usual' and the 'atypical' forms previously recognised. This is referred to as the 'fluoride resistant' form. It may occur together with the 'usual' enzyme in individuals of genotype $E_1^u E_1^f$, or together with the 'atypical' enzyme in individuals of genotype $E_1^a E_1^f$. It may also occur alone as in the genotypes $E_1^f E_1^f$ and $E_1^f E_1^s$. The 'fluoride resistant' enzyme exhibits less activity than the 'usual' enzyme, but more than the 'atypical' enzyme, so the different types in which it occurs show on average differing levels of total serum cholinesterase activity.

5.2.4. Multiple allelism causing 'continuous' variation in activity

The four alleles (E_1^u, E_1^a, E_1^s and E_1^f) give rise to ten different genotypes which are listed in table 5.4. Their identification may require not only inhibitor tests with dibucaine and fluoride, but also detailed family studies. On average they differ from one another in the levels of activity they display, but there is considerable variation between individuals of any one type due to other factors. So the distribution of levels of activity in a large population containing the whole array of genotypes is effectively continuous.

Markedly prolonged paralysis following suxamethonium occurs as a regular phenomenon in three of these genotypes $E_1^a E_1^a$, $E_1^a E_1^s$ and $E_1^s E_1^s$. $E_1^a E_1^f$, $E_1^f E_1^s$ and $E_1^f E_1^f$ individuals also show some increase in sensitivity to the drug, though in different degrees.

The alleles E_1^u, E_1^a and E_1^f evidently determine structurally distinct forms of

TABLE 5.4

Serum cholinesterase genotypes and phenotypes. Summary of inhibition characteristics with dibucaine (DN) and fluoride (FN) and of relative activities of serum cholinesterase in ten different genotypes. The values quoted are rounded off means based on the experience of several laboratories. The 'relative activities' are based on assays using benzoylcholine as substrate under standardised conditions. Such 'relative activities' will in general vary with the substrate and conditions used for the assay.

Genotype	Enzymes present	Inhibition characteristics DN	FN	Relative activity
$E_1^u E_1^u$	'Usual'	80	60	100
$E_1^u E_1^f$	'Usual'+'fluoride resistant'	75	50	85
$E_1^u E_1^a$	'Usual'+'atypical'	60	50	75
$E_1^u E_1^s$	'Usual'	80	60	70
$E_1^f E_1^a$	'Fluoride resistant'+'atypical'	50	30	60
$E_1^f E_1^f$	'Fluoride resistant'	65	30	55
$E_1^a E_1^a$	'Atypical'	20	20	50
$E_1^f E_1^s$	'Fluoride resistant'	65	30	30
$E_1^a E_1^s$	'Atypical'	20	20	25
$E_1^s E_1^s$	—	—	—	0

the enzyme protein. Possibly these only differ by single aminoacid substitutions, but so far it has not proved possible to characterise the differences precisely in structural terms. The enzymes differ in their kinetic properties with various substrates and inhibitors, but they appear to be very similar in other respects (e.g. electrophoretic mobilities and molecular size). The altered levels of activity observed in the different genotypic combinations of these alleles can probably be attributed largely to their different kinetics with the substrates used for the assay, rather than to any differences in rates of synthesis or in stability.

It is worth noting that the observed degree of reduction in the level of activity in a particular genotype, e.g. $E_1^a E_1^a$, compared with that of the common or 'normal' genotype $E_1^u E_1^u$ may vary considerably according to the substrate used in the particular assay procedure. It will also be affected by the actual substrate concentration adopted for the assay and the other conditions. This is because the kinetics of these enzymes are complex, and kinetic variations from substrate to substrate do not necessarily parallel one another in the 'usual', 'atypical' and 'fluoride resistant' forms.

There are also marked differences in the temperature activity relationships of these variants (King and Dixon 1969, King and Morgan 1970, 1971). There appears to be a characteristic temperature activity curve for each enzyme phenotype and differences from the 'usual' form are most pronounced in those phenotypes associated with suxamethonium sensitivity (fig. 5.5). These findings emphasise the complexity of such a parameter as 'level of enzyme activity' when it is used to compare different genotypes.

It has also become evident that the so-called 'silent' phenotype with a complete or nearly complete deficiency of the enzyme is heterogeneous. In some cases no serum cholinesterase activity or cross-reacting protein can be demonstrated at all despite the use of extremely sensitive methods (both enzymic and immunochemical), while in others trace amounts of activity can be consistently demonstrated (Hodgkin et al. 1963, Goedde et al. 1965, Gutsche et al. 1967, Gaffney and Lehmann 1969, Altland and Goedde 1970, Rubinstein et al. 1970, Scott 1973). The findings suggest that two or more different mutants are included in the class of E_1^s alleles, but exactly how they exert their effects is not known.

In most populations this 'silent' phenotype is extremely rare, so that the particular E_1^s alleles concerned are evidently quite uncommon. However in a somewhat isolated group of Eskimos living in Alaska, as many as 1–2% of the population have been found to show this gross deficiency of serum cholinesterase activity (Gutsche et al. 1967), and nearly 25% of the population

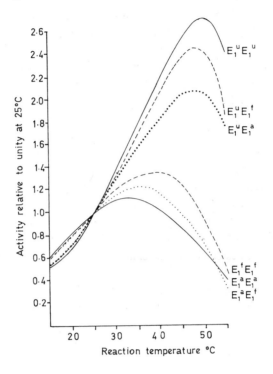

Fig. 5.5. Relative activities of serum cholinesterase from individuals of different genotypes when the assay is carried out at different temperatures (from King and Morgan 1971)

are evidently heterozygous for E_1^s alleles involved (Scott et al. 1970). Quite unexpectedly it was found (Scott 1973) that in this population of Eskimos two different E_1^s alleles are segregating, one resulting in an apparently complete deficiency of the enzyme, and the other in a deficiency which is not quite complete (2–4% of the usual).

5.2.5. *A second locus* – E_2

There is yet a further source of genetically determined variation in serum cholinesterase activity. This was discovered by electrophoresis, which does not discriminate the various enzyme types discussed above (Harris et al. 1962).

When serum cholinesterase is examined electrophoretically several isozymes (known as C_1, C_2 and C_3 and C_4) are regularly observed. C_1, C_2 and C_3 are minor components and contribute little to the total activity, most of which is derived from C_4 (fig. 5.6). However in about 10% of people in European and certain other populations, an extra isozyme called C_5 is seen as well (Harris et al. 1962, 1963a; Simpson 1966, Robson and Harris 1966). The presence of C_5 in serum appears to be determined by a gene at a locus which is separate and not closely linked to the locus at which the alleles E_1^u, E_1^a etc. occur (Harris et al. 1963c, Simpson 1966). The amount of serum cholinesterase activity attributable to C_5 varies considerably from individual to individual carrying the appropriate gene. However sera from C_5^+ individuals (i.e. those with the extra isozyme) have on average about 25% higher levels of activity than sera from those without it (so-called C_5^- individuals). But the variation is wide in both groups and the two distributions overlap to a very considerable extent.

The mode of action of the gene determining the C_5 component is not understood. It apparently leads to the appearance of an electrophoretically distinct isozyme whose properties are in other respects similar to those of the other serum cholinesterase isozymes, and which appears to be affected in the same way by the alleles E_1^u, E_1^a etc.

Thus the genetical factors concerned in determining the level of serum cholinesterase are complex. At least two chromosomal loci are involved.

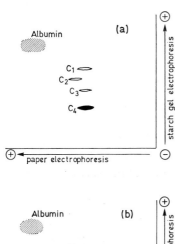

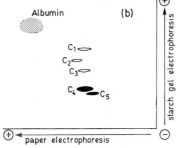

Fig. 5.6. Diagrams showing separation of serum cholinesterase isozymes by two dimensional filter paper/starch gel electrophoresis at pH 8.6 (Harris et al. 1962). (a) C_5- serum; (b) C_5+ serum. Isozyme C_4 is present in both (a) and (b), and accounts for most of the serum cholinesterase activity in (a) and a high proportion (ca. 65-90%) in (b). Isozymes C_1, C_2, and C_3 are minor components present in both (a) and (b). Isozyme C_5 is the additional component present in (b). Note that the electrophoretic mobilities towards the anode in starch gel are in the order $C_1 > C_2 > C_3 > C_4 > C_5$ but in filter paper the order is $C_2 > C_1 = C_3 = C_4 > C_5$. This is because the isozymes differ from each other both in molecular charge and also in molecular size. The molecular sizes of the various isozymes are in the order $C_4 \simeq C_5 > C_3 > C_2 > C_1$ (Harris and Robson 1963, La Motta et al. 1970).

At one locus (E_1), alleles determining both qualitative and quantitative differences in the enzyme occur. At the other locus (E_2), one allele E_2^+ determines the formation of an extra serum cholinesterase isozyme C_5, while the other allele (E_2^-) appears to be functionally inactive. The level of activity observed in any one person will depend on the particular alleles present at these two loci as well as no doubt on a variety of extraneous non-genetical factors, both in healthy individuals and in patients with particular types of disease. The level of serum cholinesterase for example is usually depressed in liver disease, presumably because of diminished formation of the enzyme protein.

5.3. Glucose-6-phosphate dehydrogenase (G-6-PD)

5.3.1. G-6-PD deficiencies

The discovery of inherited variation of glucose-6-phosphate dehydrogenase (G-6-PD) followed the finding that a significant proportion of American Blacks develop an acute haemolysis when they receive the synthetic anti-malarial drug, primaquine (Hockwald et al. 1952). It was shown that the haemolytic response to the drug was due to an intrinsic red cell abnormality (Dern et al. 1954), and that this was a specific deficiency of the enzyme glucose-6-phosphate dehydrogenase (Carson et al. 1956).

Glucose-6-phosphate dehydrogenase catalyses the oxidation of glucose-6-phosphate to 6-phosphogluconate, and this is accompanied by the concomitant reduction of the coenzyme NADP to NADPH. The reaction represents the first step in the oxidation of glucose via the so-called pentose shunt pathway (see fig. A1 p. 369). Under normal conditions this pathway accounts for only a small proportion of the glucose utilised by the red cell. It does however serve to maintain the intracellular concentration of the reduced coenzyme $NADPH_2$, and it seems likely that it is a failure in this which accounts through other reactions, particularly the maintenance of glutathione in the reduced form, for the haemolytic crises observed when G-6-PD deficient subjects receive primaquine and other drugs.

Assays of red cell G-6-PD levels in Black populations reveal a striking sex difference (fig. 5.7). In healthy males two clearly distinct classes of individuals can be recognised; those with normal levels of the enzyme, and those who are deficient. Although in each group there is considerable variation from one individual to another, the two distributions hardly overlap. The deficient group has levels of red cell G-6-PD varying around 15% of the average found in the non-deficient group. A quite different situation is found in females. Here a more or less continuous variation in levels of the enzyme in different individuals is observed. All values may be found from the normal levels characteristic of the non-deficient males to the low levels characteristic of the deficient males.

The significance of the sex differences became clear when family studies showed that the variations are determined by alleles at a locus on the X chromosome (Childs et al. 1958). Males have only one X chromosome and so may either carry the normal allele or a mutant allele causing the deficiency, but not both. Females have two X chromosomes. They may be homozygous for the normal allele and have normal levels of the enzyme like normal males; homozygous for the mutant allele and have a marked deficiency of the

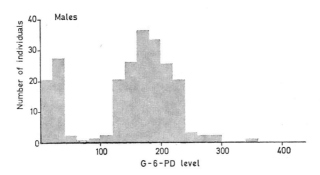

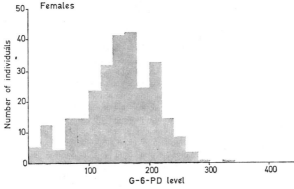

Fig. 5.7. Distribution of red cell glucose-6-phosphate dehydrogenase activity levels in 235 male and 284 female Nigerians (data of Nance 1967).

enzyme like deficient males; or heterozygous, in which case they have levels of the enzyme which are on the average almost exactly intermediate between the average level of the normal males and the average level of the deficient males. Since considerable variation in level of enzyme activity among heterozygous females occurs, the distribution of values overlaps at one extreme with the distribution of homozygous deficient values, and at the other with the distribution of homozygous normal values. Consequently it is often not possible to classify the female unequivocally into one or other of the three postulated genotypes simply in terms of the level of the enzyme in red blood cells.

It was soon recognised that other drugs besides primaquine may induce haemolytic crises in G-6-PD deficient subjects (table 5.5). These include sulfonamides such as sulfapyridine and sulfanilamide, other antibacterial agents such as nitrofurantoin, other antimalarials such as pentaquine, and so on. It was also discovered that a form of G-6-PD deficiency is the underlying cause of the disease known as favism, which has long been recognised as a not infrequent condition in certain population groups living in the Middle East and some Mediterranean countries (Zinkham et al. 1958, Larizza et al. 1958, Szeinberg et al. 1958). It manifests as an acute haemolytic anaemia which follows the ingestion of fava beans, a not uncommon feature of the diet in these areas.

TABLE 5.5

Drugs and other agents that can cause clinically significant haemolysis in G-6-PD deficient people (from WHO Technical Report Series No. 366, 1967).

Acetanilid	Nitrofurazone (Furacin)
Phenylhydrazine	Nitrofurantoin (Furadantin)
Sulfanilamide	Furazolidone
Sulfacetamide	Furaltodone (Altofur)
Sulfapyridine	Quinidine
Sulfamethoxypyridazine (Kynex)	Primaquine
Salicylazosulfapyridine (Azulfidine)	Pamaquine
Thiazosulfone	Pentaquine
Diaminodiphenylsulfone	Quinocide
Trinitroluene	Naphthalene
	Neosalvarsan
	Fava Beans

5.3.2. The common Black and Mediterranean G-6-PD variants: Gd B, Gd A,
Gd A-, and Gd Mediterranean

Like the form of G-6-PD deficiency occurring in Blacks, the G-6-PD deficiency found in Mediterranean and Middle East populations is determined by a mutant gene on the X chromosome. A different mutation is however involved, and there are certain characteristic differences between the two conditions. In particular the level of the enzyme in red cells of affected males with the so-called 'Mediterranean' type of G-6-PD deficiency is usually about 3 or 4% of the normal level, whereas in the 'Negro' type it is on average about 15%. Also a significant reduction in G-6-PD level may be demonstrated in white cells (Ramot et al. 1959) and in other tissues in the Mediterranean type, whereas in the type seen in Blacks such a reduction is either not found or is very slight (Marks et al. 1959). Thus in terms of the level of enzyme activity the Mediterranean type is a relatively more severe abnormality.

Another difference between the two kinds of deficiency was brought to light by electrophoretic studies. These studies also revealed a further kind of variation in G-6-PD which is relatively common in Black populations, but is not associated with marked enzyme deficiency (Boyer et al. 1962, Kirkman and Hendrickson 1963).

When G-6-PD from different healthy Black subjects is examined by electrophoresis, two separate forms may be separated (fig. 5.8). The slower

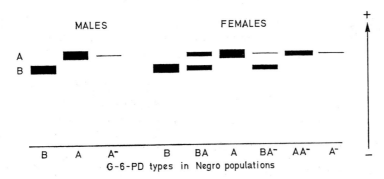

Fig. 5.8. Diagram showing electrophoretic components of red cell G-6-PD in Blacks of different phenotypes.

moving form at *p*H 8.6 is called B and the faster one A. Males may have an enzyme with the B mobility or with the A mobility but not both. Furthermore in those who are G-6-PD deficient the enzyme present almost invariably shows the A mobility. Thus three distinct classes of male, each of which is relatively common among Blacks, may be distinguished from the level of G-6-PD activity present in the red cells and from its electrophoretic properties. These three phenotypes are referred to as Gd B, Gd A and Gd A-. Gd B individuals show normal levels of G-6-PD activity and the B electrophoretic mobility. Gd A individuals show normal levels of activity and the A electrophoretic mobility. Gd A- individuals show deficient enzyme activity (about 15% of normal) and A electrophoretic mobility.

The relative frequencies of these three phenotypes vary somewhat among different Black populations. For example, among male Nigerians (Yoruba) about 56% were found to be Gd B, about 22% were Gd A, and about 22% were Gd A- (Porter et al. 1964). Among male Blacks living in the U.S.A. typical figures are Gd B: 60–70%, Gd A: 15–20%, Gd A-: 10–15%. Family studies indicate that these phenotypes are determined by three distinct alleles occurring at a single locus on the X chromosome. They have been designated Gd^B, Gd^A and Gd^{A^-}. Among Black females six different genotypes occur, and the corresponding phenotypes are illustrated diagrammatically in fig. 5.8. Three of these are similar to the three phenotypes seen in males and correspond to the homozygous genotypes $Gd^B Gd^B$, $Gd^A Gd^A$ and $Gd^{A^-} Gd^{A^-}$. The other three correspond to the heterozygous genotypes $Gd^B Gd^A$, $Gd^B Gd^{A^-}$ and $Gd^A Gd^{A^-}$.

In the so-called Mediterranean type of G-6-PD deficiency, the enzyme which is present has the B electrophoretic mobility, as does the enzyme in normal individuals in these populations. The allele causing the Mediterranean type of deficiency is at the same locus as the common alleles occurring in Black populations. It has been designated $Gd^{Mediterranean}$

The four alleles, Gd^B, Gd^A, Gd^{A^-} and $Gd^{Mediterranean}$, appear to determine four structurally distinct forms of the G-6-PD enzyme protein (see table 5.6, pp. 168–169). The so-called B enzyme must be regarded as the normal type, because it is by far the commonest and occurs in all populations. The others possibly differ from it simply by single aminoacid substitutions. Evidence on this point has been obtained from structural studies in the case of the A enzyme (determined by the Gd^A allele). Here it appears that a particular asparagine residue present in the B enzyme is replaced by an aspartic acid residue (Yoshida 1967). So far it has not proved possible to carry out similar structural investigations on the Gd A- or the Gd Mediterranean types of the enzyme because of the difficulty of isolating them in pure form in adequate amounts. However, comparisons of the properties of these various enzymes have revealed a number of other interesting differences and similarities.

The A enzyme, although it differs from the normal B enzyme electrophoretically, has been found to resemble it closely in many other respects. The two enzymes appear to have essentially the same K_m for G-6-P and NADP, similar pH optima, and similar rates of thermal denaturation. Thus the structural alteration, while causing an electrophoretic difference, does not apparently lead to any marked differences in catalytic activity or in stability. There may however be minor differences not detectable by present techniques because assays of G-6-PD activity in red cells in large numbers of Gd B and Gd A individuals have shown that there is small difference in the average level of activity. The mean level in Gd A individuals appears to be about 10% less than in Gd B individuals, though there is a very considerable overlap in the two distributions.

The A- enzyme (determined by the Gd^{A-} allele) has the same electrophoretic mobility as the A enzyme and is very similar both to it and the B enzyme in a variety of kinetic and other properties (Kirkman 1959, Marks et al. 1961). It differs markedly, however, in its stability *in vivo*. This has been demonstrated by comparing enzyme activities in relatively 'young' and relatively 'old' red cells (Yoshida et al. 1967, Piomelli et al. 1968). The cells were fractionated into 'age groups' according to their specific gravity by centrifugation in a density gradient. It was found that in the younger cells (with a high proportion of reticulocytes) the level of G-6-PD activity in Gd A- individuals is almost as high as in Gd B individuals. However, in relatively old cells there is a very marked difference, the older cells from Gd B individuals showing much more G-6-PD activity than comparable cells from Gd A- individuals. It has been estimated from these studies that the half life of the enzyme in red cells from normal Gd B individuals is about 62 days, whereas in Gd A- individuals the half life of the enzyme in red cells is only about 13 days (Piomelli et al. 1968). So the enzyme deficiency observed in Negro Gd A- individuals is evidently due to the more rapid denaturation of the A- enzyme protein *in vivo*. A sample of red cells used in the ordinary assay of the enzyme in an individual consists of course of a mixture of cells of all ages. The level of activity seen in normal Gd B individuals or in Gd A- individuals represents an average of the relatively higher values present in the younger cells, and the lower values in the older ones.

This age differential largely explains why no marked deficit of G-6-PD activity is seen in white cells from Gd A- individuals. Red cells normally have a life span of more than 100 days, but they lose their nuclei and their ability to synthesise proteins at an early stage. Consequently the G-6-PD enzyme protein, which is progressively reduced in quantity by denaturation,

TABLE 5.6

Some variants of glucose-6-phosphate dehydrogenase.

Variant	Red cell activity (% of normal)	Electrophoretic mobility relative to normal (pH 7.0 or pH 8.6)	K_m for G-6-P (μM)	K_m for NADP (μM)
Normal (Gd B)	100	—	50-78	2.9-4.4
1. Hektoen	400	fast (pH 6.5)	n	n
2. Madison	100	slow	?	?
3. A	90	fast	n	n
4. Baltimore-Austin	75	slow	n	n
5. Madrona	70-80	slow	reduced (32)	n
6. Ibadan-Austin	72	slow	n	n
7. Barbieri	40-60	fast	increased	increased
8. Kerala	50	slow	reduced (23)	reduced (1.5)
9. Tel-Hashomer	25-40	slow	reduced (30-40)	?
10. Athens	25	slow	reduced (16-19)	n
11. *Chicago	9-26	normal	n	n
12. Seattle	8-21	slow	reduced (15-25)	reduced (2.4-2.8)
13. A-	8-20	fast	n	n
14. Canton	4-24	fast	reduced (20-36)	reduced (2.0-2.4)
15. West Bengal	9	slow	reduced (31)	increased (6.6)
16. * Ohio	2-16	fast	slightly increased	slightly increased
17. * Oklahoma	4-10	normal	increased (127-200)	increased (20)
18. * Duarte	8.5	normal	n	n
19. Mediterranean	0-7	normal	reduced (18-26)	reduced (1.2-1.6)
20. * Alberquerque	1	normal	increased (115)	increased (11)
21. * Eyssen	0	slow	?	?

* variant associated with congenital haemolytic disease.

Utilisation of substrate analogue 2-dG-6-P relative to G-6-P	Thermostability relative to normal	pH activity curve	Incidence	References
<4%	—	truncate	Commonest type in all populations	
n	n	n	rare	Dern et al. (1969)
?	?	?	rare	Nance and Uchida (1964)
n	n	n	Common in Blacks	Boyer et al. (1962) Kirkman et al. (1964a)
n	n	n	rare	Long et al. (1965)
n	?	n	rare	Hook et al. (1968)
n	n	n	rare	Long et al. (1965)
?	n	?	rare	Marks et al. (1962)
increased (7.4)	n	biphasic	rare	Azevedo et al. (1968)
n	n	slightly biphasic	rare	Ramot and Brok (1964)
increased (10-15)	slightly reduced	slightly biphasic	Common in Greece	Stamatoyannopoulos et al. (1967)
n	much reduced	n	rare	Kirkman et al. (1964b)
increased (7-11)	n	slightly biphasic	rare	Kirkman et al. (1965)
n	n	n	Common in Blacks	Boyer et al. (1962), Kirkman et al. (1964a)
increased (4-15)	slightly reduced	biphasic	Common in S.E. Asia	McCurdy et al. (1966)
n	n	n	rare	Azevedo et al. (1968)
n	much reduced	?	rare	Pinto et al. (1966)
n	reduced	narrow peak	rare	Kirkman et al. (1964a)
increased (5.4)	much reduced	narrow peak	rare	Beutler et al. (1968)
increased (23-27)	reduced	biphasic	Common in Mediterranean countries and Middle East	Kirkman et al. (1964c)
n	much reduced	narrow peak	rare	Beutler et al. (1968)
?	much reduced	?	rare	Boyer et al. (1962)

n: characteristic within normal range

is not replaced by newly synthesised enzyme. But white cells which are nucleated are capable of continuing enzyme synthesis. Thus the effect of the more rapid decay of the A- enzyme is much less pronounced in circulating white cells than in circulating red cells, and may indeed be barely detectable.

The Gd Mediterranean enzyme, although it has the same electrophoretic mobility as the B enzyme, differs in a number of other properties (Kirkman et al. 1964a). The K_m values for G-6-P and also for NADP are both appreciably lower than the corresponding values obtained for the B enzyme. Another difference is in the rate of denaturation at elevated temperatures, which is significantly increased in Gd Mediterranean. There are also differences in pH optima. A particularly striking peculiarity is in its ability to utilise 2-deoxy-glucose-6-phosphate at about 25–30% of the rate at which it oxidises G-6-P. The distinctive peculiarities in the properties of the enzyme presumably reflect in different ways its altered structure. They do not, however, *per se* account for the very low levels of activity seen in the Gd Mediterranean phenotype. This is evidently due, as in the case of the A-enzyme, to a decreased stability of the enzyme protein *in vivo*. But here the effect is even more profound. The rate of decay is evidently extremely rapid, and even though younger red cells show a somewhat less severe deficiency of the enzyme than older cells, the reduced level of enzyme activity, even in reticulocytes, is nevertheless quite marked (Piomelli et al. 1968). Also a significant G-6-PD deficit is found in white cells in Gd Mediterranean individuals (Ramot et al. 1959).

The Gd A- type of G-6-PD deficiency is, as we have seen, relatively common among Blacks, and the Gd Mediterranean type is relatively common in non-Black populations living in the Middle East and Southern Europe. Most individuals affected with either of these forms of enzyme deficiency are quite healthy and do not show any ill effects unless they happen to take one or other of the various drugs to which they are sensitive or, in the case of the Gd Mediterranean type, unless they happen to eat fava beans. Under these circumstances they are liable to develop an acute haemolytic reaction. But the severity of this reaction is very variable. Thus some Gd Mediterranean individuals appear to be able to eat the fava bean without serious consequences while others, after similar amounts, show a rapid and severe response. The severity of the response may also vary from time to time in the same individual. The details of the cause or causes of these variations are not known. They may depend on the form in which the bean is eaten, or on the season of the year (Belsey 1973). They may in part be genetically determined (Stamatoyannopoulos et al. 1966, Sartori 1971) and could perhaps have to do with differences in the manner in which the active principle of the bean is metabolised.

5.3.3. *Other G-6-PD variants*

The discovery of these common G-6-PD variants in Black and Mediterranean populations was followed by extensive surveys of a wide variety of populations in different parts of the World. Also the methods for characterising such variants were considerably improved and standardised (WHO Scientific Group 1967). As a result more than 80 distinct variant forms of the enzyme have now been identified, each apparently determined by a different allele at the locus on the X-chromosome which codes for the structure of the enzyme. These variants differ from each other and from the Gd B, Gd A, Gd A- and Gd Mediterranean forms of the enzyme, in one or more of their qualitative characteristics (e.g. electrophoretic mobility, K_m for glucose-6-phosphate, K_m for NADP, thermostability, utilisation of 2-deoxyglucose-6-phosphate, utilisation of deamino NADP, pH activity curve, etc.). Some of these characteristics for a number of the different variants are shown in table 5.6, and it is apparent that quite a variety of properties must be examined for the positive identification and characterisation of a particular variant.

Extensive tabulations of the different variants with their distinctive characteristics are given by Yoshida, Beutler and Motulsky (1971), and Beutler and Yoshida (1973), and summarise the literature until the end of 1972. Most of the alleles which determine these variants are relatively rare and appear to be irregularly distributed in different population groups. But certain of the alleles occur with an appreciable frequency in particular populations. For example an allele which determines the variant known as Gd Canton is relatively common in populations originating from Southern China, and Gd Athens is relatively frequent in Greece.

The level of glucose-6-phosphate dehydrogenase activity found in individuals with different variants varies very widely (table 5.6). In some the red cell activity falls within the normal range, in others it is reduced to a moderate extent, and in others there is a marked reduction. The reason for the reduced enzyme activity where this occurs, differs no doubt from variant to variant. In some, decreased stability of the enzyme protein is probably the main cause, as it is with the Gd A- and Gd Mediterranean variants. In others the low activity may be mainly due to alteration of the catalytic efficiency of the enzyme caused by its altered structure. In still other cases the deficit may be due to a reduced rate of synthesis. It is also possible, of course, that more than one of these different sorts of effect may be important in determining the level of activity observed in any specific instance.

Thus the different variants encompass a very .wide range of activity level.

In this connexion a very unusual variant (Gd Hektoen) is of particular interest, because it is associated with a level of activity which is some four-fold greater than that normally found (Dern 1966, Dern et al. 1969). The increased activity was observed both in red cells and leukocytes, and also in fibroblasts grown in tissue culture. Quantitative studies using immuno-logical neutralisation techniques indicated that the specific enzyme activity of Gd Hektoen was the same as that of the normal enzyme Gd B (Yoshida 1968, Dern et al. 1969). Since the degradation rate of the variant *in vivo* was also found to be similar to that of the normal, it was concluded that the increased activity was a consequence of an increased rate of synthesis (Dern et al. 1969). The enzyme protein was also shown to differ from Gd B by a single aminoacid substitution, a histidine in the normal enzyme being replaced by a tyrosine in the variant (Yoshida 1970). Thus in this instance, a mutation resulting in a structural alteration in the enzyme protein appears also to have resulted in an increased rate of synthesis.

As has already been noted a number of the variants are associated with a marked reduction of the level of enzyme activity in red cells. In some cases this does not lead to haemolytic disease, except where there is some pre-cipitating exogenous factor such as a particular drug, the ingestion of fava beans, or some intervening infection. However there are also other variants which are characteristically associated with chronic forms of haemolytic disease, even in the absence of any precipitating exogenous factors. More than twenty different variants of this sort have been identified (Yoshida et al. 1971, Beutler and Yoshida 1973), some of which are shown in table 5.6. If the whole series of variants, each of which causes a marked deficiency of the level of activity in the red cells, are considered together it is found that there is a poor correlation between the red cell activity level as assayed by conventional methods and the presence or absence of chronic haemolytic disease (Yoshida 1973). Some variants which are not associated with chronic haemolytic disease show apparently lower red cell activities by standard assay methods than other variants which are associated with chronic haemolytic disease. Such apparent anomalies are probably attri-butable to the particular kinetic characteristics of the enzymes and the manner in which these affect red cell metabolism *in vivo*.

The point can be illustrated by a comparison of the variants Gd Oklahoma and Gd Mediterranean (Kirkman 1968). With conventional assay procedures haemolysates of Gd Oklahoma subjects show slightly greater activity than those of Gd Mediterranean subjects (table 5.6). Yet the former abnormality is associated with chronic haemolytic disease and the latter is not. The probable explanation for the difference is that the K_m of Gd Oklahoma with glucose-6-phosphate is very much greater than the corresponding K_m of Gd Mediterranean. In the standard assay system the concentration of glucose-phosphate is kept relatively high so as to saturate the enzyme and ensure maximal reaction velocities. However, in the red cell the concentration of glucose-6-phosphate is very much less and is probably close to or lower than the K_m values. As can be seen from fig. 5.9 (Kirkman 1968), the reaction rate achieved in the red cell by Gd Oklahoma may well be very much less than that achieved by Gd Mediterranean, even though the relative levels of activity as determined in an assay system with high glucose-6-phosphate concentrations are in the opposite direction.

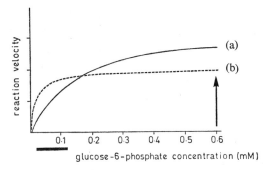

Fig. 5.9. Plots of reaction velocities at different glucose-6-phosphate concentrations of two G-6-PD variants (Kirkman 1968). (a) Gd Oklahoma (K_m = 140 μM) – continuous line; (b) Gd Mediterranean (K_m = 18 μM) – broken line. The curves are constructed so that the relative rates with standard assay concentrations of glucose-6-phosphate (indicated by the arrow) are proportional to the specific activities of haemolysates of the two variants. The solid bar indicates the actual range of intracellular concentration of glucose-6-phosphate.

Similar high K_ms with glucose-6-phosphate are found for Gd Ashdod (Ramot et al. 1969) and Gd Freiberg (Weinreich et al. 1968), both of which are associated with chronic haemolytic disease, and yet show levels of red cell G-6-PD activity appreciably greater than in certain other variants which do not give rise to chronic haemolysis.

Another group of variants giving rise to chronic haemolytic disease, exemplify a different type of kinetic abnormality (Yoshida 1973, Yoshida and Lin 1973). These variants, Gd Manchester, Gd Alhambra and Gd Tripler, show only a moderate enzyme deficiency by conventional red cell assay (more than 20% of normal activity) and they do not have unusually high K_m values for glucose-6-phosphate; they are however inhibited to an unusual degree by NADPH and also by ATP, in concentrations such as normally occur in the red cell. Consequently they cannot generate sufficient NADPH in the red cell to maintain an adequate level of reduced glutathione, and this is apparently the immediate cause of the haemolytic process.

5.3.4. G-6-PD and the Lyon hypothesis

Since the locus determining G-6-PD is on the X chromosome it is represented twice in cells of the female (XX) but only once in cells of the male (XY). Nevertheless on the average the level of G-6-PD activity found in normal females is much the same as that in normal males (Marks 1958). Furthermore the same average level of activity is observed in individuals who for one reason or another possess abnormal numbers of X chromosomes e.g. males with the Klinefelter syndrome XXY and females with the XXX syndrome, provided that they do not also carry one of the alleles determining G-6-PD deficiency (Grumbach et al. 1962, Harris et al. 1963b). Thus in general the number of G-6-PD genes present in an individual does not appear to influence the level of activity obtained.

This phenomenon which appears to involve the equalisation of the effects of X linked genes in the two sexes is sometimes referred to as dosage compensation. It probably occurs with respect to most loci on the X chromosome which are not concerned with characteristics peculiarly relevant to sexual differentiation. A hypothesis aimed at providing a general explanation of this phenomenon at least in mammalian species was put forward by Lyon (1962) and also Beutler et al. (1962). It suggests that in any cell of the female organism only one of the two X chromosomes present is functionally active. This may be the X chromosome derived from the father or the X chromosome derived from the mother, and in general one of these will be active in some cells and the other in other cells. In the male it is assumed that the single X chromosome present is functional in all cells. Consequently only a single dose of any gene on the X chromosome will be functionally active in each somatic cell of both females and males and dosage compensation would thus be achieved. It is supposed that the inactivation of one or the other of the two X chromosomes present in the cell of the female occurs more or less at random and at a fairly early stage of embryological development. It is also supposed that once it has occurred in any one cell the same X chromosome continues to be inactivated in all the daughter cells subsequently derived from it. Thus according to this hypothesis a female may be regarded as a mosaic. In approximately half of her cells only the X chromosome derived from her father will be functionally active, while in the other cells only the X chromosome of her mother is active.

A direct way of testing the hypothesis is to consider the situation in a female heterozygous for two alleles at a locus on the X chromosome which determines the structure of a particular enzyme or protein. In some cells the functionally active chromosome should carry one allele and in others the other allele, so that some cells should synthesise one form of the enzyme or protein while the other cells should only synthesise the other form. A single cell should not contain both sorts of enzyme or protein. Investigation of heterozygotes for alleles determining G-6-PD variants have shown that this in fact appears to be the case.

A particularly convincing demonstration of the phenomenon was obtained in experiments using cells grown in tissue cultures started from small explants of skin from heterozygous Black females of G-6-PD genotype $Gd^B Gd^A$ (Davidson et al. 1963). When the G-6-PD present in clones derived from individual cells in the culture were examined electrophoretically it was found that some clones showed only the A enzyme, and others only the B enzyme but none showed both (fig. 5.10). Furthermore in a tissue culture derived from skin from a single heterozygote both A clones and B clones could be obtained. Thus it appeared that two distinct cell populations were present, one synthesising only the A enzyme and the other only the B enzyme.

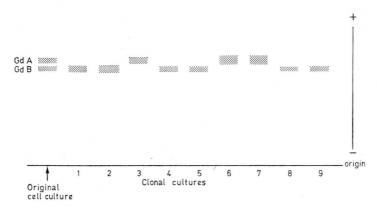

Fig. 5.10. Diagram illustrating the electrophoretic separation of glucose-6-phosphate dehydrogenase components in sonicates of tissue cultured cells from a heterozygous female of genotype $Gd^A Gd^B$ (redrawn from Davidson et al. 1963). The original cell culture shows the two G-6-PD components Gd A and Gd B. However the clones derived from single cells of this culture show either Gd A or Gd B but not both.

Similar experiments were also carried out in tissue cultures derived from females heterozygous for the Mediterranean type of G-6-PD deficiency genotype $Gd^BGd^{Mediterranean}$. Here it was found that two types of clone could be clearly distinguished. In one the level of enzyme activity was essentially the same as in tissue culture cells derived from normal males of genotype Gd^B, and in the other the level of enzyme activity was grossly reduced and equivalent to that found in tissue culture cells obtained from males of genotype $Gd^{Mediterranean}$. Thus again two biochemically distinct cell populations were present and both could be identified in tissue cultures derived from a single heterozygous individual.

Studies carried out on red cells from Negro females heterozygous for G-6-PD deficiency (Gd^BGd^{A-}) indicate that here also two biochemically distinct populations probably occur (Beutler and Baluda 1964). Evidence pointing to the same general conclusion has also been obtained by studies on the electrophoretic patterns of the enzyme derived from small skin biopsies and from single tumours (leiomyomas) in heterozygotes of G-6-PD genotype Gd^BGd^A (Linder and Gartler 1965a, b). Thus in the case of the G-6-PD locus the postulate that in any one cell of the female only one of the two alleles present is functionally active appears to be correct.

However one interesting exception to this general rule has been found. This concerns the oocyte cells of the ovary. It has already been noted (p. 37) that although glucose-6-phosphate dehydrogenase has a dimeric subunit structure, one finds that in electrophoretic studies of the enzyme in haemolysates, normal tissue extracts, and uncloned tissue cultured cells of female heterozygotes of genotype Gd^AGd^B, only the two forms Gd A and Gd B are seen. One does not see an intermediate isozyme representing a hybrid containing both the A and B type polypeptide chains because the A and B polypeptides are synthesised separately in different cells. Oocytes are however an exception, since a hybrid isozyme is indeed found in the heterozygote (fig. 5.11), though it is not seen in other cells of the ovary or in other tissues (Gartler et al. 1972). This implies that in oocyte cells the polypeptide products of both alleles are synthesised and combine together at random to give both the homodimeric forms A and B and also the heterodimeric hybrid. Thus in the oocyte both X-chromosomes are evidently active.

Other enzyme proteins have also been shown to be coded by loci on the X-chromosome. In several cases a specific deficiency of the enzyme is the cause of a particular inherited disorder (table 5.7), whose familial distribution shows that it is determined by a gene locus on the X-chromosome. In the cases of hypoxanthine guanine phosphoribosyl transferase, α-galacto-

(a)

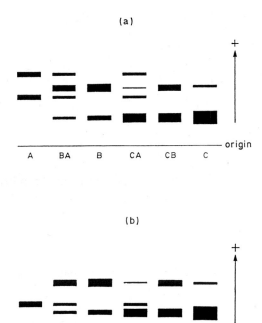

(b)

Fig. 5.12. Diagram showing electrophoretic patterns of different red cell acid phosphatase phenotypes. Starch gel electrophoresis at *p*H 6.0; (a) with citrate–phosphate buffer system; (b) with phosphate buffer system. Note that the two type A isozymes have a different relative mobility to the two type B and the two type C isozymes in the different buffer systems. For details of methods see Hopkinson and Harris (1969).

present. In type B the faster, more anodic isozyme shows much more activity than the slower isozyme, whereas in type C the opposite is the case. The isozyme patterns seen in the heterozygous types BA, CA and CB are more complex and represent simple mixtures in roughly equal proportions of the isozymes present in the two corresponding homozygous types. There is no indication of hybrid isozyme formation in the heterozygotes.

Studies on the pattern of substrate specificity and on the kinetics of the acid phosphatases of the different types have not revealed any striking differences (Scott 1966, Luffman and Harris 1967). Also the various isozymes in the different types all appear to be of the same molecular size, about 15,000 (Luffman and Harris 1967, Fisher and Harris 1971a). However significant differences in thermostability have been demonstrated (fig. 5.13). In general the relative thermostabilities of the isozymes of the different types are in

TABLE 5.8

The distribution of red cell acid phosphatase phenotypes observed in 440 English families
(Harris et al. 1968).

Type of mating	Number of matings	A	BA	B	CA	CB	C	Total
A × A	13	22	—	—	—	—	—	22
A × BA	50	65	48	—	—	—	—	113
A × B	27	—	65	—	—	—	—	65
A × CA	6	4	—	—	5	—	—	9
A × CB	10	—	5	—	14	—	—	19
BA × BA	94	40	91	54	—	—	—	185
BA × B	109	—	106	96	—	—	—	202
BA × CA	16	14	9	—	6	4	—	33
BA × CB	16	—	10	7	6	10	—	33
B × B	55	—	—	141	—	—	—	141
B × CA	12	—	12	—	—	20	—	32
B × CB	24	—	—	33	—	25	—	58
CA × CB	5	—	2	—	2	3	1	8
CB × CB	3	—	—	0	—	4	1	5
Totals	440	145	348	331	33	66	2	925

the order C > B > A, and within any type the faster (i.e. more anodic) isozyme appears to be somewhat less stable than the slower isozyme (Luffman and Harris 1967, Fisher and Harris 1969, 1971b). The fast and slow isozymes in a single homozygous phenotype (e.g. A or B) also differ in their kinetic characteristics and their pH activity profiles (Fisher and Harris 1971b).

The nature of the structural differences between the acid phosphatases determined by the several alleles is not known. One may suppose that the different alleles arose one from another by single mutational steps. Each presumably codes for a distinctive polypeptide chain, the differences between which may be no more than single aminoacid substitutions. Presumably also the two isozymic proteins which appear to be determined by each allele, both contain the same characteristic polypeptide chain, since they both seem to be modified in a characteristic way by the mutational differences.

There is evidence that the fast and slow isozymes apparently determined by a single allele may be conformational isomers (p. 82). It was found that

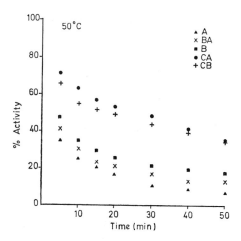

Fig. 5.13. Comparison of thermostabilities of different red cell acid phosphatase types. Mean percentage residual red cell acid phosphatase activity of haemolysates of different types held at 50 °C for varying periods of time (Luffman and Harris 1967).

after separation by ion-exchange chromatography (Fisher and Harris 1971a), each of the separated isozymes was partially converted to the other (Fisher and Harris 1971b). The results were consistent with the idea that the two isozymes produced by a single allele represent different conformations of a single polypeptide chain, each relatively stable, but with a standard free energy difference small enough to allow interconversion to occur at $+4\,°C$ in a matter of days, so that eventually an equilibrium mixture containing both isozymes is achieved. If so, such an equilibrium is probably present in the normal red cell, and the different relative staining intensities of the fast and slow bands observed in the A, B and C phenotypes after electrophoresis may reflect differences in the position of the interconversion equilibrium.

Besides the six red cell acid phosphatase phenotypes discussed above, several others have also been recognised (Giblett and Scott 1965, Karp and Sutton 1967). Three of these known as RA, RB and RC represent heterozygous combinations of another allele ACP_1^R with ACP_1^A, ACP_1^B or ACP_1^C. The ACP_1^R allele is very rare, if it occurs at all, in White populations; but has an appreciable frequency among Blacks (about 0.01 in U.S.A. Blacks). It appears to be particularly common in certain Black populations in South Africa where in some populations it attains frequencies as high as 0.2 (Jenkins and Corfield 1972). In contrast the ACP_1^C appears to be virtually absent in these populations and it has a much lower frequency in Blacks than in Whites. It is of interest that the level of activity attributable to ACP_1^R is relatively low, being about the same as that due to ACP_1^A (Jenkins and Corfield 1972).

5.4.2. Quantitative differences

Thus each of the alleles appears to determine structurally distinct forms of the enzyme. These structural differences are also associated with quantitative differences in the total amount of acid phosphatase activity produced. This was shown by assaying the acid phosphatase activity in red cells derived from individuals of the different types (Spencer et al. 1964).

Some typical findings are summarised in table 5.9. Although a great deal

TABLE 5.9

Average red cell acid phosphatase activity in individuals of different types. (Spencer et al 1964b.)

Type	Number of individuals tested	Mean activity	Standard deviation
A	33	122.4	16.8
BA	124	153.9	17.3
B	81	188.3	19.5
CA	11	183.8	19.8
CB	26	212.3	23.1

The activity is expressed as μmoles of p-nitrophenol liberated from p-nitrophenyl phosphate in 30 min at 37 °C, per gram of haemoglobin present in haemolysate.

of variation in the activity levels between individuals of the same acid phosphatase type occurs, nevertheless significant differences in average level of activity between the types are readily demonstrable. On average red cells of type B individuals show about 50% more activity than those of type A individuals, while type BA individuals have intermediate levels. Similarly the activity of red cells of type CB individuals is on average greater than those of type CA or type B.

Using this kind of data one may examine the question as to whether the quantitative effects of the three alleles are additive in a simple way or not. If they are additive one would expect the following relationships to be true:

$$\tfrac{1}{2}\bar{A} + \tfrac{1}{2}\bar{B} = \bar{B}\bar{A} \qquad \text{(a)}$$
$$\text{and} \quad \bar{C}\bar{A} - \tfrac{1}{2}\bar{A} = \bar{C}\bar{B} - \tfrac{1}{2}\bar{B} \qquad \text{(b)}$$

where $\bar{A}$, $\bar{B}\bar{A}$, $\bar{B}$ etc. are the mean values for the various types. It will be seen that the results given in table 5.9 support the idea of additivity rather well. Thus

$$\frac{1}{2}\bar{A} + \frac{1}{2}\bar{B} = 155.35 \text{ units} \qquad (a)$$
$$\bar{B}\bar{A} = 153.9 \text{ units}$$
$$\bar{C}\bar{A} - \frac{1}{2}\bar{A} = 122.6 \text{ units} \qquad (b)$$
$$\bar{C}\bar{B} - \frac{1}{2}\bar{B} = 118.15 \text{ units}$$

These results are of course consistent with the electrophoretic findings that the isozyme patterns in the heterozygous types are essentially those to be expected from simple mixtures in equal proportions of the two correspondingly homozygous types. Essentially similar findings have been obtained by Modiano et al. (1967), Shinoda (1967), Jenkins and Corfield (1971) and Eze et al (1973).

The precise way in which the structural differences between the acid phosphatases determined by the different alleles are related to these quantitative differences in activity remains to be elucidated. It is perhaps, at least in part, a consequence of the differences in enzyme stabilities, since the order of relative stabilities $C > B > A$ found in thermostability experiments is the same as the order of relative activities in red cells of the different types.

One point of some general significance which emerges from these studies is the following. If one determines red cell acid phosphatase activities in a series of randomly selected individuals in the general population one obtains a continuous unimodal distribution without any obvious discontinuities. It is not in fact dissimilar in form to the distributions often obtained when other enzymes are examined quantitatively in randomly selected populations. Yet in this case it is apparent that the overall distribution represents a summation of a series of separate but overlapping distributions corresponding to each of the discrete phenotypes (fig. 5.13). Furthermore much of the variance of the overall distribution can be attributed simply to the effects of the three alleles (Harris 1966, Eze et al. 1973). This suggests that many other examples of quantitative enzyme variation which are apparently continuous and unimodel may have a similar simple underlying basis. Indeed several cases have been analysed in the same way, for example the continuously distributed quantitative variation in red cell peptidase A (Sinha et al. 1970) has been shown to be largely attributable to two common alleles (Lewis 1973) and so also has the continuously distributed quantitative variation of glutamate-pyruvate transaminase (Chen et al. 1972).

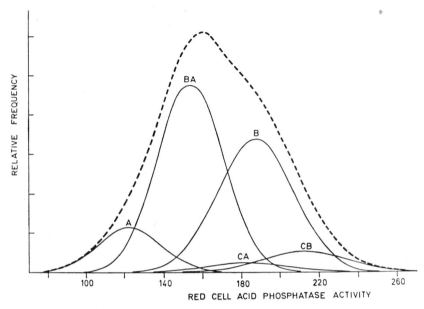

Fig. 5.14. Distribution of red cell phosphatase activities in the general population (broken line) and in the separate phenotypes. The curves are constructed from the data in table 5.9, using the frequencies of the different phenotypes as found in the English population.

These examples emphasise the difficulty pointed out earlier (p. 149) of analysing quantitative variation in the absence of qualitative methods for discriminating between discrete phenotypes.

CHAPTER 6

The inborn errors of metabolism

6.1. Garrod and the concept of 'inborn errors of metabolism'

It has been recognised for many years that there are a large number of metabolic diseases in which the characteristic clinical, pathological and biochemical abnormalities can be attributed to the congenital deficiency of a specific enzyme, which in turn is due to the presence of a particular abnormal gene. Such conditions are usually called 'inborn errors of metabolism', a term which was first used by A. E. Garrod more than sixty-five years ago.

Garrod's basic concept of the pathogenesis of these disorders (Garrod 1909) was mainly derived from studies on the rare condition known as alkaptonuria. His classical work on this abnormality has provided an elegant and simple model for the interpretation of a great variety of different inherited diseases subsequently discovered.

Alkaptonuria is characterised by the urinary excretion in massive quantities of the substance homogentisic acid, which is not normally found in urine. Alkaptonurics excrete several grams of homogentisic acid daily, and this is continuous and lifelong. It is a very striking abnormality because although the urine has a normal colour when passed, it rapidly goes black on standing due to the oxidation of the homogentisic acid present. The disorder is indeed often first recognised in infancy because of the characteristic staining of the diaper. Alkaptonurics are usually quite healthy though in later life they are particularly prone to develop a form of arthritis known as ochronosis, apparently caused by the deposition in cartilage and other connective tissue of a pigment derived from homogentisic acid.

Garrod noted that when homogentisic acid is fed to alkaptonuric subjects it is excreted quantitatively in the urine, whereas when given to normal subjects it appears to be readily metabolised. He also showed that its excretion in alkaptonurics could be augmented by feeding increased protein, and that this was due to the presence in protein of the aromatic aminoacids phenylalanine and tyrosine, which when given alone also enhanced the homogentisic acid

187

output. The excretion of homogentisic acid was also increased by certain derivatives of phenylalanine and tyrosine which could be plausibly regarded as intermediates in their catabolism.

From such metabolic experiments Garrod inferred that homogentisic acid, although it had never been detected in tissues, was itself a normal intermediate in the catabolism of phenylalanine and tyrosine, and that in alkaptonuric subjects the essential defect was a failure in its degradation, due to the lack of a necessary enzyme. Thus he supposed that in normal individuals homogentisic acid occurs at most only in trace amounts because it is broken down virtually as rapidly as it is formed. Whereas in alkaptonurics it cannot be broken down further, so it tends to accumulate in the cells of the liver where this metabolic process mainly occurs, leaks into the circulation and is excreted into the urine in large quantities.

At that time the crucial piece of evidence namely a direct demonstration of the specific enzyme deficiency was not available. In fact this was not obtained until some 50 years later when it became possible to assay in liver

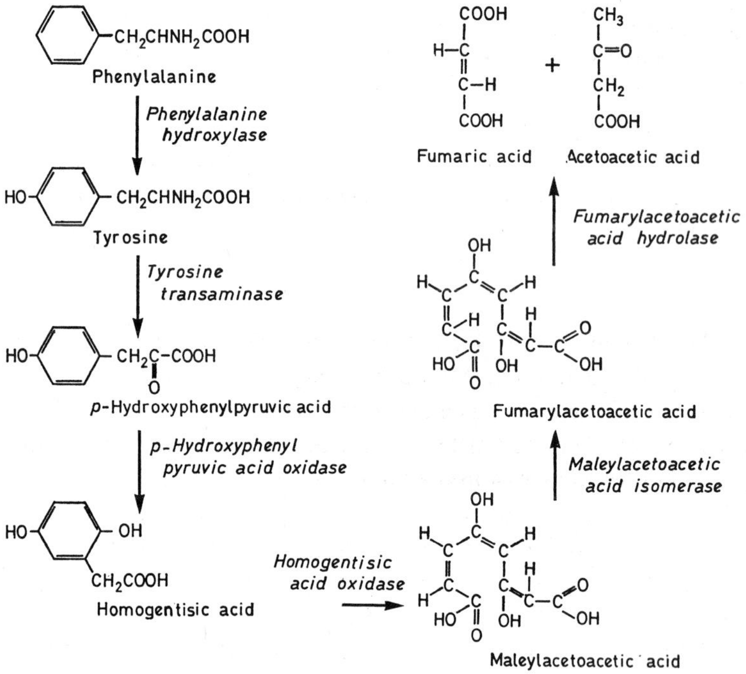

Fig. 6.1. Enzymatic steps in the oxidation of phenylalanine and tyrosine to acetoacetic acid. (From La Du 1966.)

biopsies from alkaptonuric subjects each of the series of enzymes concerned in the oxidation of tyrosine (fig. 6.1). It was found that all were present in normal amounts except for homogentisic acid oxidase, the enzyme which catalyses the conversion of homogentisic acid to maleylacetoacetic acid (La Du et al. 1958). This was not detectable (table 6.1). Thus Garrod's classical interpretation of the biochemistry of alkaptonuria was fully substantiated.

TABLE 6.1

Activity of tyrosine oxidation enzymes in biopsy specimens of alkaptonuric and non-alkaptonuric liver (La Du et al. 1958).

Enzyme	Enzyme activity	
	Non-alkaptonuric liver	Alkaptonuric liver
1. Tyrosine transaminase	3.6	3.2
2. p-Hydroxyphenylpyruvic acid oxidase	6.7	4.6
3. Homogentisic acid oxidase	26.8	<0.0048
4. Maleylacetoacetic acid isomerase	960	780
5. Fumarylacetoacetic acid hydrolase	29	22

Activity of enzymes 1, 2, 3 and 5 expressed as μmoles of substrate oxidized per hour by 0.1 g wet weight of liver. Activity of enzyme 4 expressed as Δ log O.D. per hour per 0.1 g wet weight of liver.

The other striking feature of alkaptonuria to which Garrod drew attention was its familial distribution. Although a very rare condition it was often found among more than one member of the family. Frequently two or more of a group of brothers and sisters would be affected, while their parents, children and other relatives appeared quite normal. Furthermore the parents of alkaptonurics were often blood relatives, usually first or second cousins. The pedigrees were quite characteristic, and Garrod had little hesitation in concluding that they implied an hereditary or genetical basis for the condition. He consulted Bateson, one of the earliest geneticists, who pointed out that the situation could be readily explained in terms of the then recently rediscovered laws of Mendel. The pedigrees were exactly those to be expected if alkaptonuria was determined by a rare recessive Mendelian factor, or as we should now say gene. The affected individuals could be presumed to be homozygous for the abnormal gene. This was in fact the first example of the so-called 'recessive inheritance' to be recognised as such in man.

Thus Garrod interpreted alkaptonuria as being caused by the congenital deficiency of a particular enzyme due to the presence in double dose of an abnormal Mendelian factor or gene. An important implication of the idea was that the normal allele of this gene must in some way be necessary for the formation of the enzyme in the normal organism. This was the first clue to the now well established generalisation that genes exert their effects in the organism by directing the synthesis of enzymes and other proteins.

Garrod viewed the inborn errors as conditions in which the specific enzyme deficiency effectively blocked at a particular point a sequence of reactions which form part of the normal course of metabolism. As a result metabolites immediately preceding the block would accumulate, and metabolites subsequent to the block would not be formed. The various biochemical, pathological and clinical manifestations of the condition could be regarded as secondary consequences of this primary metabolic defect. These secondary changes might be complex and widespread, and would depend in general on the nature and the biochemical effects of the metabolites which tended to accumulate or whose formation was restricted.

A large number of different disorders which can be explained in these general terms are now known. They are listed in Appendix I, pp. 368–396. Although a specific enzyme deficiency has been demonstrated in each of these conditions, the underlying nature of the defect is understood in very few. In some cases it may represent the synthesis of a structurally altered enzyme protein which has defective catalytic properties. In other cases a structurally altered enzyme protein which is extremely unstable and so is rapidly broken down in the tissues, may be the cause of the enzyme deficiency. In still other cases there may be a specific reduction or complete failure in the synthesis of the enzyme protein.

The enzymes involved in these different disorders are very diverse and are concerned with many aspects of metabolism. The metabolic disturbances and clinical abnormalities which result also vary widely. They range from conditions which may be effectively lethal in early life (e.g. maple syrup urine disease), through those that can produce a permanent disability such as mental retardation (e.g. phenylketonuria) or chronic haemolytic disease (e.g. pyruvate kinase deficiency), to those which are comparatively benign (e.g. alkaptonuria) or apparently harmless (e.g. fructokinase deficiency).

The examples discussed below illustrate something of the variety of biochemical and metabolic disturbances that may follow from such specific enzyme deficiencies, and the clinical abnormalities that can ensue.

6.2. *Phenylketonuria*

Phenylketonuria is among the most common of the 'inborn errors' which give rise to severe clinical disability, and since its discovery by Fölling in 1934 it has been studied extensively. It was, prior to the development of an effective therapy, characteristically associated with a marked degree of mental retardation, and in most institutions for the mentally retarded about 0.5–1 % of the patients suffered from this disorder. In European populations the abnormality is probably present in about 1 in 15,000 newborn infants (Levy 1973).

The deficient enzyme is phenylalanine 4-hydroxylase (Jervis 1953, Mitoma et al. 1957, Wallace et al. 1957, Kaufman 1958, Friedman et al. 1973) which in the normal individual occurs in the liver and catalyses para-hydroxylation of the aminoacid phenylalanine to give tyrosine. Phenylalanine is continuously being produced from the normal breakdown of tissue protein, and from the digestion of dietary protein (fig. 6.2). Its conversion to tyrosine in the liver

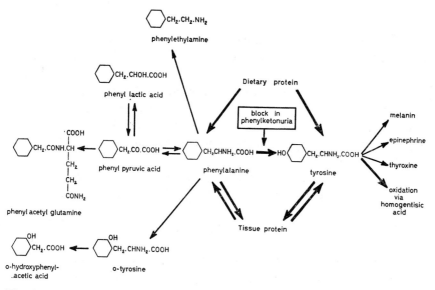

Fig. 6.2. Metabolic pathways involving phenylalanine, showing the site of the metabolic block in phenylketonuria.

is the first step in its catabolism and if this is blocked it accumulates intracellularly and appears in high concentrations in the body fluids. The level
of phenylalanine in blood serum in phenylketonuria is generally more than
thirty times normal, and there is an increased excretion of the aminoacid
in urine. The level of phenylalanine in the cerebrospinal fluid is also considerably elevated.

The very high concentrations of phenylalanine that occur result in various
secondary biochemical disturbances (fig. 6.2). One set of reactions involve
changes in the side chain of phenylalanine (Jervis 1950). Large amounts of
phenylpyruvic acid (Fölling 1934), phenyllactic acid (Zeller 1943) and phenylacetic acid are formed. The phenylacetic acid is subsequently conjugated
with glutamine to give phenylacetyl glutamine (Woolf 1951). These substances have a low renal threshold and are excreted in the urine in considerable
quantities. It was indeed the presence of phenylpyruvic acid in the urine that
first led to the identification of phenylketonuria and gave it its name.

Another derivative of phenylalanine formed in abnormal amounts is
o-hydroxyphenyl-acetic acid (Armstrong et al. 1955). Ortho-hydroxylation of
phenylalanine (or phenylpyruvic acid) may occur in the normal individual
(Tashian 1959) but is quantitatively insignificant compared with para-
hydroxylation of phenylalanine to give tyrosine. When however the main
pathway is blocked the product of the minor pathway is formed in increased
amounts. Phenylethylamine is yet another derivative of phenylalanine not
apparently formed in appreciable amounts in normal individuals but
produced in considerable quantities from phenylalanine in phenylketonurics
(Oates et al. 1963).

Abnormalities in tryptophan metabolites are also characteristic features
of the biochemical upset. There is an increased excretion of indole acetic
acid, indole lactic acid (Armstrong and Robinson 1954) and indole pyruvic
acid (Schreie and Flaig 1956). These metabolites are probably mainly of
intestinal origin, being absorbed after their formation by bacterial action on
non-absorbed tryptophan. The absorption of tryptophan is inhibited by
elevated phenylalanine concentrations (Scriver and Rosenberg 1973). There
is also a diminished excretion of 5-hydroxyindole acetic acid (Pare et al.
1957) and a reduced concentration of 5-hydroxytryptamine in the blood.
All these abnormalities are corrected on a phenylalanine restricted diet.

Another interesting effect is a small but significant reduction in the formation of the pigment melanin from tyrosine. This causes the hair and skin
of phenylketonuric patients to be slightly less pigmented than that of their
normal sibs. It appears to be due to a partial inhibition of the enzyme

tyrosinase by the high concentrations of phenylalanine which occur. Phenylalanine acts as a competitive inhibitor of the tyrosine–tyrosinase system *in
vitro* (Miyamoto and Fitzpatrick 1957, Boylen and Quastel 1962), and it has
been demonstrated that darkening of new grown hair in the phenylketonuric
may be achieved either by increasing tyrosine intake considerably (Snyderman et al. 1955), or by severely restricting phenylalanine intake (Armstrong
and Tyler 1955).

There is usually no tyrosine deficiency in phenylketonuria because
adequate amounts of tyrosine are present in the diet. But it is of interest to
note that while tyrosine is not an indispensible dietary ingredient (i.e. an
'essential' aminoacid) in the normal individual, it becomes one in phenylketonuria. This is because in the normal organism tyrosine is readily formed
from phenylalanine, while in phenylketonuria this is not the case.

Physical development is not seriously affected by the metabolic disorder,
but mental development is markedly impaired. The patients usually have a
severe degree of intellectual retardation. The majority are graded as idiots
(I.Q. less than 20) and the remainder nearly all as imbeciles (I.Q. less than
50), though some individuals with a lesser degree of impairment do occur.

The exact manner in which the damage to the brain is brought about is not
understood. While it is reasonable to suppose that phenylalanine itself, or
one of the other substances which are present in unusual concentrations in
the body fluids may either by inhibiting certain enzyme systems, or by
blocking particular transport processes, so modify the intracellular milieu in
the brain that its normal biochemical development is impeded, the exact
causal relations involved are still obscure. Among the specific compounds
that may be implicated are phenylethylamine (Oates et al. 1963) which is
thought to be a neurotoxic agent and which is probably formed in abnormal
amounts in the brain from phenylalanine because the appropriate L-amino
acid decarboxylase is normally present. Another substance which may be
significant in this connection is 5-hydroxytryptamine (serotonin) which is
present in less than normal amounts in phenylketonuric blood (Pare et al.
1957).

Treatment of the condition has been mainly directed to restricting the
phenylalanine content of the diet. Since phenylalanine is an essential aminoacid which is required for normal protein synthesis and growth, it cannot be
eliminated from the diet entirely. However diets can be constructed which
contain only enough phenylalanine to allow normal growth, but little or no
excess. When these are fed to phenylketonuric patients the concentration of
phenylalanine in the body fluids is brought down to normal or near normal

levels, and the other biochemical abnormalities consequent on the high phenylalanine concentrations disappear (Armstrong and Tyler 1955).

Sufficient time has now elapsed since the widespread introduction of phenylalanine restricted diets for the treatment of phenylketonuria, to make it abundantly clear that the treatment when properly applied is very effective. It prevents the development of severe mental retardation which used to be a characteristic feature of the condition, and children treated in this way can achieve intellectual levels within the normal range (for recent literature reviews see Knox 1972, Scriver and Rosenberg 1973, Levy 1973). However for successful results it is essential that the dietary treatment should be initiated shortly after birth, since it appears that most of the brain damage in phenylketonuria occurs during the first four months of life.

The introduction of dietary therapy for phenylketonuria, coupled with the recognition that it had to be introduced very early in life if it were to be effective, led to the development of screening techniques designed to be applied to all newborn infants, so that phenylketonuria could be diagnosed shortly after birth. The most satisfactory methods depend on determining the blood level of phenylalanine either by bacterial inhibition, fluorimetric assay or by paper chromatography, applied to dried blood spots on filter paper, or on blood taken into heparinised capillary tubes (for review see Levy 1973). An important point about these screening procedures is that the blood sample should preferably not be collected before about the sixth or seventh day of life. The reason for this is that in foetal life, the phenylketonuric does not have a grossly elevated blood level of phenylalanine, because the aminoacid is normally metabolised effectively in the mother's liver. Consequently several days usually elapse after birth before the blood phenylalanine level becomes obviously abnormal. Phenylketonuric infants are liable to be missed if the screening test is applied too soon.

Maternal phenylketonuria: Of the phenylketonuric females born prior to the widespread introduction of dietary therapy, only a small proportion became pregnant and had liveborn children. However when the children of such individuals are studied, it is found that although as a rule they do not show the classical metabolic disorder characteristic of phenylketonuria since the great majority are heterozygotes, they nevertheless nearly all exhibit developmental abnormalities and mental retardation which are attributable to the high levels of blood phenylalanine to which they were continuously exposed during foetal life (Mabry et al. 1966, Howell and Stevenson 1971).

These findings will of course become of particular importance in the management of pregnancy in phenylketonuric females who were born subsequent to the introduction of effective dietary control. There is still uncertainty as to how long the strict dietary control applied in early infancy to phenylketonurics need be continued without leading to mental deterioration. It has, for example, been suggested that relaxation of dietary control and even a normal diet can be safely allowed in later childhood and subsequently (Hornes et al. 1962, Solomons et al. 1966). However the adverse effects produced in the foetal life of high maternal blood phenylalanine levels, makes it clear that strict dietary control in pregnancy will continue to be necessary.

Other hyperphenylalaninaemias: The widespread introduction of screening programmes for phenylketonuria in which the blood level of phenylalanine is measured in newborns a few days after birth, led to the discovery of certain infants whose blood levels of phenylalanine are higher than in normal individuals and heterozygotes for the phenylketonuria gene (p. 232) but are not as high as those usually seen in classical phenylketonuria. In general the blood levels of phenylalanine in these infants range between 6 and 18 mgm per 100 ml, (normal $<$ 2 mgm per 100 ml, classical phenylketonuria $>$ 20 mgm per 100 ml), and the values subsequently often persist at these intermediate levels. This class of individuals is still poorly characterised, but it seems that it includes more than one distinct entity, and these are each distinct from classical phenylketonuria. Taken together these various types of hyperphenylalaninaemia appear to occur with a frequency which is on average about one third of that of phenylketonuria itself (Levy 1973), though in some newborn surveys the two frequencies are about the same. Comparisons between different surveys are difficult, because differences in the criteria used to define hyperphenylalaninaemia have varied (Levy 1973).

Various classifications of these disorders have been proposed (see Hsia 1970, Scriver and Rosenberg 1973, Levy 1974), but the precise delineation of the conditions included in the so-called group of 'hyperphenylalaninaemias' which are distinct from classical phenylketonuria is still obscure. Some individuals with what has been called 'persistent hyperphenylalaninaemia' evidently manage quite well on a normal diet and are not mentally retarded (Levy et al. 1971). There are also atypical cases of phenylketonuria in which although a restricted phenylalanine diet is necessary, the amount of phenylalanine which can be tolerated is significantly greater than in classical phenylketonuria (Kennedy et al. 1967, Scriver and Rosenberg

1973). In addition there is what has been called 'transient phenylketonuria in which the hyperphenylalaninaemia does not persist, though in which a phenylalanine restricted diet is necessary for a period, but subsequently there can be a return to a normal diet without the reappearance of the hyperphenylalaninaemia (O'Flynn et al. 1967, Castells et al. 1968, Hsia 1970).

These various conditions appear to be familial and usually run true to type in any one family. But the genetics is still obscure. Perhaps they are due to different alleles at a single phenylalanine hydroxylase locus, but other possibilities have certainly not been excluded. Assays of phenylalanine hydroxylase in liver biopsy material have been carried out in a few cases, (Justice et al. 1970, Kang et al. 1970, Hsia 1970). Demonstrable activity was found, but it was much reduced compared to that present in control liver material, though not as grossly deficient as in liver from patients with classical phenylketonuria.

6.3. *Galactosaemia*

Galactosaemia is an inborn error of carbohydrate metabolism in which there is a specific inability to metabolise the hexose sugar galactose (Townsend et al. 1951, Komrower et al. 1956). Galactose is an important constituent of an infant's diet because it is a component of the disaccharide lactose which is the main carbohydrate present in milk. Thus an infant affected with this disorder who is fed on milk in the ordinary way will receive considerable quantities of galactose. The consequences are generally severe. The infant fails to thrive, weight gain is slow, mental development is retarded, the liver becomes enlarged and eventually cirrhotic, and cataracts develop in the lens. Death in infancy is not uncommon if the condition is not recognised. If however such an infant is placed on a diet entirely free of galactose a dramatic improvement in the physical condition takes place. In fact it seems probable that provided the treatment is started early enough and great care is taken rigorously to exclude galactose from the food, growth and development may take place normally. If however commencement of treatment is delayed, then some degree of liver damage, cataract and mental impairment may persist because irreversible changes have already taken place.

Galactose enters the main stream of carbohydrate metabolism via a series of reactions which result in its conversion to glucose-1-phosphate (fig. 6.3). It first reacts with ATP to give galactose-1-phosphate. The enzyme concerned is galactokinase. The galactose-1-phosphate then reacts with the nucleotide uridine diphosphoglucose to give glucose-1-phosphate and uridine diphosphogalactose. The enzyme here is galactose-1-phosphate uridyl transferase. Uridine diphosphoglucose can then be regenerated from uridine diphosphogalactose by the enzyme uridine diphosphogalactose-4-epimerase, a reaction which requires the coenzyme NAD. In galactosaemia galactose-1-phosphate uridyl transferase activity is virtually absent (Kalckar et al. 1956, Isselbacher et al. 1956). Galactokinase and the epimerase occur in normal amounts as do other enzymes concerned in carbohydrate metabolism.

When galactose is present in the diet, the enzyme deficiency results in an intracellular accumulation of galactose-1-phosphate (Schwarz et al. 1956) and an abnormally high level of galactose itself in the body fluids. The blood galactose is considerably elevated after a galactose containing meal, and declines only slowly. Because of the high blood levels, large amounts of galactose appear in the urine. The elevated galactose concentrations in the body fluids also lead to the formation of the sugar alcohol galactitol in abnormal amounts (Wells et al. 1964, 1965).

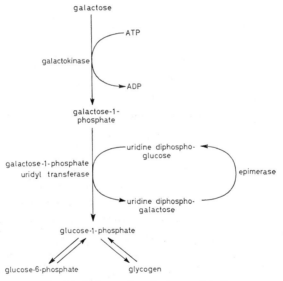

Fig. 6.3. Pathways in galactose metabolism.

Most of the pathological consequences that occur in galactosaemia can be attributed to the high intracellular concentrations of galactose-1-phosphate that develop. This is thought to produce pathological effects by inhibiting other enzyme reactions in carbohydrate metabolism which involve phosphorylated intermediates. The liver and brain damage as well as the general failure to thrive are probably secondary consequences of this. A notable feature is a tendency to hypoglucosaemia due to a decreased hepatic output of glucose.

The development of cataracts of the lens, which is a typical feature of the disease, is however probably due to the high concentrations of galactose in the body fluids and the resultant formation of abnormal quantities of galactitol (Gitzelmann et al. 1967). In this connexion it is interesting to compare the pathological changes which occur in galactosaemia due to galactose-1-phosphate uridyl transferase deficiency with those found in another inborn error of galactose metabolism, where there is a gross deficiency of galactokinase, but normal levels of the uridyl transferase and the epimerase. In galactokinase deficiency (Gitzelmann 1967), high blood levels of galactose occur when milk is fed, and galactitol is produced in abnormal amounts. Galactose-1-phosphate however is not formed. The outstanding clinical feature is the development of severe lens cataracts from an early age. In other respects the affected individuals may be healthy and develop normally. This is in marked contrast to the situation in classical galactosaemia where the transferase is deficient. Evidently cataract formation which occurs in both conditions is due to galactose accumulation and excessive galactitol formation, whereas the other severe pathological features seen in the uridyl transferase deficiency but not found in galactokinase deficiency are consequent on galactose-1-phosphate accumulation. Galactitol appears to be readily formed in lens and is probably not further metabolised. Its accumulation in this tissue may lead to cataract formation by causing over-hydration and electrolyte imbalance.

6.4. Isozyme deficiencies and tissue differences

Often more than one gene locus is concerned in determining the molecular structure of a particular enzyme. Lactate dehydrogenase (pp. 52–60) and phosphoglucomutase (pp. 60–69) which have already been discussed are typical examples of this kind of phenomenon. Each locus codes for a distinctive polypeptide chain and as a result several structurally different molecular forms or isoenzymes of the enzyme occur. Also it is not infrequently found that the relative amounts of the different isozymes vary considerably from tissue to tissue, presumably because of variations in expression of the several gene loci in different cell types.

In such cases, a mutation at one of the loci may result in a deficiency of some isozymes but not of others. Furthermore if tissues or organs differ in the isozymes they normally contain, then this will be reflected in the biochemical and clinical manifestations of a particular mutant. The characteristic and sometimes rather unexpected features of many inherited diseases are often explicable in terms of such isozymic differences in the normal organism.

6.4.1. Aldolase deficiency in hereditary fructose intolerance

At least three structurally distinct forms of the enzyme aldolase have been shown to occur (Penhoet et al. 1966, 1967, 1969, Lebherz and Rutter 1969). These are known as aldolase A, aldolase B and aldolase C. Aldolase A occurs in most tissues and is the only form present in muscle. Aldolase B is the predominant form in liver and also occurs in kidney. Aldolase C occurs in brain and heart. These isozymic proteins are tetramers and evidently differ in the structures of their characteristic polypeptide subunits, each of which is presumably determined by a separate gene locus. The tissue differences seem to reflect differences in the relative amounts of synthesis of the polypeptide products of the different loci. Thus in liver the polypeptide determined by the 'B' locus is formed in much greater amounts than that determined by the 'A' locus, and the 'C' locus is probably virtually inactive. In muscle on the other hand practically all the enzyme formed appears to be a product of the 'A' locus. In tissues where two different polypeptides occur (e.g. A and B, or A and C) hybrid tetrameric isozymes are formed.

Each of the aldolase isozymes (Rutter et al. 1968) catalyses both of the following reactions:

(1)   Fructose diphosphate $\rightleftharpoons$ dihydroxyacetone phosphate+glyceraldehyde-3-phosphate

(2)   Fructose-1-phosphate $\rightleftharpoons$ dihydroxyacetone phosphate+glyceraldehyde.

But they differ from one another in their detailed kinetics. This is particularly apparent in comparisons of the relative reaction rates obtained when fructose diphosphate (FDP) and fructose-1-phosphate (F-1-P) are used separately as substrates. Thus with muscle extracts which contain virtually only aldolase A the FDP:F-1-P activity ratio found is about 50:1, whereas with liver extracts where the predominant form is aldolase B the FDP:F-1-P activity ratio is about 1:1.

Hereditary fructose intolerance (Chambers and Pratt 1956, Froesch et al.

1957, Froesch 1972) appears to be specifically an inborn error of fructose metabolism. Individuals with the abnormality, whether they are in infant or adult life, remain healthy and symptom free as long as they do not take any food containing fructose. The ingestion of fructose however, whether as the free form or in the disaccharide sucrose, causes immediate deleterious effects the main symptoms being attributable to a severe hypoglucosaemia which develops. The condition is usually first recognised in infancy, since the symptoms are likely to become manifest as soon as breast feeding is supplemented or replaced by feeds which contain added sucrose. There is often a rapid deterioration in the child's condition, and the disorder may prove fatal if it is not recognised and sucrose promptly removed from the diet (fig. 6.4).

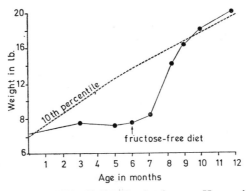

Fig. 6.4. Weight chart of a child with fructose intolerance. He was breast fed for only two days and then put on a dried milk preparation with added sucrose. There was virtually no weight gain over the next few months, and his condition deteriorated. He only began to improve and gain weight when a fructose-free diet was instituted (Black and Simpson 1967).

Fructose enters the main stream of carbohydrate metabolism, by first being converted to fructose-1-phosphate with ATP and the enzyme fructokinase (fig. 6.5). The fructose-1-phosphate is then cleaved by aldolase (reaction 2 above). In hereditary fructose intolerance a considerable reduction in aldolase activity in liver but not in muscle has been demonstrated (Hers and Joassin 1961, Froesch et al. 1963, Nordmann et al. 1968). Furthermore the reduction of activity in liver is much more pronounced when fructose-1-phosphate is used as substrate for the enzyme assay, than when fructose diphosphate is used. Thus in one study (table 6.2) it was found that with fructose-1-phosphate as substrate the aldolase activity in liver specimens

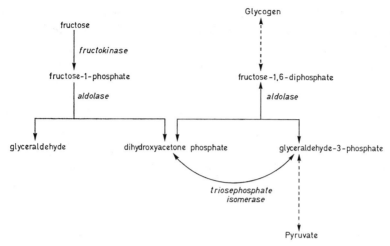

Fig. 6.5. Pathways in fructose metabolism.

TABLE 6.2

Aldolase activity in liver biopsy specimens from patients with fructose intolerance and from appropriate controls (Hers & Joassin 1961).

Subjects		Liver aldolase activity (μmoles substrate utilised per min per g tissue)			Ratio of activities FDP:F-1-P
		With fructose diphosphate (FDP)		With fructose-1-phosphate (F-1-P)	
Patients with fructose intolerance	1	3.1		0.5	6.2:1
	2	2.5		0.4	6.2:1
	mean		2.8	0.45	6.2:1
Controls	1	7.2		8.0	0.90:1
	2	15.4		15.8	0.97:1
	3	8.5		9.1	0.93:1
	4	11.0		12.4	0.89:1
	5	14.9		13.7	1.09:1
	mean		11.4	11.8	0.96:1

from the affected patients was only about 4% of that found in comparable controls. With fructose diphosphate as substrate the liver aldolase activity was about 25% of the control levels. The FDP:F-1-P activity ratios for the patients were around 6:1 compared with control values of about 1:1. Similar results have been obtained in other studies (Froesch 1972), and it is clear that a consistent feature of the disease is a marked reduction in liver aldolase F-1-P activity, a moderate reduction in liver FDP activity, and a considerable rise in the FDP:F-1-P activity ratio. In contrast to this, studies on muscle aldolase show no abnormality; activity levels with both FDP and F-1-P are similar to those in the controls and the FDP-F-1-P ratio is unaltered.

These findings are most simply accounted for by a specific deficiency of aldolase B. The residual aldolase activity in liver is probably mainly due to aldolase A, the characteristic polypeptide subunit of which is normally formed in liver but only in small quantities (Nordmann et al. 1968). It is also possible that there may be some compensating synthesis of aldolase A as a response to aldolase B deficiency. In muscle, where only the aldolase A polypeptide is normally formed, no aldolase deficiency occurs.

When fructose is administered, the aldolase B deficiency results in an intracellular accumulation of fructose-1-phosphate and an abnormal elevation of fructose in the body fluids. The blood fructose levels are high and fructose is excreted in the urine. The toxic effects of fructose in these circumstances are almost certainly attributable to the grossly increased concentrations of fructose-1-phosphate in the cells of the liver, because in another abnormality of fructose metabolism due to fructokinase deficiency (Schapira et al. 1961) no clinical disorder occurs at all, even when large amounts of fructose are given. Here high blood fructose levels develop as in aldolase deficiency, but there is no accumulation of fructose-1-phosphate. The precise manner in which fructose-1-phosphate exerts its toxic effects in hereditary fructose intolerance is not known, but it seems likely that it does so by inhibiting other enzymes involved in carbohydrate metabolism in liver, particularly those concerned in glycogen breakdown and the maintenance of a normal blood sugar level.

In effect the specific aldolase abnormality in hereditary fructose intolerance results only in defective metabolism of ingested fructose. Glycolysis in muscle is not affected because aldolase A activity is normal. Glycolysis and gluconeogenesis in liver are not seriously disturbed because the enzyme activity necessary for fructose diphosphate cleavage or synthesis is sufficient for normal requirements provided fructose is omitted from the diet and therefore no secondary disturbances due to fructose-1-phosphate accumulation occur.

6.4.2. *Pyruvate kinase deficiency*

Pyruvate kinase catalyses the conversion of phosphoenolpyruvate to pyruvate, a key step in glycolysis which is coupled to the generation of ATP:

2-phosphoenolpyruvate $+$ ADP $\rightleftarrows$ pyruvate $+$ ATP.

Two or more distinct isozymic forms of pyruvate kinase have been found to occur (Bigley et al. 1968, Osterman and Fritz 1973, Imamura and Tanaka 1972a, b). They differ in their kinetic behaviour, as well as in various physical properties, and they can also be distinguished immunochemically. They are presumably determined by separate gene loci. In man one of these forms (PK I) has been found only in red cells and in liver. Another (PK III) is more widely distributed and has been found for example in liver, kidney and leucocytes, but it is not present in red cells (Bigley et al. 1968).

Many examples of a form of chronic haemolytic anaemia, apparently due to a specific deficiency of red cell pyruvate kinase, have been identified (Valentine et al. 1961, Bowman and Procopio 1963, Grimes et al. 1964, Keitt 1966, Tanaka and Valentine 1968). The enzyme deficiency causes a major disturbance of red cell glycolysis with impairment of energy supplies and shortening of the average life of the red cells. The severity of the anaemia may vary considerably between patients in different families. In some cases it is mild and fully compensated (Nixon and Buchanan 1967), but often the functional defect and consequent anaemia is very severe. It appears that there are a number of distinct mutant alleles which can cause the pyruvate-kinase deficiency, and do so to different degrees, some of the patients being homozygous and others heterozygous for two different abnormal alleles (Paglia et al. 1972). In some cases the defective enzyme has been shown to have altered kinetic properties (Boivin and Galand 1967, Paglia et al. 1968, Paglia et al. 1972). The red cells of pyruvate kinase deficient individuals characteristically show high levels of 2,3-diphosphoglycerate (2,3 DPG), phosphoenol pyruvate (PEP) and 3-phosphoglycerate (3 PG), while pyruvate lactate and ATP concentrations are usually lower than normal (see fig. A.1, p. 369).

Even though the level of pyruvate kinase in the red cells is grossly reduced in these cases, the level of the enzyme in leucocytes is quite normal. So the defect is evidently peculiar to the red cell type of isozyme. The other tissue in the body in which the red cell type of isozyme occurs is liver, but here it is present together with the isozymic form also found in leucocytes. Thus one would expect to observe some degree of deficiency of liver pyruvate kinase activity in patients with the red cell deficiency. This has indeed been demonstrated (Bigley and Koler 1968), and it was also shown that the reduction in the total level of activity in the liver was due to the reduction of the isozymic form which also occurs in the red cell, whereas the other isozyme was unaffected. Evidently in such circumstances the amount of activity contributed by the unaffected isozyme is adequate to maintain normal function, so that no significant metabolic disturbance in the liver is apparent.

6.5. *Partial enzyme deficiencies and their metabolic consequences. The urea cycle enzymes*

The classical idea of an inborn error of metabolism was a situation where the specific enzyme concerned was absent so that the metabolic pathway it subserved was completely blocked. Metabolites immediately preceding the block would be expected to accumulate while the normal products of the reaction would not be formed. As more and more examples have been studied it has become clear that in many inherited metabolic diseases although there is certainly a specific enzyme deficiency, the loss of activity is not complete. Consequently some formation of the normal reaction products goes on, and the metabolic pathway is therefore only partially blocked. Nevertheless the restriction on the formation of the products of the reaction as well as the accumulation of metabolites immediately preceding the partial block are often sufficient to give rise either directly or indirectly to serious clinical abnormalities. In other cases of course although some distortion of the normal metabolic situation may be evident, no clinical abnormality ensues except perhaps under special conditions of metabolic stress.

According to the degree of the enzyme deficiency and the particular nature of the metabolic pathway involved a wide range of different effects may occur. One particular type of situation is of special interest because the biochemical findings may at first sight seem somewhat anomalous. This is where the product of a reaction or reaction sequence appears to be formed at normal or near normal rates, but where the occurrence of grossly increased concentrations of certain intermediary metabolites in the reaction sequence indicates that a quite severe metabolic block is nevertheless present, and indeed a marked though partial deficiency of the appropriate enzyme may be demonstrable. This type of situation can often be explained in terms of an increased reaction velocity produced by elevation in substrate concentration. Although the specific enzyme may be much reduced compared with the normal, the increased concentration of its substrate which this reduction inevitably produces may be sufficient to raise the reaction velocity, that is the rate of formation of the product, to normal or near normal levels.

The general point is illustrated by the biochemical findings (Shih and Efron 1972) in the series of rare disorders involving specific deficiencies of enzymes in the Krebs–Henseleit cycle of reactions (fig. 6.6). This is the well-

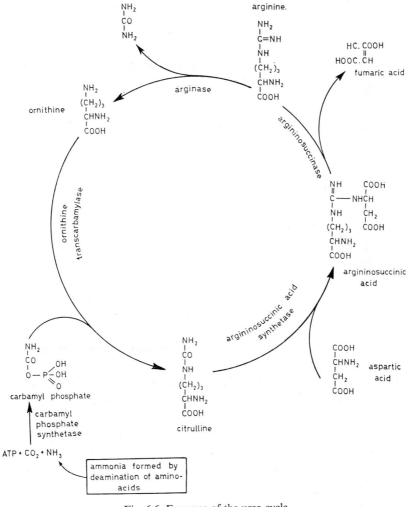

Fig. 6.6. Enzymes of the urea cycle.

known reaction sequence by which urea, the principal nitrogenous excretory product of the organism, is produced. Disorders due to specific defects in each of the enzymes involved in the cycle have now been identified.

Argininosuccinase: One condition called argininosuccinicaciduria, is due to a deficiency of the enzyme argininosuccinase (argininosuccinate lyase) which cleaves argininosuccinate to give arginine and fumarate (Allan et al. 1958, Westall 1960). Argininosuccinate, an intermediate in the urea cycle and not normally present in more than trace amounts, appears in considerable quantities. Its level in serum is elevated and it is excreted in the urine in large amounts. There is also some increase in the serum level of citrulline, its immediate precursor in the cycle. Blood ammonia levels, which may be normal in the fasting state, tend to rise significantly after meals containing any substantial amount of protein. But the blood urea level is not reduced below normal levels. Most patients show some degree of mental retardation (Moser et al. 1967, Shih and Efron 1972) and may exhibit other neurological abnormalities such as fits.

Urea is mainly formed in liver, and a marked and quite specific deficiency of argininosuccinase activity in liver samples from patients with this condition has been demonstrated. But although less than 5% of the normal activity of the enzyme (Miller and McLean 1967) is apparently present, there is evidently still sufficient activity to allow urea synthesis to proceed at a more or less normal rate. Presumably, in the normal, the concentration of the intermediate metabolite argininosuccinate is very low and far from saturating the enzyme. When the enzyme is defective the argininosuccinate concentration tends to rise and levels are reached which give a reaction velocity similar to that occurring in the normal state with much lower concentrations of substrate. Consequently the rate of urea production is not significantly reduced, but there is a considerable alteration in the concentration of those intermediates in the pathway which precede argininosuccinate cleavage. The concentration of argininosuccinate itself is greatly increased, and there is also some increase in the levels of its precursors in the cycle. Immediately following a heavy protein meal these effects are accentuated.

Argininosuccinate synthetase: In another condition, known as citrullinaemia (McMurray et al. 1963), the defective enzyme is argininosuccinate synthetase. There is a markedly increased level of citrulline in the blood plasma, and this aminoacid is excreted in large amounts in the urine. Elevation of blood ammonia also occurs, especially after a protein meal. But area production still goes on. In one well-studied case the blood urea levels remained consistently within normal limits (McMurray et al. 1963, Mohyuddin et al. 1967). In another, however, they were somewhat lower than normal so there was evidently some restriction in the rate of urea formation (Morrow 1967). Presumably in the former case the increased intracellular concentration of citrulline was sufficient to give with the deficient enzyme a more or less normal reaction velocity. Whereas in the second case this was not quite achieved.

These observations are of interest because in the normal subject argininosuccinate synthetase is thought to be the rate limiting enzyme of the urea cycle (Miller and McLean 1967). However, even this rate limiting step appears to have a considerable functional reserve, because calculations based on the observed levels of activity of the enzymes in the cycle indicate that a much greater rate of urea formation is possible than is actually required under normal circumstances. But presumably the margin by which argininosuccinate synthetase may be reduced without slowing the whole cycle and diminishing the rate of urea formation significantly is likely to be less than with the other enzymes of the cycle. It may be reached in certain cases of citrullinaemia and not in others.

Citrullinaemia results in neurological damage and mental retardation (McMurray et al. 1963, Morrow 1967) probably because of the toxic effects of ammonia which tends to be elevated in the blood, particularly after high protein meals. A fulminating form of the disorder with very severe manifestations occurring immediately after birth and a very rapidly fatal course has also been described (Van der Zee et al. 1971, Wick et al. 1973).

As has been noted previously (p. 147), a deficiency of a particular enzyme may arise in several different ways. In some cases it may be due to the synthesis of a structurally abnormal protein which is less efficient catalytically than its normal counterpart, or which is less stable so that it is broken down excessively rapidly. In other cases the deficiency may be due to a reduction in the actual rate of synthesis of the enzyme protein. Thus the deficiency state could represent a situation in which essentially normal amounts of the enzyme protein are present but its catalytic activity is defective; or a situation in which the catalytic properties of the enzyme are not significantly altered, but there is a true reduction in the actual amount of enzyme protein present. But whatever the precise cause of the enzyme deficiency, it is possible, provided that the deficiency is incomplete, for the product of the reaction to be formed at a normal or near normal rate. This will in general occur if, in the normal organism, the enzyme is not saturated by the concentration of substrate that normally exists, and if in the abnormal individual the increase in concentration of substrate which develops is sufficient to raise the reaction velocity to normal levels before the enzyme present is fully saturated.

In most cases of enzyme deficiencies in the urea cycle, the nature of the enzyme defect has not been determined. However, in one instance, a case of citrullinaemia, it was shown that the deficiency was most probably due to the synthesis of an abnormal enzyme with grossly altered catalytic properties (Tedesco and Mellman 1967). In this case it was found that the apparent Michaelis constant (K_m) of argininosuccinate synthetase with citrulline as substrate was at least twenty-five times greater than that found for the enzyme in appropriate normal controls. The Lineweaver–Burk plots obtained are

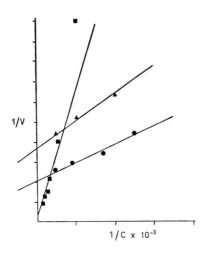

Fig. 6.7. Lineweaver–Burk plots derived from measurements of reaction velocities (V) of argininosuccinate synthetase at various concentrations (C) of citrulline (Tedesco & Mellman 1967). The enzyme was studied in extracts of fibroblast cells grown in tissue culture and derived from a patient with citrullinaemia, and from appropriate controls. *Key:* ■: enzyme from cells of citrullinaemic patient; ▲ and ●: enzymes from cells of two different control subjects.

shown in fig. 6.7. It will be seen that provided a sufficiently high intracellular concentration could be reached, the velocity of the reaction with the abnormal enzyme might be essentially the same as in the normal individual where the intracellular concentrations of citrulline are very low. In fact in this particular case this degree of compensation was probably not quite achieved *in vivo*, since the blood urea levels of the patient were somewhat less than normal. This was perhaps because the very considerable intracellular concentration of citrulline in the liver cells that would have been required to obtain a normal rate of urea formation could not be reached because of leakage of the citrulline from the liver cells into the body fluids. However, one can see that in other cases, if the elevation in K_m were not quite so severe, an essentially normal rate of urea formation might well occur.

Ornithine carbamoyl transferase: In another disorder involving the urea cycle, ornithine carbamoyl transferase (ornithine transcarbamylase) is the defective enzyme. The cardinal feature is a gross elevation of blood ammonia. The blood urea is usually in the normal or low normal range. There appears to be a clear difference between the degree of enzyme deficit and the severity of the clinical signs and symptoms in males as opposed to females. This is so even in the same family and these findings together with the pattern of inheritance observed in various pedigrees suggest that ornithine transcarbamylase is determined by a gene located on the X-chromosome (Campbell et al. 1971, Scott et al. 1972, Short et al. 1973).

In males, a gross deficiency of hepatic ornithine transcarbamylase activity ($< 0.5\%$ of normal) occurs and this is associated with severe hyperammonaemia and a clinical picture of rapid progressive deterioration and death within a few days (Campbell et al. 1973, Saudubray et al. 1973). In contrast, female patients have been found to show less marked reductions in hepatic ornithine transcarbamylase activity, and although these were associated with hyperammonaemia and corresponding clinical manifestions such as vomiting, lethargy, seizures and sometimes coma, the patients have not usually succumbed during the first few days of life (Russell et al. 1962, Hopkins et al. 1969; Levin et al. 1969a, Corbeel et al. 1969, Matsuda et al. 1971, Sunshine et al. 1972). Several have died in later infancy or childhood but others survive and may remain quite well provided their dietary protein intake is restricted.

Taken as a whole the findings are consistent with the idea that in any one family the affected males carried a mutant allele on their single X-chromosome and the affected females were heterozygous for the mutant. Presumably in heterozygous females the overall enzyme deficiency varies quite widely, probably because of variations in the proportion of cells in which the 'active' X-chromosome carries the mutant allele (see p. 242), and only those with the more severe deficits show marked clinical abnormality.

Carbamyl phosphate synthetase: Severe hyperammonia due to a specific deficiency of carbamyl phosphate synthetase has also been described, death occurring in infancy after four days in each of the affected sibs in one family (Gelehrter and Snodgrass 1974) and later in infancy in others (Hommes et al. 1969, Freeman et al. 1970).

Arginase: Deficiency of arginase has been reported in three sisters by Terheggen et al. (1969, 1972). They showed elevated blood arginine and ammonia levels which were diminished by restricting dietary protein. There was marked mental retardation and neurological damage.

In these various disorders of the urea cycle, the primary metabolic disturbance is in the liver, but the principal clinical abnormalities that ensue are largely attributable to damage to the central nervous system. A severe degree of mental retardation is a common feature. It appears probable that the neurological damage is mainly caused by the toxic effects of elevated blood ammonia levels, which tend to occur in each of these conditions particularly after heavy protein meals. This may be particularly important in infancy and early childhood when rapid maturation of the brain is taking place. So though in most of the cases it appears that the overall rate of production of urea is not markedly altered, severe clinical abnormality occurs because of the secondary effects of the accumulation of metabolites prior to the partial block in the sequence of reactions.

6.6. *Glycogen diseases*

Certain enzymes are concerned with the synthesis or breakdown of complex macromolecules, and specific deficiencies of such enzymes can, as the case

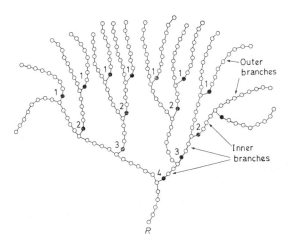

Fig. 6.8. Model of a segment of a glycogen molecule (after Cori 1954). Open circles: glucose residues in α-1,4 linkage; black circles: glucose residues in α-1,6 linkage. R = reducing end group. There are four tiers of branch points (glycogen has at least seven). Inner chains are terminated by branch points in adjacent tiers; outer chains by a branch point and by the non-reducing terminal glucose residue.

may be, result in a deficit or an accumulation of the macromolecule concerned. They may also in some cases lead to the occurrence of macromolecules with a quite unusual structure. Such phenomena are well illustrated by the series of inborn errors involving the polysaccharide glycogen.

Glycogen is the main form in which carbohydrate is stored in the animal body. Although it is present in the cells of most tissues it occurs in particularly large amounts in liver and muscle. It is a polydisperse polymer with an average molecular weight of between 2.5 and 4.5 million and is made up from only one type of building block, α-D-glucose. It has a multi-branched tree-like structure (fig. 6.8). Most of the glucose residues are joined in chains by α-1,4 linkages. But at the branch points α-1,6 linkages occur. The outer chains which end in non-reducing terminal glucose residues tend to be longer than the inner chains which occur between two branch points. These outer chains usually consist of 7 to 10 glucose units and may constitute approximately 50% of the macromolecule. In normal individuals glycogen molecules are

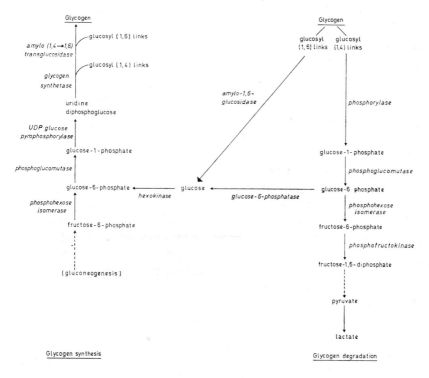

Fig. 6.9. Enzymatic steps in the synthesis and breakdown of glycogen. (Lysosomal degradation by α-1,4 glucosidase, referred to on p. 215, is not shown.)

constantly being degraded and resynthesised to various degrees according to the immediate metabolic requirements. Thus the amount and to some extent the size and structure of the glycogen molecules present will depend at any one time on the nutritional state of the individual.

Quite a number of different enzymes are involved in the synthesis and degradation of glycogen and the main pathways are indicated diagrammatically in fig. 6.9. A variety of rare disorders each due to a specific deficiency of one or another of these enzymes have been recognised (table 6.3). In some cases the defect has been found to involve reactions concerned with the synthesis of the macromolecule. But more often the defect appears to involve some particular step in glycogen degradation and results in the intracellular accumulation of glycogen in various tissues thus giving rise to what is often referred to as a 'glycogen storage disease'.

6.6.1. Defect in glycogen synthesis

Glycogen synthesis involves the successive addition of glucose units in α-1,4 linkage to the ends of the most peripheral glycogen chains. First, uridine diphosphoglucose (UDP glucose) is formed from glucose-1-phosphate and uridine triphosphate by the action of UDP glucose pyrophosphorylase, and then the glucose unit in the UDP glucose is transferred to the growing end of the chain by the action of glycogen synthetase. The reactions may be written as follows:

(1)   Glucose-1-phosphate+UTP → UDP glucose+pyrophosphate
(2)   UDP glucose+[glucosyl (1,4)]$_n$ → UDP+[glucosyl (1,4)]$_{n+1}$

When a peripheral chain of a growing glycogen molecule reaches an appropriate length (perhaps seven or more residues) as a result of successive transfers of glucosyl units from UDP glucose, the so called 'brancher' enzyme (amylo 1,4 → 1,6 transglucosidase) transfers a string of several glucosyl units from their α-1,4 attachment on the chain to an α-1,6 attachment on the same or another unit on the chain, and the α-1,6 linkage so formed constitutes a new branch point.

A characteristic abnormality in glycogen synthesis is due to a deficiency of the 'brancher' enzyme (Illingworth and Cori 1952, Brown and Brown 1966). This results in the formation of a glycogen with quite abnormal structure. It has much fewer branch points, and the outer and inner chains are abnormally long. The glycogen formed resembles amylopectin, the branched polysaccharide of starch and the disease has therefore sometimes

TABLE 6.3

Glycogen diseases (for references see text).

	Eponym	Enzyme defect	Glycogen structure	Main clinical manifestations
1.	von Gierke's disease (Cori type I)	Glucose-6-phosphatase	Normal	Enlargement of liver and kidneys; hypoglycaemia; acidosis
2.	Pompe's disease (Cori type II)	α-1, 4-glucosidase (lysosomal)	Normal	Enlargement of heart; cardio-respiratory failure
3.	Forbe's disease (Cori type III)	Amylo-1, 6- glucosidase (debrancher enzyme)	Abnormal; very short outer chains	Enlargement of liver; moderate hypoglycaemia and acidosis
4.	Andersen's disease; amylopectinosis (Cori type IV)	Amylo-(1,4 → 1,6)-transglucosidase (brancher enzyme)	Abnormal; long inner and outer chains with very few branch points	Cirrhosis of the liver
5.	McArdle's disease (Cori type V)	Muscle phosphorylase	Normal	Muscle cramps on exercise
6.	Hers' disease (Cori type VI)	Liver phosphorylase	Normal	Enlargement of liver; moderate hypoglycaemia and acidosis
7.	—	Phosphorylase kinase	Normal	Enlargement of liver
8.	—	Phosphofructokinase	Normal	Muscle cramps on exercise

'Cori types' based on classification proposed by G. Cori (1954, 1957).

been referred to as 'amylopectinosis'. The abnormal glycogen is very much less soluble than normal glycogen and tends to be precipitated in the tissues. Cirrhosis of the liver (Andersen 1956) is a characteristic feature of the disorder, and this has been attributed to a tissue reaction due to the abnormal glycogen being treated as if it were a foreign body.

6.6.2. *Defects in glycogen mobilisation*

Various enzymes are concerned in the sequential breakdown of glycogen. The situation is however complicated by the fact that an enzyme concerned with a particular step may not be the same in all tissues. Thus the successive cleavage of the α-1,4 links in the outer chains of the macromolecule is brought about by 'phosphorylase' and leads to the formation of glucose-1-phosphate. But the phosphorylase enzyme protein in muscle is different from that in liver, and a mutation may affect one and not the other. In one form of phosphorylase deficiency, known as McArdle's disease, only muscle phosphorylase is affected and the glycogen accumulation is confined to this tissue (McArdle 1951, Schmid et al. 1959). Abnormalities confined only to liver phosphorylase have also been found (Hers 1959, Koster et al. 1973). These phosphorylase deficiencies are probably an even more heterogeneous group because an apparent deficiency of liver phosphorylase can also be due to a deficiency of the enzyme phosphorylase kinase which is necessary for the conversion of the inactive dephospho form of phosphorylase to its active state (Hug et al. 1966, Huijing 1967).

Phosphorylase cleaves the α-1,4 glucosyl linkages of glycogen but cannot attack the α-1,6 linkages at the branch points. This requires the so-called 'debrancher' enzyme, amylo 1,6 glucosidase. A specific deficiency of this enzyme results in the accumulation of an abnormal glycogen with another type of unusual structure (Forbes 1953, Illingworth et al. 1956). There is an increased proportion of branch points (i.e. α-1,6 linkages), and the outer chains of the macromolecule tend to be much shorter than normal, particularly so in the fasting state. The peculiar structure is due to the fact that phosphorylase which is present in normal amounts can degrade the outer chains of the macromolecules but its action ceases as the outer tiers of branch points are approached. On the other hand lengthening of the chains by glycogen synthetase action and formation of new branch points by the brancher enzyme can still proceed in the normal way.

The main product of glycogen degradation is glucose-1-phosphate which is produced by the sequential action of phosphorylase on α-1,4 linkages. The glucose-1-phosphate is readily converted by the enzyme phosphoglucomutase to glucose-6-phosphate, and this may be metabolised via several different pathways. In the liver it is in large part hydrolysed by glucose-6-phosphatase to give rise to free glucose, which passes into the circulation, and is the main metabolic source of blood glucose. Glucose-6-phosphatase besides being present in liver also occurs in kidney, but is not found in muscle. A deficiency of this enzyme (Cori and Cori 1952) causes a characteristic form of glycogen storage disease known as Von Gierke's disease (Von Gierke 1929). In this condition glycogen accumulates in both liver and kidney which become very enlarged, but there is no accumulation in muscle. Marked hypoglycaemia is a characteristic feature, and growth tends to be severely retarded.

In muscle, glucose-6-phosphate is largely metabolised via the glycolytic pathway, most of the energy for muscular activity being derived from the breakdown of glycogen to lactate by this sequence of reactions. Phosphofructokinase which catalyses the phosphorylation of fructose-6-phosphate to give fructose-1,6-diphosphate is one of the enzymes involved, and a form of glycogen storage disease involving the gross deficiency of phosphofructokinase has been described (Tarui et al. 1965, Layzer et al. 1967). Here there is an accumulation of fructose-6-phosphate and glucose-6-phosphate as well as of glycogen itself. Clinically, a marked weakness and stiffness occurring in the muscles on vigorous or prolonged exertion is a notable feature.

There is another form of glycogen storage disease which is of special interest because it was found to be due to the deficiency of an enzyme not originally thought to be involved in glycogen degradation at all, and because it has proved to be the prototype of a number of other inherited abnormalities in which the intracellular accumulation of some complex macromolecule is the chief feature (see p.216). The condition is called Pompe's disease. Glycogen accumulation occurs in most tissues in the body, but it is particularly marked in the heart which becomes greatly enlarged (di Sant'Agnese et al. 1950). The patients seem to be normal at birth but the abnormality is rapidly progressive and they frequently die before the age of one year. Apart from the gross accumulation of glycogen, carbohydrate metabolism as such does not seem to be impaired, and all the enzymes known to be involved in the main pathways of glycogen degradation occur in normal amounts.

This was a very puzzling problem for many years, but eventually it was shown (Hers 1962) that the disease is in fact due to the deficiency of an α-1,4 glucosidase which normally occurs as one of a number of different hydrolytic enzymes in the cytoplasmic organelles known as lysosomes. This glucosidase, like other lysosomal enzymes, has a characteristically low pH optimum (about pH 4.0). It can hydrolyse maltose, linear oligosaccharides, and the outer chains of glycogen to give glucose. In Pompe's disease, electronmicroscopy of the liver has shown that most of the glycogen present is segregated in large vacuoles not seen in other forms of glycogenosis (Baudhuin et al. 1964), although some glycogen is also freely dispersed in the cytoplasm as occurs normally. The vacuoles appear to be lysosomes which are grossly enlarged and distended by glycogen accumulation. Other lysosomal enzymes are present in Pompe's disease, so that the lysosomal α-1,4 glucosidase deficiency appears to be specific. It seems probable that in the normal cell fragments of glycogen are constantly being taken up by the lysosomes and degraded. In Pompe's disease the uptake of glycogen fragments goes on but their degradation is blocked by the absence of the glucosidase, so that the organelles become grossly swollen with glycogen. This in turn leads to progressive degenerative changes in the cells. It is not clear why the effect although general is particularly marked in heart muscle.

6.7. The lysosomal storage diseases

A large number of different hydrolytic enzymes occur in most tissue cells. Many of them appear to be segregated within lysosomes and are kept apart from the rest of the contents of the cell by the thin membrane which encloses these organelles and which effectively prevents leakage of the hydrolases into the surrounding cytoplasm (De Duve 1963, Dingle and Fell 1969). Characteristically these lysosomal hydrolytic enzymes have rather low pH optima, usually between pH 3.0 and pH 6.0. Each has its own specificity directed at a particular type of linkage, and together they appear to be capable of degrading to their constituent residues a wide variety of different complex macromolecules which may be polysaccharide, lipid, protein or nucleic acid in nature. Such macromolecules may originate from the components of the cell itself or may be taken up from the outside by the process of endocytosis. Thus the lysosomes can be regarded as representing a kind of intracellular digestive system.

Hers (1965) following his discovery that the accumulation of glycogen in Pompe's disease is due to a specific deficiency of α-1,4 glucosidase, suggested that perhaps many other obscure storage diseases might have an analogous pathology. That is to say, they might be due to specific deficiencies of one or another of the many acid hydrolases which occur in lysosomes. This suggestion proved to be correct, and a large number of different rare inherited 'storage' diseases each apparently due to the deficit of a single specific lysosomal hydrolase have now been identified (Hers and Van Hoof 1973). Some of them are listed in table 6.4.

Characteristically there occurs in each of these conditions the progressive accumulation of particular kinds of complex macromolecule in the cells of various tissues. The substances which accumulate are sequestrated within the cells in lysosomal vacuoles and give rise to typical electron micrographic appearances. These substances represent intermediates in the turnover of compounds which form a normal integral part of cellular structure. They are normally degraded by being taken up into lysosomes where they are then broken down in a stepwise manner by various acid hydrolases with specificities directed at the particular linkages they contain. The central defect in each of these 'storage' diseases is a block at some point in this normal process of degradation due to the specific deficiency of one of the hydrolases concerned in the overall process. Progressive accumulation of the partially hydrolysed products occurs because while synthesis of the parent substance can still go on, degradation cannot proceed further than the step normally catalysed by the deficient enzyme. The lysosomes consequently become increasingly distended by the partially degraded materials.

As a rule the enzyme deficiency is present in most tissues of the body. But the amount of material accumulated and also its nature, may vary considerably from one tissue to another according to the distribution and normal rates of synthesis and breakdown of the parent substances in different cells; and also according to the potentialities of different cells for endocytosis. Furthermore, the rate of accumulation depends on the degree and character of the enzyme deficit. Very often the enzyme activity though much reduced is not completely absent and the rate of accumulation of the storage substance in different tissues will be affected by the amount of residual activity.

The clinical signs and symptoms of the various disorders are largely a function of the tissue sites affected and the rates of accumulation of the storage compounds. Often the affected individuals appear normal at birth though the enzyme deficiency is present, and the clinical signs of the disease may not become apparent for several months, years, or in some cases decades. The subsequent progressive deterioration appears to parallel the increasing accumulation of the storage substances.

6.7.1. Mucopolysaccharidoses

It is customary to classify these disorders according to the nature of the substances which accumulate. One series of conditions, for example, is referred to as mucopolysaccharidoses, because the principal storage compounds are acid mucopolysaccharides. Several of these disorders are listed

TABLE 6.4

Some lysosomal 'storage' diseases.

Eponym	Enzyme deficiency	'Storage' substances	References to enzyme studies
1) Hurler syndrome*	α-iduronidase	heparan and dermatan sulphates	Matalon and Dorfman 1972, Bach et al. 1972
2) Scheie syndrome*	α-iduronidase	heparan and dermatan sulphates	Wiesmann and Neufeld 1970, Bach et al. 1972
3) Hunter syndrome*	sulphoiduronate sulphatase	heparan and dermatan sulphates	Bach et al. 1973
4) Sanfilippo syndrome A*	heparan sulphate sulphatase	heparan sulphate	Kresse and Neufeld 1972
5) Sanfilippo syndrome B*	N-acetyl-α-glucosaminidase	heparan sulphate	O'Brien 1972, Figura and Kresse 1972
6) β-Glucuronidase deficiency*	β-glucuronidase	dermatan sulphate	Hall et al. 1973, Sly et al. 1973
7) Generalised gangliosidosis**	GM$_1$-β-galactosidase	GM$_1$ ganglioside; keratan sulphate	Okada and O'Brien 1968, Van Hoof and Hers 1968, Dacremont and Kint 1968
8) Tay-Sachs' disease**	N-acetyl hexosaminidase A	ganglioside GM$_2$	Okada and O'Brien 1969, Sandhoff 1969, Tallman et al. 1972
9) Sandhoff's disease**	N-acetyl hexosaminidase A *and* B	ganglioside GM$_2$	Sandhoff 1969, Okada et al. 1972
10) Ceramide lactoside lipidosis**	ceramide lactoside-β-galactosidase	ceramide lactoside	Dawson and Stein 1970

TABLE 6.4 (continued)

Some lysosomal 'storage' diseases.

Eponym	Enzyme deficiency	'Storage' substances	References to enzyme studies
11) Gaucher's disease**	glucosylceramidase	glucosylceramide (glucosylcerebroside)	Brady et al. 1965, 1966a, Patrick 1965, Brady 1968
12) Metachromatic leucodystrophy**	aryl-sulphatase A	ceramide galactose-3-sulphate (sulphatide)	Austin et al. 1963, Jatzkevitz and Mehl 1969, Ratazzi et al. 1973
13) Krabbe's disease**	galactosylceramidase	galactosylceramide (galactocerebroside)	Suzuki and Suzuki 1970, 1971
14) Fabry's disease**	α-galactosidase	ceramide trihexoside	Brady et al. 1967a, b, Kint 1970, Romeo and Migeon 1970
15) Niemann–Pick disease**	sphingomyelinase	sphingomyelin (ceramide phosphoryl choline)	Brady et al. 1966a, Sloan et al. 1969
16) α-Fucosidosis	α-fucosidase	fucose rich acid muco-polysaccharides and glycolipids	Van Hoof and Hers 1968
17) Mannosidosis	α-mannosidase	mannose rich oligosaccharides	Ockerman 1967, 1969
18) Wolman's disease	'acid' lipase	cholesteryl esters and triglycerides	Patrick and Lake 1969, Lake and Patrick 1970
19) Pompe's disease	α-1,4-glucosidase	glycogen	Hers 1962

* Mucopolysaccharidoses; ** sphingolipidoses (also see fig.6.12). For further details and references see Appendix 1.

in table 6.4 and the substances that accumulate in these particular conditions are partially degraded acid mucopolysaccharides derived from dermatan sulphate (chondroitin sulphate B) and heparan sulphate (heparitin sulphate) (Dorfman and Matalon 1972).

Dermatan sulphate and heparan sulphate are normal though minor constituents of most connective tissues. Both are however quantitatively important in the blood vessels and heart valves, and dermatan sulphate is a major component of the skin. They normally occur as large polymers, consisting of a protein core with extensive carbohydrate branches. In connective tissue they are bound to the various other proteins, such as collagen, which

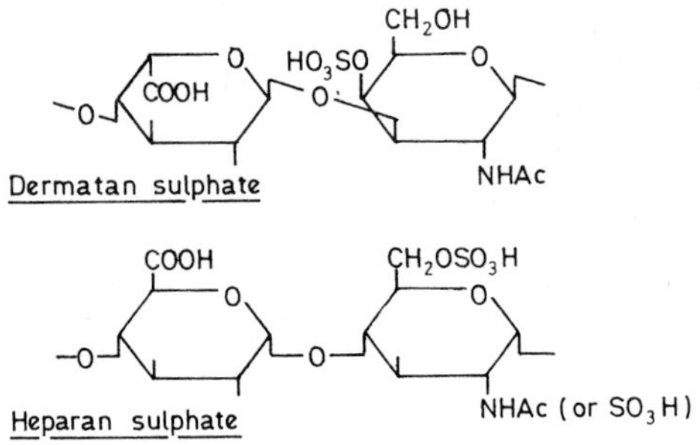

Fig. 6.10. Repeating units in the polysaccharide chains of dermatan sulphate (L-iduronic acid and N-acetylgalactosamine-4-sulphate); and of heparan sulphate (D-glucuronic acid and N-acetylglucosamine-6-sulphate).

make up the intercellular matrix. The carbohydrate chains consist of alternating uronic acid and sulphated hexosamine units (fig. 6.10). In dermatan sulphate the repeating dimer is L-iduronic acid linked to N-acetyl-galactosamine-4-sulphate. In heparan sulphate it is D-glucuronic acid linked to N-acetylglucosamine which is either sulphated in the 6-position or has N-sulphate replacing the N-acetyl. The chains however are not entirely regular. Thus a few of the L-iduronic acid residues in dermatan sulphate may be replaced by glucuronic acid, and L-iduronic acid residues may occur in the heparan sulphate chains. There may also be irregularities in the distribution of the sulphate groups. In both cases the carbohydrate chains are attached to the protein core by a xylose residue at the end of the carbohydrate chain linked to a serine residue in the protein.

In the partially degraded fragmented forms of heparan and dermatan sulphates which accumulate in the various mucopolysaccharidases, the protein core is missing. A few aminoacids are attached to some of the carbohydrate chains while other fragments have neither aminoacids nor the region of the xylose-serine linkage. The protein part of the macromolecule is evidently fully digested but the carbohydrate chains are only partially cleaved. The abnormal amount of acid mucopolysaccharide excreted in urine in these cases is also made up of such partially degraded materials. In the Hunter, Hurler and Scheie syndromes they are derivatives of both heparan sulphate and dermatan sulphate though in differing relative amounts. In the Sanfillipo syndromes (A and B) partially degraded heparan sulphate occurs, and in β-glucuronidase deficiency it is partially degraded dermatan sulphate which accumulates.

These patterns of acid mucopolysaccharide accumulation are the consequence of the different specific enzyme deficiencies. In both the Hurler and Scheie syndromes the defective enzyme is α-iduronidase (Matalon and Dorfman 1972, Bach et al. 1972); in the Hunter syndrome a sulphoiduronate sulphatase is deficient (Bach et al. 1973); in Sanfillipo A a heparan sulphate sulphatase is deficient (Kresse and Neufeld 1972); and in Sanfillipo B the defective enzyme is N-acetyl-α-glucosaminidase (O'Brien 1972, Figura and Kresse 1972).

6.7.2. Sphingolipidoses

Another major group of lysosomal storage diseases are often referred to as the sphingolipidoses because they accumulate different members of the class of substances known as sphingolipids (Volk and Aronson 1972). These substances have in common the compound known as ceramide (fig. 6.11), which consists of a long chain aminoalcohol (usually sphingosine) linked to a long chain fatty acid. They each have a further moiety which is linked to carbon-1 of the sphingosine portion of ceramide and which may contain quite a number of different residues.

For example, in the subclass of sphingolipids known as gangliosides the

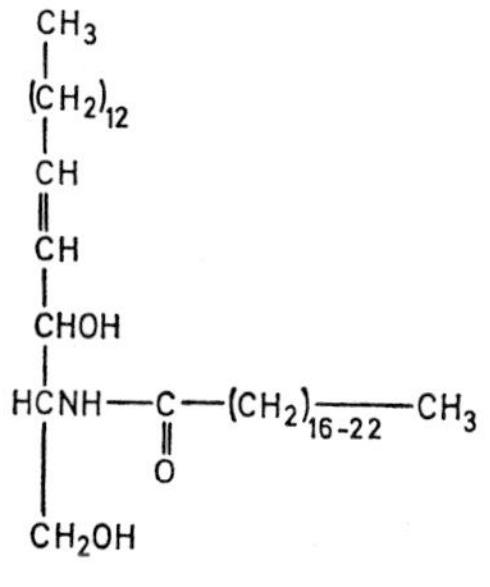

$$
\begin{array}{l}
CH_3 \\
| \\
(CH_2)_{12} \\
| \\
CH \\
\| \\
CH \\
| \\
CHOH \\
| \\
HCNH-\underset{\underset{O}{\|}}{C}-(CH_2)_{\overline{16-22}}-CH_3 \\
| \\
CH_2OH
\end{array}
$$

Fig. 6.11. Ceramide (N-acylsphingosine).

moiety attached to the ceramide consists of a polysaccharide chain with one
or more N-acetylneuraminic acid (sialic acid) residues attached to the sugars.
Gangliosides are important constituents of the cells of the grey matter of the
brain, but only occur in relatively small amounts in non-neural tissues. In
the so-called gangliosidoses accumulation of such substances in the brain
is the characteristic feature (O'Brien 1972). Typical examples are generalised
gangliosidosis in which the ganglioside known as GM_1 accumulates, and
Tay-Sachs disease in which ganglioside GM_2 accumulates (fig. 6.12). In the
normal subject, gangliosides are degraded by the sequential removal of sugar

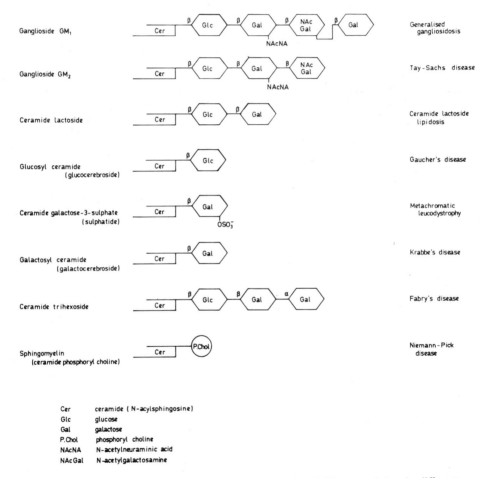

Fig. 6.12. Diagram of the structure of the main sphingolipid accumulating in different
spingolipidoses.

and N-acetylneuraminic acid residues from the molecules. GM_1 is an intermediate in this process and it accumulates in general gangliosidosis because of a deficiency of the β-galactosidase normally concerned with cleaving the terminal galactose residue from the end of the carbohydrate chain (Okada and O'Brien 1968, Van Hoof and Hers 1968, Dacremont and Kint 1968). GM_2 is the normal product of this reaction. This accumulates in Tay-Sachs disease because of a specific deficiency of the N-acetylhexosaminidase which would normally cleave the now terminal N-acetylgalactosamine residue (Kolodney et al. 1969, Sandhoff and Jatzkewitz 1972, Tallman et al. 1972). The progressive destruction of neuronal cells due to intracellular accumulation of these substances accounts for the profound cerebral degeneration which characterises both generalised gangliosidosis and Tay-Sachs disease.

Another subclass of sphingolipids, the so-called neutral glycosphingolipids, have a mono- or oligo-saccharide moiety attached to ceramide, but do not contain N-acetylneuraminic acid. These substances are widely distributed in the body and blocks at different steps of their sequential degradation account for a variety of distinct lysosomal storage diseases. Examples of substances which accumulate in some of these disorders are shown in fig. 6.12. Further subclasses of sphingolipids are represented by sphingomyelin (ceramide phosphorylcholine) which accumulates in Niemann-Pick's disease, and sulphatide (ceramide-galactose-3-sulphate) which accumulates in metachromatic leucodystrophy.

6.7.3. Tissue variations

Lysosomal hydrolases are in general specific for a particular type of linkage rather than for a particular substrate. Consequently the substances that accumulate in any one condition may be heterogeneous and there may be marked differences from one tissue to another according to the nature of the parent substances principally degraded at different sites.

A striking illustration of this is provided by the findings in generalised gangliosidosis. Here as mentioned earlier there is massive accumulation of the ganglioside GM_1 resulting in progressive neurological degeneration. However this is associated with accumulation of acid mucopolysaccharide in cells of the reticuloendothelial system in liver, spleen, bone marrow and lymph nodes (Suzuki 1968). Marked enlargement of the liver and spleen is a characteristic finding and there are also bony abnormalities. The particular mucopolysaccharide which accumulates appears to be partially degraded keratan sulphate (Suzuki et al. 1969). In this substance the carbohydrate chains mainly consist of alternating residues of N-acetylglucosamine and galactose. Evidently the β-galactosidase which is deficient, is normally concerned both in the degradation of gangliosides in the brain and in the degradation in the cells of the reticuloendothelial system of keratin sulphate derived from connective tissue. It is apparent from this example that the formal classification of lysosomal storage diseases into mucopolysaccharidoses, sphingolipidoses and so on, cannot be clear cut and is in some cases quite arbitrary. Heterogeneity of the accumulated material appears indeed to be a general phenomenon in most of these disorders although one particular substance often predominates.

A rather common finding in many of these conditions is that the specific deficiency of one lysosomal enzyme may be associated with a considerable non-specific increase in the levels of activity of other lysosomal enzymes, and this is often very marked in tissues where the accumulation of storage substance is most prominent. Thus in liver from patients with generalised gangliosidosis Van Hoof and Hers (1968) noted a marked elevation of the activities of α-fucosidase, α-galactosidase and N-acetylhexosaminidase in association with the virtual absence of β-galactosidase. Similar findings have been reported in many of the other conditions and the phenomenon appears to be a quite general one. Its nature is not entirely clear, but it is thought to be the consequence in some way of the proliferation and enlargement of the lysosomes which occurs in cells where the storage process is actively proceeding. However, some lysosomal enzyme activities appear to remain within normal limits when others are considerably elevated, and the various enzymes appear to be affected to different degrees in different conditions. Occasionally the activity of a particular lysosomal enzyme not apparently primarily affected in the particular disorder may be significantly reduced. This is so for example with regard to β-galactosidase in Hurler's syndrome in certain tissues (Van Hoof and Hers 1968, MacBrinn et al. 1969).

6.7.4. Genetic heterogeneity

Family studies indicate that most of the disorders listed in table 6.4 are inherited as autosomal recessives, the exceptions being Fabry's disease and Hunter's syndrome which are sex-linked. However, the systematic investigation of many of these conditions has uncovered a considerable degree of previously unsuspected genetical heterogeneity.

In some cases clinical syndromes formerly regarded as representing the same disease entity have been found to include two or more quite distinct enzyme abnormalities determined by mutant genes at different loci. For example, the Sanfillipo syndrome has been found to include two separate entities (Kresse et al. 1971), Sanfillipo A due to a deficiency of heparan sulphate sulphatase (Kresse and Neufeld 1972), and Sanfillipo B due to a deficiency of N-acetyl-α-glucosaminidase (O'Brien 1972, Figura and Kresse 1972). Since both these enzymes are normally concerned in the degradation of the carbohydrate chains of the mucopolysaccharide heparan sulphate, it is perhaps not surprising that the clinical and pathological findings in the two conditions should be very similar, and in fact it appears that they are indistinguishable on clinical grounds alone (McKusick 1972).

In other cases a deficiency of the same enzyme and accumulation of the same storage substance or substances are found in association with clinically distinct syndromes, which because they run true to type in separate sibships appear to be determined by different mutant genes. One example of this are the so-called adult and infantile forms of Gaucher's disease (Brady 1968, Fredrickson and Sloan 1972). Adult Gaucher's disease the commoner form, is a chronic slowly progressive disorder characterised by enlargement of the spleen, bone pain and pathological fractures. Infantile Gaucher's disease is a very much more severe and rapidly progressive condition, in which marked cerebral degeneration occurs in addition to the features shown by the adult form. The patients usually die in the first year or two in contrast to those with adult Gaucher's disease who live very much longer, usually well into adult life. In both conditions glucosylceramide accumulates because of a deficiency of glucosyl ceramidase. Much of this glucosylceramide appears to be derived from neutral glycosphingolipids in red cell stroma which is taken up into reticuloendothelial cells, as the circulating red cells age, and is there degraded. The enzyme defect in infantile Gaucher's disease is evidently much more profound than in the adult condition, since the rate of accumulation of glucosylceramide appears to be so much more rapid, but the exact nature of the difference is obscure.

The so-called infantile and juvenile forms of generalised gangliosidosis illustrate the same kind of difference in the rate of development of the clinical abnormalities (O'Brien 1972). In the infantile form, signs of psycho-motor deterioration are present at birth or shortly after and the condition runs a rapid course with death before the second year. In the juvenile form, mental and motor development may appear normal in the first year of life but shortly thereafter signs of deterioration appear and death usually comes in the first decade. In both conditions there is a gross deficiency of β-galactosidase resulting in accumulation of GM_1 ganglioside in the brain and partially degraded keratan sulphate in the viscera. But presumably the enzyme defect is functionally more severe in the infantile form so that the rate of accumulation of the storage substances is greater.

A particularly striking example of this general phenomenon came to light when it was discovered that in both Hurler's syndrome and Scheie's syndrome there was a gross deficiency of the same enzyme, α-iduronidase (Wiesmann and Neufeld 1970, Bach et al. 1972). Until then the conditions had been regarded as quite different mucopolysaccharidoses because of their marked clinical differences (McKusick 1972). In Hurler's syndrome mental deterioration, severe stunting of growth with bony deformities, gross enlargement of the liver and spleen and clouding of the corneae are characteristic features which become apparent in infancy or early childhood and usually lead to death by the age of ten. In contrast, Scheie's syndrome is a very much milder disorder. Intellectual and physical development appears normal and the patients live well into adult life. The main disabilities arise from clouding of the corneae and relatively minor abnormalities of the extremities. It has been suggested that the mutant genes which determine these two very different clinical disorders are allelic and affect in different degrees the functional efficiency of α-iduronidase (McKusick et al. 1972). Another syndrome with clinical features which can be regarded as intermediate between those of the Hurler and Scheie syndromes and which is also attributable to α-iduronidase deficiency, has also been identified. It is suggested that this represents a compound heterozygote, the affected individuals having both one allele which in homozygotes would result in the Hurler syndrome, and one allele which in homozygotes would result in the Scheie syndrome (McKusick et al. 1972).

Where a deficiency of a particular enzyme is associated with the occurrence of two or more clinically and genetically distinct conditions, one would anticipate that there are differences in the degree and character of the residual enzyme activity in the several disorders, due to different mutations having brought about different alterations in the structure of the enzyme protein with differing functional consequences. So far however the nature of such presumed functional differences has remained obscure in most cases. A major difficulty in elucidating these problems is that for technical reasons the *in vitro* enzyme studies have perforce been carried out using artificial substrates containing the same linkage as that whose hydrolysis in the natural substrate is defective in the particular condition. In practice the use of the true natural substrates for such studies is restricted because of their very limited availability and because of the relative insensitivity of the assay procedures so far developed. Although the same linkage may be hydrolysed by a particular enzyme in both the natural substrate and in artificial substrates, the kinetics of the process may be very different. Thus assay results obtained with artificial substrates such as p-nitrophenyl- or 4-methyl-umbelliferyl-glycosides which are commonly used for such studies may only partially reflect the actual functional defect which occurs *in vivo* with the natural substrates.

Another difficulty in elucidating this heterogeneity is that many of the enzymes involved appear to occur in two or more isozymic forms whose structural relationships are imperfectly understood, and these may be affected differentially in the various disorders. This point is well illustrated by the extensive studies on the isozymes of N-acetylhexosaminidase which have been carried out in relation to Tay-Sachs disease and its variants. These conditions are characterised by the accumulation of the ganglioside GM_2 (fig. 6.12) in the brain due to an inability to cleave the terminal N-acetyl-β-galactosamine residue from the carbohydrate chain of the ganglioside. The deficient enzyme is usually referred to as N-acetylhexosaminidase because it is capable of hydrolysing β-N-acetylglucosaminyl- as well as β-N-acetylgalactosaminyl-linkages. Two distinct isozymes are normally present in all tissues except red cells (Robinson and Stirling 1968, Okada and O'Brien 1969, Sandhoff 1969). They are referred to as hexosaminidase A and hexosaminidase B, and they can be readily separated by ion-exchange

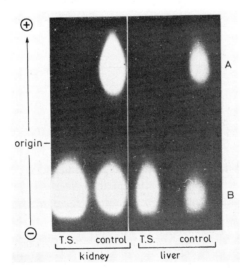

Fig. 6.13. Electrophoretic appearance of hexosaminidase A and B in kidney and liver from a normal subject and from an individual with Tay-Sachs' disease (T.S.).

chromatography, isoelectric focussing or electrophoresis. Hexosaminidase A is more acidic and also more thermolabile than hexosaminidase B, but the isozymes appear to be similar in immunological characteristics (Carrol and Robinson 1973, Srivastava and Beutler 1972, 1973) and in molecular size (Sandhoff and Wässle 1971). They also resemble one another very closely in their kinetic properties at least with artificial substrates (Sandhoff and Wässle 1971, Sandhoff and Jatzkewitz 1972).

In classical Tay-Sachs disease (also referred to as GM_2 gangliosidosis Type 1) hexosaminidase A is grossly deficient (Okada and O'Brien 1969, Sandhoff 1969). However hexosaminidase B is not reduced (fig. 6.13). It appears to be present in normal or slightly increased amounts in most tissues (e.g. liver and spleen) and it is very considerably increased in brain. In a less common variant form of the disease, now referred to as GM_2 gangliosidosis Type 2 or Sandhoff's disease, a very different situation obtains. Here hexosaminidase A and also hexosaminidase B are grossly deficient (Sandhoff 1969, Okada et al. 1972). A third form of the disorder known as GM_2 gangliosidosis Type 3 or juvenile Tay-Sachs disease has also been identified. In this condition hexosaminidase A is deficient but to a lesser degree than in classical Tay-Sachs disease, and hexosaminidase B is not reduced (Suzuki and Suzuki 1970, Okada et al. 1970).

Classical Tay-Sachs disease (GM_2 gangliosidosis Type 1) and Sandhoff's disease (GM_2 gangliosidosis Type 2) are clinically and pathologically very similar disorders despite the presence of hexosaminidase B in the former and its virtual absence in the latter (O'Brien 1972). Both conditions are characterised by rapidly progressive neurological degeneration becoming clinically apparent in the first few months of life. In both conditions the ganglioside GM_2 is the principal storage substance, although increased amounts of its asialo derivative are also found. In juvenile Tay-Sachs disease the same substances accumulate, but evidently at a slower rate because neurological damage does not become clinically apparent till between two and six years of age.

The molecular relationships between hexosaminidase A and hexosaminidase B are not yet understood. There is therefore uncertainty about the nature of the primary effects of the mutant genes which give rise to the different types of GM_2 gangliosidosis. One hypothesis is that the isozymes are multimeric and that there is one polypeptide subunit which is common to both hexosaminidase A and hexosaminidase B, and another polypeptide subunit which is present in A but not in B. The two polypeptides would be coded at separate gene loci so that a mutation at the first could be the cause of Sandhoff's disease and a mutation at the second could be the cause of classical Tay-Sachs disease (Robinson and Carroll 1972, Srivastava and Beutler 1973). Another hypothesis is that hexosaminidase A may be normally formed from hexosaminidase B and that the mutant gene in classical Tay-Sachs disease may in some way cause this conversion to be blocked (Tateson and Bain 1971).

The considerable elevation of hexosaminidase B in brain in classical Tay-Sachs disease is probably a relatively non-specific phenomenon attributable to proliferation and enlargement of the lysosomes consequential on the marked ganglioside accumulation (O'Brien 1972). Other lysosomal enzymes such as β-glucosidase, β-galactosidase and acid phosphatase are also increased in brain. In liver and spleen where lysosomal ganglioside storage is not prominent, hexosaminidase B is not so markedly elevated.

The failure of hexosaminidase B to bring about cleavage of the terminal N-acetylgalactosamine residue of GM_2 ganglioside in classical Tay-Sachs disease despite its presence in elevated amounts in the brain remains however a puzzling finding. The observations on this isozyme in Tay-Sachs disease have mainly been made using artificial substrates and it is likely that the isozyme behaves very differently with the natural substrate, GM_2 ganglioside. Thus brain extracts from classical Tay-Sachs disease have been found to show an increase in the total level of N-acetylhexosaminidase activity when this is assayed with artificial substrates presumably because of the considerably elevated amounts of hexosaminidase B, but with radioactively labelled GM_2 ganglioside as substrate a significant reduction in the total hexosaminidase activity is observed (Tallman et al. 1972).

Yet a further form of GM_2 gangliosidosis indistinguishable clinically from Tay-Sachs disease has been found by Sandhoff and Jatzkewitz (1972) who refer to it as the 'AB' variant. Here both the A and B isozymes as detected with artificial substrates are present and indeed show enhanced activity in brain. But presumably they are incapable of hydrolysing the natural substrate, GM_2, *in vivo*.

Another mutant affecting hexosaminidase A but with quite peculiar properties has also been identified (Navon et al. 1973, Vidgoff et al. 1973). Certain apparently healthy individuals were found, who showed no A activity with the use of artificial substrates. These electrophoretic isozyme patterns were indistinguishable from those seen in Tay-Sachs disease, yet there was presumably no abnormal accumulation of GM_2 in the brain. Family studies showed that these individuals were heterozygous for the usual type of Tay-Sachs allele and also for another allele, which it was inferred determines a variant of the A isozyme which is incapable of hydrolysing the artificial substrates but is able to hydrolyse the natural substrate, GM_2 *in vivo*.

6.7.5. *Tissue culture studies*

Most lysosomal enzymes may be demonstrated in fibroblast cells grown in tissue culture and this *in vitro* system has proved to be particularly useful in experimental investigations of various lysosomal storage diseases. The work of Neufeld and her colleagues using tissue cultured cells from patients with different kinds of mucopolysaccharidosis is of special interest (Neufeld and Fratantoni 1970). This work was initiated before the various specific enzyme deficiencies had been identified, and indeed before it was clear that the accumulation of acid mucopolysaccharides in these disorders is primarily due to a failure in degradation rather than to excessive synthesis. By adding radioactive sulphate to the medium in which the fibroblasts were growing it was possible to follow the processes of synthesis, secretion and degradation of the intracellular mucopolysaccharides. Comparisons of the rates of these various processes as they occur in fibroblasts from patients with the Hurler and Hunter syndromes and in fibroblasts from normal individuals, showed that the rates of synthesis, secretion and accumulation are not affected in Hurler or Hunter cells, but the rates of degradation are markedly reduced (Fratantoni et al. 1968).

The stored mucopolysaccharide in the abnormal cells was found to turn over with a half life of several days or even a week compared to about eight hours for normal fibroblasts. The effect is illustrated by the experimental results shown in fig. 6.14. Following the addition of radioactive sulphate to the medium, labelled intracellular mucopolysaccharide in normal cells increases in the first day or two and then reaches a plateau. With Hurler and Hunter cells, in contrast, the labelled mucopolysaccharide goes on accumulating progressively. If the mucopolysaccharide in the cells is pre-labelled, then the rate of decay may be followed (fig. 6.14). In Hurler and Hunter fibroblasts this is much slower than in controls. Similar results were also obtained with cells from other sorts of mucopolysaccharidosis.

A remarkable discovery was made using this general technique (Fratantoni et al. 1968). It was found that if Hurler and Hunter fibroblasts were grown together in the same culture medium, the progressive mucopolysaccharide accumulation which occurred when they were grown separately was prevented (fig. 6.14). It turned out that each produced a so-called 'corrective factor' which when taken up by the other type of cell enabled mucopolysaccharide degradation to proceed at a normal rate. The corrective factor for Hurler cells was shown to be present in medium in which Hunters' cells had been grown alone, and was also present in medium in which normal fibroblasts were grown. Similarly the corrective factor for Hunter's fibroblasts was present in medium in which either normal or Hurler cells had been grown (Fratantoni et al. 1968, 1969). Similar specific corrective factors were demonstrated for other mucopolysaccharidoses in the same way (Neufeld and Cantz 1971).

In general it appeared that each of the conditions lacked a specific factor which could be supplied by cells from either normal individuals or individuals with a different type of mucopolysaccharidosis or other disease. An exception to this general rule was the finding that no cross correction occurred between Hurler cells and cells from individuals with the Scheie syndrome, indicating a deficiency of the same factor in both these conditions (Wiesmann and Neufeld 1970). Eventually it was shown that these various corrective factors are in fact the actual enzymes which are specifically deficient in each of the disorders.

The uptake by cells deficient in a particular lysosomal enzyme of the enzyme when supplied in the extracellular medium appears to be a general

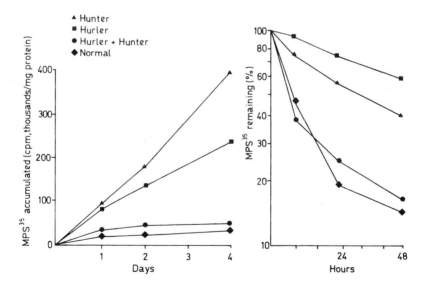

Fig. 6.14. Patterns of accumulation (left panel) and decay (right panel) of radioactively labelled sulphated mucopolysaccharide in normal, Hurler and Hunter fibroblasts, and also a mixture of Hurler and Hunter fibroblasts in the same culture (from Neufeld and Cantz 1971).

phenomenon and has been demonstrated in a number of conditions other than the mucopolysaccharidoses (e.g. metachromatic leukodystrophy, Porter et al. 1971; Wolman's disease, Kyriakides et al. 1972). The pheno-menon has of course important therapeutic implications, because it opens up the possibility that the progressive accumulation of storage substances in different lysosomal diseases might be considerably retarded if the appropriate enzyme could be supplied to the extracellular fluids in the correct form and on a continuing basis.

6.8. Heterozygotes

6.8.1. Partial enzyme deficiencies in heterozygotes

Most of the inborn errors of metabolism are inherited as 'recessive' abnor-malities. That is to say, affected individuals with the typical clinical and metabolic features of the disease usually appear to be homozygous for the particular abnormal gene, while the corresponding heterozygotes with one dose of the abnormal gene and one of its normal allele are generally quite healthy. There is a characteristic familial distribution. The disorder occurs on average in one in four of the sibs of affected individuals. It is however only rarely seen in their parents, children and other relatives. If the abnormality is sufficiently uncommon a significant increase in the incidence of cousin marriage in the parents of affected individuals may be observed.

Although apparently quite healthy the heterozygotes usually have a partial deficiency of the enzyme and often exhibit a minor metabolic disturb-ance qualitatively similar to that found in the affected patients. Such disturb-ances are usually relatively slight. They can best be studied by comparing the healthy parents and children of affected individuals with a randomly selected series of appropriate controls. These parents and children will, with rare exceptions due to fresh mutations, be heterozygotes. The randomly selected controls will if the abnormality is uncommon, almost entirely consist of homozygotes for the normal allele.

Extensive studies on these lines have been carried out on heterozygotes for the gene which in homozygotes causes phenylketonuria. Blood phenylal-anine levels in patients with phenylketonuria are considerably elevated and are usually about thirty or more times greater than those found in control subjects. In the parents of phenylketonurics there is also a significant elevation of blood phenylalanine (Hsia et al. 1957, Knox and Messinger 1958), but it is very slight, and on average the blood phenylalanine levels in the

and some of them are listed in table 6.5. In virtually every case the heterozygote appears to have a partial deficiency of the enzyme which is grossly
reduced or is undetectable in the abnormal homozygote. As a general rule
the average level of enzyme activity in the heterozygotes is intermediate
between the very low levels seen in the abnormal homozygote and the levels
found in randomly selected controls. Where for example the enzyme is
completely or almost completely absent in the affected homozygotes, the
values found in the heterozygotes are usually about 50% of those found in
normal homozygotes. Thus there often seems to be a simple gene dosage
relationship. Two doses of the normal allele in the homozygote lead to the
formation of twice as much enzyme as one dose of the allele in heterozygotes,
and there is usually no obvious compensation in activity of the normal allele
for the defective activity of the normal allele in the heterozygote. It is, however, possible that there are exceptions to this general rule, because in some
instances the limited data so far available suggest that the average level of
activity seen in heterozygotes may be greater (e.g. in the Swiss type of
acatalasia, Aebi et al. 1968) or smaller (e.g. in orotic aciduria, Fallon et al.
1964; and in homocystinuria, Goldstein et al. 1973) than would be expected
from a simple dosage relationship. Detailed analysis of these apparently
unusual situations would be of obvious interest.

Although the average level of enzyme activity in heterozygotes is generally
significantly less than that in normal homozygotes, it should be noted that
there is always considerable variation about the means. The two distributions
usually overlap so that it is not always possible to identify the heterozygote
unequivocally by determinations of enzyme level. This variation is often
largely due to extraneous non-genetic factors, and the discrimination between
heterozygotes and normal homozygotes can often be improved by their
identification and elimination from the test system. However, it is probable
that in some cases the variation is at least in part genetic in origin. It may, for
example, arise because there are in fact several different so-called 'normal'
alleles, each resulting in a distinct average activity level within the normal
range, or different 'abnormal' alleles giving different degrees of deficiency.
Another possibility is that variation in genes at other loci may affect the
overall level of the particular enzyme.

The fact that heterozygotes for genes determining most inborn errors of
metabolism are usually perfectly healthy, implies that in the corresponding
normal homozygotes the amount of enzyme present is well in excess of that
required for ordinary metabolic function. Reduction to as much as half of
its normal level may have no obvious pathological effects, so there is clearly

TABLE 6.5

'Inborn errors' in which partial enzyme deficiencies have been demonstrated by *in vitro* studies on tissues from clinically unaffected heterozygotes.

Condition	Enzyme	Tissue	References
1. Histidinaemia	Histidase (histidine deaminase)	Skin	La Du et al. (1962), Holton (1965)
2. Homocystinuria	Cystathionine synthetase	Liver, phytohaemagglutinin stimulated lymphocytes	Finkelstein et al. (1966) Goldstein et al. 1973
3. Maple syrup urine disease	Branched chain ketoacid decarboxylase(s)	Leucocytes	Goedde et al. (1966b), Dancis et al. (1965), Goedde and Keller (1967)
4. Argininosuccinic aciduria	Argininosuccinase	Erythrocytes	Tomlinson and Westall (1964)
5. Hexokinase deficiency	Hexokinase (erythrocyte isozyme)	Erythrocytes	Valentine et al. (1967)
6. Phosphohexose isomerase (glucose phosphate isomerase deficiency)	Phosphohexose isomerase (glucose phosphate isomerase)	Erythrocytes	Baughan et al. (1968), Paglia et al. (1969)
7. Triosephosphate isomerase deficiency	Triosephosphate isomerase	Erythrocytes and leucocytes	Schneider et al. (1965), Valentine et al. (1966)
8. Pyruvate kinase deficiency	Pyruvate kinase	Erythrocytes	Tanaka et al. (1962), Tanaka and Valentine (1968)
9. Diphosphoglycerate mutase deficiency	Diphosphoglycerate mutase	Erythrocytes	Schröter (1965)
10. 'Glucose-6-phosphate dehydrogenase deficiency (primaquine sensitivity, favism etc.)	Glucose-6-phosphate dehydrogenase	Erythrocytes, fibroblasts grown in tissue culture	See pp. 162–174

11. Galactokinase deficiency	Galactokinase	Erythrocytes	Gitzelmann (1967)
12. Galactosaemia	Galactose-1-phosphate uridyl transferase	Erythrocytes, leucocytes and fibroblasts grown in tissue culture	Kirkman and Bynum (1959), Donnell et al. (1960), Hugh-Jones et al. (1960), Russell and De Mars (1967)
13. Andersen's disease (glycogen storage disease type IV)	Amylo (1, 4 → 1, 6) transglucosidase	Leucocytes	Legum and Nitowski (1969)
14. Acatalasia	Catalase	Erythrocytes	Nishimura et al. (1959), Aebi et al. (1964), Aebi (1967) Scott (1960)
15. Congenital methaemo-globinaemia	NADH diaphorase	Erythrocytes	
16. Orotic aciduria	Orotidine-5'-phosphate decarb-oxylase *and* orotidine-5'-phosphate pyrophosphorylase	Leucocytes, fibroblasts grown in tissue culture	Smith et al. (1961), Fallon et al. (1964), Krooth (1964)
17. Hypophosphatasia	Serum alkaline phosphatase	Serum	Currarino et al. (1957), Rathbun et al. (1961)
18. Erythrocyte glutathione peroxidase deficiency	Glutathione peroxidase	Erythrocytes	Necheles et al. (1969)
19. Tay-Sachs' disease	N-acetylhexosaminidase A	Serum, leucocytes, fibroblasts in culture, tears	O'Brien et al. (1970), Okada et al. (1971), Kaback and Zeigler (1972), Carmody et al. (1973) Singer et al. (1973)
20. Von Gierke's disease	Glucose-6-phosphatase	Thrombocytes	Soyama et al. (1973)
21. Lesch-Nyhan disease	Hypoxanthine guanine phosphoribosyl transferase	Fibroblasts in culture, hair follicles	Rosenbloom et al. (1967), Migeon et al. (1968), Salzmann et al. (1968), Migeon (1971), Gartler et al. (1971), Silvers et al. (1972)

a considerable functional reserve. While this is no doubt true for many and perhaps most enzymes, it may not be so for all. If the level of an enzyme without such a degree of functional reserve were reduced, perhaps to half of its normal value, then this might lead to obvious pathological consequences and clinical disorder. Heterozygotes for a gene causing a gross deficiency of such an enzyme are therefore likely to exhibit some characteristic clinical disorder. Such a disease would be expected to show a typical familial distribution. It would appear to be transmitted directly from parent to child in successive generations, and on average about half the children of matings between an affected patient and a normal individual would be affected. In other words it would appear to be inherited as a so-called 'dominant' abnormality. The homozygotes for the abnormal gene may have a much more severe form of the disorder than the heterozygotes. However they would constitute only a small proportion of all cases, and indeed if the gene were rare, would quite possibly never be observed at all.

6.8.2. Acute intermittent porphyria: a 'dominant' inborn error of metabolism

Acute intermittent porphyria is an example of one condition which is probably due to a partial enzyme deficiency occurring in heterozygotes (Marver and Schmid 1972, Meyer et al. 1972). It has been recognised for a long time that the disorder, unlike most other inborn errors of metabolism, has an autosomal 'dominant' mode of inheritance, and that the affected individuals are heterozygous. Quite extensive pedigrees have indeed been constructed, showing how the condition appears to be transmitted from parent to child over several generations (Waldenström 1957), though not all the presumed heterozygotes have exhibited clinical or biochemical abnormalities. The disorder is relatively uncommon and so far no homozygotes have been identified. Indeed the homozygous state may well be lethal in foetal life. The condition is sometimes referred to as the Swedish type of porphyria because a considerable number of examples have been identified in Sweden and extensively studied (Waldenström 1937, 1957, Waldenström and Haeger-Aronson 1967). However it is not confined to Scandinavia and indeed appears to be quite widely distributed in other populations of European origin and perhaps elsewhere.

The clinical manifestations are extremely variable, but characteristically there are periodic attacks of a peculiarly severe intestinal colic, and neurological disturbances which result in irregularly distributed though often widespread paralysis and in some cases mental confusion. The acute attacks vary in severity and either the abdominal or neurological features of the disease may predominate. The symptoms usually do not occur until adult life, and they often appear to have been precipitated by the taking of certain drugs, particularly barbiturates but also sulphonamides and steroids (see also p.250).

A characteristic biochemical finding is the appearance in the urine of grossly abnormal quantities of δ-aminolaevulinic acid and porphobilinogen, which are normal intermediates in the biosynthesis of haem (fig. 6.16). These substances are formed in excessive quantities in the liver, particularly during acute attacks. They leak into the circulation and are excreted in the urine. In between attacks a moderately increased excretion of these metabolites is usually present, and increased amounts of these substances may also be

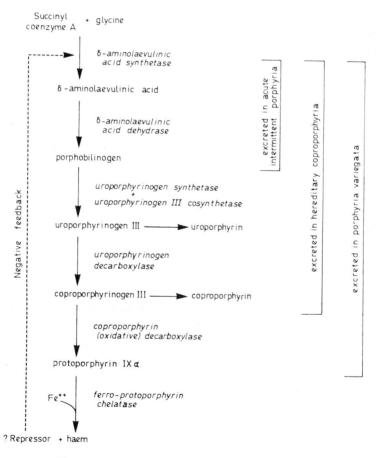

Fig. 6.16. Biosynthesis of haem (from Meyer et al. 1972).

found in symptom-free heterozygotes who have no previous history of an acute attack, though this is not invariably so. The increased quantities of δ-aminolaevulinic acid and porphobilinogen which are produced, are apparently mainly formed in liver cells, and not in the haemopoietic cells of the bone marrow where much haem synthesis is normally occurring. The immediate cause of excessive production of these metabolites in liver appears to be an increased formation of the enzyme δ-aminolaevulinic acid synthetase (fig. 6.16), since considerable elevations of the activity of this enzyme have been found in liver samples from patients with acute intermittent porphyria (Tschudy et al. 1965, Nakao et al. 1966, Dowdle et al. 1967).

Two other inherited forms of hepatic porphyria, porphyria variegata and hereditary coproporphyria, closely resemble acute intermittent porphyria in certain of their features. In particular they appear to be inherited as autosomal 'dominants'; they are characterised clinically by acute episodes of neurological dysfunction often precipitated by barbiturates or other drugs; the acute attacks are associated with an excessive urinary excretion of δ-aminolaevulinic acid and porphobilinogen; and they show increased hepatic activity of δ-aminolaevulinic acid synthetase (Marver and Schmid 1972). However they differ from acute intermittent porphyria in certain of their other clinical and biochemical manifestations; in particular by the urinary and foetal excretion of endogenously formed porphyrins (fig. 6.16). They are presumably determined by mutant genes at other loci.

The primary enzyme defect in acute intermittent porphyria is most probably in the enzyme uroporphyrinogen synthetase (Meyer et al. 1972). This enzyme is rate limiting in the conversion of porphobilinogen to uroporphyrinogen, but requires the presence of a second enzyme, uroporphyrinogen III co-synthetase to form the natural intermediate in haem biosynthesis, uroporphyrinogen III. A significant reduction of uroporphyrinogen synthetase in liver samples from patients with acute intermittent porphyria was demonstrated by Strand et al. (1970). Meyer et al. (1972) subsequently showed that this apparently specific enzyme deficiency also occurred in red cells in patients with a clinical history of acute intermittent porphyria, though not at the time in an acute phase. The same defect was also found in some of their immediate relatives who had not yet shown any clinical abnormality. Furthermore it was found that such individuals when given an oral dose of δ-aminolaevulinic acid showed an abnormal metabolic response resembling that seen in the acute phase of acute intermittent porphyria.

On average the level of activity of uroporphyrinogen synthetase in hetero-zygotes for the gene which determines acute intermittent porphyria is probably about half that in normal individuals. However it varies somewhat from one individual to another, and this may in part account for the variability in the clinical manifestations. The variation in clinical severity is also dependent on whether or not particular drugs have been taken and on their dosage.

The elevation of δ-aminolaevulinic acid synthetase which is a charac-teristic feature of the disorder is probably a secondary phenomenon. The synthesis of this enzyme which is the first and rate controlling enzyme in haem biosynthesis, appears to be normally regulated by a negative feed back system involving haem (see Marver and Schmid 1972) and it is likely that this is affected in liver by the partial block in haem biosynthesis due to reduced uroporphyrinogen synthetase. It has also been shown by a variety of experimental studies in laboratory animals (Marver and Schmid 1972) and in tissue culture (Granick 1966), that increased synthesis of δ-amino-laevulinic acid can be induced by a number of different substances, some of which are known to precipitate acute episodes of intermittent porphyria in man. Such induction of increased δ-aminolaevulinic acid synthetase occurring in individuals who already have a partial deficiency of uroporphyrinogen synthetase may well so overload the system as to result in the observed metabolic and clinical abnormalities. However it is not yet clear exactly how the various secondary biochemical manifestations of the disorder, such as the excessive formation of δ-aminolaevulinic acid and porphobilinogen give rise to the observed clinical effects.

Nor have the primary enzyme defects in porphyria variegata or hereditary coproporphyria been identified, although it seems likely that as in acute intermittent porphyria they produce abnormalities affecting the biosynthetic pathway leading to haem. The excessive synthesis of δ-aminolaevulinic acid synthetase in these disorders is probably a secondary phenomenon, as in acute intermittent porphyria.

6.8.3. *Heterozygotes in X-linked inborn errors of metabolism*

Partial enzyme deficiencies in heterozygotes are found not only in inborn errors of metabolism determined by mutants at loci on one or other of the autosomal chromosomes, but also in conditions determined by mutants at loci on the X-chromosome. However because of the phenomenon of X-chromosome inactivation in females (see p. 174), the fundamental basis of the situation is usually quite different in the two cases. In autosomal conditions all the cells which produce the enzyme in the normal organism will have a partial deficiency in the heterozygote. In heterozygotes for X-linked conditions however, some of the cells will produce normal amounts of the enzyme because the normal allele is on the 'active' X-chromosome, while other cells will have a gross deficiency of the enzyme because here the 'active' chromosome will be carrying the mutant allele. Consequently whether or not the partial enzyme deficiency present in heterozygotes for X-linked conditions results in overt clinical abnormality is likely to depend on different factors from those involved in autosomal conditions.

Although on average half the cells in heterozygotes will have an 'active' X with the mutant allele, the actual proportion of such cells will vary in different tissues of the same individual, and also in the same tissue in different individuals, and this may have an important effect on the severity and degree of the metabolic upset and on any consequent clinical abnormalities.

The disorder due to ornithine carbamoyl transferase deficiency (p. 209) illustrates the general point. The abnormality is evidently due to a mutant at an X-chromosomal locus. In affected males there is an almost complete deficiency of the enzyme and this results in a rapidly fatal condition. In heterozygous females the clinical abnormality occurs, but is less profound and rather variable so that while some females show a moderately severe disorder there are others who appear to be clinically unaffected. Presumably this variation largely depends on the actual proportion of liver cells with the mutant allele on the 'active' X-chromosome.

The clinical manifestations of an X-linked mutant may also be influenced by two types of effect which can occur in cell populations where some cells lack a particular enzyme which is present in normal amounts in other cells. One of these involves the process of cell selection in which those cells with the normal allele on the 'active' X-chromosome are favoured because they do not have the metabolic defect present in the other cells (e.g. Nyhan et al. 1970, see p. 178). The other involves the process which has been called 'metabolic co-operation', in which it appears that enzyme molecules are transferred from cells containing the enzyme to those lacking it, provided they are immediately contiguous (Subak-Sharpe et al. 1969, Cox et al. 1970). The transfer of lysosomal enzymes from one cell to another via the extra-cellular fluid (see p. 231) is another possible mechanism of 'metabolic co-operation'.

6.9. *Defects in active transport systems*

The passage of small molecular weight substances in and out of cells across cell membranes appears to depend to a considerable extent on a variety of more or less specific active transport systems. For example absorption of foodstuffs by the organism requires the transport of the various substances which are the products of gastric and intestinal digestion across the mucosal cells of the jejunum and small intestine. The processes involved are often highly specific either for only a single substance or for a small number of related substances, and they require an energy supply which is generated metabolically in the cell. Similar specific active transport systems are also known to occur in the cells lining the renal tubules. In the kidney, blood plasma is constantly being filtered by the glomeruli, so that the glomerular filtrate contains essentially all the small molecular weight constituents of blood plasma. As the fluid passes down the renal tubules, selective reabsorption of many of these substances across cells lining the renal tubule takes place. In this way many of the small molecular weight constituents of plasma are retained by the organism while others are excreted in the urine. There are also a variety of other active transport systems in the organism which are concerned with maintaining concentration gradients of particular substances across the membranes of different cells, so producing a characteristic and more or less constant intracellular distribution of metabolites.

A number of genetically determined and quite specific defects of such active transport systems have been identified. Their precise molecular basis has not yet been clarified but they seem to involve defects of particular enzymes or 'carrier' proteins, which are essentially analogous to the enzyme deficiencies which result in blocks in intermediary metabolism in the more classical forms of inborn errors of metabolism. The best studied abnormalities involve defects in the active transport of aminoacids, but specific transport defects of other substances such as glucose (as in renal glycosuria) are also known to occur and no doubt many more remain to be identified.

6.9.1. *Cystinuria*

This condition was first recognised in the last century as a disorder characterised by the tendency to form renal stones composed almost entirely of the aminoacid cystine. Individuals so affected were found to excrete cystine continuously in large amounts in the urine, and the abnormality was originally included by Garrod among the conditions which he called 'inborn errors of metabolism'. At that time and indeed for many years afterwards it was thought that the disorder was a block at some point in the normal catabolism of cystine leading to accumulation of the amino-acid, and hence to its excretion in abnormal quantities in the urine. Eventually however it was shown that this is not the case, and that the abnormal urinary excretion was essentially due to a defect in the renal tubular reabsorption of cystine from the glomerular filtrate (Dent and Rose 1951). Blood plasma levels of cystine were found not to be elevated as would be expected if accumulation were occurring, but were lower than normal. Also it emerged that cystine

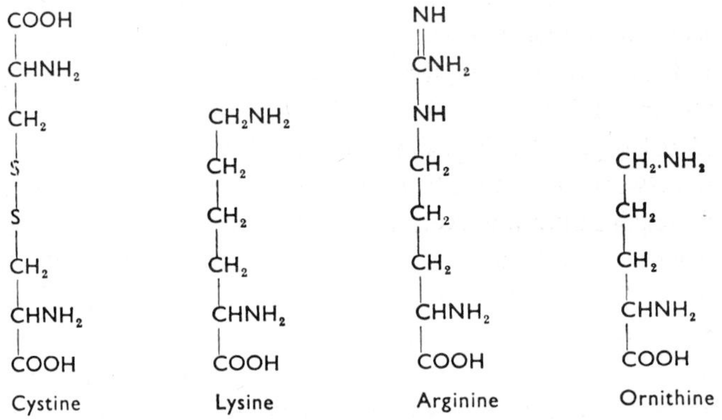

Fig. 6.17. Aminoacids characteristically excreted in excess in cystinuria.

was not the only aminoacid involved. The dibasic aminoacids lysine, arginine and ornithine (fig. 6.17), were also found to be continuously excreted in large amounts by cystinuric patients, and the renal clearances of these aminoacids like that of cystine were greatly increased (Dent et al. 1954, Arrow and Westall 1958, Doolan et al. 1957). Thus there appeared to be a specific defect in renal tubular transport of four aminoacids normally present in blood plasma, cystine, lysine, arginine and ornithine, but not of others.

Subsequently a similar abnormality in active transport was also shown to occur in the mucosal cells of the small intestine in cystinuric patients. It results in a reduction in the rate of absorption of these aminoacids from the gut. This was first demonstrated by feeding experiments in the whole organism (Milne et al. 1961), but the specific character of the abnormality was subsequently investigated by *in vitro* studies on the uptake of these and other amino-acids by small pieces of jejunal mucosa obtained by biopsy from cystinuric and control subjects (McCarthy et al. 1964, Thier et al. 1964, 1965). These experiments confirmed the existence of a common transport defect for the four aminoacids in cystinuria. They also demonstrated competitive inhibition of the uptake of one aminoacid in this group of aminoacids by another.

It is of interest that neither the defective rates of intestinal absorption, nor the abnormal excretion in large quantities of these four aminoacids in the urine, usually lead to any nutritional abnormality in growth or development. Evidently the usual protein intake is well in excess of minimal requirements. All the clinical features of the disorder are in fact due to cystine calculus formation and the consequent obstruction of the renal tract with secondary kidney damage. The daily urinary excretion of the four aminoacids in cystinuric subjects on average amounts to about 2.0 g lysine, 1.0 g arginine, 0.75 g cystine and 0.4 g ornithine. Cystine happens to be relatively insoluble, and in urine between pH 5.0 and pH 7.0 it is kept in solution only to the extent of about 0.3–0.4 g per litre (Dent and Senior 1955). In patients who may be excreting between 0.5 and 1.0 g per day, the urinary concentration will frequently reach saturation levels, particularly at night when the urine passed is most concentrated. The cystine therefore tends to come out of solution, and this leads to calculus formation. Lysine, arginine and ornithine are, however, all very soluble and therefore do not form calculi. Because of variations in fluid intake and other physiological differences, the propensity of different cystinuric patients to form calculi, even when they are excreting similar quantities of cystine, appears to be very variable. Some may go for years without trouble, others develop renal symptoms and damage in early life.

Detailed genetical analysis of cystinuria was originally carried out by measuring the quantities of the different aminoacids in urine samples from patients and their relatives. It soon became evident that the condition was genetically heterogenous (Harris and Warren 1953, Harris et al. 1955a, b). In one group of families, individual members showed either a grossly abnormal excretion of cystine, lysine, arginine, ornithine, or their excretion of these aminoacids was quite normal. The familial distribution indicated that the affected individuals were most probably homozygous for an abnormal autosomal gene. The presumed heterozygotes were indistinguishable from homozygotes for the normal allele. This condition was therefore called 'recessive' cystinuria.

In other families a quite different situation was found. Three types of individuals could be identified; those with a grossly abnormal excretion of cystine, lysine, arginine and ornithine; those with a moderately increased output of cystine and lysine but little or no increase in arginine and ornithine excretion; and those with normal excretion rates. Their occurrence in the families made it apparent that individuals with greatly increased output of all four aminoacids were likely to be homozygous for an abnormal gene, while individuals of the intermediate type with moderately increased cystine and lysine excretions were heterozygotes. This condition was referred to as 'incompletely recessive' cystinuria.

Thus it seemed that at least two different abnormal genes could cause cystinuria. The two sorts of homozygote showed on average a similar degree of abnormality in urinary aminoacid excretion and were indistinguishable phenotypically. However, while heterozygotes for one of the genes showed no obvious abnormality, heterozygotes for the other displayed a partial defect in renal tubular aminoacid reabsorption. The quantities of cystine and lysine excreted in heterozygotes of the 'incompletely recessive' type are very variable though closely correlated. Occasionally the cystine levels may be sufficiently high as to lead to calculus formation. But this is unusual, whereas in both sorts of abnormal homozygotes it is a frequent occurrence.

Studies of the active uptake of these aminoacids in jejunal biopsy material obtained from different patients have revealed further complexity in the genetics (Rosenberg et al. 1966, Rosenberg 1966). Three distinct types of

TABLE 6.6

Distinction between three types of cystinuria based on the degree of active transport of cystine, lysine and arginine observed by *in vitro* studies on jejunal biopsies from homozygotes, and on the urinary excretion of cystine and lysine observed in the corresponding heterozygotes.

	Active intestinal transport in homozygotes			Excretion of cystine and lysine in the urine in heterozygotes
	Cystine	Lysine	Arginine	
Controls	7.0 ± 1.4	11.2 ± 1.6	28.3 ± 1.3	Normal
Cystinurics				
Type I	1.1 ± 0.2	1.0 ± 0.3	0.9 ± 0.2	Normal
Type II	2.4 ± 0.2	1.0 ± 0.2	Not tested	Increased
Type III	4.1 ± 2.8	4.2 ± 3.0	6.6 ± 3.3	Increased

The degree of intestinal active transport is expressed as a distribution ratio. A value of 1.0 indicates the absence of active transport of the particular aminoacid. Values greater than 1.0 indicate the occurrence of active transport of varying degrees (for details see Rosenberg et al. 1966, Rosenberg 1966).

homozygotes have been defined by such investigations (table 6.6). Type I corresponds to the homozygote for so called recessive cystinuria, because the corresponding heterozygotes show no abnormality in urinary aminoacid excretion. Types II and III correspond to homozygotes for what was previously called 'incompletely recessive' cystinuria. In Type I virtually no uptake of the specific aminoacids can be detected in the jejunal material. In Type II there is slight activity, and in Type III there is a significant degree of uptake though on average less than in non-cystinuric controls. Types II and III also differ in the degree of aminoaciduria shown by the corresponding heterozygotes. In Type II this is somewhat more marked than in Type III.

Family studies support the idea that at least three distinct abnormal genes are indeed involved. Each appears to affect the active transport process in a specific way. Furthermore, in several instances (Rosenberg 1967, Morin et al. 1971) it has been possible to show by family studies that certain individuals with gross excretion of all four aminoacids are probably heterozygous for different combinations of these abnormal genes (e.g. I-II, I-III and II-III). This suggests that the genes may be allelic and each alters but in a different manner the structure of a single enzyme or carrier protein. However, the nature of this postulated protein is not known, and it may well be that if it contained two non-identical polypeptide chains its functional properties could be modified by mutations at more than one locus. One point of particular interest, though at present unexplained, is that the gene for Type I cystinuria, which apparently in homozygotes produces the most marked transport deficit in the jejunum, does not in heterozygotes lead to any obvious renal abnormality; whereas the genes for Type II and Type III cystinuria, which have apparently a less profound effect in the jejunum in homozygotes, cause significant renal abnormalities in heterozygotes. Also the renal defect, as indicated by the abnormal aminoacid excretion, does not obviously differ between the three types of homozygotes.

6.9.2. Other aminoacid transport defects

Another genetically determined transport defect has been shown to involve specifically the aminoacids glycine, proline and hydroxyproline (Scriver 1968, Rosenberg et al. 1968). All three aminoacids are excreted in abnormally large amounts in the urine of affected homozygotes evidently because of a specific abnormality in renal tubular reabsorption (table 6.7). There appear however to be no untoward clinical consequences. In heterozygotes an increased renal clearance and an abnormal excretion of only glycine is found. The findings suggest that in the tubular reabsorption of aminoacids from the

TABLE 6.7

'Hereditary renal iminoglycinuria'. Estimated percentage tubular reabsorption of proline, hydroxyproline and glycine from glomerular filtrate in heterozygotes and homozygotes (Scriver 1967).

	Percentage renal tubular reabsorption from glomerular filtrate		
	Normal	Heterozygote	Homozygote
Proline	99.8	99.8	80
Hydroxyproline	100	100	65
Glycine	93	84	65

glomerular filtrate in the kidney, these three aminoacids share a common transport system. It seems that some defect in an enzyme or 'carrier' protein specific to the pathway is likely to be the cause of the particular abnormality. Defective intestinal absorption of proline has also been observed in this condition (Goodman et al. 1967).

A further and somewhat larger group of aminoacids appear to share a common transport pathway in the cells of the renal tubules and probably also in the intestinal mucosa. They include alanine, serine, threonine, asparagine, glutamine, valine, leucine, isoleucine, phenylalanine, tyrosine, tryptophan, histidine and citrulline. They are all excreted in excess in the urine in a condition known as Hartnup disease, in which there is presumably some defect in this common transport system.

Hartnup disease (Jepson 1965) was originally described under the title 'hereditary pellagra-like skin rash with temporary cerebellar ataxia, constant renal aminoaciduria and other bizarre biochemical features' (Baron et al. 1956). This is an apt description, though the symptomatology is somewhat variable from case to case. The other biochemical features referred to are the urinary excretion in varying amounts of certain tryptophan derivations, notably indoxyl sulphate and indolylacetic acid. These are probably derived from excessive bacterial breakdown of tryptophan in the gut consequent on its delayed absorption (Milne et al. 1961). It is possible also that the various clinical features of the disease may be caused by toxic products of bacterial degradation of tryptophan or other aminoacids in the gut, but their detailed causation is not understood. Individuals with Hartnup disease, although variable in symptomatology, all show the same very characteristic pattern of aminoacid excretion. Their distribution in families indicates that they are homozygous for a rare autosomal gene, but so far no detectable abnormality has been found in presumptive heterozygotes.

Thus these various abnormalities define three different aminoacid transport systems in the kidney and probably also in the gut. These are concerned with distinct groups of aminoacids which comprise cystine, lysine, arginine and ornithine; glycine, proline and hydroxyproline; and the large group of mainly mono-amino mono-carboxylic aminoacids excreted in excess in Hartnup disease. Each of these systems can evidently be specifically blocked (at least partially) by an appropriate mutation. There is however, evidence for the kidney that other systems concerned with the transport of some of these aminoacids also occur (Rosenberg et al. 1967, Scriver 1967, Scriver and Hechtman 1970). These systems may be specific for only one (or some) of the aminoacids in the three group transport systems mentioned above, and are probably affected by mutations at other loci. They probably differ in their capacities and in their kinetics, so that any particular aminoacid may, to a greater or less extent according to the particular conditions, be actively transported by more than one system. Thus a variety of effects can be expected from different mutations.

6.10. *'Inborn errors' of drug metabolism*

6.10.1. *Enzyme deficiencies and pharmacological aberrations*

It has long been recognised that the administration to certain individuals of particular drugs in normal doses sometimes results in a markedly abnormal response, which may have very undesirable clinical consequences. Certain of these so-called drug idiosyncrasies have been shown to be due to specific inherited enzyme defects, essentially similar to those occurring in the more classical types of inborn errors of metabolism.

The enzyme involved may be concerned in the normal metabolism of the drug, and its defective action can result in the drug, or one of its pharmacologically active derivatives, persisting at much higher concentrations in the body than is normally the case, so that the effects of a standard dose are excessively prolonged. A well known example of this, which has already been discussed (pp. 150–162), is the abnormally prolonged period of respiratory paralysis which occurs in certain individuals when the drug suxamethonium is administered to obtain muscular relaxation. Serum cholinesterase is the defective enzyme, and it has been shown that the excessive sensitivity to the drug is due to particular alleles which result either in the synthesis of an abnormal form of the enzyme with unusual kinetic properties, so that it is not capable of hydrolysing the suxamethonium at an appreciable rate under the conditions that are obtained after the drug's administration, or which result in a complete deficiency of the enzyme. In these circumstances the drug persists in the unhydrolysed and pharmacologically active form at relatively high concentrations for a prolonged period.

In other cases the specific enzyme defect may be the cause of the drug idiosyncrasy because it sets up an abnormal metabolic situation in particular cells, so that they are not capable of dealing effectively with certain secondary pharmacological effects of the drug which under normal metabolic conditions are not harmful. Examples of this kind of phenomenon are provided by the series of structurally abnormal forms of glucose-6-phosphate dehydrogenase which in different ways give rise to a marked deficiency of glucose-6-phosphate dehydrogenase activity in red cells (pp. 162–174). In the commoner types of glucose-6-phosphate dehydrogenase deficiency there is generally no clinical abnormality under ordinary conditions, even though red cell metabolism must be in some degree disturbed. However, when particular drugs, such as the antimalarial compound primaquine, or certain sulfonamides (see also table 5.5, p. 164) are administered, the unusual metabolic state of the red cells becomes accentuated and premature destruction causing haemolytic anaemia ensues.

The rare metabolic diseases, acute intermittent porphyria, porphyria variegata and hereditary coproporphyria (see p. 238) provide further examples of how individuals with particular genetic constitutions may react

TABLE 6.8

Some drugs known or reported to have precipitated attacks of acute intermittent porphyria (Elder et al. 1972).

Antipyrine	Isopropyl dipyrone
Barbiturates	Methyldopa
Chlordiazepoxide	Meprobamate
Chlorpropamide	Oestrogens (natural and synthetic)
Diazepam	Pentazocine (Fortral)
Dichloral phenazone	Progestogens
Ergot preparations	Sedormid
Glutethemide	Succinimides
Griseofulvin	Sulfa drugs
Hydantoins	

unfavourably to the administration of certain drugs. In these cases a wide range of drugs which include barbiturates, sulfonamides and steroids (table 6.8) may precipitate severe clinical reactions, when administered in doses which for normal individuals are quite acceptable. Exactly how such drugs produce their effects in these conditions is not known. It may be a consequence of the induction of δ-aminolaevulinic acid synthetase activity which could result in a grossly abnormal metabolic disturbance in individuals who have a partial deficiency of one or another of the enzymes involved in haem biosynthesis in the liver.

6.10.2. Isoniazid inactivation

In some cases genetically determined differences in the activity of a certain enzyme may result in quite marked differences between individuals in the manner in which they metabolise a particular drug, although this may not be associated with any acute clinical consequences. An illustration of this is provided by the differences which have been found to occur in the acetylation of the drug isoniazid, which is widely used in the chemotherapy of tuberculosis. In its acetylated form, isoniazid is much less active therapeutically and is also less toxic, so that the drug is effectively inactivated by acetylation.

Shortly after the use of isoniazid in tuberculosis therapy was introduced, it

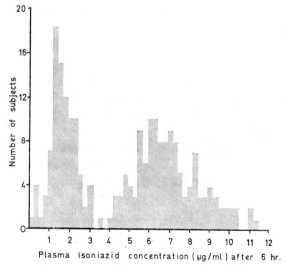

Fig. 6.18. Distribution of plasma isoniazid levels in 220 different subjects six hours after an oral dose of isoniazid (Evans et al. 1961). In this study the dose of isoniazid administered was 40 mg/kg of 'metabolically active mass' ($Wt^{0.7}$).

was found that there are considerable differences between individuals in the rate at which isoniazid is inactivated by acetylation, though in any one individual the metabolism of the drug appears to be remarkably constant (Hughes et al. 1954, Bell and Riemensnider 1957). It was then shown that people can be differentiated into two more or less sharply distinct groups according to the rate at which the inactivation proceeds. For example, if a standard dose of isoniazid is administered to a randomly selected series of individuals and the blood level measured a few hours later, the distribution of blood levels is clearly bimodal (fig. 6.18). Individuals can be readily classified into two distinct groups: 'rapid inactivators' in whom the blood level is relatively low a few hours after taking the drug, and 'slow inactivators' in which it is relatively high (Knight et al. 1959, Evans et al. 1960, 1961). In the 'rapid inactivators' a much higher proportion of the drug is excreted in the acetylated form in the urine than in the 'slow inactivators', in whom it is mainly excreted unacetylated.

Family studies demonstrate that the differences are genetically determined, and the results can largely be accounted for in terms of two common alleles, such that 'slow inactivators' represent the homozygotes for one allele, and 'rapid inactivators' are either heterozygotes or homozygotes for the other (Knight et al. 1959, Evans et al. 1960). It is probable that the rate of inactivation of the drug is somewhat more rapid in the homozygous 'rapid inactivators' than in the heterozygotes (Dufour et al. 1964).

Acetylation of isoniazid is brought about by an acetyltransferase enzyme which occurs in the liver and which is concerned in a reaction by which an acetyl group from acetyl-coenzyme A is transferred to isoniazid (fig. 6.19). Assay of acetyltransferase activity in liver samples obtained by biopsy has shown striking differences in level between 'rapid' and 'slow' inactivators, the activity being on average much higher in the former group than in the latter (Evans and White 1964). Similar results have also been obtained with autopsy specimens (Jenne 1965). Semipurified preparations of the enzymes

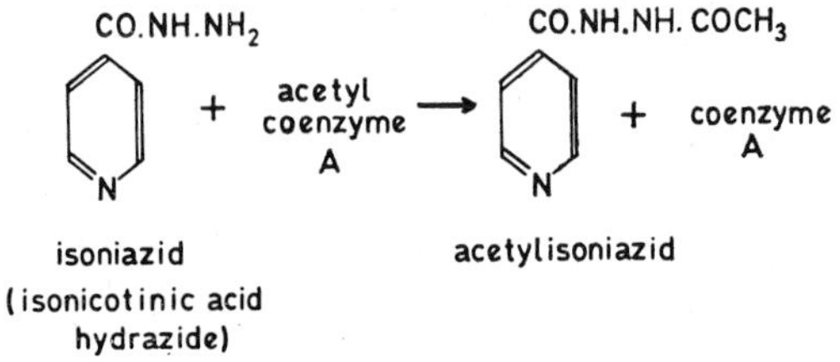

Fig. 6.19. Acetylation of isoniazid.

obtained from 'rapid' and 'slow' inactivators appeared to be closely similar in a number of properties, such as Michaelis constants and substrate specificities (Jenne 1965), which suggests that the difference between the two types may depend on the amount of the enzyme protein actually present in the liver cells, rather than on differences in its specific activity.

The enzyme has also been shown to be concerned in the acetylation of certain other drugs, such as sulfamethazine and hydralazine (Evans and White 1964) sulphapyridine (Schröder and Evans 1972) and dapsone (Gelber et al. 1971). However, other compounds normally acetylated in the body, such as sulfanilamide (Evans and White 1964) or p-aminosalicylic acid (Jenne 1965) are not apparently dealt with by the same enzyme. These *in vitro* findings correlate with the observation that subjects receiving sulfamezathine show significant differences in the proportions of the free and acetylated forms of the drug excreted in the urine according to whether they are 'slow' or 'rapid' inactivators of isoniazid (Evans and White 1964), but the same kind of differences are not found in individuals receiving sulfanilamide (Peters et al. 1965) or p-aminosalicylic acid (Jenne 1965).

. The existence of these marked individual differences in isoniazid inactivation raises the question of its significance in the therapeutic use of the drug in tuberculosis. In comparisons of large groups of patients on standardised anti-tuberculous treatment including isoniazid, significant differences have not usually been found between the results of treatment in 'rapid' and 'slow' inactivators (Evans 1963). But while there may be little or no difference when the drug dosage scheme is optimal it is likely that if it is suboptimal, for example when the isoniazid is given too infrequently, then differences in response may occur (Tuberculosis Chemotherapy Centre 1970, Ellard et al. 1972a).

Slow inactivators of isoniazid appear to be somewhat more likely than 'rapid' inactivators to develop peripheral neuropathy which is one of the main complications which may occur in prolonged isoniazid therapy, and which is evidently due to a toxic side effect of the drug (Devadatta et al. 1960, Evans and Clarke 1961). The development of peripheral neuritis as a complication of isoniazid treatment is, however, now rare because it can be prevented by the simultaneous administration of pyridoxine.

Differences between 'rapid' and 'slow' inactivators in the toxic or therapeutic effects of other drugs acetylated by the same enzyme have not yet been very extensively studied. However in the use of phenelzine (β-phenylethylhydrazide) in the treatment of depression, it appeared that severe adverse effects of the drug were somewhat more common among 'slow'

inactivators of isoniazid than among 'rapid' inactivators, though the therapeutic response appeared to be much the same in the two groups (Evans
et al. 1965). Some differences have also been noted between rapid and slow
inactivators with respect to the toxic effects of hydralazine used in hypertension (Perry et al. 1970). Differences in response to dapsone in the treatment of leprosy have not been observed (Ellard et al. 1972b).

It is of some interest that the proportion of 'rapid' and 'slow' inactivators
of isoniazid has been found to vary quite widely in different human populations. Among Europeans and Blacks about half the population appear to
be 'slow' inactivators (Evans et al. 1960), whereas among Japanese (Sunahara
et al. 1961) the proportion is much less – about 10%. Thus among Europeans
and Blacks it appears that the allele causing a deficiency of the enzyme must
be more than twice as common as the allele responsible for relatively high
enzyme activity (table 6.9). This of course is in striking contrast to the
situation with genes causing analogous enzyme deficiencies in the more
classical types of 'inborn errors of metabolism' which are, in general,
extremely infrequent.

Some allele frequencies in different populations are given in table 6.9.

TABLE 6.9

Frequencies of alleles determining 'rapid' and 'slow' inactivation of isoniazid in different populations.

Population	No. tested	Allele determining 'rapid' inactivation	Allele determining 'slow' inactivation	References
U.S.A. Whites	200	0.28	0.72	Evans et al. (1960)
U.S.A. Blacks	91	0.27	0.73	
U.S.A. Whites	105	0.24	0.75	
U.S.A. Blacks	116	0.27	0.73	Dufour et al. (1964)
Japanese	209	0.71	0.29	
Japanese	1,808	0.66	0.34	
Ainu	86	0.69	0.31	Sunahara et al. (1961)
Korean	65	0.67	0.33	
Thai	108	0.46	0.54	
Burmese	121	0.39	0.61	Smith and Kyi (1968)
Indian (South)	321	0.22	0.78	
Indian (Madras)	529	0.18	0.82	Tuberculosis Chemotherapy Centre (1970)
Alaskan Eskimos	157	0.54	0.46	Scott et al. (1969)
Alaskan Indians	47	0.38	0.62	

6.11. '*Inborn errors*' *with increased enzyme activity*

The great majority of inborn errors of metabolism are due to deficiencies of specific enzymes. But sometimes a mutation may result directly in the increased activity of a specific enzyme, and this, by disturbing the regulation of particular metabolic pathways, may give rise to overt pathological consequences.

An interesting example of one such abnormality is provided by a mutant allele determining a structurally variant form of the enzyme phosphoribosyl-pyrophosphate synthetase (Becker et al. 1973a, b). This enzyme is concerned in the synthesis of 5-phosphoribosyl-1-pyrophosphate (PRPP) from ATP and ribose-5-phosphate. The PRPP formed is a substrate in the PRPP amidotransferase reaction which is the first reaction committed to *de novo* purine synthesis, and which is probably rate limiting for this pathway (see fig. A.10 p. 388). PRPP is also a substrate in the phosphoribosyl transferase reactions which constitute the so-called salvage pathway of purine nucleotide synthesis.

The abnormality in PRPP synthetase activity was discovered in the course of studies on patients with gout due to overproduction of uric acid (Becker et al. 1973a). Two brothers with this syndrome were found to have levels of PRPP synthetase activity in their red cells and also in fibroblast cells grown in tissue culture, which were more than twice as high as those found in control subjects or in other patients with gout due to uric acid overproduction. A daughter of one of the affected patients also showed elevated levels of PRPP synthetase, though she did not have gout or hyperuricaemia. The level of other enzymes involved in purine metabolism such as hypoxanthine guanine phosphoribosyl transferase (p. 351) and adenine phosphoribosyl transferase appeared to be normal in these individuals.

Studies on intact fibroblasts showed the elevated intracellular PRPP synthetase levels were associated with an actual increase in rate of production of PRPP. This could account *in vivo* for the excessive rate of uric acid formation which was estimated in the patients to be four or fivefold that in normal individuals.

Studies on the molecular basis of this enzyme abnormality showed that it was due to the synthesis of a structurally altered enzyme protein with increased specific activity (Becker et al. 1973b). Thus immunological studies demonstrated that the actual quantity of enzyme protein present was similar to that in normal individuals, but that the specific enzyme activity per molecule of enzyme was about 2.5 to 3.0 times greater than normal.

Electrophoretic studies showed that the abnormal enzyme protein differed from the normal in its electrophoretic mobility. The family studies suggest that the abnormality segregates as a 'dominant' characteristic, but it is not yet clear whether the mutant allele occurs at a locus on the X-chromosome or on one of the autosomal chromosomes.

Other enzyme variants in which a structurally altered enzyme protein is associated with increased enzyme activity have been reported. One example is Gd Hektoen which is a variant of glucose-6-phosphate dehydrogenase (p. 172). Another is a variant of serum cholinesterase known as E. Cynthiana (Yoshida and Motulsky 1969, Reys and Yoshida 1971). But in neither of these cases does the increased enzyme activity appear to give rise to any untoward metabolic or clinical consequences.

There are other cases, where enhanced activity of a particular enzyme has been found to occur in an inborn error of metabolism, but the increased activity appears to be a secondary consequence of a primary enzyme deficiency involving some other enzyme. One example of this is the enhanced activity of δ-aminolaevulinic acid synthetase observed in acute intermittent porphyria (see p. 240). Another is the increased activity of adenine phosphoribosyl transferase and other enzymes found in red cells in the Lesch-Nyhan syndrome. This condition is due to a virtually complete deficiency of the enzyme hypoxanthine-guanine phosphoribosyl transferase (Seegmiller et al. 1967, Kelley 1968), which results in a major upset in purine metabolism. One feature of the disorder is an increase in the intracellular concentration of PRPP in red cells (Greene et al. 1970). The increased PRPP concentration has the effect of enhancing the stability of adenine phosphoribosyl transferase and consequently this enzyme shows an elevated level of activity, because it is less rapidly degraded than is normally the case (Rubin et al. 1969, Greene et al. 1970). Other enzymes showing increased activity in red cells from patients with the Lesch-Nyhan syndrome are inosinic acid dehydrogenase (Pehlke et al. 1972), orotate phosphoribosyl transferase and orotodylic decarboxylase (Beardmore et al. 1973). The enhancement of activity of these enzymes in the red cells is apparently also a secondary consequence of the primary deficiency of hypoxanthine guanine phosphoribosyl transferase deficiency. But the precise mechanism is not yet clear. However it is noteworthy that the enhanced activity is not observed in leucocytes.

The various lysosomal diseases furnish further examples of increases in the activity of a number of enzymes occurring as a secondary consequence of a primary enzyme deficiency. The matter has been discussed on p. 244.

CHAPTER 7

The blood group substances

7.1. The ABO blood groups

The first inherited antigenic differences to be recognised as such in man were discovered by Landsteiner at the beginning of the present century. He showed that when suspensions of red blood cells obtained from different people are mixed with blood serum obtained from other people clear-cut differences in reaction are observed. In some cases there is marked agglutination or clumping of the red cells. In other cases the red cells remain unaffected. The agglutination is due to the binding of particular antigenic substances present on the surface of the red cells with specific antibodies (immunoglobulins) present in the serum. By cross-agglutination tests using red cells and sera from normal healthy individuals, it was found possible to classify people into four distinct groups in terms of two antigenic specificities (A and B). Some people (group O) have neither of these specificities, others have only one (group A or group B), while still others have both (group AB). The corresponding serum antibodies are called anti-A and anti-B, and their occurrence in sera of individuals of the four groups are indicated in table 7.1.

TABLE 7.1

The ABO blood groups.

Blood group	Antigenic specificities on red cells	Antibodies in serum
O	—	anti-A and anti-B
A	A	anti-B
B	B	anti-A
AB	A and B	—

257

These findings laid the foundation for modern blood transfusion, by defining who would be compatible donors of blood for particular recipients. Among Europeans about 47% of people are group O, about 42% group A, about 8% group B and about 3% group AB. However, the relative frequencies of the four groups varies from population to population.

Early studies on the familial distributions and the population frequencies of these four groups showed that they were inherited, and led to the hypothesis that they are determined by three allelic genes; allele *A* determining A specificity, allele *B* determining B specificity, and allele *O* being inactive. According to this, group O individuals are all homozygous *OO* and group AB individuals all heterozygous *AB*. But group A individuals may be either homozygous *AA* or heterozygous *AO*; and group B individuals may be either homozygous *BB* or heterozygous *BO*. Later work substantially confirmed this interpretation, and extended it in certain ways. For instance it was shown that by using appropriate antisera group A could be subdivided into two relatively common types A_1 and A_2, evidently determined by separate alleles, and also a number of rare alleles resulting in slightly modified specificities were identified (for detailed review see Race and Sanger 1968).

In the red cell, the substances which carry these A and B antigenic specificities are firmly bound and cannot be extracted from the stroma with water or salt solutions. Active preparations may however be obtained by extraction with ethanol. Substances with similar properties and specificities have also been shown to occur elsewhere in the body (Hartman 1941), notably in membranes of the endothelial cells which line the cardiovascular system (Szulman 1964, 1966). These substances are usually referred to as 'alcohol soluble' group specific substances. This is to distinguish them from another class of substances which carry the same antigenic specificities evidently determined by the same alleles (*A, B*, etc.), but which are water soluble. The so called 'water soluble' group specific substances occur in large amounts in mucous secretions, notably in saliva and the mucus of the gastrointestinal tract. They are usually demonstrated and assayed by absorption techniques. That is by showing that they have the capacity when mixed with an appropriate antiserum, to combine specifically with blood group antibody so that agglutination fails to take place upon subsequent addition of red cells of the corresponding type.

The 'alcohol soluble' group specific substances extracted from red cells appear to be glycosphingolipids (Koscielak 1967, Hakomori and Strycharz 1968). They are complex macromolecules containing a carbohydrate moiety joined through sphingosine to fatty acids. The 'water soluble' group specific substances on the other hand are glycoproteins (Kabat 1956, Morgan 1967), containing a high percentage of carbohydrate. Thus there are two quite different classes of macromolecule which exhibit similar or identical antigenic specificities determined by the same alleles. The antigenic specificities appear to be a reflexion of the structural arrangement of certain sugars on the more superficial parts of the carbohydrate moieties of the macromolecules. These groupings are probably very similar if not identical in the two classes of macromolecules.

In fact most of the work on the chemical structures which underlie these antigenic specificities has been carried out on the so called 'water soluble' group specific substances. This is because until very recently these were the only form of blood group substance which could be isolated in a satisfactory state and in adequate quantities for structural studies.

The most potent sources of water soluble blood group substances among the normal secretions of the body are saliva and gastric juice. Meconium, the first stool of the newborn is also rich in these substances. However a particularly useful source of material for the isolation of relatively large amounts of these substances from single individuals has proved to be fluid obtained from ovarian cysts (Morgan and van Heyningen 1944). These fluids accumulate in the cysts over long periods of time, and large volumes often containing several grams of active group specific material may be obtained from a single cyst.

Purified blood group substances obtained from such secretions turn out to be high molecular weight glycoproteins. The molecular weights range from 3×10^5 to 1×10^6, and it seems that a preparation having specific blood group activity obtained even from a single individual and a single secretion, may contain a family of macromolecules varying somewhat in overall size and perhaps in composition, though they are no doubt very closely related in structure. Such substances usually contain about 85% carbohydrate and 15% aminoacids. The full details of their molecular organisation are not yet known, but their general properties and degradation products suggest that they are made up of a large number of relatively short oligosaccharide chains which are covalently attached at intervals to a polypeptide backbone. On the assumption that the specific substances usually have a molecular weight around 500,000, and that the carbohydrate chains are composed of seven or eight sugar units, there would appear to be some 300 carbohydrate chains, each with a non-reducing end group (Morgan and Watkins 1969).

The carbohydrate moiety of the macromolecules contains (fig. 7.1) a hexose, D-galactose, a methyl pentose, L-fucose and two aminosugars, N-acetyl-D-glucosamine and N-acetyl-D-galactosamine. The nine carbon

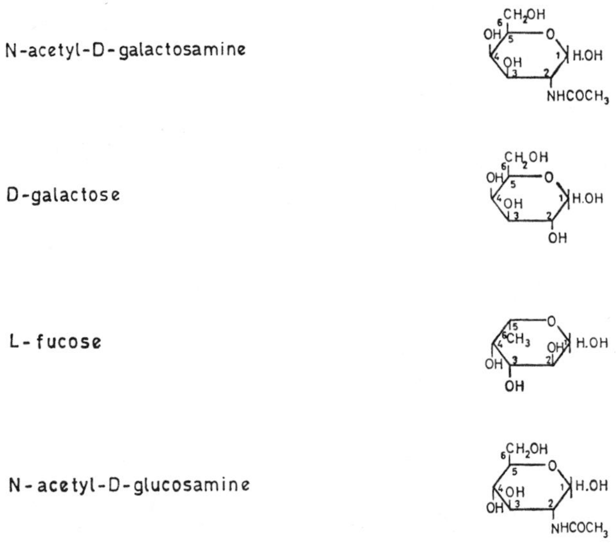

N-acetyl-D-galactosamine

D-galactose

L-fucose

N-acetyl-D-glucosamine

Fig. 7.1. Sugars which occur in the blood group substances.

sugar, N-acetyl-neuraminic acid (sialic acid) is also often present. The polypeptide part of the macromolecules is composed of fifteen aminoacids, and is distinctive in that four aminoacids, threonine, serine, proline and alanine make up about two thirds of the aminoacids present (Pusztai and Morgan 1963). Another notable feature is that sulphur containing amino-acids are virtually absent. The integrity of the complete macromolecule is essential for maximum serological reactivity, and the role of the polypeptide backbone appears to be that of maintaining the correct spacing and orienta-tion of the carbohydrate chains which provide in their terminal groupings the determinants of the antigenic specificity.

The terminal sequences of sugars which are now thought to confer A and B group specificity are shown diagrammatically in fig. 7.2 (Watkins 1966). The essential point is that the A specific chains terminate with an N-acetyl-galactosamine residue, while the B specific chains terminate in a galactose residue. In other respects the chains are the same, although it should be noted that both A and B specificity can occur on chains in which there may be a β-1,3 (type 1 chains) or β-1,4 (type 2 chains) link in the second position between a galactose and an N-acetylglucosamine residue. Also shown in fig. 7.2 are oligosaccharide chains which occur in many of these group specific substances and give rise to what has been called H specificity. It will be noted

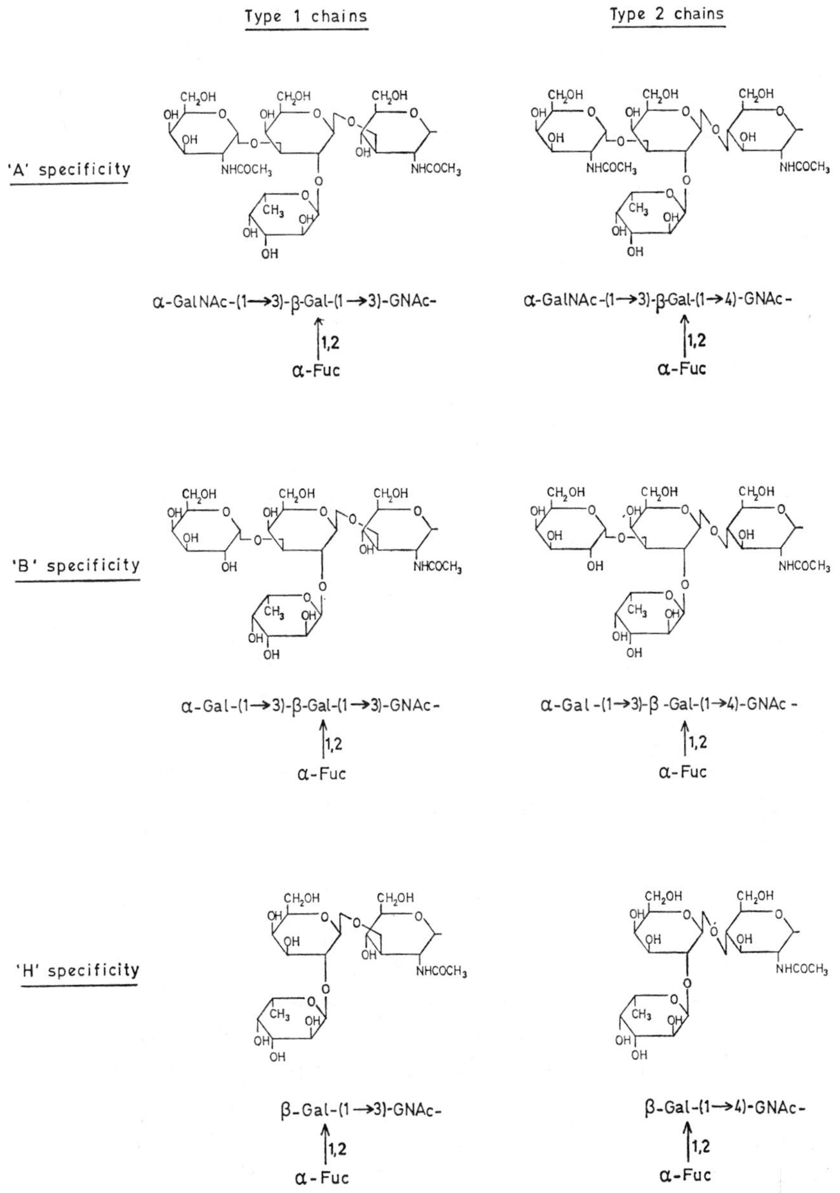

Type 1 chains   Type 2 chains

'A' specificity

α-GalNAc-(1→3)-β-Gal-(1→3)-GNAc-   α-GalNAc-(1→3)-β-Gal-(1→4)-GNAc-
            ↑1,2   ↑1,2
           α-Fuc   α-Fuc

'B' specificity

α-Gal-(1→3)-β-Gal-(1→3)-GNAc-   α-Gal-(1→3)-β-Gal-(1→4)-GNAc-
            ↑1,2   ↑1,2
           α-Fuc   α-Fuc

'H' specificity

β-Gal-(1→3)-GNAc-   β-Gal-(1→4)-GNAc-
        ↑1,2   ↑1,2
       α-Fuc   α-Fuc

Fig. 7.2. Terminal sugar sequences in polysaccharide chains of glycoproteins which confer 'A', 'B' and 'H' specificity (see text). Gal: D-galactopyranosyl; Fuc: L-fucopyranosyl; GNAc: N-acetyl-D-glucosaminopyranosyl; GalNAc: N-acetyl-D-galactosaminopyranosyl.

that the H specific chains are the same as the A and B chains except that they lack the terminal N-acetylgalactosamine or galactose residues which are necessary for these characteristic specificities.

H specificity was originally discovered when it was found that certain sera obtained from normal cattle could cause selective agglutination of human group O red cells (Schiff 1927). It was also found that a high proportion of group O individuals also had in their saliva and other secretions substances which were capable of neutralising by absorption the agglutinating effect of the cattle sera on group O red cells. Subsequently a number of other antibodies with very similar properties were obtained from a variety of different sources. These included sera of goats and chickens immunised with *Shigella shiga*, sera from the eel *Anguilla anguilla*, sera from rabbits immunised with material isolated from ovarian cyst fluids of group O individuals, and also extracts prepared from the seeds of various plants (so-called lectins). Very occasionally human sera have been found to contain an antibody with the same specificity.

At first it was supposed that these antibodies were reacting with a specific antigen determined by the blood group gene *O*. This idea was however abandoned when it became clear that substances with H specific reactivity were formed in individuals who could not be carrying the *O* gene (Morgan and Watkins 1948). In particular salivas from individuals of group AB could exhibit H specificity. These results led to the idea that there was a so-called H substance which was a precursor of substances with A and B specificities. The *A* or *B* genes or rather their immediate enzyme products by acting on this precursor caused the appearance of A or B specificities with the complete or partial elimination of H activity. In this way H activity would be expected to be most pronounced in blood group O individuals (Watkins and Morgan 1955b).

The oligosaccharide sequences (fig. 7.2) associated with A, B and H specificities were deduced by a variety of experimental approaches. The earliest indications of their nature came from studies based on the classical work of Landsteiner who showed that a simple substance with a structure closely related to or identical with the immunological determinant grouping of an antigen, can often inhibit competitively a specific antigen–antibody reaction. It was found for example that under appropriate conditions the H–anti H reaction could be inhibited by L-fucose, but not by other sugars present in the group specific glycoproteins. Similarly the A–anti A reaction could be specifically inhibited by N-acetylgalactosamine and the B–anti B reaction by D-galactose. These findings (Watkins and Morgan 1952, Morgan and Watkins

1953, Kabat and Leskowitz 1955) indicated that despite the overall similarities in the apparent composition of the A, B and H group specific substances different sugars were implicated in the immunologically specific groupings. This approach was further extended and made more specific and also more sensitive by the use as inhibitors of particular di- and trisaccharides obtained from partial hydrolysates of the different substances, and also from other sources.

Another approach came from the discovery of certain enzymes in a number of different microorganisms, which are capable of splitting terminal sugars from the ends of carbohydrate chains in the group specific substances, and at the same time altering their antigenic specificity (Iseki et al. 1953, 1959; Watkins 1956; Watkins et al. 1962; Harrap and Watkins 1964). For example particular enzymes were found which cause the preferential liberation of free D-galactose from purified preparations of B reacting substances. This change is accompanied by a loss of B reactivity and the appearance or enhancement of H reactivity which was absent or only barely detectable in the original preparation. Other enzymes lead to a liberation of N-acetylgalactosamine from A reacting glycoproteins, with the concomitant loss of A reactivity. Here also the change leads to the appearance or enhancement of H reactivity. Still other enzymes occur which lead to the liberation of L-fucose from H reacting substances, and the concomitant loss of H reactivity. Furthermore it is found (Watkins and Morgan 1955a) that the destruction of the particular group specificity by the action of these different enzymes can be specifically inhibited by the addition of the corresponding sugar to the reaction mixture.

From these types of experiment the nature of the sugars specifically concerned in the A, B and H determinants was deduced. But the detailed elucidation of the structures involved was finally dependent on the isolation and characterisation of numerous short chained oligosaccharides obtained by partial acid and alkaline hydrolysis of purified glycoproteins with different group specificities (Cote and Morgan 1956, Schiffman et al. 1962, 1964; Painter et al. 1962, Rege et al. 1964a). Determination of the structures of these fragments as well as the study of their behaviour in the inhibition and enzyme systems mentioned above eventually led to the elucidation of the oligosaccharide sequence shown in fig. 7.2.

Progress in establishing the exact nature of the group specificities of the active glycolipid materials obtained from red cells is much less advanced. However it has been shown that the carbohydrate moiety of the molecules contains the same sugars as are present in the group specific glycoproteins

(Koscielak et al. 1970). Also serological and enzymic inhibition tests indicate that here also N-acetylgalactosamine and D-galactose are the major determinant units in A and B specificity respectively (Watkins et al. 1964). So it appears that the same or very similar oligosaccharide groupings may be responsible for the group specificities of both the 'water soluble' glycoproteins occurring in the mucous secretions and the 'alcohol soluble' glycolipids present in the red cell membrane.

All these results suggest that the *A* and *B* alleles at the *ABO* gene locus act by determining the formation of specific glycosyl transferring enzymes which add either N-acetylgalactosaminosyl or D-galactosyl units to the ends of the carbohydrate chains in the final stages of the synthesis of these group specific macromolecules. One may suppose that whatever *A*, *B* or *O* genes are present, the synthesis of the macromolecule proceeds in much the same way as far as the formation of a substance containing multiple carbohydrate chains with H specificity (as shown in fig. 7.2) is concerned. Then in individuals carrying the *A* allele, and hence possessing the corresponding transferase enzyme, N-acetylgalactosamine residues will be added as the terminal units of the oligosaccharide chains. Similarly in individuals carrying the *B* allele, and so forming a specific D-galactosyl transferase enzyme, galactose is added as a terminal unit to the oligosaccharide chains. In individuals homozygous for the *O* allele, no corresponding enzyme is apparently present, and so no further addition to the carbohydrate chain takes place. The H specific grouping is left exposed and is therefore serologically detectable. Hence H reactivity is most marked in group O individuals.

This general hypothesis of the action of the *A*, *B* and *O* alleles is also consistent with finding that a single glycoprotein molecule may exhibit more than one blood group specificity. It has been shown for example that when a purified group specific preparation obtained from an AB subject is precipitated with an antiserum specific for A, both A and B activities are carried down in the precipitate (Morgan and Watkins 1956). Presumably in the synthesis of the glycoprotein macromolecules in group AB individuals, there is competition for the completion of the oligosaccharide chains. Any one chain may be completed either as an A active structure by the addition of N-acetylgalactosamine, or a B active structure by the addition of D-galactose. But since there are many chains on the same macromolecule some will be completed in one way and others in the other. It has also been shown that H activity may be exhibited by the same glycoprotein macromolecule which shows A or B activity (Watkins and Morgan 1957b). Evidently here, not all the chains available are completed by the addition of the A or B determinants, and the H reactivity is due to the presence of these chains to which no additional sugar has been added.

Direct evidence for the occurrence of specific glycosyltransferase enzymes which can be regarded as the products of the *A* and the *B* alleles at the ABO locus was subsequently obtained. Thus an α-D-galactosyltransferase was shown to occur in submaxillary glands and also in gastric mucosal material, from individuals who are group B or AB, but absent in tissues from group A or O individuals. This enzyme transfers D-galactose from uridine diphosphate galactose to oligosaccharides containing at the terminal non-reducing end the H active structure α-L-fucosyl-(1 → 2)-galactose (Ziderman et al. 1967, Race et al. 1968, Poretz and Watkins 1972). The D-galactosyl-transferase has also been found in milk from B and AB individuals but not in A or O individuals (Kobata et al. 1968b).

Similarly an α-N-acetyl-D-galactosaminyltransferase which transfers N-acetyl-D-galactosamine from uridine diphosphate N-acetyl-D-galactosamine to the same oligosaccharide acceptors has been demonstrated in submaxillary glands (Hearn et al. 1968) from group A or group AB individuals, but not from group B or group O individuals. Also in human milk and serum the N-acetyl-D-galactosaminyltransferase has been found in A and AB but not in B or O individuals (Kobata et al. 1968a, Kim et al. 1971).

Furthermore it has been found that the N-acetylgalactosaminyl transferase in serum of blood group A_1 individuals differs significantly in its kinetic properties from that in serum of blood group A_2 individuals (Schachter et al. 1973). The two alleles evidently determine qualitatively different versions of the enzyme.

Schenkel-Brunner and Tuppy (1970, 1973) made the very interesting observation that enzyme preparations from human gastric mucosa microsomes were able under appropriate conditions to confer A and B specificities on red cells. Thus an enzyme preparation from gastric mucosa of blood group A individuals, when incubated with O or B red cells in the presence of UDP-acetylgalactosamine, rendered them agglutinable by anti-A serum. Similarly after treatment with enzyme from B individuals and UDP-galactose, O and A red cells could be agglutinated by anti-B serum. When the enzymes from A and B individuals were allowed to act together with O red cells in the presence of UDP-acetylgalactosamine and UDP-galactose, the red cells could be agglutinated with both anti-A and anti-B sera.

As yet little is known about the nature of the specific glycosyltransferase enzymes determined by the *A* and *B* alleles. One would expect that the different enzyme proteins are very similar in structure and perhaps only differ by a single aminoacid substitution. However this difference is presumably sufficient to result in a clear difference in substrate specificity. The 'A' enzyme presumably has a specificity for a nucleotide diphosphate compound containing an N-acetyl-D-galactosamine grouping. Similarly the 'B' enzyme has a specificity for a substrate containing a D-galactose grouping. It is noteworthy that N-acetyl-D-galactosamine and D-galactose are structurally identical except that the N-acetyl amino group at carbon No. 2 in the former, is replaced by an hydroxyl group in the latter. A relatively small difference in enzyme protein structure may well, by altering the conformation of the active site give rise to this difference in substrate specificity. The 'inactivity' of the *O* allele might be due to the formation of an enzyme protein molecule with a slightly modified structure which renders it devoid of glycosyl transferase activity. Other possibilities are a true failure in the synthesis of the enzyme protein, or the synthesis of an extremely unstable form of the enzyme protein so that its half life is very much shorter than those of the 'A' or 'B' enzymes.

7.2. The 'secretor' and the 'H' loci

In most people the saliva and other mucous secretions contain water soluble glycoproteins with A, B or H specificities according to the individual's red cell group. However there are some individuals in whom although there is no lack of glycoproteins in their secretions, the substances present are devoid of the characteristic A, B or H specificities. In fact individuals can be divided into two sharply distinct classes in this way (Lehrs 1930, Putkonen 1930). One class, the so-called 'secretors' exhibit A, B or H specificity in their saliva and other secretions, the other class called 'non-secretors' do not. Among Europeans about 80% of people are 'secretors' and about 20% 'non-secretors'.

The 'secretor' or 'non-secretor' status of an individual is constant and is genetically determined. Family studies have shown that the dimorphism is determined by a pair of allelic genes now referred to as *Se* and *se* (Schiff and Sasaki 1932). An individual carrying the *Se* allele whether homozygous *SeSe* or heterozygous *Sese* is a 'secretor'. An individual homozygous for the other allele i.e. *sese* is a 'non-secretor'. These alleles occur at a locus which is distinct from the ABO locus, and it has been shown that the two loci are not clearly linked. They either lie well separated on the same chromosome, or occur in different chromosomes.

The *Se* allele appears to be necessary for the formation of the H specific grouping in the carbohydrate chains of the water soluble glycoproteins. If it is not present, as occurs in homozygotes *sese*, then the H specific grouping is not formed. Since this appears to be a necessary prerequisite before the A or B terminal groupings can be added to the chains, A or B specificities are also absent in non-secretors even though the individual may carry the corresponding *A* or *B* alleles.

However the alleles at this so-called 'secretor' locus only affect the synthesis of 'water soluble' group specific substances. They do not appear to influence the formation of the 'alcohol soluble' substances, since A, B and H specificities are exhibited in the ordinary way by the red cells of 'non-secretors'. Furthermore the apparent block in 'non-secretors' in the formation of the characteristic carbohydrate chains of the glycoproteins, evidently only affects the last stages of their synthesis, because as will be seen later the glycoproteins actually present in the secretions of 'non-secretors' are qualitatively very similar to those in 'secretors' and only appear to lack the terminal L-fucosyl residue characteristic of the H specific chains, and the terminal N-acetylgalactosamine or D-galactose residues which determine the A or B specificities.

There is yet another gene locus which is intimately concerned with the formation of the H specific grouping and consequently also the formation of A and B specificities. It was recognised following the discovery of certain rare individuals with a quite unusual pattern of blood group specificities (Bhende et al. 1952). The individuals in whom this peculiarity was first discovered came from Bombay, and it is consequently often referred to as the 'Bombay phenotype'. It is also called O_h. In these individuals the red cells are not agglutinated by anti-A, anti-B or anti-H, and the serum contains not only anti-A and anti-B but also anti-H in high titre. Furthermore no A, B or H specificities are detectable in saliva. The genetical basis of this unusual situation became clear when a detailed study was made of a family in which there occurred three individuals with this peculiar phenotypic pattern (Levine et al. 1955). The pedigree is shown in fig. 7.3. II_3, II_4 and II_6 are the individuals with the unusual phenotype (O_h). The essential point is that one of the children of II_6 shows B reactivity in her red cells though this antigen is apparently not present in II_6 herself or her husband who is group A. Furthermore the other child of II_6 who appeared to be a normal group O individual showed H activity in her saliva and was therefore presumably a 'secretor'. Yet her father II_5 was a 'non-secretor', and her mother II_6 also showed no A, B or H activity in her saliva and might therefore also have been considered to be a 'non-secretor'. Thus it would appear that II_6 is carrying a *B* allele at

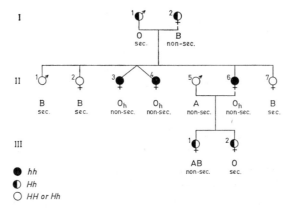

Fig. 7.3. Pedigree showing segregation of unusual blood group phenotype O_h (Bombay phenotype), and its interpretation in terms of the segregation of the rare allele h at the H locus (after Levine et al. 1955). Individuals I_1 and I_2 were first cousins. The apparent ABO types as determined by red cell agglutination tests with anti-A and anti-B are shown for each individual. Also shown are the secretor (sec.) or non-secretor (non-sec.) phenotypes as determined by standard methods on the saliva. Individuals with O_h or Bombay phenotype lack A, B and H antigens in red cells and secretions. Their sera contain the antibodies anti-A, anti-B and anti-H.

the *ABO* locus though its activity is not expressed, and that she is also carrying an *Se* allele at the *Secretor* locus though neither B or H activity is present in her secretions. The simplest explanation of the findings is that II_6 and also II_3 and II_4 who show the same peculiar phenotypic pattern are homozygous for a rare allele at yet another gene locus, and that this has the effect of suppressing the formation of B and H (and probably also A) specificities in both red cells and secretions. Further evidence that homozygosity for a rare allele is involved is provided by the fact that the parents of the affected individuals were first cousins.

Several other examples of this phenomenon have since been found, and they generally confirm the interpretation that a rare allele at a gene locus different from either the *ABO* or the 'Secretor' loci is involved. The common allele at this new locus is now referred to as *H*. It is present in the great majority of individuals and it is evidently necessary for the formation of the H specific grouping and hence the A and B specific groupings, in both the glycolipid group substances of the red cells and the glycoprotein substances of the secretions. The rare allele *h* is ineffective in this regard, so that homozygotes *hh* fail to show A, B or H specificities in either their red cells or their secretions even though the appropriate ABO alleles and also the *Se* allele may be present. In heterozygotes *Hh*, the expression of the ABO alleles or of the *Se* allele is not affected.

Thus for the development of H specificity in the glycoproteins of the secretions both the *H* allele and the *Se* allele appear to be necessary. However in the formation of H specificity in the glycolipids of the red cell, only the *H* allele appears to be essential while *Se* seems to play no significant part.

In the biosynthesis of the blood group substances the appearance of H specificity apparently depends on the addition of L-fucose in α-(1→2) linkage to a terminal galactose residue in an oligosaccharide chain so as to give the structures shown in fig. 7.2. This step is an essential preliminary to the action of the glycosyl transferase enzymes determined by the *A* and *B* alleles of the ABO locus. Presumably it requires a specific L-fucosyl transferase capable of transferring an L-fucose residue from a suitable donor substrate to the terminal galactose residue of the oligosaccharide chain. It has been suggested (Watkins 1966) that this fucosyl transferase is the specific enzyme product of the *H* allele and is not formed by the *h* allele. However for the enzyme to occur and also take part in the biosynthesis of the 'water soluble' group specific substances in the secretions, the allele *Se* must also be present. So 'water soluble' substances with H specificity will be formed in individuals who are either *HH* or *Hh*, and also *SeSe* or *Sese*, but not in those who are *hh* and *sese*. Since the 'Secretor' locus appears to play no part in the biosynthesis of the 'alcohol soluble' blood group substances, H specificity may be found in red cells even though the individual is a 'non-secretor' i.e. *sese*. It will not however occur in red cells and neither will the A or B specificities if the individual is of the rare 'Bombay' type, *hh*. On this view then, the 'Secretor' locus is regarded as regulating the activity of the *H* locus, in the biosynthesis of the group specific 'water soluble' glycoproteins of the secretions, but not in the biosynthesis of the group specific 'alcohol soluble' glycolipids.

Shen et al. (1968) and Chester and Watkins (1969) have studied the fucosyltransferases present in milk, submaxillary gland and stomach from secretors and non-secretors. They found that one of the enzymes was present in secretors but not in non-secretors. This enzyme transferred L-fucose in α-(1→2) linkage to β-galactosyl residues from GDP-L-fucose, and thus appeared to have the specificity predicted for the formation of H reactive groupings in the blood group substances.

7.3. The 'Lewis' or Le locus

The '*Lewis*' locus is a further gene locus which affects the biosynthesis and hence the antigenic specificity of the water soluble glycoproteins occurring in saliva and other mucous secretions.

Its recognition followed the discovery (Mourant 1946) that occasional individuals have in their sera an antibody which causes agglutination of red cells of some 18% of European individuals. It was subsequently found however that in as many as 90% of the population, substances which also reacted specifically with this antibody could be detected in saliva and other mucous secretions (Grubb 1951). The antigen specificity involved is now referred to as Lewis a or Le^a. Secretions or red cells which exhibit this specificity are said to be $Le(a+)$, and those which do not $Le(a-)$.

The presence or absence of Le^a specificity in secretions depends on a pair of alleles *Le* and *le* (Grubb 1951, Ceppellini 1955). Individuals carrying the *Le* allele either as homozygotes *LeLe* or heterozygotes *Lele* exhibit $Le(a+)$ specificity. Individuals who are homozygous *lele* do not. The locus is separate from and apparently not closely linked to the other loci known to be involved in determining the group specificities of glycoproteins in the mucous secretions (e.g. the ABO locus and the Secretor locus).

However rather unexpectedly it was found that the degree of Le^a reactivity present in secretions of individuals carrying the *Le* allele is markedly dependent on their 'secretor'–'non-secretor' status. In the secretions of 'non-secretors' Le^a activity is very much greater than in the secretions of 'secretors'. Also individuals who show Le^a activity in red cells are all 'non-secretors' (Grubb 1948, 1951). So there is an interaction between the effects of the alleles at the *Secretor* locus on the one hand and the alleles at the *Lewis* locus on the other.

The characteristic sequence of sugars which give rise to Le^a specificity in the glycoprotein molecules (Watkins and Morgan 1957a, 1962; Rege et al. 1964b) is shown in fig. 7.4. The critical sugar is the L-fucosyl residue attached in α-1,4 linkage to the penultimate N-acetylglucosaminyl residue in the chain. The main backbone of the carbohydrate chain is the same as that of one of the two types of chain which by the addition of an L-fucosyl residue to the terminal galactosyl unit can give rise to H specificity; and which by the further addition of an N-acetylgalactosaminyl or a D-galactosyl unit can give rise to A or B specificity respectively. Evidently in the biosynthesis of the glycoprotein molecules carbohydrate chains are formed which may be acted on not only by the *H*, *Se* and *A* and *B* genes but also by the *Le* gene. A fucosyltransferase with the appropriate properties has been demonstrated in milk (Jarkovsky et al. 1970).

In 'non-secretors' (*sese*), H reactive groupings are not formed, so that if the *Le* gene is present many carbohydrate chains are available to which Le^a specific activity can be conferred by the addition of L-fucosyl residues in the appropriate position. However in 'Secretors' (*SeSe* or *Sese*) there is presumably competition for the chains and a much smaller proportion of them are

'H' specificity

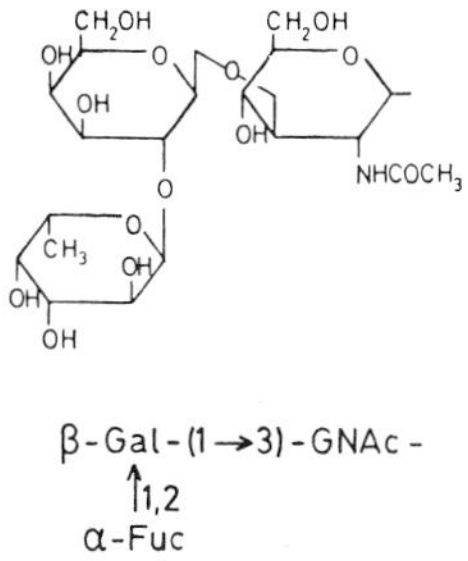

β-Gal-(1→3)-GNAc-
↑1,2
α-Fuc

'Le^a' specificity

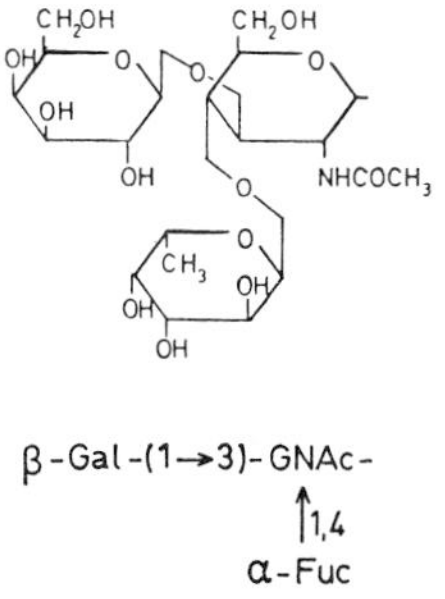

β-Gal-(1→3)-GNAc-
↑1,4
α-Fuc

'Le^b' specificity

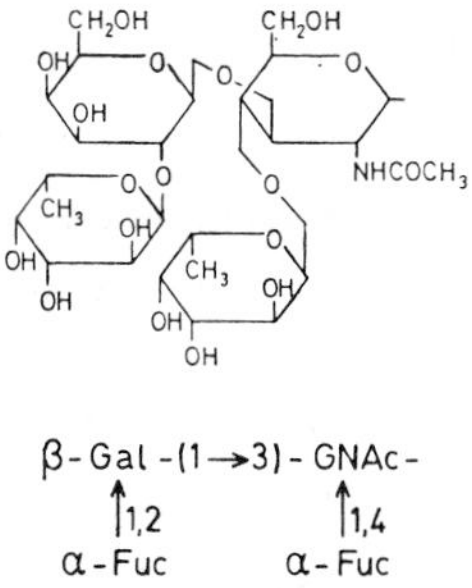

β-Gal-(1→3)-GNAc-
↑1,2 ↑1,4
α-Fuc α-Fuc

Fig. 7.4. Comparison of terminal sugar sequences in polysaccharide chains of glycoproteins which confer 'H', 'Le^a' and 'Le^b' specificities (see text). Abbreviations as in fig. 7.2.

ably competition for the chains and a much smaller proportion of them are likely to end up with only the specific Le^a grouping. Others will have the H grouping and perhaps the A or B groupings according to which genes are present. Furthermore it has been shown that some chains may acquire under the influence of the *Le* gene, an L-fucosyl residue attached to the penultimate sugar residue (N-acetylglucosamine), and also under the influence of the *H* and *Se* genes an L-fucosyl residue attached to the next sugar residue (D-galactose). This compound grouping exhibits neither Le^a reactivity or H reactivity. Instead a new specificity known as Le^b and detectable only by a quite different antiserum, is produced (Marr et al. 1967). The apparent inter-action between the Secretor and the Lewis loci, whereby much greater Le^a activity is found in 'non-secretors' (*sese*) who carry the *Le* gene, than in 'secretors', who carry the *Le* gene, is thus readily accounted for.

It seems likely that the *Le* gene does not play a specific role in the biosyn-thesis of the glycolipid group specific substances of the red cell, since it has been shown that red cell Le(a+) reactivity which is normally exhibited only by individuals who are non-secretors (*sese*) is largely if not entirely due to adsorption on the red cell of Le^a active substances formed elsewhere but which occur in relatively greater amounts in 'non-secretors' (Sneath and Sneath 1955, Marcus and Cass 1969).

7.4. The biosynthetic pathways for the group specific glyco-proteins

According to the behaviour of the red cells and the mucous secretions of different individuals with the five specific antibodies, anti-A, anti-B, anti-H, anti-Le^a and anti-Le^b it is possible to define six distinctive types of reaction as shown in table 7.2.

The fourth set of reactions shown in table 7.2 is of special interest because in these individuals although the red cells have the standard reactions with anti-A, anti-B, and anti-H according to the particular ABO genotype of the individuals, the mucous secretions fail to react with any of the antibodies. These individuals constitute about 2% of European populations and are homozygous for the 'inactive' alleles at both the '*Secretor*' and the '*Lewis*' loci. That is, they have the genotype *sese, lele*. The secretions of such individ-uals are however not devoid of glycoproteins of the kind which in other people exhibit the different group specificities and in fact such glycoproteins have been isolated and studied. They can be regarded as precursors of the

TABLE 7.2

Six types of individuals c. stinguishable on the basis of the reactions of their red cells and secretions with the antibodies anti-A, anti-B, anti-H, anti-Leᵃ and anti-Leᵇ (Watkins 1966).

Gene combination			Specificities detectable on red cells			Specificities detectable in secretions		
H locus	*Se* locus	*Le* locus	ABH	Leᵃ	Leᵇ	ABH	Leᵃ	Leᵇ
1. *HH* or *Hh* *SeSe* or *Sese* *LeLe* or *Lele*			+++	—	++	+++	+	++
2. *HH* or *Hh* *sese* *LeLe* or *Lele*			+++	+++	—	—	+++	—
3. *HH* or *Hh* *SeSe* or *Sese* *lele*			+++	—	—	+++	—	—
4. *HH* or *Hh* *sese* *lele*			+++	—	—	—	—	—
5. *hh* *SeSe* or *Sese* *LeLe* or *Lele*			—	+++	—	—	+++	—
6. *hh* *sese* *lele*			—	—	—	—	—	—

('Bombay' types: rows 5 and 6)

Strong specific activity +++; weak specific activity +; no activity —.

other group specific substances, in which the biosynthesis has been stopped before groupings giving rise to the different characteristic specificities have been added to the carbohydrate chains (Watkins 1966).

This so called 'precursor' substance (fig. 7.5) appears to have at least two sorts of carbohydrate chains (Rege et al. 1964a) and these may be joined through a branch point to a common chain linking to the polypeptide backbone (Lloyd et al. 1966, 1968a, b). Both these chains appear to have a terminal galactosyl residue at the non reducing end (table 7.3). This is joined to an N-acetylglucosaminyl residue in one chain (type 1) by a β-1,3 linkage, and in the other chain (type 2) by a β-1,4 linkage. The substance has been shown to react with antisera prepared against Type XIV pneumococcus, and this specificity can be attributed to the terminal structure of the type 2 chain.

A variety of gene loci presumably play a part in determining the biosynthesis of this so-called 'precursor' glycoprotein. They would be concerned in determining the synthesis of the various enzymes necessary for assembling the carbohydrate chains. And no doubt at least one further locus is necessary to define the sequence of aminoacids in the polypeptide backbone to which the carbohydrate chains are thought to be attached. However nothing is known about these loci because so far no genetical differences in the 'pre-

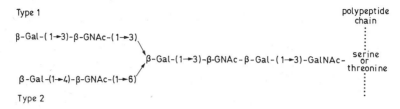

Fig. 7.5. Proposed structure of the main carbohydrate chain which carries the sugars for the serological specificity of the soluble blood group specific substances; the so-called precursor substance (Lloyd and Kabat 1968).

cursor' substance as it occurs in different individuals have been detected. The four loci *ABO*, *Secretor*, *H* and *Lewis*, about which something is known, are evidently concerned only with the later stages of the biosynthesis of their glycoproteins and appear to act only after synthesis as far as the 'precursor' has occurred.

Table 7.3 (Morgan and Watkins 1969) summarises the way in which the various genes at these several loci are thought to act in the formation of group specific glycoproteins of mucous secretions. It is supposed that if an individual carries the *Le* gene, he forms a specific fucosyl transferase enzyme which is capable of adding an L-fucose unit to the carbon No. 4 position of the subterminal N-acetylglucosamine unit in the type 1 chains of the 'precursor'. The type 2 chains are unaffected because in the corresponding N-acetylglucosamine unit the carbon no. 4 position is already substituted. If the individual carries the *H* allele at the '*H*' locus and also the *Se* allele at the '*Secretor*' locus gene, then he forms a different fucosyl transferase enzyme which is capable of adding an L-fucose unit to the carbon no. 2 position of the terminal galactose unit in either the type 1 or the type 2 chains. The prior addition of this fucose unit to form the H type structure appears to be necessary before the specific transferases thought to be determined by the *A* and *B* alleles at the ABO locus, can act. The 'A' transferase adds an N-acetylgalactosamine unit in α-linkage to the carbon no. 3 of the terminal galactose units of the H active structures. The 'B' transferase adds D-galactose in α-linkage to the same position. The alternative alleles at the various loci, for example the *h* allele at the *H* locus, the *le* allele at the '*Lewis* locus and the *O* allele at the *ABO* locus are ineffective in the biosynthetic process because they fail to lead to formation of the appropriate transferases.

Thus from the results obtained in extensive and very diverse serological, genetical and biochemical studies it has become possible to see how the group specificities of these complex glycoproteins are built up stepwise by

TABLE 7.3

Genetic control of the biosynthesis of structures with the blood group specificities A, B, H, Leᵃ, and Leᵇ (Morgan and Watkins 1969). Abbreviations as in fig. 7.2. The numerals in brackets refer to two different L-fucosyl transferases referred to in the text (pp. 269, 270).

Gene	Enzyme product	Chain-ending in precursor	Structure	Serological specificity
—	—	Type 1	β-Gal-(1→3)-GNAc--	—
—	—	Type 2	β-Gal-(1→4)-GNAc--	Type XIV
H	α-L-Fucosyltransferase (1)	Type 1	β-Gal-(1→3)-GNAc-- ↑1, 2 α-Fuc	H
		Type 2	β-Gal-(1→4)-GNAc-- ↑1, 2 α-Fuc	H
Le	α-L-Fucosyltransferase (2)	Type 1	β-Gal-(1→3)-GNAc-- ↑1, 4 α-Fuc	Leᵃ
		Type 2	β-Gal-(1→4)-GNAc--	Type XIV
H and *Le*	α-L-Fucosyltransferases (1) and (2)	Type 1	β-Gal-(1→3)-GNAc-- ↑1, 2 ↑1, 4 α-Fuc α-Fuc	Leᵇ
		Type 2	β-Gal-(1→4)-GNAc-- ↑1, 2 α-Fuc	H
H and *A*	α-L-Fucosyltransferase (1) and α-N-acetylgalactosaminyltransferase	Type 1	α-GalNAc-(1→3)-β-Gal-(1→3)-GNAc-- ↑1, 2 α-Fuc	A
		Type 2	α-GalNAc-(1→3)-β-Gal-(1→4)-GNAc-- ↑1, 2 α-Fuc	A
H and *B*	α-L-Fucosyltransferase (1) and α-D-galactosyltransferase	Type 1	α-Gal-(1→3)-β-Gal-(1→3)-GNAc-- ↑1, 2 α-Fuc	B
		Type 2	α-Gal-(1→3)-β-Gal-(1→4)-GNAc-- ↑1, 2 α-Fuc	B

the action of a series of glycosyl transferase enzymes determined by genes at several different loci. It is of some interest to note that while the action of a particular enzyme in the course of the biosynthesis of glycoproteins, may be adding the appropriate sugar unit to an oligosaccharide chain to produce a new antigenic specificity, it may at the same time abolish or cancel a specificity previously present. For example, the addition of the N-acetylgalactosamine unit to the H reactive structure leads to the appearance of A reactivity, but the loss of H reactivity.

It is also of interest that the sequential formation of the various group specificities can be effectively reversed experimentally by the stepwise

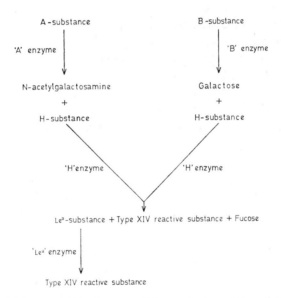

Fig. 7.6. Specificities revealed by the sequential enzymic degradation of A and B specific substances (Watkins 1966).

degradation of the glycoproteins using specific enzymes which split off sugar units one at a time. This is illustrated by experiments in which purified substances exhibiting only A or B specificities were treated successively with different enzymes capable of splitting off the characteristic sugar units associated with A, B, H and Le^a specificities (fig. 7.6, Watkins 1966). The so-called 'A' and 'B' destroying enzymes lead to a loss of A and B reactivities and the appearance of H reactivity. Then the 'H' destroying enzyme leads to the loss of the H reactivity and the appearance of Le^a and also of pneumococcal type XIV reactivity. Finally treatment with the Le^a destroying enzyme resulted in the loss of Le^a reactivity and the further enhancement of pneumococcal type XIV reactivity. In effect the various specific groupings had been stripped off leaving the so-called 'precursor' glycoprotein.

The role of the *ABO* and the *H* loci is probably much the same in the biosynthesis of the glycolipid 'alcohol soluble' group specific substances of the red cell as in the biosynthesis of the glycoprotein 'water soluble' group specific substances of the mucous secretions. However the '*Lewis*' and '*Secretor*' loci do not appear to play any part in the biosynthesis of the glycolipid substances. Exactly why this should be so, is not at present understood. The reason will perhaps emerge as the structures of the group specific glycolipid substances are elucidated.

A series of other gene loci (*Rh*, *MNS*, *Xg*, etc.) at which alleles determining inherited differences in antigenic substances which form part of the structure of red blood cells have also been identified by serological methods (Race and Sanger 1968). The study of the nature of the substances concerned, and the elucidation of the structural basis of the specific individual antigenic differences which occur, are, however, still in their early stages. One may perhaps anticipate that they will turn out to show complexities similar to those found with the classical ABO system.

CHAPTER 8

Enzyme and protein diversity in human populations

8.1. 'Common' and 'rare' alleles

A very large number of different enzymes and other proteins are synthesised in the human organism and we may presume that the primary aminoacid sequence of each of their distinctive polypeptide chains is coded in the DNA of a separate gene locus. So there must be a vast array of so-called 'structural' gene loci in the genetic constitution of every individual. Furthermore as a result of mutational events which have occurred in earlier generations there may occur at any given gene locus a series of different alleles each of which determines a structurally distinct version of the particular polypeptide chain. Thus the extent to which individual members of a population differ from one another in the structural characteristics of the enzymes and proteins they synthesise, will in general depend on the number of different alleles which are present at these various gene loci, and on the relative frequencies with which they occur.

In fact studies on a variety of different enzymes and proteins have led to the discovery of a considerable number of structurally variant forms which are genetically determined. And in this chapter we will be concerned with a consideration of their incidence and distribution in different human populations and with the kind of inferences it is possible to make from such observations about the extent of enzyme and protein diversity in general. We will also be concerned the question as to how the actual distributions of enzyme and protein variants that we observe may have been brought about.

8.1.1. How many alleles occur?

In theory, a very large number of structurally different alleles may be generated by separate mutational events within the confines of a single gene. For example from a typical gene containing a DNA sequence of say 900 bases and coding for a protein of 300 aminoacids, as many as 2,700 different alleles may be generated simply from mutational events which result in no

278

more than a single base change at some point in the sequence. This is so because each base may be changed to one of three others. In addition it is also known that other types of mutational event resulting in small deletions, duplications, frame shifts, etc. may also occur within the confines of a single gene, and each will generate its own structurally distinct mutant allele.

Now the inherited variants of different enzymes and proteins which we find among living members of our species today, must be attributed to specific gene mutations which occurred in single individuals among our ancestors in earlier generations; some living perhaps relatively recently, others in the more remote past. So the question arises as to how many of the different possible mutants which in theory could have been generated by separate mutational events at any given gene locus, actually occur among living members of the species today.

One can approach this question by thinking about the variants of haemoglobin, since this protein happens to have been investigated from this point of view, more extensively than any other.

Consider the gene which codes for the β-chain of haemoglobin. There are 146 aminoacids which are presumably coded by a DNA sequence of 438 bases. So point mutations each resulting in only a single base change could generate 1,314 different alleles. From the known aminoacid sequence and the genetic code one can classify their effects approximately as shown in table 8.1, provided one makes the simplifying assumption that each of the possible codons for any aminoacid is equally likely to occur. Because the code is degenerate, about 23.4% (307) of the mutant alleles generated in this way would not result in any change in structure of the haemoglobin. Such mutants are called 'synonymous' because the base change simply alters a codon specifying one aminoacid to another codon specifying the same aminoacid. In about 3.7% of cases (49 alleles) the single base change will result in the alteration of a codon specifying a particular aminoacid at some point in the polypeptide sequence to a so-called 'nonsense' codon which will specify chain termination. Such mutations would be expected to result in the synthesis of an abbreviated polypeptide chain which lacks the series of aminoacids normally coded by the base sequence in the gene beyond the point where the mutation occurred. In most cases such an abbreviated polypeptide is unlikely to give rise to a stable and detectable form of haemoglobin. However in some 72.9% of cases (958 alleles) a single base change will result in the alteration of a codon specifying one aminoacid to a codon specifying another and thus lead to the synthesis of an altered β-polypeptide chain which differs from the original by a single aminoacid substitution.

Because of the degeneracy of the code, the 958 mutant alleles of this sort will give rise to only 858 structurally different haemoglobin variants.

TABLE 8.1

Possible alleles which may be generated by one step single base change mutations in the gene coding for β-polypeptide chain of human haemoglobin.

Haemoglobin β-chain

Number of amino acids	= 146
Number of bases in gene	= 146 × 3 = 438
Number of possible alleles generated by single base change mutations	= 438 × 3 = 1,314

Effect of mutant	%	Number of alleles	Different variant proteins	Variants with charge change
Synonymous	23.4	307	0	0
Chain termination	3.7	49	?	?
Amino acid substitution	72.9	958	858	287
Total		1,314		

Known variants with charge change (see fig. 1.4, p.12) = 61

∴ Percentage of possible variants detected $= \dfrac{61}{287} \times 100 = 21.3\%$

Now the screening technique which has been most widely used and systematically applied to searching for haemoglobin variants is the method of electrophoresis. This will in general be expected to detect aminoacid substitutions which involve a change in a charge, i.e. substitutions of a neutral for a basic or acidic aminoacid *or* vice versa, and substitutions of a basic for an acidic aminoacid and vice versa. One can estimate that 287 of these possible 858 variants would involve substitutions of this sort (i.e. almost exactly one third). Thus on purely theoretical grounds one can expect that about 287 different electrophoretic variants might be generated by single one step base change mutations in the β Hb gene.

Actually among all the haemoglobin variants so far identified some 61 appear to have involved this type of aminoacid substitution in the β-chain (see fig. 1.4, p. 12). So we can say that of all the possible mutants of this sort which might exist, more than 20% are known to occur.

Although electrophoretic screening of haemoglobin and the necessary aminoacid sequencing of variants has now been carried out very extensively both in patients with haemolytic disease of various sorts and in random population surveys, it is clear that only a minute fraction of the world's population can have been examined in this way. So the fact that as many as 20 % of the possible examples of this type of mutant have already been found is quite remarkable. It suggests indeed that a very high proportion of all possible mutants of this sort actually exist somewhere in the species, though of course the majority are likely to be very rare.

There is no reason to think that the Hb β-gene locus is in any way peculiar in this respect. More than 80 different allelic variants of the enzyme glucose-6-phosphate dehydrogenase have already been reported (Yoshida et al. 1971, Beutler and Yoshida 1973), and numerous variants of such proteins as albumin (Weitkamp et al. 1973) transferrin (Kirk 1968), and serum α_1-antitrypsin (Fagerhol and Laurell 1970) have also been identified. Furthermore the enzyme surveys which will be discussed in more detail later have revealed the widespread occurrence of a multiplicity of allelic variants which are strictly analogous to those found for haemoglobin.

So it now appears probable that at virtually all loci coding for enzyme and protein structure a very large number of the different possible mutant alleles which may be generated by separate mutations, actually exist among living members of the species.

This leads us to the question of the frequencies of these alleles in different human populations. One may anticipate that many will be very rare, but others will occur with an appreciable frequency, and indeed may be common enough as to give rise to the well-known phenomenon usually referred to as genetic polymorphism. That is to say, a situation where the individual members of a population may be sharply classified into two or more relatively common genetically determined phenotypes, due to the occurrence of common alleles at a particular locus.

This in turn poses the question as to how the actual distributions of the different enzyme and protein variants that we observe today in human populations, have been brought about during the course of the evolution of the species.

An idea of the nature and dimensions of this rather general problem can perhaps best be obtained in the first instance by considering the main features of a few well-studied, though very different examples.

8.1.2. Haemoglobin

Haemoglobin A which constitutes most of the protein content of normal red blood cells, has been studied more extensively than any other human protein. It will be recalled that the protein is a tetramer and contains two distinct types of polypeptide chain (α and β) each of which is represented twice in the molecule. The α-chain has a sequence of 141 aminoacids, and is presumably coded by a gene containing a stretch of DNA with at least 423 bases, and the β-chain has 146 aminoacids presumably coded by a DNA sequence of 438 bases. The two gene loci are apparently well separated on the chromosomes and not closely linked.

More than 130 different variants of Hb A in which the distinctive structural peculiarity has been identified, have now been discovered. They can. be attributed to alleles at either the α- or β-gene loci. Some of these variants give rise to quite severe haematological disease, but many appear to be relatively harmless and these are probably under-represented in the total data because of the manner in which much of it has been collected, since patients with haematological abnormalities have been particularly selected for study.

Thus at both the α and the β gene loci a large number of different alleles, generated by mutational events in the remote or more recent past, exist in human populations. But these alleles vary greatly in the relative frequencies with which they occur, and in fact it is possible to classify them fairly sharply into two main groups on this basis.

The first and much the smallest group comprises those alleles which occur with a relatively high incidence in one or another human population (fig. 8.1).

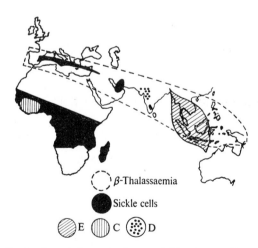

Fig. 8.1. World distribution of the major haemoglobin abnormalities. (From Lehmann et al. 1966.)

In the case of the β locus it includes the allele that determines Hb S (or sickle-cell haemoglobin) which is very common in tropical Africa where in different populations it may occur in 20% or more of all individuals; the allele determining Hb C which is more localised to W. Africa; the allele determining Hb E which is common in many populations in S.E. Asia, and the allele determining Hb D Punjab which occurs in appreciable frequency in certain populations in India (for detailed references see Livingstone 1967). Thus in each of these populations, besides the so-called normal allele at the β gene locus, at least one other allele determining a structurally distinct form of the β polypeptide chain is relatively common, so that the individual members of the population can be classified into two or more not uncommon types according to the characteristics of the haemoglobin they synthesise. These are typical examples of what is meant by genetically determined protein 'polymorphism'.

The second group of alleles determining haemoglobin variants are in contrast all extremely rare. They have turned up in an irregular manner in a wide variety of different populations, and many have so far been seen only among individual members of a single family. An idea of the incidence of these rare alleles can be obtained from the results of some surveys of population samples in England and Denmark (Huntsman et al. 1963, Liddell et al. 1964, Sick et al. 1967, and Lehmann, personal communication). The haemoglobin from large numbers of essentially randomly selected and un-related individuals was screened for electrophoretic variants. In all, out of nearly 11,000 individuals examined, 14 were found to be heterozygous for one or another allele giving rise to a structurally altered α- or β-chain. Among them 10 different variants were found, 4 turning up twice and 6 only once (table 8.2). So the gene frequencies of each of these different alleles appeared to be less than 0.0001 (i.e. $< \frac{1}{10,000}$). Thus individually they appear to be less frequent by several orders of magnitude than such alleles as the sickle cell allele in Africa or the Hb E allele in S.E. Asia which give rise to the so-called polymorphisms.

Multiple alleles determining rare haemoglobin variants appear to occur in most human populations, although the actual ones that are present probably vary from population to population. It is important to note that although these variants are individually extremely rare, one or another of them may be present in an appreciable number of people in any given population. Thus the results of Lehmann and his colleagues suggest that rather more than 1 in 1,000 individuals in Europe are heterozygous for an allele determining an electrophoretic variant of Hb A. One may plausibly assume that these represent no more than about one third of the structural variants due to single aminoacid substitutions that actually occur, the others not being detectable by electrophoresis. So it is quite likely that in Northern Europe as many as 1 in 300 people are heterozygous for an allele causing a structural variant of Hb A.

The high incidence of the specific alleles which give rise to the so-called haemoglobin polymorphisms in certain parts of the world, has generally been explained by the hypothesis that heterozygotes for such alleles are at some selective advantage compared with normal homozygotes under the environmental conditions that prevail or have prevailed in the past in these

TABLE 8.2

Rare haemoglobin variants found in electrophoretic surveys in English and Danish populations (data from Huntsman et al. 1963, Liddell et al. 1964, Sick et al. 1967, and Lehmann, personal communication 1969).

Affected polypeptide chain	Aminoacid substitution	Number of heterozygotes in total sample
α	15 Gly → Asp	2
α	53 Gly → Asp	1
α	47 Asp → His	1
β	16 Gly → Asp	2
β	43 Glu → Ala	1
β	47 Asp → Asn	1
β	69 Gly → Asp	2
β	6 Glu → Lys	1
β	26 Glu → Lys	1
β	121 Glu → Gln	2
Total heterozygotes		14

Total number of individuals examined: 10,791.

$$\text{Total incidence of heterozygotes} = \frac{14}{10,971} = \frac{1}{784}$$

Number of different rare variants $= 10$

$$\text{Separate allele frequencies} = \frac{2}{21,942} \quad \text{or} \quad \frac{1}{21,942} \text{ i.e. each} < 0.0001$$

areas. One may suppose that the primary aminoacid sequences of a protein such as haemoglobin have been evolved by natural selection and have come to be more or less optimal for the species, so that mutant alleles which alter the structure of the protein are unlikely in the great majority of cases to confer any biological advantage on individuals who carry them, and many will be in some degree deleterious. But very occasionally a mutant may appear which results in a positive selective advantage in the sense that individuals carrying the allele contribute on average more offspring to the next generation than normal homozygotes, at least under certain environmental conditions. Such an allele will, in general, tend to increase in frequency in successive generations providing the particular environmental conditions which favour it persist. However, if the specific advantage is peculiar to the heterozygous state, so that the heterozygote contributes more to the next generation than either type of homozygote, an equilibrium situation may eventually be established in which the relatively greater contribution of the variant allele to the next generation by the heterozygote is in effect balanced by the relatively reduced contribution from the homozygote. Such a situation is usually referred to as a balanced polymorphism.

The sickle-cell polymorphism in Africa provides the clearest example of this general type of phenomenon (Allison 1954, 1964). Homozygotes with sickle-cell disease tend to die in early life and very few survive to adult life and have children. So the particular sickle-cell alleles carried by homozygotes tend to be selectively eliminated from the population in each successive generation. Yet the sickle-cell allele is extremely common in many populations living in Africa, and since the possibility of a special and exceptionally high rate of mutation can be excluded (Vandepitte et al. 1955), the remarkable incidence of this particular allele in these populations could only have come about and subsequently been maintained in the face of the marked selective pressure against the sickle-cell homozygotes, if the heterozygotes have enjoyed a significant selective advantage over the normal homozygotes.

There is indeed a not inconsiderable body of evidence which indicates that such a selective advantage in fact exists for sickle-cell heterozygotes in Africa, and is due to their better chance of surviving to adult life in areas where malaria (*P. falciparum*) is an important cause of morbidity and mortality. Table 8.3, for example, summarises data collected in different places on the incidence of sickle-cell heterozygotes in children dying from malaria. Among the 104 deaths due to malaria, only one occurred in a sickle-cell heterozygote although about 23 might have been expected if mortality from malaria had been the same in sickle-cell heterozygotes as in other members of

TABLE 8.3

Incidence of sickle-cell trait among African children whose deaths could be attributed to malaria (Allison 1964).

Locality	Deaths due to malaria	Number with sickle-cell trait	Incidence of sickle-cell trait in the population	Expected number with sickle-cell trait if no selective differential	References
Uganda (Kampala)	16	0	0.16	2.6	Raper (1956)
Congo (Leopoldville)	23	0	0.235	5.4	Lambotte-Legrand, J. and C. (1958)
Congo (Luluaborg)	23	1	0.25	5.7	Vandepitte (1959)
Ghana (Accra)	13	0	0.18	2.3	Edington and Watson-Williams (1964)
Nigeria (Abadan)	29	0	0.24	7.0	Edington and Watson-Williams (1964)
Totals	104	1		23.0	

these populations. Data from morbidity surveys point in the same direction. Table 8.4 shows the results of a study in which all the patients admitted to a children's ward in Kampala (Uganda) in a given period were classified according to whether or not they had the sickle-cell trait (Raper 1956). The outstanding finding is the absence of any sickle-cell heterozygotes among the patients admitted with cerebral malaria. In patients with uncomplicated malaria however, the incidence of the sickle-cell trait was not noticeably different from that in patients with other conditions. Similar data have also been obtained elsewhere, and they indicate that sickle-cell heterozygotes are significantly less liable to the development of cerebral malaria.

Other evidence indicating that sickle-cell heterozygotes, particularly in childhood, are less severely affected by malaria has come from comparisons of the degree of *P. falciparum* parasitaemia in sickle-cell heterozygotes and in other individuals in areas where the sickle-cell allele is common and malaria is endemic (for detailed review see Allison 1964). It has also been shown that in general a high frequency of the sickle-cell allele is only found in populations living in regions where malaria is, or was until recently, endemic, or in population groups such as the Blacks in the U.S.A. whose ancestors came from such regions.

TABLE 8.4

Incidence of sickle-cell trait amongst 818 consecutive admissions to a children's ward at Kampala (thirty-one patients with sickle-cell anaemia admitted during this period are not included). (After Raper 1956.)

Disease group	Total	Number with sickle-cell trait	Incidence of sickle-cell trait
Miscellaneous	186	25	0.13
Pneumonia	118	18	0.15
Upper respiratory infections	59	13	0.22
Diarrhoea and vomiting	106	25	0.24
Poliomyelitis	26	4	0.15
Tuberculosis	37	8	0.22
Meningitis (purulent)	26	5	0.19
Malnutrition	77	11	0.14
Hookworm anaemia	30	2	0.07
Typhoid fever	17	6	0.35
Malaria (a) Uncomplicated	83	13	0.16
(b) Cerebral	47	—	0.00
(c) Blackwater fever	6	—	0.00
Total admissions	818	130	0.16

In general it appears that the malarial parasite, *P. falciparum*, flourishes less well in individuals who are heterozygous for the sickle-cell allele, and who have a mixture of both Hb A and Hb S in their red cells, than in individuals whose red cells contain only Hb A. This evidently results in sickle-cell heterozygotes having a reduced susceptibility in childhood to the development of the more serious complications of malaria and to a reduced mortality, so that where malaria is a major cause of death in early life, sickle-cell heterozygotes have a better chance of surviving to become adults and so to contribute offspring to the next generation. This differential susceptibility to the effects of malaria appears to be most marked in early life. Later on it tends to be obscured by the development of acquired immunity.

The precise biochemical reasons for this differential susceptibility to the effects of malaria are still rather uncertain. Also it is difficult to evaluate quantitatively in any particular population the degree of selective advantage enjoyed by the sickle-cell heterozygotes. However, it is possible to estimate theoretically the selective differential that would be required to maintain an allele at a given frequency in a population by the selective survival of hetero- zygotes. A simple model (Penrose 1954, 1963) of a population in equilibrium with balanced polymorphism is shown in table 8.5. It will be seen that if homozygotes (*aa*) for a particular allele always die in early life so that they contribute nothing to the next generation, the allele may nevertheless be maintained with a frequency as high as 0.1 if the heterozygotes (*Aa*) have a selective advantage of about 11% over the 'normal' homozygotes (*AA*).

TABLE 8.5

Population model illustrating balanced polymorphism due to selective advantage of the heterozygote (Penrose 1954, 1963).

Geno-types	Relative frequencies at conception (zygotic genotypes)	Relative fitness	Relative frequencies among parents (parental genotypes)	Example assuming $p = 0.9$, $q = 0.1$, and genotype *aa* is lethal in early life		
	(i)	(ii)	(iii) = (i)×(ii)	(i)	(ii)	(iii)
AA	p^2	$1-k/p^2$	$p^2(1-k/p^2)$	81	0.988	80
Aa	$2pq$	$1+k/pq$	$2pq(1+k/pq)$	18	1.111	20
aa	q^2	$1-k/q^2$	$q^2(1-k/q^2)$	1	0.000	0
All types	1	—	1	100	—	100

The three genotypes *AA*, *Aa* and *aa* differ in their relative fitnesses, that is, their chances of becoming parents of the next generation. However, the frequencies *p* and *q* of the two alleles *A* and *a* remain the same from one generation to the next. Random mating is assumed.
The example shows the relative fitnesses of individuals of genotypes *AA* and *Aa* which would be required to maintain a balanced polymorphism with $p = 0.9$ and $q = 0.1$, assuming that individuals of genotype *aa* die in early life (i.e. fitness = 0). The incidence of the heterozygote among the parents is 20%.

In such circumstances some 20% of the adult population would be hetero- zygotes. This kind of model makes it plausible to suppose that the observed differences in morbidity and mortality from malaria in different parts of Africa are, or have been in the past, sufficient to produce a selection differen- tial of the required order of magnitude.

The polymorphisms of other haemoglobin variants, such as Hb C in W. Africa and Hb E in S.E. Asia, have also been attributed to a selective advantage of the heterozygotes due to differential susceptibility to malaria. However, although the geographical distribution of these polymorphisms is very suggestive, there is as yet little direct evidence to support the idea. The same is true for the thalassaemia polymorphisms, where again the likely importance of malaria as a selective agent is indicated by the geographical distributions of the alleles. Thus in Italy (Bianco et al. 1952) a relatively high frequency (0.05–0.10) of one or other of the β-thalassaemia alleles is particularly marked in the Ferrara area, Sardinia and the South, and it is in these parts of the country that endemic malaria is, or has been until recently, a particularly important cause of mortality. Similar correlations between the distributions of thalassaemia and malaria are also seen elsewhere. One may note that Cooley's disease or 'thalassaemia major' which occurs in individuals homozygous for these alleles is in general a very severe condition and considerably reduces the chances of survival of affected subjects to adult life, so that for the thalassaemia alleles which cause this condition to have reached the high frequency observed in many areas, it seems necessary to postulate a quite significant selective advantage for the heterozygotes.

The evidence for the possible role of malaria in these various haemoglobin polymorphisms has been reviewed in detail by Livingstone (1971).

8.1.3. Glucose-6-dehydrogenase (G-6-PD)

More than eighty different variant forms of G-6-PD have now been discovered (pp. 162–174). They differ from the normal form of the enzyme and from one another in such properties as electrophoretic mobility, Michaelis constants, thermostability and pH optima, and it seems very probable that most or all of them are due to single aminoacid substitutions in the protein, similar to those found in the haemoglobins. They are apparently determined by a series of alleles at a gene locus on the X chromosome.

The incidence and distribution of these alleles in different populations is in a number of respects similar to that found with the haemoglobin alleles. Many are evidently rare, but some have an unusually high incidence in particular populations and give rise to characteristic polymorphisms. For instance, besides the allele Gd^B which determines the normal form of the enzyme, two other alleles, Gd^{A-} and Gd^A, both occur in many African populations with gene frequencies of around 0.2, though they are rare or absent elsewhere. The variant protein determined by Gd^{A-} causes the well-known African form of G-6-PD deficiency which is the basis of primaquine sensitivity and certain other adverse drug reactions. However, apart from this drug idiosyncrasy individuals carrying this allele appear to be in other respects quite healthy. The other common variant in Blacks determined by the allele Gd^A is associated with only a very slight reduction in enzyme level, and this apparently is harmless.

In many populations living in Southern Europe and the Middle East, a different sort of G-6-PD polymorphism occurs due to the high incidence of the allele $Gd^{Mediterranean}$. This determines another striking form of G-6-PD deficiency, and it predisposes to the haemolytic disease known as favism which may occur when affected individuals eat fava beans, a common feature of the diet in this part of the world. There are probably also other G-6-PD alleles which occur commonly in particular areas, for example Gd^{Canton} in S.E. Asia and Gd^{Athens} in Greece, although their distributions have not yet been worked out in detail.

Because populations which have a high incidence of one or another form of G-6-PD deficiency come from areas in which malaria is or has been in the past a major cause of mortality, it has been suggested (Motulsky 1964) that here, as in the case of the sickle-cell gene, malaria may have been an important selective agent in determining the prevalence of particular G-6-PD alleles (e.g. Gd^{A-} in Black populations and $Gd^{Mediterranean}$ in Southern European and Middle Eastern populations). The malaria parasite might proliferate less well in individuals whose red cells were G-6-PD deficient and therefore in some degree metabolically abnormal. There is some, though as yet not very extensive or completely convincing evidence (Gilles et al. 1967, Luzzatto et al. 1969, Livingstone 1971, Bienzle et al. 1972) to suggest that these alleles may indeed confer some selective advantage in terms of malarial morbidity or mortality. But one must also note that the Gd^A allele which, though as prevalent as the Gd^{A-} allele in Africa and similarly rare or absent elsewhere, does not result like the Gd^{A-} in a marked enzyme deficiency, and would be expected to alter the metabolism of the red cell hardly at all.

8.1.4. The haptoglobin variants

The various haptoglobin types (ch. 3, pp. 91–99) provide another example of protein variation which has been studied very extensively. There are two sorts of polypeptide chains, α and β, and most of the variants which have been observed can be attributed to alleles at the α gene locus. The findings here however are in striking contrast to those obtained with haemoglobin and G-6-PD because there are at least three alleles (Hp^{1S}, Hp^{1F} and Hp^2) which are common and widespread throughout the world (Kirk 1968). In European and African populations all three are found, though with differing frequencies, and in Asiatic populations Hp^{1S} and Hp^2 both have a significant incidence, though Hp^{1F} may be rare (Shim and Bearn 1964). As with haemoglobin and G-6-PD a number of other very rare alleles at the haptoglobin loci (both α and β) have been shown to occur.

A point of special interest about the haptoglobin polymorphism is that it is possible to infer from the structural differences in the protein something about the origin of the alleles (Smithies et al. 1962b). The α polypeptides determined by Hp^{1F} and Hp^{1S} each contain 83 aminoacids, and differ only in a single one (Black and Dixon 1968). The α polypeptide determined by Hp^2 is nearly twice as long (142 aminoacids) and appears to represent an end to end fusion of the hp1Fα and the hp1Sα polypeptides with a sequence of 24 residues missing at the site of fusion. It presumably originated as the result of a mutational event involving a chromosomal rearrangement in an individual who happened to be heterozygous for Hp^{1F} and Hp^{1S}. In other words, the new allele probably arose in a population already polymorphic for the Hp^{1F} and Hp^{1S} alleles. Furthermore, the peculiar structure of the polypeptide results in a rather characteristic polymerisation of the haptoglobin molecule which is readily detected by starch gel electrophoresis, and since this effect has not been seen in haptoglobins in other species including higher apes (Parker and Bearn 1961), it seems quite likely that the mutational event giving rise to the Hp^2 allele occurred only after the separation of the human line. Nevertheless it has apparently spread throughout the species and today is the commonest of the three alleles in most human populations (fig. 8.2).

Thus one appears to be observing an evolutionary change in the gross structure of a specific protein, and one might anticipate that the new structure conferred some distinctive selective advantage. Yet it is difficult to see from what is known about the differences in the properties of the common haptoglobin types exactly what this might be.

Unfortunately the exact significance of haptoglobin in normal function is still very unclear (Sutton 1970). It has the characteristic property of binding tightly and specifically with free haemoglobin, a reaction which might have some significance in connexion with the conservation of iron in the body. It has been shown (Nyman 1959) that the haptoglobin content of serum as measured by its haemoglobin binding capacity differs on average between the three types, Hp 1-1, Hp 2-1 and Hp 2-2, roughly in the ratio 135:110:85, although there is much overlap between the distributions (fig. 8.3). Also the haptoglobin–haemoglobin complex appears to be a substrate, for the liver

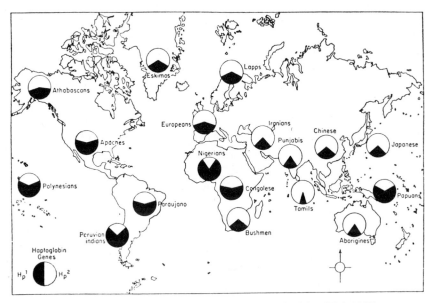

Fig. 8.2. World distribution of the *Hp*¹ and *Hp*² alleles (Kirk 1968).

enzyme α-methenyl oxygenase which is thought to be concerned in bile pigment formation (Nakajima et al. 1963), and it has been reported that the Hp 2-2 haptoglobin type is somewhat more effective in this respect than Hp 2-1 or Hp 1-1 (Nakajima 1963). But these effects have not yet been shown to lead to any obvious metabolic differences, and as far as is known, individuals of the different haptoglobin types are all equally healthy. It may also be noted that there appear to be certain as yet poorly defined haptoglobin alleles which apparently lead to a gross reduction and, in some cases, virtual absence of serum haptoglobin, and at least one of these alleles is apparently quite common in Black populations (Giblett and Steinberg 1960), though not elsewhere. However, individuals in whom such a genetically determined deficiency of haptoglobin has been demonstrated do not appear to exhibit any other obvious phenotypic peculiarities which might be selectively significant.

In general then, it seems necessary to assume that such selective differences as may occur between the various common haptoglobin types are very slight, or that they were for some unknown reason much more significant in the past but have been minimised and rendered trivial by subsequent changes in the environment.

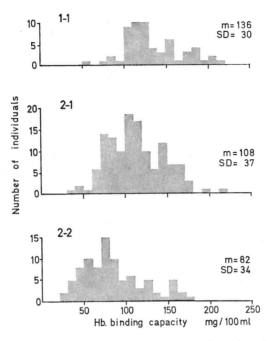

Fig. 8.3. Distributions of haemoglobin binding capacities of sera from individuals of the common haptoglobin types Hp 1-1, Hp 2-1 and Hp 2-2. The mean values (m) and standard deviations (SD) for each type are also shown. (Data of Nyman 1959.)

8.1.5. Phosphoglucomutase

It has been shown that three separate and unlinked loci determine the complex patterns of phosphoglucomutase isozymes which occur in human tissues (ch. 2, pp. 60–69). Quite a number of different electrophoretic variants determined by alleles at each of these loci have been discovered. But the incidence and distribution of these alleles varies markedly from one locus to another.

At locus PGM_1 two common alleles occur and give rise to a polymorphism which appears to occur throughout the species. Among Europeans the frequencies of the two alleles PGM_2^1 and PGM_2^2 are about 0.76 and 0.24 respectively, and very similar frequencies have been observed in various other ethnic groups, including various African and Oriental populations. Indeed the polymorphism appears to be remarkably uniform throughout the species. A number of variants due to other alleles at this locus have also been discovered, but each of these is extremely rare.

Polymorphism due to the occurrence of two common alleles at locus PGM_3 is also widespread. But here the relative incidence of the two alleles is very different in different populations. In Europeans PGM_3^1 is nearly three times as frequent as PGM_3^2 (PGM_3^1, 0.74; PGM_3^2, 0.26), whereas among Africans the situation is effectively reversed, PGM_3^2 being more frequent than PGM_3^1 (PGM_3^1, 0.4; PGM_3^2, 0.6).

At locus PGM_2 yet a further type of situation is observed. Here the great majority of individuals in all populations show the same electrophoretic phenotype and are apparently homozygous for the common allele PGM_2^1. A number of different electrophoretic variants have indeed been demonstrated, but in most populations they are very rare.

These phosphoglucomutase variants were discovered in the course of an electrophoretic screening programme deliberately aimed at searching for common polymorphic differences (see p. 295). The individuals studied were in the main normal and healthy, and there was no indication that the common variant types of the enzyme were associated with any marked functional differences which might be of selective significance. It seems therefore that if such differences do occur, they are probably very subtle and relatively small in magnitude.

8.2. Polymorphism

8.2.1. Enzyme surveys

The examples discussed above illustrate something of the degree of allelic variation that may occur at different gene loci. It seems that at some loci although multiple alleles may be demonstrable, there is one allele which can be regarded as the standard or normal form and is almost universally present while all the others are extremely rare. At other loci (e.g. the β-haemoglobin locus and the G-6-PD locus) although a standard allele occurs and is recognisable as such, there are in some populations but not in others alleles which are present in sufficient frequency as to give rise to a common polymorphism. At still other loci (e.g. the α-haptoglobin locus and the PGM_1 and PGM_3 loci) polymorphism is the rule. Two or more alleles occur relatively frequently and are widely distributed in many different populations. Indeed in certain cases there appears to be no valid reason for regarding one allele rather than another as the so-called normal or standard form.

A question which obviously arises is what is the relative incidence of these various situations among gene loci in general, and in particular how often do polymorphisms occur. Do the haemoglobin, G-6-PD, haptoglobin and phosphoglucomutase polymorphisms represent special and perhaps rather unusual forms of variation not typical of enzymes or proteins in general, or are they examples of a relatively common phenomenon? It would clearly be of great interest to know how many of the very large number of proteins and enzymes which are formed in the human organism exhibit this sort of variation. Since the structure of each protein is presumed to be determined by at least one gene locus, we are in effect asking at what proportion of this vast array of gene loci do two or more relatively common alleles occur in different human populations.

With the advent of electrophoretic methods capable of being applied to a wide range of different enzymes it became possible to approach this general problem in a direct manner. In principle one could examine a number of different arbitrarily chosen enzymes in a randomly selected series of individuals from any given population and find out how often individual differences could be detected. By family studies one could then work out the genetical basis of different variants, and so determine the population frequencies of the alleles involved. In this way one could obtain at least a minimal estimate of the proportion of gene loci at which two or more common alleles occurred. This general approach has in fact been pursued in studies on human populations (Harris 1966, 1969, Harris and Hopkinson 1972) and similar investigations have been carried out on naturally occurring populations of many other animal species (Hubby and Lewontin 1966, Lewontin and Hubby 1966, Selander and Kaufman 1973). In the event it has turned out that genetically determined enzyme polymorphism is a very much more common phenomenon than had previously been anticipated.

A polymorphic locus can for convenience be defined as one at which the most commonly occurring allele in the particular population has a frequency which is less than 0.99. This means that at least 2% of the individual members of the population will be heterozygous at that locus. In many cases of course the proportion of heterozygotes is much greater than this. Adopting this definition it appears from a variety of population surveys in man that about 30% of loci coding for enzyme structure exhibit polymorphisms which are detectable by electrophoresis (Harris and Hopkinson 1972). Since there are no doubt other polymorphisms involving alleles which determine differences in enzyme structure, not detectable by electrophoresis, the overall incidence of enzyme polymorphism is presumably greater than this.

8.2.2. *The average degree of heterozygosity*

The widespread occurrence of enzyme and protein polymorphisms implies that any given individual is likely to be heterozygous at many different gene loci, and it is of some interest to enquire what the average degree of such heterozygosity may be. That is, the proportion of gene loci in a single individual at which there are likely to occur two different alleles, each specifying a structurally distinct version of a particular enzyme or protein.

TABLE 8.6

Enzyme polymorphism in Europeans (Harris and Hopkinson 1972).

Number of loci screened	71
Number of loci showing electrophoretic polymorphism (i.e. >0.02 heterozygotes)	20
Percentage of polymorphic loci	28.2
Average heterozygosity per locus (detected electrophoretically)	0.067

A recent calculation is summarised in table 8.6. It is concerned with individuals of European origin and is based on electrophoretic enzyme surveys (Harris and Hopkinson 1972). In all, data on the enzymic products of some 71 different gene loci was available. Of these 71 loci, 20 showed electrophoretic polymorphism, i.e. 28%. By summing the observed values for heterozygosity at each locus, and dividing by the total number of loci (i.e. 71), one obtains an estimate of the average heterozygosity per locus for common alleles which result in electrophoretic differences. This is 0.067. So this particular set of data suggests that in man any single individual is likely to be heterozygous at about 7% of his loci coding for enzyme structure, for common alleles giving rise to electrophoretic differences. Of course this is likely to be an underestimate of the true average heterozygosity because only electrophoretic variants were taken into account.

Essentially similar results have also been obtained from electrophoretic surveys of enzymes and proteins in natural populations of a variety of other animal species. Data on nearly fifty different species have now been reported (table 8.7), and it has become increasingly clear that electrophoretic polymorphism is a widespread phenomenon. And as the data accumulate it is becoming possible to make informative comparisons between species. Selander and Kaufman (1973) for example have drawn attention to the fact

TABLE 8.7

Average heterozygosity per locus in different animal species, for alleles giving rise to electrophoretic enzyme and protein differences. For detailed references see Selander and Kaufman (1973).

Organism	No. of species	No. of loci	Average heterozygosity per locus	
			Mean	Range
Invertebrates				
Drosophila	19	11–33	0.145	0.05–0.22
Field cricket	1	20	0.145	—
Land snails	3	17	0.207	0.14–0.25
Horseshoe crab	1	25	0.097	—
Total	24		0.1507	
Vertebrates				
Fish (Tetra)	1	17	0.112	—
Lizards	4	15–29	0.058	0.05–0.07
Sparrow	1	15	0.059	—
Rodents	14	18–41	0.056	0.01–0.09
Seal	1	19	0.030	—
Man (European)	1	71	0.067	—
Total	22		0.0584	

that among the animal species which have been studied, the average heterozygosity per locus appears to be on average about two and a half times greater in invertebrates than vertebrates (table 8.7). This is evidently due to a greater proportion of polymorphic loci and also a greater average number of common alleles per locus. On the other hand, within each of these two groups of species, the average heterozygosities in different species are quite similar considering the variety of populations studied and the variations in the techniques which have been applied.

8.2.3. The uniqueness of the individual

An important point which emerges from these general considerations is the very considerable degree of individual diversity in enzyme and protein make-up which must actually occur among the individual members of human populations. A rough idea of the extent of this can be obtained by considering together some of the enzyme and protein polymorphisms that have already been shown to exist in a single population. Relevant data on twelve different enzymes which are polymorphic among Europeans are given in table 8.8. Since the several distinct phenotypes which make up each of these polymorphisms appear to occur independently of those of the others, it is clear that a very large number of different combinations of phenotypes must occur among individual members of the population. By combining the figures given in column 3 of the table, one finds that the most commonly occurring

TABLE 8.8

Enzyme individuality in the English population. Data on 12 enzymes (15 loci).

Enzyme	Number alleles with frequency >0.01	Frequency of commonest phenotype	Probability of two randomly selected individuals having the same phenotype	References
Red cell acid phosphatase	3	0.42	0.32	pp. 180–184
Phosphoglucomutase				
locus PGM_1	2	0.57	0.46	pp. 60–67
locus PGM_3	2	0.55	0.46	pp. 60–67
Placental alkaline phosphatase	3	0.39	0.29	pp. 39–40
Liver acetyl-transferase	2	0.53	0.50	pp. 151–154
Serum cholinesterase				
locus E_1	2	0.96	0.92	pp. 150–160
locus E_2	2	0.90	0.82	pp. 160–162
Adenylate kinase	2	0.92	0.85	Fildes and Harris (1966)
Adenosine deaminase	2	0.90	0.82	Spencer et al. (1968)
Phosphogluconate dehydrogenase	2	0.96	0.92	Parr (1966)
Alcohol dehydrogenase				
locus ADH_2	2	0.94	0.86	Smith et al. (1971, 1972)
locus ADH_3	2	0.48	0.38	Smith et al. (1971, 1972)
Glutamate-pyruvate transaminase	2	0.50	0.38	Chen and Giblett (1971)
Esterase-D	2	0.82	0.70	Hopkinson et al. (1973)
Malic enzyme (mitochondrial form)	2	0.48	0.42	Cohen and Omen (1972)
Combined	—	0.0017	0.00017	

combination of types will be found in only about 0.17% of the population. Furthermore, from column 4 one can show that the chance that two randomly selected individuals in the population would have exactly the same combination of types is only about 1 in 6,000. Thus quite a high degree of individual differentiation in enzyme and protein makeup can be demonstrated with what is only a very limited series of examples. This must surely represent only the tip of the iceberg, and one may plausibly conclude that in the last analysis every individual will be found to have his own unique enzyme and protein constitution.

One might expect that this uniqueness of the individual in his basic enzyme and protein makeup is likely in a great variety of ways to be reflected in his physical, metabolic and physiological characteristics, and also in the degree of his susceptibility to the development of particular sorts of disease or other abnormalities. However so far it has not proved possible to analyse this question in any systematic way, and for the most part we have very little certain information about the possible functional effects of different combinations of polymorphic phenotypes occurring in different individuals.

However in quite a number of the enzyme polymorphisms, significant differences in the level of enzyme activity between the commonly occurring phenotypes which make up the particular polymorphism have been demonstrated. Examples of this are seen in the polymorphisms involving glucose-6-phosphate dehydrogenase (p. 162); liver acetyl transferase (p. 251); red cell acid phosphatase (p. 180); placental alkaline phosphatase (p. 39); serum cholinesterase (p. 150); 6-phosphogluconate dehydrogenase (Parr 1966); glutamate-pyruvate transaminase (Chen et al. 1972); peptidase A (Sinha et al. 1970, Lewis 1973); and adenylate kinase (Modiano et al. 1970, Rapley and Harris 1970).

A particular interesting example of such activity differences between the phenotypes of a polymorphism is provided by alcohol dehydrogenase (locus ADH_2). The so-called 'atypical' phenotype which is found in about 6% of individuals in European populations (Smith, Hopkinson and Harris 1971) has greatly enhanced activity compared with the so-called 'usual' phenotype. There are also marked differences in the pH optimum of the enzyme between the enzymes of the two phenotypes and also differences in kinetic, inhibition and other characteristics (Von Wartburg et al. 1965, Von Wartburg and Schürch 1968, Smith et al. 1973). In Japan, however this phenotype which from the European studies was called 'atypical', appears to be the common or usual form, since it has been found in some 90% of a random population sample (Fukui and Wakasugi 1972). So far, no overt metabolic differences have been identified between individuals with these very different alcohol dehydrogenase phenotypes. But Fukui and Wakasugi (1972) suggest that the much higher incidence of the so-called 'atypical' phenotype in Japan as compared with Europe, might perhaps account for the higher incidence of chronic alcoholism in Europe.

In all, some thirty different enzyme polymorphisms have now been identified in human populations (Appendix 2, p. 297). In about half the cases significant differences in activity levels between individuals of the different phenotypes have been demonstrated. Many of the others have not been studied in detail from this point of view.

Occasionally the common variants of a particular enzyme which make up a polymorphism may exhibit a qualitative difference in their substrate specificity rather than just a quantitative change in activity. A probable example of this is provided by the well known ABO blood group polymorphism (p. 257). Here it seems that the *A* and *B* alleles result in characteristic antigenic differences because they determine glycosyl transferase enzymes with qualitatively different substrate requirements. The 'A' enzyme specificity is directed to the transfer of an N-acetylgalactosamine residue from a nucleotide diphosphate compound to the end of a particular type of oligosaccharide chain, while the 'B' enzyme specifically transfers a D-galactose residue. It is of interest that individuals with different phenotypes of the ABO polymorphism appear to differ to some degree in their susceptibility to the development of peptic ulceration and gastric carcinoma (McConnell 1966, Vogel 1970), and possibly also other conditions such as atherosclerosis (Kingsbury 1971), thromboembolic disease (Mourant et al. 1971) and coronary disease (Medalie et al. 1971). However the causal connexions between the enzyme differences in the several phenotypes and the apparent differences in susceptibility to these conditions are not known. Presumably they are very indirect.

8.3. *Rare alleles*

In searching for polymorphism in any given population, the number of different unrelated individuals examined need usually amount to no more than a hundred or two, because the alleles involved are by definition relatively common. But to study rare alleles very much larger samples are required.

The point is illustrated by the findings in the case of one of the loci which determines the enzyme phosphoglucomutase (PGM_1). In the early studies the three common phenotypes were identified and were shown to be due to

two common alleles with frequencies in European populations of about
0.76 and 0.24; the homozygous phenotypes occurring in about 56% and 6%
of the population respectively and the heterozygous phenotype in about
36% of the population. Subsequently in the course of examining this enzyme
in a much larger number of individuals a series of rare phenotypes were
identified, which could be attributed to the occurrence of rare alleles in
heterozygous combination with one or the other of the two common alleles.
Fig. 8.4 illustrates the findings in a sample of 10,333 unrelated Europeans.
Since the locus is autosomal this means that in effect 20,666 alleles were
screened. Five different rare alleles were identified, one of which was found
four times, two three times and two only once, so their individual frequencies
were each less than 0.0002. Their incidence is in fact not dissimilar from the
incidence of the different rare alleles at the α- and β-haemoglobin loci
discussed earlier (p. 283).

Similar rare alleles have of course also been discovered in the course of
electrophoretic surveys of other enzymes and it is of obvious interest to try
and obtain a general picture of the incidence of such rare alleles over a wide
range of loci.

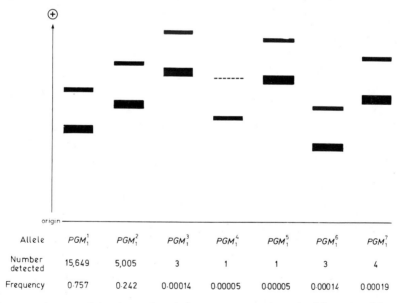

Allele	PGM_1^1	PGM_1^2	PGM_1^3	PGM_1^4	PGM_1^5	PGM_1^6	PGM_1^7
Number detected	15,649	5,005	3	1	1	3	4
Frequency	0.757	0.242	0.00014	0.00005	0.00005	0.00014	0.00019

Fig. 8.4. Diagram of the electrophoretic isozyme pattern determined by seven different
alleles at the phosphoglucomutase PGM_1 locus. The incidence of the different alleles
found in the course of screening 10,333 unrelated Europeans (i.e. 20,666 alleles) is indicated.
Data of Harris et al. (1974).

To this end, Harris, Hopkinson and Robson (1973) assembled the data which had been collected by one laboratory over a period of about ten years. It was found that reasonably reliable information on 43 different enzyme loci was available. The number of unrelated individuals examined (all Europeans) varied for the different enzymes from just a few hundred up to more than 10,000, but the average per locus was about 2,500, and since the loci are probably all or nearly all autosomal this implies that on average the products of about 5,000 genes at each locus had been screened for electrophoretic variants. For the purpose of the analysis a rare allele was defined as one which had a gene frequency of less than 1 in 200 in the general population. In all, 56 different ones had been identified, and more than 80% of these appeared to have frequencies of less than 1 in 1,000, so the majority of these alleles were indeed extremely uncommon.

The tabulations of the detailed findings over the 43 loci are rather extensive (Harris et al. 1973), so in order to illustrate the main features of the data the results in just 12 of them have been summarised in table 8.9. The first line for example gives the findings for the phosphoglucomutase locus (PGM_1) which have already been mentioned (fig. 8.4). The number of unrelated individuals tested was 10,333, i.e. 20,666 genes. Five different rare alleles were identified, and these occurred in 12 individuals who were heterozygous for one or another of them. So the combined incidence of rare heterozygotes was 1.16 per 1,000. At this locus there are also two common alleles, and the heterozygosity in the population due to these is 0.366 or 366 per 1,000.

In the surveys one or more rare alleles were detected at just over half the 43 loci studied. Obviously the chance of detecting such rare events will depend on the size of the sample. And indeed the number of individuals examined was found to be on average much lower among the loci where no rare alleles were detected (1,300), than among the others (4,023). Also at several of the loci, although no rare variants were detected in these particular surveys, such variants have in fact been found in other studies. So in general it seems not unlikely that such rare alleles actually occur at virtually all loci.

It is obvious that estimates of the frequencies of individual alleles must inevitably be very imprecise because most of them are extremely rare. It is more practical therefore in comparing different loci to consider the combined incidence of heterozygosity due to the rare alleles at each of the loci (penultimate column in table 8.9). These heterozygosities per locus

TABLE 8.9

Incidence of rare alleles determining electrophoretic enzyme variants in unrelated Europeans. Further details and also detailed data on 31 other loci are given in Harris, Hopkinson and Robson (1973).

Enzyme	Locus	No. of unrelated individuals tested	No. of different rare alleles	No. of rare heterozygotes	Heterozygotes for rare alleles (per 1,000)	Heterozygotes for common alleles (per 1,000)
Phosphoglucomutase	PGM_1	10,333	5	12	1.16	366
Phosphoglucomutase	PGM_2	10,333	3	7	0.68	—
Glutamate-oxalate transaminase 'soluble'	GOT_S	1,195	2	2	1.67	—
Glutamate-oxalate transaminase 'mitochondrial'	GOT_M	1,195	0	0	—	33
Adenylate kinase	AK	6,760	1	1	0.15	77
Nucleoside phosphorylase	NP	1,542	2	2	1.30	—
Pyrophosphatase	PP	2,190	0	0	—	—
Alkaline phosphatase-placental	PL	3,244	10	76	23.43	502
Peptidase-B	$PEP\text{-}B$	7,041	3	15	2.13	—
Triosephosphate isomerase	TPI	1,750	2	2	1.17	—
Superoxide dismutase	SOD_A	11,237	1	7	0.62	—
Inosine triphosphatase	ITP	641	0	0	—	—

were found to vary considerably, extending over a 150-fold range even if one considers only those loci at which rare alleles were in fact discovered. To some extent this variation undoubtedly occurs for technical reasons, because it is known that there is much variation in the discriminative power of the electrophoretic methods available for the different enzymes. Nevertheless it seems likely that the variation does to a significant degree reflect real differences between loci.

A case in point is locus *PL* which determines placental alkaline phosphatase (p. 39) and which shows a quite remarkable degree of allelic variation. In the course of examining placentae from 3,244 individuals, evidence for as many as 14 different alleles was obtained. Ten of these could be counted as rare alleles according to the definition and they account for as many as 23 heterozygotes per 1,000 individuals. The number of different rare alleles found in this particular case and the degree of heterozygosity to which they give rise, makes one suspect that at this locus some unusual process is responsible either for generating the allelic diversity or maintaining it at this high level.

Despite this variation in heterozygosity from locus to locus, it is nevertheless of some interest to estimate the average heterozygosity per locus due to rare alleles, over the full range of loci including those where no rare variants at all were discovered (table 8.10). This average is 1.76 per 1,000 for all the loci. But it is reduced to 1.14 per 1,000 if we exclude *PL*. Thus the data as a whole suggest that on average for any single locus one may expect that between 1 and 2 individuals per 1,000 will be heterozygous for a rare allele determining an electrophoretic variant. If we suppose for example that

TABLE 8.10

Average heterozygosity per locus for rare alleles determining electrophoretic enzyme variants. For data see Harris, Hopkinson and Robson (1973). *PL* is the locus for placental alkaline phosphatase.

	Number of loci	Total No. of genes screened	Total No. of rare alleles	Heterozygotes per locus per 1,000 individuals
All loci	43	231,508	204	1.76
All loci excluding *PL*	42	225,020	128	1.14

there are say 30,000 loci coding for enzyme structure, each of us is likely to be carrying at least 30 such alleles, and it is extremely improbable that any two of us will, with the exception of monozygotic twins, have exactly the same combination.

It is also of interest to ask whether the so-called 'polymorphic' loci and 'non-polymorphic' loci differ in the incidence of rare alleles. The average heterozygosity per locus due to rare alleles is 2.61 per 1,000 for the 13 'polymorphic' loci in the data and 1.16 for the 'non-polymorphic' loci (table 8.11). This is a considerable difference, but it is entirely accounted for by the locus *PL*. If this is removed the two values are almost the same.

The alleles which determine these rare electrophoretic enzyme variants must have originated by mutations occurring in single individuals in earlier generations. The question therefore arises as to how many of the rare

TABLE 8.11

Average heterozygosity per locus for rare alleles determining electrophoretic enzyme variants at polymorphic and non-polymorphic loci. For detailed data see Harris, Hopkinson and Robson (1973). *PL* is the locus for placental alkaline phosphatase.

	Number of loci	Total No. of genes screened	Total No. of rare alleles	Heterozygotes per locus per 1,000 individuals
Non-polymorphic loci	30	135,750	79	1.16
Polymorphic loci	13	95,758	125	2.61
Polymorphic loci excluding *PL*	12	89,280	49	1.10

variants observed in such surveys are the consequence of fresh mutations. That is to say mutations which occurred in the germ line of one or other of the parents of individuals found to have the variant.

In the course of this work (Harris et al. 1973) systematic studies on the families of individuals showing rare variants had been carried out whenever this was practicable, and an extensive body of pedigree data had been assembled. In fact no clear example of a fresh mutation was found. But of course not all the family data is informative, because sometimes critical individuals could not be tested. However in 77 unrelated individuals who were heterozygous for one or another of these rare variants both the parents had been tested. And in each case, either the father or the mother of the individual with the variant showed the same variant. So none of these cases was due to a fresh mutation.

One can therefore conclude that the great majority of the rare variants observed in these population samples were not the products of fresh mutations, and the fraction of all rare variants attributable to fresh mutations must be very small – probably less than 1–2%.

By combining the data for the rare alleles and the common alleles when these occurred it was possible to obtain an estimate for the heterozygosity at each of the loci studied in these surveys. Of course at some loci no variants were found at all, but even here the size of the population sample screened gives one an idea of the possible upper limit of the amount of heterozygosity. Fig. 8.5 shows the distribution on a log scale of the heterozygosities over the whole series of loci. It shows two interesting features. The first is the very wide variation in the values that occur. At one extreme there are loci with heterozygosities of around 0.5. At the other, there are loci with heterozygosities

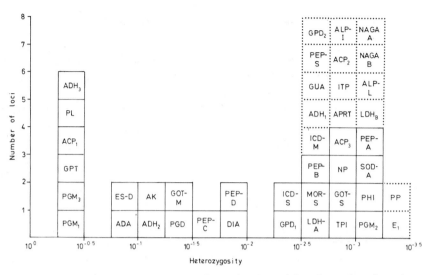

Fig. 8.5. Distribution of heterozygosities for alleles determining electrophoretic variants at 43 loci (from Harris, Hopkinson and Robson 1973). Each square represents a locus, the symbol for which is indicated. At some of the loci (indicated by dotted lines) no heterozygotes were detected in the surveys, but to include them in the distribution they have been placed in the position where they would have been if in fact one heterozygote had been observed. The true heterozygosities for these loci must therefore lie somewhat to the right of the position shown.

Key: *ADA*, adenosine deaminase; *ACP₁*, *ACP₂*, *ACP₃*, acid phosphatase loci; *ADH₁*, *ADH₂*, *ADH₃*, alcohol dehydrogenase loci; *AK*, adenylate kinase; *ALP-I*, intestinal alkaline phosphatase; *ALP-L*, liver alkaline phosphatase; *APRT*, adenine phosphoribosyl transferase; *DIA*, NADH diaphorase; *E₁*, serum cholinesterase; *ES-D*, esterase D; *GPD₁*, *GPD₂*, α-glycerophosphate dehydrogenase loci; *GOT-S*, *GOT-M*, glutamate-oxaloacetate transaminase loci; *GPT*, glutamate-pyruvate transaminase; *GUA*, guanase; *ICD-S*, *ICD-M*, isocitrate dehydrogenase loci; *ITP*, inosine triphosphatase; *LDH_A*, *LDH_B*, lactate dehydrogenase loci; *MOR-S*, NAD malate dehydrogenase (soluble); *NAGA-A*, *NAGA-B*, hexosaminidase loci; *NP*, purine nucleoside phosphorylase; *PEP-A*, *PEP-B*, *PEP-C*, *PEP-D*, *PEP-S*, peptidase loci; *PGD*, phosphogluconate dehydrogenase; *PGM₁*, *PGM₂*, *PGM₃*, phosphoglucomutase loci; *PHI*, phosphohexose isomerase; *PL*, placental alkaline phosphatase; *PP*, inorganic pyrophosphatase; *SOD-A*, superoxide dismutase; *TPI*, triosephosphate isomerase.

of 0.001 or less. So the heterozygosities over this series of loci vary over at least a 500-fold range.

The second feature of interest is that the distribution tends to bimodality. Thus it appears to provide objective evidence for the view that in any population there is a real dichotomy between so-called polymorphic loci (i.e. those with high heterozygosities) and so-called monomorphic loci (i.e. those with low heterozygosities).

However it should be noted that the distribution is concerned only with heterozygosity attributable to alleles which determine electrophoretically distinct forms of the corresponding enzymes. There are, we must presume, many other alleles at these loci which do not result in a variant with an altered electrophoretic mobility (vide infra) and at present we have very little information about the heterozygosity attributable to such alleles at individual loci, or how it may vary from locus to locus.

8.4. *Alleles not detected by electrophoresis*

Although electrophoresis is a most powerful general technique for screening populations in order to detect allelic variants, estimates both of the number of different alleles and also of the amount of heterozygosity at various loci obtained by this procedure, are inevitably underestimates of the true values. There are a number of different reasons why this must be so.

1) Because the genetic code is degenerate some single base change mutations will simply alter a codon specifying a specific aminoacid in a protein to another codon specifying the same aminoacid. So they produce no change in the structure of the protein. Such mutations are said to be synonymous.

At present there is virtually no information about the possible occurrence of such synonymous alleles in human populations because there has been no general method for detecting them. Fitch (1973) has however made an interesting observation which suggests a possible approach to the problem. He points out that three rare haemoglobin variants with different aminoacid substitutions in the β 67 position normally occupied by valine have been reported. They are Hb M Milwaukee with glutamic acid at β 67 (Gerald and Efron 1961), Hb Bristol with aspartic acid at β 67 (Steadman et al. 1970), and Hb Sydney with alanine at β 67 (Carrell et al. 1967). But if these substitutions are correctly identified and if one supposes that both the glutamic acid substitution and the aspartic acid substitution each arose by single base change mutations then it follows that some normal β-alleles must have a purine and others a pyrimidine in the third position of the valine codon (see the genetic code fig. 1.5, p. 14). If this is so the two synonymous β-alleles are each likely to be relatively common and the situation would therefore represent an example of polymorphism. The alanine substitution in Hb Sydney could have been derived from either allele.

2) It has been pointed out already that only a proportion of mutants resulting in single aminoacid substitutions will be detectable by electrophoresis, because many of them will cause no alteration in charge (p. 16). On theoretical grounds one would expect only about one-third of the possible protein variants with single aminoacid substitutions to show a change in electrophoretic mobility (p. 280). A number of 'no charge' substitutions have in fact been identified among the haemoglobin variants. They were however mainly found in studies on individuals selected by the fact that they had some form of haemolytic disease attributable to an abnormal haemoglobin, usually an unstable one (p. 24). Those not resulting in clinical abnormality generally pass undetected. The same of course will be true for enzyme proteins. Though even here where detailed structural studies are far less advanced, occasional examples have been identified. For instance the so-called 'atypical' and 'fluoride resistant' variants of serum cholinesterase (p. 150) show significant changes in their kinetics and other properties, which enable them to be readily identified by screening, and imply structural changes in the enzyme protein. However they cannot be detected in routine electrophoretic surveys. In general therefore although such non-electrophoretic enzyme variants are known to occur we have at present very little information about their incidence at different loci.

In this connexion some interesting observations have been made by Boyer et al. (1972) in the course of aminoacid sequencing studies on the haemoglobins of various primates. Their results suggest that among the commoner variants which occur, neutral aminoacid substitutions are about five times more frequent than electrophoretically detectable substitutions.

3) In electrophoretic surveys it is usually assumed for operational reasons that variants which are indistinguishable electrophoretically are attributable to the same allele. But this is not necessary so, since two quite different aminoacid substitutions in an enzyme can produce the same change in electrophoretic mobility. So the number of alleles recorded in such surveys must be minimal estimates.

4) It is important to remember that the electrophoretic systems which have been developed for enzyme screening, vary considerably in their discriminative power from enzyme to enzyme. It is therefore quite probable that at some loci, because of the inadequacy of the technique, electrophoretic variants which actually occur have not been detected.

Indeed in some cases even where the discriminative power of a particular technique was thought to be quite good, further improvements have led to the discovery of previously unsuspected allelic differences. One example of this is peptidase A (p. 32). Originally it was thought that most individuals in European populations were homozygous for a single allele which determined the phenotype designated Pep A 1 (Lewis and Harris 1967). Subsequently, however, it was found that by modifying the electrophoretic method this phenotype could be differentiated into three different types which were called Pep A 1, Pep 8-1 and Pep A 8 (Lewis 1973). It emerged that there is in fact a common polymorphism in European populations involving at least two different alleles *Pep A*^{1} and *Pep A*^{8}.

Another example is the case of the haptoglobin 1-1 phenotype (p. 92). This was originally attributed to homozygosity for a single allele. Subsequently, however, when it became possible to examine the separated α-polypeptide chains it was found that there were two electrophoretically different types distinguished by a single aminoacid substitution (lysine or glutamic acid at α 54), and determined by two common alleles *Hp*^{1F} and *Hp*^{1S}. Evidently the charge difference in the separated polypeptide chains is sufficiently concealed in the fully constituted protein that it cannot be recognised by routine electrophoresis.

The *p*H at which an electrophoretic separation is carried out is not surprisingly an important factor in determining the discriminative power of a particular system. This is well illustrated by the placental alkaline phosphatase polymorphism (p. 39). Here it was found that the isozyme products of the allele *PL*^{1} could not be distinguished from those of allele *PL*^{3} at *p*H 6.0, though they were readily separable at *p*H 8.6. Routine electrophoretic work carried out on this enzyme both at *p*H 6.0 and *p*H 8.6 has been important in improving the discrimination between the various phenotypes which occur. Even so there is no doubt that the incidence of many rare alleles at this locus is underestimated in population surveys (Robson and Harris 1966, Donald and Robson 1973).

5) Underestimation of alleles in electrophoretic surveys must also have occurred for another and perhaps rather less obvious reason. The electrophoretic techniques depend on the use of specific staining reactions designed to detect the activity of the particular enzyme under investigation. Consequently a variant will only be detected in heterozygous individuals, if it has retained the specific catalytic activity characteristic of the enzyme. It is known however, that certain mutant alleles result in a complete loss or a very marked reduction in the enzymic activity. This may be because the presumed aminoacid substitution has affected the catalytic site, or has rendered the enzyme molecule excessively unstable so that it decays very rapidly, or perhaps because it has reduced its rate of synthesis. In a hetero-

zygote for such an allele, one would therefore only see the electrophoretic band corresponding to the normal allele with which it is in heterozygous combination, and although this may be somewhat weaker than usual, the methods are in general so insensitive to quantitative variation that in virtually all cases the anomaly would be missed.

Evidence for the occurrence of heterozygotes for mutants resulting in such activity loss – so-called 'silent' or 'null' alleles – has indeed been obtained at a number of the loci which have been studied by these electrophoretic techniques. But this evidence has mainly come from the anomalous segregation patterns which are occasionally produced in families in which such an allele is segregating. Some examples are, red cell acid phosphatase (Herbich, Fisher and Hopkinson 1970); phosphoglucomutase *PGM*₁ (Brinkmann et al. 1972); phosphogluconate dehydrogenase (Parr 1966, Parr and Fitch 1967); and haptoglobin (Harris, Robson and Siniscalco 1958, Matsunaga 1962, Prokop and Dietrich 1968, Cook et al. 1969). It has however not yet proved possible to obtain satisfactory estimates of their incidence from population surveys, even where these have been combined with quite extensive family studies.

Occasionally the enzyme deficiency state in homozygotes for 'silent' or 'null' alleles has been detected in the course of electrophoretic surveys (e.g. phosphogluconate dehydrogenase, Parr and Fitch 1967). However more extensive information on the incidence of such alleles has come from a quite different approach – screening for inborn errors of metabolism – and this is considered in the next section.

One may note that although the mutations giving rise to such 'silent' or 'null' alleles no doubt include many cases where there has been a single base change in the gene resulting in a single aminoacid substitution in the protein, other kinds of mutations may produce the same effect. Thus chain termination mutants (p. 16), deletions (p. 112), frame shifts and a variety of other types of mutational change may result in what is effectively a 'null' or 'silent' allele, because they lead to a loss of the specific enzyme activity.

8.5. *Incidence of alleles determining inborn errors of metabolism*

The inborn errors of metabolism are conditions mainly inherited as autosomal recessives, in which there is some characteristic metabolic disturbance due to a gross deficiency of a specific enzyme (ch. 6, p. 187). Many inborn errors of metabolism have now been identified and the important point for the present discussion is that the mutant alleles which determine them are effectively the same as the so-called 'null' or 'silent' alleles discovered as a result of the anomalous segregation in particular families, of enzyme phenotypes detected electrophoretically.

One well known condition of this sort is phenylketonuria (p. 191). Here there is a gross deficiency of the enzyme phenylalanine hydroxylase and this results in a block in phenylalanine metabolism, which if left untreated causes severe mental defect. It is however possible to treat the condition very effectively by feeding a phenylalanine restricted diet. But to do this it is necessary to diagnose the condition very shortly after birth, because if the introduction of the diet is delayed irreversible brain damage will occur. To cope with the situation it was necessary to devise simple but specific methods for screening all newborn infants in order to pick out those with phenyl-ketonuria. And as a result of such surveys, good estimates of the incidence of this disorder have become available. Furthermore this general approach has been extended to the screening of the newborn for other inborn errors of metabolism which like phenylketonuria can be identified by the presence of very abnormal concentrations of particular low molecular weight metabo-lites in blood or urine. So estimates of the population incidence of a number of other rare conditions due to specific enzyme deficiencies are also being obtained.

The most extensive data presently available is that collected by Levy and his colleagues in Massachusetts (Levy 1973) where the great majority of all newborn infants are now routinely screened for a series of different metabolic disorders. Some of their findings are summarised in table 8.12. The number of newborn screened for the various conditions varies from about 350,000 up to over a million, and the estimated incidence of the different conditions at birth varies from about 1 in 15,000 to less than 1 in 300,000. The conditions listed are probably all autosomal recessives, so approximate estimates of the gene frequencies can be obtained by taking the square roots of the case frequencies, and from these one can obtain estimates of the incidence of heterozygotes at each of the loci. It seems that about 11% of the population must be heterozygous for an allele which determines one or other of these conditions.

For technical reasons it has not yet proved possible to examine the enzymes involved in this series of conditions by the electrophoretic techniques. Nevertheless, it is of some interest to compare the levels of heterozygosity obtained by the two different approaches. In fact, the heterozygosities per locus for these rare deficiency alleles are on average several times greater

TABLE 8.12

Incidence of certain metabolic disorders among newborn infants in Massachusetts (Levy 1973).

Disorder	Total screened	Number detected	Incidence	Estimated frequency of heterozygotes (per 1,000)
Phenylketonuria	1,012,017	66	1: 15,000	16
Cystinuria	350,176	23	1: 15,000	16
Hartnup disease	350,176	22	1: 16,000	16
Histidinaemia	350,176	20	1: 17,500	15
Argininosuccinic acidaemia	350,176	5	1: 70,000	8
Galactosaemia	588,827	5	1: 118,000	6
Cystathioninaemia	350,176	3	1: 117,000	6
Maple-syrup-urine disease	872,660	5	1: 175,000	5
Homocystinuria	480,271	3	1: 160,000	5
Hyperglycaemia (non-ketotic)	350,176	2	1: 175,000	5
Propionic acidaemia (ketotic hyperglycinaemia)	350,176	1	<1: 350,000	3
Hyperlysinaemia	350,176	1	<1: 350,000	3
Rickets vit. D dependent (with hyperaminoaciduria)	350,176	1	<1: 350,000	3
Fanconi syndrome	350,176	1	<1: 350,000	3

than those obtained for the rare alleles determining functionally active enzyme variants detected in electrophoretic surveys (see tables 8.9 and 8.10). This is at first sight a very paradoxical result because in general one would expect that alleles causing gross enzyme deficiencies would be those which were under the most pressure from natural selection, and therefore the least common. This discrepancy however is probably mainly due to the sampling procedure involved in such surveys for metabolic disease. One does not know how many other metabolic disorders due to deficiency alleles at other loci there might be, which could have been picked up in the screening tests but which were not, simply because the allele frequencies in the population were too low. The findings do suggest however that there are likely to be many other so-called deficiency or 'null' alleles at different loci with frequencies similar to those observed for the rare alleles detected in the electrophoretic surveys.

What is needed is some general procedure analogous to electrophoresis which would enable us to detect such alleles in heterozygotes in population surveys. It has of course been shown for many of the inborn errors of metabolism that a partial deficiency of the particular enzyme can be demonstrated in heterozygotes (p. 235). However in most cases the variation of enzyme levels in normal individuals and in heterozygotes and the consequent overlapping of the distributions effectively precludes any attempt to determine the incidence of the particular alleles by random population survey, based simply on assays of levels of enzyme activity, except perhaps in rather special cases such as the screening of Ashkenazi Jews for Tay-Sachs heterozygotes (Kaback and Zeiger 1972).

8.6. The causes of allelic diversity

8.6.1. Natural selection and random genetic drift

The incidence and distribution of the enzyme and protein variants which are observed in natural populations can be considered as representing the consequences of the operation over many previous generations of three main kinds of process.

(1) Mutation, which results in the generation of new alleles. Mutations appear to occur in an essentially random manner, though the average rates at which different types of spontaneous mutation such as single base substitutions, deletions, gene fusions and so on, take place no doubt vary according to the nature of the mechanisms involved.

(2) Natural selection, which tends to eliminate those alleles which reduce the biological 'fitness' of individuals who carry them, and tends to cause the spread of alleles which increase 'fitness'. In this context 'fitness' is measured by the relative contributions which individuals of different genotypes make to the next generation. Individuals of a given genotype will, for example, be on average less 'fit' if they are more prone to die in early life so that fewer survive to become parents, or if their effective fertility is reduced for some other reason. In general, the effects of a particular allele on 'fitness' will differ according to whether it is present in the heterozygous or homozygous state, and it is important to note that alleles which are relatively uncommon will mainly occur in the population in heterozygotes.

(3) Chance effects or random genetic drift, due to the fortuitous character of the sampling process determining which of the gametes (sperm and ova) produced by members of one generation happen to give rise to the new individuals of the next.

It should be noted that, quite apart from the question as to whether a particular mutant is relatively deleterious and so tends to be eliminated by natural selection, or confers some kind of selective advantage and so tends to spread, the odds against any new mutant allele persisting in a population for many generations are very considerable. The new allele will on average only be transmitted to half the children of the individual who first receives it. So there is a distinct chance that it will not be transmitted to the next generation, and the chances of its being lost are compounded in successive generations. In a reasonably large stable population where each pair of parents is on average replaced by two children who become parents in the next generation, the probability (because of chance effects alone) that a new mutant will still be present after, say, 15 generations is only about 1 in 9 (Fisher 1930). The odds in favour of the persistence of a mutant are greater if the population happens to be increasing in numbers when it appears, and are less if the population is declining. But in general the majority of new mutant alleles that occur are likely to be eliminated in the course of the next five to twenty generations in a more or less random manner.

However despite this steady elimination of the majority of the mutant alleles that have occurred, many must have persisted as is indicated by the very extensive degree of allelic variation which we observe in living populations.

There is, as we have seen, increasing evidence that at most loci coding for enzyme or protein structure a large number of different alleles exist in human populations. Furthermore only a very small proportion of these are evidently attributable to fresh mutations which occurred in the immediately preceding generation. At any particular locus most of the alleles which occur are quite rare. There must of course always be one allele which is common. However it turns out that at many loci there are two or more. This is the basis of the phenomenon known as polymorphism, and in fact accounts for much of the individual diversity that we observe among members of the species.

The recognition in recent years of this extraordinary degree of allelic variation in natural populations, and in particular of the widespread occurrence of polymorphism has led to a very fierce controversy about its evolutionary origins. The argument concerns the relative importance of natural selection and random genetic drift and there are two main and essentially opposite positions.

One – which we can call the classical position – maintains that the allelic distributions we observe today are primarily the consequence of the action of natural selection over the course of many previous generations, on the flux of new mutant alleles which are being continuously generated at a low but steady rate. It is argued that most mutants are in some degree deleterious, because a single random alteration in the structure of a complex structure such as an enzyme protein will usually impair its function. Consequently the majority of mutant alleles will be kept at a low frequency because of the pressure of natural selection. However, occasionally, mutants will occur which in particular genotypic combinations are selectively advantageous and consequently tend to spread. These give rise to the polymorphisms.

Thus the polymorphisms are on this view regarded as the consequences of differential selection. A number of different kinds of polymorphism have been envisaged. In some cases it is thought the heterozygotes for the alleles involved in the polymorphism are in the particular environmental circumstances fitter than either type of homozygote. This results in a so-called 'balanced' polymorphism which is in stable equilibrium at particular allele frequencies, and this can in principle be maintained indefinitely provided the environmental situation on which the heterozygous advantage depends does not alter. If, of course, it changes then the allele frequencies will also change. The sickle-cell haemoglobin polymorphism in Africa (p. 285) is the most convincing example of this type of situation in man.

Polymorphism without heterozygous advantage may occur if the environmental situation in a certain area favours one type of homozygote, while in another area another type of homozygote is at an advantage. Provided some degree of migration occurs between the different areas, a cline of allele frequencies will develop and local populations in the cline will be polymorphic. Another interesting possibility is known as frequency-dependent selection. If either of two homozygotes are fitter when they are rare than when they are common, then polymorphism will be set up at intermediate allele frequencies. This might occur for example if certain parasites or microorganisms were best adapted to attack the commonest phenotypes of the host. Other polymorphisms may represent intermediary stages in gene evolution in which one allele is progressively replacing another because it confers a selective advantage, and will eventually become fixed as the standard type in the species.

However whatever the details of a particular situation, the classical view suggests that the co-existence of two or more common alleles in a population is in general the consequence of differential selection on the phenotypes they produce. Such differential selection may, of course, have resulted in the development of a polymorphism in the past, and may no longer operate today because of alterations in the environmental circumstances in which the population now lives. However because changes in allele frequencies only occur slowly over very many generations, the polymorphism may still occur even if it is no longer maintained by differential selection.

A quite different explanation of the matter has, however, been put forward. This suggests that the alleles which determine the enzyme and protein variants commonly observed in natural populations, and in particular those which make up the polymorphisms, are selectively neutral or near neutral, and that their frequencies and distribution are simply the consequence of random genetic drift (Kimura 1968a, b, King and Jukes 1969, Kimura and Ohta 1971a).

It is argued that for any particular enzyme or protein, many of the possible amino acid substitutions which can occur because of random mutations, have little or no effect on the properties of the protein, and so the particular alleles concerned will not influence the biological fitness of individuals who carry them. Such 'neutral' alleles like other alleles will in most cases be lost from the population a few generations after they arise by mutation. But in any population a number of them will always be present, representing as it were a balance between the generation of new alleles by mutation and their extinction by random drift. The particular 'neutral' alleles present in a population at any given time will therefore represent a random collection of all the 'neutral' alleles which might be generated by separate mutations. Many will be relatively rare but some which happened by chance to have persisted in the population through many generations may become quite common, and it suggested that it is these which determine the enzyme and protein polymorphisms. Occasionally by chance one such allele after very many generations may become so common that it effectively replaces the allele which had earlier been the commonest form throughout the species. In this way the standard aminoacid sequence of the protein in the species would be altered by the aminoacid substitution so that in effect a step in molecular evolution will have occurred.

8.6.2. *Molecular evolution*

The evidence which Kimura has suggested provides the strongest argument in favour of the neutral mutation-random drift hypothesis comes in fact not from data on enzyme and protein polymorphism within species, but from calculations on the rates of molecular evolution of homologous proteins over a range of different species (Kimura 1969).

The aminoacid sequences of certain proteins such as cytochrome c and haemoglobin have now been determined in many different organisms, and from the number of aminoacid differences between the same protein in different species it is possible to construct phylogenetic trees showing the paths of evolutionary descent.

For example, the aminoacid sequence of the cytochrome c obtained from some thirty different species have been determined (Nolan and Margoliash

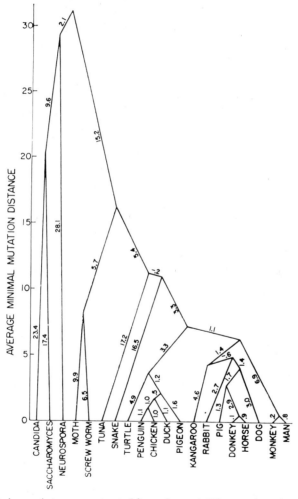

Fig. 8.6. Phylogenetic tree reconstructed from observed differences in amino acid sequence of cytochrome c in different species (from Fitch and Margoliash 1967). Each number on the figure is the estimated minimal number of nucleotides that would need to be altered in order for the gene for one cytochrome to code for the other.

1968). In the sequence of the hundred or so aminoacids which occur in each of these proteins, about thirty positions appear to be invariant, and the others differ to varying extents in the aminoacid substituents which occur. These aminoacid differences may be assumed to have arisen during the course of evolution by single base change mutations, and by comparison of the aminoacid differences between each pair of proteins it is possible to estimate the minimal number of mutations which would have been required for their divergence from their nearest common ancestor. From such data it is then possible to construct the most likely phylogenetic tree (fig. 8.6) which would have given rise with the minimum number of mutational changes, to the aminoacid sequences in the proteins as they are found to occur in the different species to-day (Fitch and Margoliash 1967). The evolutionary tree so constructed turns out to resemble quite closely the phylogenetic relationships obtained in the orthodox way using morphological and other standard taxonomic criteria. In this case however the phylogenetic tree is entirely deduced from the aminoacid sequences of a single polypeptide chain and can therefore be taken to represent the evolution of a single gene.

Using estimates derived from palaeontology of the time that has elapsed since various groups of species diverged from a common ancestor, it is possible from such data to estimate and compare the rates of molecular evolution by aminoacid substitution in particular proteins and in different lines of evolutionary descent (King and Jukes 1969, Kimura 1969). Thus for cytochrome c the average rate of aminoacid substitution per aminoacid site per year appears to have been about 4×10^{-10}. Not surprisingly the

TABLE 8.13

Average rates of aminoacid substitutions in mammalian evolution (King and Jukes 1969).

Protein	Average rate of aminoacid substitution per aminoacid site per year
Insulin* (A & B chains)	3.3×10^{-10}
Cytochrome c	4.2×10^{-10}
Haemoglobin α-chain	9.9×10^{-10}
Haemoglobin β-chain	10.3×10^{-10}
Ribonuclease	25.3×10^{-10}
Fibrinopeptide A	42.9×10^{-10}

* Guinea pig insulin omitted (see p. 332).

rate varies from protein to protein. For example, the rate for haemoglobin appears to have been about two and a half times greater than that for cytochrome c, and the rate for fibrinopeptide A is about four times that for haemoglobin (table 8.13).

But the crucial point which emerges according to Kimura (Kimura 1969, Kimura and Ohta 1971a, b) is that for any particular protein there is a remarkable uniformity in the rate of aminoacid substitution along widely divergent lines of evolutionary descent.

The point is illustrated by calculations on the rates of aminoacid substitution in the α- and β-chains of haemoglobin (fig. 8.7, table 8.14). If one compares for example the haemoglobin α-chain of the carp with that of man, one finds that there are 140 equivalent aminoacid positions (excluding deletions or insertions amounting to 3 positions), and of these 68 are occupied by different aminoacids and 72 by the same aminoacids. From this

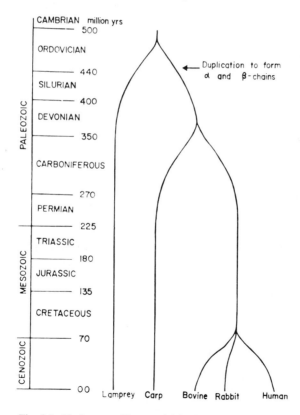

Fig. 8.7. Phylogeny of haemoglobin (from Kimura 1969).

one can estimate the mean number of substitutions per aminoacid site over the whole evolutionary period in the two different lines of descent (0.665, table 8.14). From palaeontological evidence it appears that the divergence from a common ancestor of the lines of descent leading to the carp and man occurred about 375 million years ago. By dividing the mean number of aminoacid substitutions per site in the evolutionary period by twice the time since the divergence from a common ancestor occurred, one obtains an estimate of rate aminoacid substitution/per aminoacid site/per year. This is 8.9×10^{-10}.

A similar calculation can be made by comparisons of the α-chains of other pairs of species, by comparisons of the β-chains, and by comparisons of the α- and β-chains (table 8.14). The divergence of the various mammalian species from a common ancestor is taken to be about 80 million years ago,

TABLE 8.14

Summary of some comparisons of the rates of aminoacid substitution in the evolution of the haemoglobin α- and β-chains between different pairs of vertebrate species (Kimura 1969).

Comparison	Average K_{aa}	$2T \times 10^8$	Average rate of substitution per aminoacid site per year (k_{aa}) $k_{aa} \times 10^{-10}$
α carp vs α human, mouse, rabbit, horse and bovine	0.665	7.5	8.9
α human vs α horse, pig, sheep and bovine	0.141	1.6	8.8
α mouse vs α human, horse, bovine, pig, rabbit and sheep	0.175	1.6	10.9
β human vs β horse, pig, sheep and bovine	0.190	1.6	11.9
β rabbit vs α human, mouse, rabbit, horse, bovine and carp	0.829	9.0	9.2
β human vs α human, mouse, rabbit, horse, bovine and carp	0.799	9.0	8.9
β human vs α human	0.766	9.0	8.5

K_{aa} = mean number of substitutions per aminoacid site over the whole evolutionary period. Estimated from p_d, the proportion of all aminoacid sites which are different between two polypeptide chains. $K_{aa} = -2.3 \log_{10}(1 - p_d)$.
T = number of years that have elapsed since divergence from a common ancestor.
k_{aa} = rate of substitution per aminoacid site per year, i.e. $K_{aa}/2T$.

and the time since the gene duplication occurred which started the divergence of the α- and β-chains is taken as about 450 million years ago. As will be seen the values obtained from the various comparisons are remarkably similar to one another. Thus for the α and β-chains of haemoglobin the rate of aminoacid substitution/per aminoacid site/per year along very different lines of evolutionary descent appears to have been close to 10^{-9}.

Kimura argues that this apparent uniformity of evolutionary rate in a protein can only be plausibly accounted for, by supposing that the evolutionary changes are due to the fixation in different species of particular mutations which while resulting in aminoacid substitutions in the protein cause no significant change in its functional properties and which are therefore selectively neutral. The term non-Darwinian evolution (King and Jukes 1969) has sometimes been used to describe such a process.

Only a proportion of all the possible mutants that may arise in any particular gene are assumed to be neutral. So the differences in rates of aminoacid substitution in different proteins such as haemoglobin and cytochrome C (table 8.13) can be explained by supposing that a different fraction of mutants are neutral in each case, according to the structural and functional characteristics of the different proteins.

If the hypothesis is correct then the rate of aminoacid substitution in a particular protein determined from comparisons of the aminoacid sequences in different species and from their time of evolutionary divergence, is the same as the mutation rate to neutral alleles of the gene determining the particular protein in members of any of the species (Kimura 1968). This important point can perhaps be most easily seen by the following simple argument (Smith 1970). Suppose that in a species at a given time the population size is N, and the mutation rate of a particular gene to neutral alleles is u. Then there are $2N$ genes in this population and therefore $2Nu$ newly arising neutral mutations. If we now travel far enough into the future we will find that by the process of random sampling all the genes then present will be descended, with or without further mutation from a single one of the original $2N$ genes. Hence each gene has a probability of $1/2N$ of ultimately being established or fixed. Since a newly arising neutral allele makes the same contribution to fitness as any other allele, it too has a probability of $1/2N$ of being established. Hence the number of new neutral genes which arise and which are ultimately established is $2Nu \times 1/2N = u$. Thus for the α- and β-genes of haemoglobin this argument suggests that the mutation rate to neutral alleles determining aminoacid substitutions is about 10^{-9}/per aminoacid site/per year. This is equivalent to about 1.4×10^{-7}/per gene/per year assuming 140 aminoacid sites per polypeptide chain. It will be noted that the hypothesis implies that the mutation rate to neutral alleles per unit time is the same in different species. But species differ in their generation times and usually mutation rates are expressed per gene per generation. Thus according to the hypothesis the mutation rates to neutral alleles per gene/per generation of the α- and β-haemoglobin genes would be about 2.8×10^{-6} in man assuming an average generation time of 20 years, but only about 7×10^{-8} per gene/per generation in the mouse assuming that here the generation time is on average about 6 months.

The hypothesis that the differences in aminoacid sequence of a particular protein in different species are simply the consequence of neutral mutations which have been fixed by random genetic drift, is of course exactly the opposite of the classical view which would ascribe the observed evolutionary changes to the operation of natural selection. On this view the structure of a particular protein in a given species has been selected because it is best adapted to meet the particular features of the external environment in which the species live, and also to meet the special internal environment which depends on the activities of all the other genes present.

However whether these differences between species are the products of neutral mutations, or whether they are the result of natural selection, it is clear that they must have occurred in any given case by the progressive replacement of a pre-existing gene by a particular mutant which has spread through the species and eventually become fixed as the characteristic type. So at any one time one might expect to find evidence for such partially completed replacements occurring. The obvious candidates for this are of course the enzyme and protein polymorphisms. And indeed Kimura and Ohta (1971a) have argued that the common enzyme and protein variants observed in population surveys, simply represent a phase of molecular evolution by neutral mutations.

8.6.3. The 'neutral' gene hypothesis: the question of functional equivalence

An important implication of the neutral mutation-random drift hypothesis is that the structurally different variants of any particular enzyme or protein polymorphism must be functionally equivalent. It should make no essential difference to the organism whether it has one or the other.

There certainly appear to be cases where this proposition is just not true. The best studied example is the sickle-cell haemoglobin polymorphism in Africa. Sickle-cell haemoglobin in certain of its properties, such as solubility in the deoxygenated state, is functionally very different from normal haemo-

globin. And in view of all the evidence now available about this polymorphism (p. 285), it can hardly be argued that the sickle-cell allele achieved
its high incidence in African populations purely as a consequence of random
genetic drift, or that it represents a mutation on its way to fixation in the
species. Advocates of the neutral mutation-random drift hypothesis would
no doubt accept this. But they would argue that the sickle-cell case represents
an exception rather than the rule.

So the controversy is not simply about whether or not all polymorphisms
are due to neutral alleles, but about how many. Neutralists would expect
that the great majority of polymorphisms involve neutral alleles, while
selectionists would consider that this would rarely if ever be the case. But
clearly all degrees of intermediate position are possible, and it is not difficult
to envisage a role both for natural selection and random genetic drift in
determining the many different polymorphisms that occur, though it is
certainly difficult to obtain critical evidence in particular cases.

The search for functional differences between the enzyme phenotypes
which make up the different polymorphisms is however obviously important.
For technical reasons this is in most cases a difficult problem, and in man it
has only been possible to attack it at a rather simple level. This has been to
try and determine in any particular enzyme polymorphism whether there
are significant differences between the level of enzyme activity in individuals
of different phenotypes. Such differences have indeed been demonstrated in a
number of cases (p. 299), and at present there is evidence for this in about
half of the known enzyme polymorphisms in man. This would not be expected
on the neutralist view. But it is still rather inconclusive evidence because we
know that enzyme assays *in vitro* may for various reasons be poor in-
dicators of enzyme function *in vivo*. And so far there has been very little
work done on the question as to whether these activity differences are
reflected in metabolic differences, which would be expected if they were
functionally significant.

The question of possible functional differences between the structurally
distinct versions of particular proteins such as haemoglobin and cytochrome c in different species is also important to the argument, since in its
simplest form the neutral mutation-random drift hypothesis of molecular
evolution implies that the structural differences which they exhibit do not
have significant effects on their functional properties.

It might be supposed according to the hypothesis that if one could entirely
replace the haemoglobin present in human red cells by haemoglobin obtained
from say the mouse or the rabbit, or even the carp, no functional disability

would result. Such an experiment is not of course feasible, but essentially the same kind of comparison may be provided in the particular cases where the long term evolutionary products of a gene duplication occur together in a single species. For example Kimura (1969) has argued that the rates of aminoacid substitution in the evolution of the α- and β-chains of human haemoglobin since they originated from a common ancestral gene by duplication, has been essentially the same as the rates of aminoacid substitution occurring in the subsequent divergence of evolutionary lines leading to structurally different α-chains in different species, and also in the lines leading to the structurally different β-chains. This is considered as strong evidence for the view that the various differences between these polypeptide chains are the consequences of neutral mutations. If so one might have expected that the α- and β-chains of haemoglobin would be functionally interchangeable. That this is not the case however is readily seen from a comparison of the properties of the normal Hb A tetramer $\alpha_2\beta_2$ and the abnormal tetramer Hb H which has four normal β-chains but no α-chain (Jones et al. 1959) and which is formed in individuals in whom α-chain synthesis is specifically restricted (p. 143). Haemoglobin H is very much less stable than Hb A and readily precipitates. It has an oxygen affinity ten times greater than the Hb A, and its oxygen dissociation curve shows no evidence of heam–haem interaction (Benesch, Ranney, Benesch and Smith 1961). This and other evidence makes it clear that the molecular evolution of the α- and β-chains has been shaped to a significant degree by natural selection, and that the emergence of the tetrameric form of haemoglobin with two α- and two β-chains is not purely a fortuitous consequence of neutral mutations and random drift.

Slightly more than 50% of the homologous sites in the human α- and β-chains differ in their aminoacid substituents and in addition there are several sites represented in one chain but not in the other which are presumably the consequence of the fixation of mutations causing deletion or additions to the ancestral DNA sequences. It remains to be determined to what extent the specific differences at the various sites contribute to the marked overall difference in the functional properties of the two chains. If it should turn out that only a small proportion of the substitutions were significant in this respect then the neutral mutation-random drift hypothesis would not be too seriously eroded. But even so the question of the deletions and accretions arises. These must also have arisen as mutations in single individuals and subsequently spread through the species and become fixed. Was this a matter of random drift or of selection?

The β- and γ-haemoglobin chains which are presumed to have derived from a somewhat more recent gene duplication raise another interesting point. They are both formed in the normal individual but at different times, so that while the principle haemoglobin in the red cells of the foetus is Hb F $(\alpha_2\gamma_2)$, in post-natal life it is Hb A $(\alpha_2\beta_2)$. The two chains differ in about one quarter of their aminoacid substituents. If these differences are all to be attributed to neutral substitutions, then it would seem necessary to conclude that the developmental shift from γ- to β-chain synthesis at birth is also fortuitous and of no functional significance.

But Hb F and Hb A have in fact been shown to differ in their functional properties. The difference turns on the facility with which they bind the phosphate ester 2,3-diphosphoglycerate, a substance which is present in similar concentrations in foetal and adult red cells and which has the effect on binding to haemoglobin of reducing the oxygen affinity (Benesch and Benesch 1969) so that the oxygen dissociation curve is shifted to the right. Hb F binds 2,3-diphosphoglycerate less readily than Hb A (De Verdier and Garby 1969, Tyuma and Shimizu 1970) and consequently the oxygen dissociation curve of foetal red cells is to the left of that of adult red cells. This difference appears to be appropriate to the physiological requirements of oxygen uptake and delivery in the foetus on the one hand, and in the adult on the other. This functional difference between Hb A and Hb F must depend on differences in the aminoacid sequences of the β- and γ-chains. One of the critical residues in the binding of 2,3-diphosphoglycerate appears to be the histidine at position 143 in the β-chain (Arnone 1972) which is replaced by serine in the γ-chain. It is not clear however, how far the other aminoacid differences between the two chains may play a part in causing the difference in oxygen affinities of the two forms of haemoglobin.

8.6.4. *The random drift versus differential selection controversy: some further aspects*

(1) Differential selection implies that phenotypes differ from each other either in viability or in effective fertility. Some fifty different enzyme and protein loci have now been shown to exhibit polymorphism in one or more major human populations (Appendix 2, p. 397). But in only very few cases is there any clear evidence for differential selection between the common phenotypes. One example of course is the sickle-cell haemoglobin polymorphism in Africa. Others are the α- and β-thalassaemia polymorphisms in S.E. Asia and in certain Mediterranean countries. In each of these cases individuals who are homozygous for the mutant suffer from a severe disease

which usually precludes their survival to adult life and hence their making any significant contribution to the next generation. But the alleles which determine these conditions occur with quite high frequencies in the particular populations. The loss of the alleles in each generation due to the failure of the homozygotes to pass them on cannot be being made up fresh mutations, since this would imply an implausibly high mutation rate to these particular alleles and also require that such high rates of specific mutations occurred only in certain populations and not in others. So it is necessary to conclude that the loss of mutant alleles by the defective homozygotes is made up by an increase in the fitness of the heterozygotes compared with normal homozygotes. This differential selection between the heterozygotes and normal homozygotes may be assumed to occur, or have occurred in the past, only in populations living in particular environmental circumstances, and it appears that endemic malaria has in these particular cases probably been the most important factor. In the case of the sickle-cell heterozygotes direct evidence for their selective survival from severe malaria has indeed been obtained (pp. 285–287).

However in the great majority of enzyme and protein polymorphisms so far identified, there is as yet little direct evidence to suggest that the several common phenotypes which occur differ in either viability or effective fertility. This of course is what would be expected on the neutralist view. However selectionists can reasonably contend that the degree of selective differential which would have been needed to produce and maintain most polymorphisms would be too small to be demonstrable by presently available techniques. The case of the sickle-cell polymorphism is an extreme one because here the mutant homozygote is virtually lethal, and the selective advantage of the heterozygote over the normal homozygote was probably about 10–15% (p. 288). Even so a clear demonstration of the selective differential was in practice quite difficult to achieve. In most other polymorphisms since the difference in viability or effective fertility of the two sorts of homozygote is clearly very much less, the required selective advantage of the heterozygote would be very much smaller.

In order to tackle the problem some kind of demographic approach would appear to be required. The aim would be to categorise individuals in one or more populations in terms of the various common allelic differences that are known, and then search for differences between them in the main parameters involved in selection, such as mortality and morbidity rates at various ages, and fertility. Also by family analysis investigate possible disturbances in segregation ratios, and so on. Such data, although they may give only indirect information about specific selective factors in relation to particular polymorphisms, should in principle provide an assessment of the magnitude of any selective effects that are actually occurring in the given environmental situations. And this, of course is fundamental to the whole problem. Such surveys are, however, extremely hard to mount on a scale which is both sufficiently large as to be likely to yield significant results, and yet sufficiently detailed and exact in the determination of the various demographic parameters as to yield precise answers. So far, although much suggestive information has been obtained, the results of even the most sophisticated surveys of this sort (e.g. Morton 1964, Neel and Salzano 1967, Reed 1968) have mainly served to emphasise the considerable difficulties of obtaining unequivocal answers in this field. The blood group polymorphisms provide the most extensive data available, yet even here it appears that selective differences of less than about 5% cannot be detected at present by this approach. Such a differential would certainly be very high in terms of population genetics theory.

An inherent source of uncertainty arises from the fact that the environments in which human populations live today, or even in the last few generations, are very different in important aspects from those in which they lived in the past. In particular the incidence and age distribution of mortality and morbidity and its main causes have changed, and are changing profoundly. So what may have been important selective agents in the past, and may well have shaped many of the polymorphisms that we see today, may now be of only minor or no significance. We are only looking at what is inevitably a changing situation over a very narrow period of time. Furthermore, as a general rule we have no means of knowing whether in any particular polymorphism we are dealing, as is often assumed, with a situation close to stable equilibrium due to heterozygous advantage, or with the steady increase of one particular allele at the expense of another, or with its progressive disappearance.

A further important source of difficulty in such investigations is that in general one would expect that selection will be directed at complex phenotypic characteristics dependent on many different enzymes and proteins acting together. It may therefore be necessary to compare the relative fitness of multiple combinations of phenotypes of different enzymes and proteins, rather than rely simply on comparisons between the common phenotypes for each enzyme or protein taken separately. This is likely to enhance greatly the difficulty of obtaining unambiguous evidence for selective differences in terms of viability or effective fertility.

Another approach is to try and find out whether particular alleles render individuals more or less susceptible to the development of particular disorders or disabilities, especially common ones. The general method is to compare the incidence of the allele in individuals affected by the particular condition with the overall incidence in the population of which they are a part. A now well established example of this kind of effect is the association of the ABO blood groups with certain gastro-intestinal disorders (for detailed references see McConnell 1966, Vogel 1970). Blood group A individuals are somewhat more susceptible to gastric cancer than group O individuals. Group O individuals are more susceptible to peptic ulceration than group A. However, the effects are quite small, and the natural history of the diseases in question is not such as to suggest that the particular associations so far discovered could have been the main source of the selective differential which is presumed to have established and maintained this very widespread polymorphism. More recently, associations with atherosclerosis (Kingsbury 1971), thromboembolic disease (Mourant et al. 1971) and coronary disease (Medalie et al. 1971) have been reported, and these might prove to be of more selective significance. However it is not known exactly how these different blood group antigens, or the enzyme differences which apparently determine them, influence susceptibility to these particular diseases. The causal relations involved are quite obscure and presumably very indirect, and the specific associations between the polymorphic types and these particular diseases could hardly have been predicted from the known characteristics of the enzymes or the antigens. This may well be commonly the case for other polymorphisms and other diseases, and if so the search for such associations is likely to be peculiarly difficult. There are a great variety of different common diseases, including many sorts of acute and chronic infections, which could in theory have been important in establishing different polymorphisms, and there are an increasing number of polymorphisms which might be tested for such disease associations. But if there is no particular reason to expect one sort of association rather than another in a specific case, the discovery of significant associations is likely to be somewhat fortuitous.

(2) When the allele frequencies of particular polymorphisms are determined in different human populations it is found that there are considerable differences from one locus to another in the amount of variation between populations. At one extreme there are certain loci at which the common alleles appear to have similar frequencies in most of the populations which

have been studied (e.g. phosphoglucomutase locus PGM_1). At the other extreme there are loci at which the frequencies of different alleles vary very widely from population to population (e.g. glucose-6-phosphate dehydrogenase). Indeed certain loci appear to be polymorphic in some populations, and 'monomorphic' in others.

Such variations can be attributed to selection by assuming that where the allele frequencies throughout the world are very similar, the particular environmental circumstances which determine the selective differential are world wide; whereas when the allele frequencies vary widely, the environmental circumstances determining the selective differentials are peculiar to some populations and do not occur in others. But these similarities and differences of the allelic frequencies at the various loci in different subpopulations of the species can also be accounted for in terms of random genetic drift.

However while natural selection may be expected to operate differently for each locus and each allele at a locus, random genetic drift would not discriminate in this way. Using this principle Lewontin and Krakauer (1973) have devised a statistical method which they claim enables one to test whether the variance of the observed allele frequencies at a series of polymorphic loci in a number of different subpopulations of a species departs significantly from that expected, assuming that the distributions are simply a consequence of random genetic drift. Applying this test to published data on the blood group polymorphisms over an extensive range of different human populations they found a significant deviation from expectation, a possibility suggested earlier by Cavalli-Sforza (1966). The results would suggest the action of natural selection in determining at least some of the blood group polymorphisms. As yet it has not been possible to apply this test to the human enzyme and protein loci on a world scale because the available data on sufficient loci in a wide enough range of different populations is so far insufficient. However Nevo (1973) has applied Lewontin and Krakauer's test to data on a series of enzyme and protein polymorphisms as they occur in a number of populations of pocket Gophers, and a marked deviation from the expectation assuming random drift was obtained. Thus here again there appears to be evidence for differential selection having been responsible for at least some of the polymorphisms.

(3) Occasionally a group of individuals from one population have migrated elsewhere and founded a new community which has subsequently expanded so that a population emerges whose genes are largely derived from a small

number of founders. The alleles which happen to have been carried by the original founders are not likely to be exactly representative of the alleles in the population from which they came because of sampling effects. Consequently the allele frequencies at various loci in the new population will be expected to differ to a greater or lesser degree from those of the original population. Furthermore if one of the 'founders' happened by chance to be heterozygous for a rare allele then this might eventually be and establishment. This type of phenomenon is often referred to as the 'founder effect' and is in essence a special case of random genetic drift.

Essentially the same phenomenon may occur without migration, if the numbers of an established population are severely reduced by some epidemic or other disaster and then subsequently increase again. The sample of alleles which happen to be carried by the survivors and so form the gene pool from which the population is reconstituted is likely to differ in some degree from that of the original population. It may well include alleles originally very rare but which fortuitously become relatively common after the reestablishment of the population from the survivors of the disaster.

The 'founder effect' provides the most plausible explanation for a number of cases where a particular allele, sometimes quite deleterious, occurs with a high frequency in some relatively small and reproductively somewhat isolated group, while it is extremely rare elsewhere in the world.

A striking example is the case of the allele which determines the severe and usually lethal, autosomal recessive disease hereditary tyrosinosis, which has an unusual incidence in French-Canada but is extremely rare elsewhere. Laberge (1969) found that in certain relatively isolated French-Canadian populations living in northern Quebec, the incidence of heterozygotes for this allele might be as high as 3–4%. Pedigree studies suggested that one of the founders of the population who were immigrants to Quebec from France in the 17th century was probably heterozygous for this allele.

The relatively high incidence of the autosomal dominant metabolic disorder, porphyria variegata (p. 24〇), among the descendents of the Dutch settlers in South Africa furnishes another example of the 'founder effect' (Dean 1963). There were about 40 original Dutch settlers and their wives who came to South Africa in the latter part of the 17th century, and genealogical studies suggest that the present cases of porphyria variegata in South Africa can all be traced back to one of them. Dean (1969) estimated that by 1969 there were about 9,000 individuals with porphyria variegata in South Africa, and that the incidence in the total white population was about 1 in 400, and higher among the descendents of the original Dutch settlers. These individuals represented about the 12th to the 16th generation since the introduction of the gene.

A high incidence of a particular allele in certain relatively discrete populations, and its extreme rarity elsewhere has been found to be a not unusual phenomenon, and can probably in most cases be accounted for by the 'founder effect' or random drift. One example is a type of albumin variant which was found in as many as 25 % of members of a group of NorthAmerican Indians known as the Naskapi and in several closely related tribes (Melartin and Blumberg 1966). This particular variant is probably different from the numerous other albumin variants which have been discovered (Weitkamp et al. 1973), and which in any case occur among Europeans and other major population groups with frequencies of less than about one in 1,000. Another illustration is the relatively high incidence among Ashkenazi Jews of the gene which determines the autosomal recessive metabolic disorder, xyloketosuria (congenital pentosuria p. 371) and which appears clinically to be quite harmless. It has been estimated that the incidence of xyloketosuria in Israel among Jews whose ancestors came from eastern Europe is about 1 in 5,000, indicating an allele frequency of about 0.014 (Mizrahi and Ser 1963). It was not found in Israeli Jews who originated elsewhere. Similarly the condition has only been found in the U.S.A. among Jews who came, or whose ancestors came, from Eastern Europe (Hiatt 1972). The disorder is extremely rare in other populations, if indeed it occurs at all.

The unusually high incidence of a complete or nearly complete deficiency of serum cholinesterase (p. 159) among certain Eskimo populations is another apparent example of the same sort of thing. However the recent finding of Scott (1973) that there are probably two different alleles determining this deficiency both occurring with quite high frequencies among Eskimos, but both extremely rare elsewhere, is unexpected and makes one wonder whether random drift can be the whole explanation.

The remarkable incidence among Ashkenazi Jews of an allele which determines Tay-Sachs disease (p. 227) has also been ascribed to random genetic drift (Livingstone 1969, Chase and McKusick 1972). The allele which in the homozygous state results in this severe and lethal disease, appears to occur with frequencies of between 0.01 and 0.02 in Ashkenazi Jews whose ancestors came from eastern Europe (Aronson and Volk 1962, Kaback and Zeiger 1972) and the highest frequencies appear to occur among Jews whose antecedents came from North East Poland, the neighbouring area of Lithuania and the adjacent portion of Byelorussia. Alleles determining Tay-Sachs disease are at least ten times less frequent in other populations

(O'Brien 1972), and quite possibly different mutant alleles are involved here in causing the disease. However although the remarkable population distribution of this lethal condition can be formally ascribed to random drift (Livingstone 1972), it has been argued that heterozygous advantage may have been a significant factor in the early spread of this allele in Jewish populations (Myrianthopoulos and Aronson 1966, Myrianthopoulos et al. 1971, Knudson 1973). The genes determining adult Gaucher's disease and also Niemann-Pick disease also have elevated frequencies among Ashkenazi Jews, and both these conditions like Tay-Sachs disease are disorders of sphingolipid metabolism (p. 221). While the founder effect may well result in more than one different mutant acquiring a high frequency in a given population, the functional relatedness of the genes should be randomly distributed. Thus the finding of three different mutants each affecting sphingolipid metabolism, with relatively high frequencies in Ashkenazi Jews, seems too much of a coincidence and suggests that some kind of heterozygous advantage exists or once existed (Knudson 1973). Relative resistance of heterozygotes for the Tay-Sachs gene towards tuberculosis (Myrianthopoulos and Aronson 1972), or towards plague (Knudson and Kaplan 1962) have been suggested as possible selective factors.

(4) An important plank in the argument for the neutral mutation-random drift hypothesis is the evidence pointing to a uniform rate of aminoacid substitution in the evolution of a particular protein along different lines of descent (p. 319). Not surprisingly the question has been raised as to how general this apparent uniformity in the rate of amino acid substitutions actually is.

One apparent exception has been found in the case of insulin. The number of aminoacid differences between guinea pig insulin (A- and B-polypeptide chains) and other mammalian insulins is very much larger than between other pairs of mammalian insulin (Smith 1966). Indeed the aminoacid differences between guinea pig insulin and other mammalian insulins appear to be larger than the differences between insulin from the chicken and the other mammals, and similar in number to those between certain fish insulins and mammalian insulins. The findings appear to imply a much higher rate of molecular evolution along the line of descent to the guinea pig, than along the lines of descent to other animals (King and Jukes 1969). It is difficult to know what to make of this remarkable exception, but it does seem to indicate the operation of natural selection.

Further analyses of the molecular evolution of the α- and β-chains of haemoglobin, cytochrome c and fibrinopeptide have also been carried out and indicate that a general uniformity of the rates of molecular evolution in these cases is also open to question. For example the rates for the α- and β-chains of haemoglobin in the descent of higher primates appears to have been significantly slower than in the descent of other mammals (Barnabas et al. 1971, Goodman et al. 1971). Also detailed comparisons of the rates of apparent nucleotide substitutions during the evolution of the α- and β-chains of haemoglobin, cytochrome c and fibrinopeptide A over a range of 24 species including humans and fish gave evidence for differences in the relative rates for these proteins, both within particular segments of the evolutionary tree, and also between divergent lines of descent for the total rate over all four proteins (Langley and Fitch 1973, 1974). Some modification of either the definition or the generality of the neutral mutation-random drift hypothesis seems necessary to accomodate these results, and various possibilities have been suggested (e.g. Fitch and Markowitz 1970, Ohta 1973). At present however it is not clear how far the evidence for non-uniformity in rates at present available, erodes the main hypothesis.

(5) Kimura and Crow (1964) showed that assuming random genetic drift the proportion of homozygotes in a population at equilibrium is given by $F = 1/4N_e u + 1$ where N_e is the effective size of the population and u is the mutation rate to neutral alleles. Thus the proportion of heterozygotes in the population is given by

$$1 - F = \frac{4N_e u}{4N_e u + 1}.$$

It follows that in any one population the variation in heterozygosities between loci will be a function of the mutation rates to neutral alleles at different loci. The greater the mutation rate the greater will be the heterozygosity. This is illustrated in fig. 8.8.

In the surveys of electrophoretic variants in the European population discussed earlier (p. 305) it was found that the heterozygosities varied widely over the range of loci. Some loci showed high heterozygosities, around 0.5, and others very low heterozygosities, 0.001 or less. Now in this population we do not know what the effective population size or number actually is. However it must be the same for all loci. Consequently if the alleles involved in determining these electrophoretic variants are indeed neutral then the 500-fold range of heterozygosities observed implies a similarly wide range

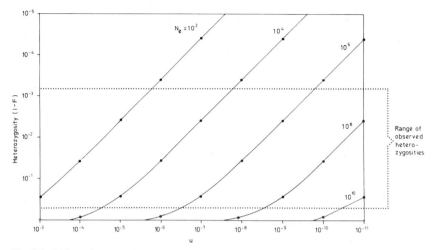

Fig. 8.8. Values for 1-F, the expected proportion of heterozygotes, for various values of the effective population number, N_e, and the mutation rate to neutral alleles, u (from Harris, Hopkinson and Robson 1973, based on Kimura and Crow 1964).

in mutation rates to neutral alleles giving electrophoretic variants over this series of loci (Harris, Hopkinson and Robson 1973). Furthermore one may note that if the population were an ideal population at equilibrium as assumed theoretically in the derivation of Kimura and Crow's formula, then as can be seen in fig. 8.8 the relationship between mutation rate and heterozygosity would be effectively linear over a considerable range of possible population sizes and mutation rates. So the distribution of heterozygosities shown in fig. 8.5 (p. 306) would in essence be a distribution of mutation rates. It is unlikely that the population is in fact in the ideal equilibrium state because it is known that it has increased in size considerably over the last twenty generations or so. But nevertheless if the neutral gene hypothesis is correct and can be applied to these observations, the heterozygosities must be a function of the mutation rates and a very wide range of mutation rates to this class of alleles over this series of loci has to be inferred.

This point is of some general significance because according to the neutral mutation-random drift hypothesis, the rate of aminoacid substitution in molecular evolution is directly related to the mutation rate to neutral alleles (p. 321). So the variation in heterozygosities between different loci within a species should be reflected in the variation in rates of molecular evolution of different proteins as deduced from comparisons of the amino acid sequences of homologous proteins over a series of different species.

It is hardly feasible yet to examine the possible correlation between heterozygosities within species and the rates of molecular evolution for a series of polypeptide chains determined by different loci, because there is too little data. But no doubt appropriate data for such an analysis will become available in due course and could provide one way of testing the general validity of the hypothesis.

(6) An intriguing feature of the neutral mutation-random drift hypothesis as applied to rates of molecular evolution, is that it implies that the mutation rate to neutral alleles at any one locus would be the same per unit of time for different species (p. 321). But species vary in their generation times, and mutation rates are usually expressed per gene per generation. Thus according to the hypothesis the mutation rate per generation to a particular class of alleles at a given locus in man might be about 40 times the mutation rate per generation to the same class of alleles for the homologous locus in the mouse and about 500 times the equivalent mutation rate in *Drosophila*, assuming average generation times of twenty years in man, six months in the mouse and two weeks in *Drosophila*. Enzyme or protein variants detected by electrophoresis can be reasonably regarded as representing the same class of mutations in the different species. So if one could determine the mutation rates for electrophoretic variants of homologous enzymes and proteins in the different species this should provide a direct test of the hypothesis.

Unfortunately good estimates of such mutant rates are very difficult to achieve because of the very large numbers of individuals who need to be examined, and very few studies have as yet been carried out. Tobari and Kojima (1972) in an experiment in *Drosophila melanogaster* in which the enzyme products of ten loci were examined electrophoretically in a very considerable number of flies found three mutations and estimated the average mutation rate over all the loci as 4.5×10^{-6} per gene per generation. This estimate must obviously have a large sampling error. In man the only study at all comparable is that of Harris, Hopkinson and Robson (1973). They found no mutants in what was estimated to be 133,478 alleles screened. The alleles were at 43 different loci, a few of which were probably homologous to those examined by Tobari and Kojima in *Drosophila melanogaster*. It was estimated from these results that the upper limit of the average mutation rate over the whole series of loci was unlikely to be greater than 2.24×10^{-5} per gene per generation. Clearly much more data on both species would be required to make a satisfactory comparison, but since a difference of two orders of magnitude is predicted by the theory for the mutation rates in *Drosophila* and man, such a test might be feasible in the future.

(7) A variety of other arguments, based on both experimental results in a variety of species and on the implications of and developments in the mathematical theory, have been deployed both for and against the neutral mutation-random drift hypothesis. The problem can hardly be said to have been resolved. However the key questions at issue have become clarified, and it is now becoming easier to see what kinds of information will bear critically on the hypothesis. Relevant contributions on different aspects of the problem which have not already been mentioned are contained in the following: Sved et al. 1967, Boyer et al. 1969, Prakash et al. 1969, Smith 1968, 1970, Richmond 1970, Clarke 1970a, b, Boyer 1971, Crow 1972, Haigh and Smith 1972, Yamazaki and Murayama 1972, Kimura and Ohta 1971, 1973, Wade Cohen et al. 1973, Ayala and Anderson 1973, Cavalli-Sforza 1973, and the series of papers in the symposium 'Darwinian, neo-Darwinian and non-Darwinian evolution' edited by Lecam, Neyman and Scott 1972.

8.7. *The extent of allelic variation: summarising conclusions*

The information we have about the extent and character of allelic variation in human populations has been mainly obtained by two very different lines of investigation. One of these has been directed to the study of particular inherited diseases, with the object of identifying a defect in a specific enzyme or protein as the cause of the clinical and pathological abnormalities that occur. The other approach has been by random population surveys of mainly healthy individuals in a search for individual differences in various enzymes or proteins. At one time it was usually tacitly assumed that there is in any population a clear line of distinction between individuals who could be regarded as 'normal' and those who are 'abnormal', but it is now clear that this is only a convenient operational distinction and in fact there is a continuous gradation between health and disease. The two different approaches to the problem of genetic variation, that is by the study of the so-called 'normal' population and by the study of so-called 'abnormal' individuals with disease states has emphasised the continuity of the distribution, but has at the same time shown how it is possible to analyse it in terms of the discrete effects of specific alleles.

As a result of the extensive data which have been obtained by these very different approaches, a general picture of the character of inherited variation of enzymes and proteins in human populations is beginning to appear. It is perhaps useful to summarise the principle generalisations which seem to be emerging.

1) A very large number of different enzymes and proteins are formed in the human organism, and the aminoacid sequences of their constituent polypeptide chains, are each defined by the sequence of bases in a gene at a particular gene locus. It now seems likely that at the great majority and perhaps all of these loci many different alleles occur among individual members of the species. These alleles have been generated by separate mutations occurring in single individuals in earlier generations. Many of the alleles result in the synthesis of a structurally altered polypeptide chain, but some may result in its complete deficiency.

2) At any given locus the majority of alleles which exist are quite rare, and mainly occur in the heterozygous state. However at an appreciable proportion of loci (probably at least 30%) there are two or three alleles each of which is quite common and because of the discrete phenotypes they produce in different common combinations they give rise to the phenomenon known as polymorphism.

3) Because of the remarkable degree of allelic variation which exists, many different combinations of alleles at various loci can occur in different individuals, and this gives rise to an enormous range of diversity between individuals in their enzyme and protein makeup. In fact it is unlikely that any two members of the species with the exception of monozygotic twins are exactly alike in the details of their enzyme and protein constitutions. Much of this diversity comes from different combinations of the common alleles over the whole range of polymorphic loci. But a significant fraction must also derive from the rare alleles, since all individuals are probably heterozygous for a number of these at different loci, and no two individuals are likely to have exactly the same combination.

4) The enzyme or protein products of the series of alleles which occur at any given locus vary in their structures and hence in their properties, and these differences may be reflected functionally. In some cases the alteration in the functional properties of the enzyme or protein is sufficiently marked that it gives rise in either the homozygous or heterozygous state to overt clinical abnormality, that is an inherited disease. But in other cases the functional alteration is less profound and may only give rise to overt clinical effects under certain environmental circumstances or when the allele in either the heterozygous or homozygous state occurs in combination with particular

alleles at other gene loci. In still other cases functional change may be minimal or possibly absent.

Thus the genes which are said to cause inherited disease can be regarded simply as extreme examples of the allelic variation which is ubiquitous throughout the species and is indeed responsible for the inherited variation which is observed in virtually all 'normal' phenotypic characteristics.

CHAPTER 9

Gene mutations and inherited disease

A recent catalogue of inherited abnormalities (McKusick 1971) lists more than one thousand distinct clinical syndromes each of which can be plausibly attributed to the effects of a single abnormal gene. They vary greatly in manifestations and severity. Some are present at birth or appear shortly after; others may not become apparent till middle or late life. Some are inevitably progressive and fatal; others give rise to only minor disability. Any organ or tissue may be affected in some degree or another, and often quite characteristic and specific pathological changes are demonstrable. Thus this great variety of disorders encompasses virtually every branch of medicine. Judging from the rate at which new examples are currently being described in the medical literature, it seems that many more must still remain to be identified.

It is usual to classify these conditions according to whether they are inherited as so-called 'dominant' or 'recessive' characteristics, and according to whether the abnormal gene concerned is located on one or other of the twenty-two autosomal chromosomes or is sex linked, that is, located on the X or Y chromosomes. Among those so far characterised, somewhat more than half can be classified as autosomal dominant disorders and nearly forty per cent as autosomal recessive. The remainder (about 8%) are mainly X linked recessive conditions. As yet no certain example of a disease state attributable to a specific abnormal gene located on the Y chromosome has been identified.

The essential point about the so-called 'autosomal dominant' conditions is that virtually all the clinically affected individuals are heterozygous. They carry one dose of the abnormal gene which comes from one parent, and one dose of its functionally normal allele which comes from the other. Because most of the abnormal genes which give rise to such 'dominant' disorders are rare, the homozygous state has generally not been observed. It would be expected, however, that this would usually be represented by a much more

339

severe clinical disorder than that seen in the affected heterozygotes and quite probably often be lethal in early life.

In 'autosomal recessive' disorders, the clinically affected individuals are often homozygous and carry two doses of the abnormal gene, one derived from each of the parents. Heterozygotes with one dose of the abnormal gene and one of the functionally normal allele appear in most circumstances to be quite healthy. There may, however be two or more different abnormal genes which can occur at a particular gene locus, each producing a distinctive 'recessive' disorder in the homozygous state. Individuals heterozygous for two such alleles usually exhibit a disorder similar to what is seen in the two corresponding homozygous conditions, and if these differ in their characteristics or severity the 'double' heterozygote will generally show intermediate features. A well-known example of this is sickle-cell–haemoglobin C disease.

In the so-called 'X linked recessive' disorders, the clinical disorder occurs predominantly in males. Males having only one X chromosome will, if this carries the abnormal gene, manifest the disorder, whereas females having two X chromosomes only show the disorder if both carry the abnormal gene. If as is often the case an abnormal gene causing such a disorder is rare, the condition may never have been observed in females at all.

9.1. The molecular pathology of inherited disease

If the general theory that genes produce their effects by directing the synthesis of proteins is in its main essentials correct, then it should be possible in each of these different conditions to trace the particular constellation of clinical abnormalities that are observed, back to the effects of some specific enzyme or protein defect resulting from a single gene mutation. Indeed, a full account of the pathology of such a disease should in principle start with the details of the alteration in the base sequence of the DNA brought about by the mutation; show in what way this has modified the synthesis of the specific enzyme or protein; proceed to elucidate the secondary biochemical consequences that ensue; and finally show how these give rise to the clinical signs and symptoms that are observed.

In fact, for the great majority of inherited diseases such a complete exposition is still very far from realisation. The characteristic features that we observe in any one disorder must often represent the consequences of a very complex chain of phenomena involving interactions at many different levels of the biochemical and physiological organisation of the organism. So far it is only in relatively few conditions that it has been possible to piece together the details of even some of the steps in the causal sequence of events.

9.1.1. Defining mutational changes at the DNA level

It is not yet feasible to define the primary genetic abnormality in any particular disease by examining directly the base sequence of the abnormal gene. However, where this determines an abnormal protein which can be isolated and whose structural defect can be identified, it is often possible to deduce with some precision the likely nature of the underlying abnormality in the DNA. The haemoglobinopathies provide an extensive series of examples (see ch. 1). In sickle-cell disease, for example, the various clinical and pathological features of the disorder can be attributed to the synthesis of an abnormal haemoglobin which differs from its normal counterpart in only a single aminoacid. At the sixth position in the β polypeptide chain, valine replaces glutamic acid. Since the normal β-chain contains 146 aminoacids, and each aminoacid is coded by a sequence of three bases in the DNA, we can argue that the gene determining this polypeptide chain contains a length of DNA 438 bases long, and that the mutation giving rise to the sickle-cell allele involved the sixth triplet in the sequence, that is the 16th, 17th and 18th bases. Furthermore we can reasonably infer from what has been discovered about the genetic code and about the general nature of mutations which give rise to single aminoacid substitutions, that the specific base in the triplet which has been changed is the 17th (adenine for thymine on one of the two complementary strands of the DNA).

Similar localisations of the primary defect within the gene can be made in a number of other haemoglobin diseases. In most of these there is a single aminoacid substitution in the protein, and as with sickle-cell haemoglobin the mutation can usually be pinpointed to a specific base alteration. One interesting exception is Hb Constant Spring (p. 115) where there is an elongation of the polypeptide chain due apparently to an alteration of a single base in the chain terminating codon.

However, in some cases other types of protein abnormality have been demonstrated, and these imply that the mutational change in the structure of the gene must have been of a different type. For instance in the Lepore haemoglobins (pp. 104–109) the abnormal polypeptide chain has an amino-acid sequence which, in its first part, is the same as that of the δ polypeptide chain characteristically found in normal Hb A$_2$, but which later assumes the sequence seen in the latter part of the normal β-chain of Hb A. This is most simply accounted for by supposing that a sequence which comprises the distal part of the δ-chain gene and the proximal part of the adjacent β-chain locus has been lost from the chromosome. Since the abnormal polypeptide chain which is now defined by the new gene contains 146 aminoacids, as do both the normal β-chain and the normal δ-chain, we may infer that the primary genetic abnormality is a deletion of a chromosomal segment involving a stretch of DNA at least 438 bases long. There are also other abnormal haemoglobins characterised by the loss of a portion of the polypeptide sequence are (p. 112). For example, in Hb Freiburg a single aminoacid is missing from the β-chain and this implies a loss of three consecutive bases in the corresponding gene. In Hb Gun Hill a sequence of five aminoacids is missing indicating a deletion of fifteen bases.

Clearly different kinds of mutational events can result in the abnormal genes which cause disease. Little is known about the nature of the phenomenon (or phenomena) by which single base alterations come about. But deletions can probably arise in at least two different ways. One involves unequal crossing over between homologous chromosomes following mispairing at meiosis. This is almost certainly the cause of the Lepore deletions, and may well also account for the other haemoglobin deletions that have been identified. The phenomenon is particularly likely to occur where, as a result of some previous localised duplication of the genetic material, very similar DNA sequences lie close together on the same chromosome. The more extensive the regions of homology, the greater is the probability of mispairing, so that presumably some genes are likely to be much more prone to this type of mutational event than others. Another way in which deletions originate is when by chance two chromosomal (or chromatid) breaks occur more or less simultaneously and this is followed by aberrant reunions, so that an intervening segment is lost if the breaks occur on the same chromosome, or there is a translocation of material from one chromosome to another associated with some loss if different ones are involved.

The data obtained from structural studies on abnormal haemoglobins make it clear that both mutations involving single base changes and mutations involving deletion can lead to clinical abnormality, and presumably the same is true for other proteins and enzymes. But it is not possible as yet to assess what might be the relative importance of these or other types of mutational change as causes of the great variety of inherited diseases that are observed. Present information is derived entirely from situations where it has been possible to isolate and characterise structurally the abnormal protein, and these represent a very limited and probably somewhat biased sample of inherited disorders taken as a whole. Even in some of the haemoglobin diseases, for instance the thalassaemias (pp.133–146) it does not appear that studies of the protein structure *per se* can provide specific information about the nature of the mutational change in the DNA. In these cases it seems that investigations in depth of the abnormality in the mechanism of protein synthesis, as well probably as structural studies at the RNA level will be necessary before insight into the type of mutation involved will be obtained.

9.1.2. Molecular properties of protein variants

In conditions where a specific abnormal protein or enzyme can be identified, the characterisation of its physico-chemical properties is an important step in elucidating the pathological process. In the case of sickle-cell haemoglobin, for instance, the critical phenomenon is its dramatically reduced solubility in the deoxygenated state. This is presumably a consequence of altering a very small region of the surface of the protein molecule by substitution of the hydrophobic valine for a hydrophilic glutamic acid residue (p. 18). It is a striking fact that although many other types of aminoacid substitution occurring in different parts of the haemoglobin molecule have been identified, the very marked change in solubility occurring in sickle-cell haemoglobin still remains unique to this substitution. The altered solubility is the reason for the morphological changes seen in the red cells when they are exposed to low oxygen pressure, the so-called sickling phenomenon. *In vivo* similar deformation of the red cells tends to occur in the venous side of the circulation particularly in the small veins and venous capillaries, leading to increased blood viscosity. In turn these effects can result in localised thrombosis and tissue damage. Also the deformed red cells are more readily destroyed than normal ones, and so chronic anaemia occurs and this induces other secondary pathological changes. Thus, although many of the details are still obscure, it is possible to envisage the main sequence of events which give rise to the clinical syndrome of sickle-cell anaemia. The initial mutational alteration involves only a very small change in the DNA of the gene, but its effects are progressively amplified first by the subtle alteration in haemoglobin structure which modifies its solubility properties, and then by the effects this has on the characteristics of the circulating red cells, so that eventually a complex pattern of pathological changes is produced.

One of the physical properties of a protein which may be significantly altered by a slight modification in structure is its stability. If an abnormal protein is markedly less stable than its normal counterpart, then the rate at which it is denatured *in vivo* is likely to be much increased, and the loss of functional activity which results can have important pathological consequences. For example, several inherited forms of severe chronic anaemia have been shown to be associated with abnormal haemoglobins whose most striking feature is their instability (pp. 24–26). They undergo much more rapid denaturation in the red cell than normal haemoglobin and this is evidently the main cause of the various pathological effects that are observed. Several abnormal enzyme proteins, for instance the glucose-6-phosphate dehydrogenase variant Gd, Mediterranean (pp. 165–170), also exhibit decreased stability and here again such secondary biochemical and clinical disturbances that ensue can be largely attributed to this effect. Probably many other inherited disorders have a similar causal basis.

Instability is likely to be brought about by any change in the primary structure of a protein which significantly distorts its normal three dimensional conformation. Where this is due to a single aminoacid substitution the severity of the effect will depend on the chemical properties and size of the side chain of the aminoacid which is substituted, and also the precise site of the substitution. However, a number of quite different substitutions occurring at different sites in the protein molecule may well have essentially the same consequences as far as their effect on the stability of the protein is concerned, and thus give rise to the same pathological process. Consequently a variety of different mutations may cause a series of distinct conditions which are, however, in all respects other than the primary structure of the abnormal protein, indistinguishable from one another. It may also be noted that different small deletions within a gene may, by leading to the absence of one or several aminoacids and consequent shortening of the corresponding polypeptide chain, each result in marked distortion of three-dimensional structure with much reduced stability of the protein. Larger deletions of course, by causing even greater abbreviation of the polypeptide, would often fail to lead to the appearance of a recognisable protein at all. Similar effects could also result from those mutations causing single base alterations where the base change alters a base triplet coding for an aminoacid to one coding for chain termination.

In the study of the properties of abnormal enzyme proteins, investigation of the kinetics of their catalytic activity is a matter of obvious interest. Changes for example in affinity of an enzyme protein for substrate or coenzyme which would be reflected by altered kinetics, may clearly be important causal factors in the development of a clinical disorder. Examples are the altered kinetics of argininosuccinate synthetase in citrullinaemia (pp. 207–208), of the 'atypical' form of serum cholinesterase associated with suxa-

methonium sensitivity (pp. 151–155), and of the glucose-6-phosphate dehydrogenase variant Gd Oklahoma (pp. 172–173) which causes a particular
form of chronic haemolytic anaemia. In each of these cases the Michaelis
constants (K_m) with respect to the enzyme substrate have been shown to
be significantly elevated and the magnitude of the effect seems to be sufficient
to account for the pathological consequences observed in these conditions.

One would expect that quite a number of different single aminoacid substitutions in an enzyme protein would either by causing an alteration in the
conformation or the chemical structure of the active site, result in one way or
another in a change in the kinetic parameters. The same substitution might
also lead to an alteration in one or more of the other physico-chemical
properties of the enzyme protein, for example its molecular stability. It is of
obvious importance, therefore, in the elucidation of the pathology of a
particular condition to assess the relative significance of such different effects.
An interesting illustration is provided by the glucose-6-phosphate dehydrogenase variant, Gd Mediterranean. Although the details of the structural
alteration in the abnormal enzyme protein are not yet known, several significant changes in its properties have been recognised. It is much less stable
than its normal counterpart, the Michaelis constants, with respect to both the
substrate glucose-6-phosphate and the coenzyme NADP are lower than
normal, and it shows an increased facility to utilise the substrate analogue,
2-deoxyglucose-6-phosphate. The clinical disorder, favism, with which this
abnormal enzyme is associated, almost certainly occurs because of the very
low level of the enzyme activity present in the abnormal red cells, and this
can be largely accounted for by the marked instability of the enzyme protein.
The altered kinetics are probably of only minor or no significance in the
development of the pathological process. Indeed, reduced Michaelis constants for substrate and coenzyme would be expected to be associated with
enhanced activity *in vivo*.

9.1.3. Metabolic and clinical consequences

Many different mutations may lead to the specific deficiency of a particular
enzyme, either by causing the synthesis of an abnormal enzyme protein with
altered kinetics or stability, or by causing a true reduction in the rate of
synthesis of the enzyme protein, or a complete failure in synthesis; and in
quite a number of inherited diseases it has been possible to identify such a
specific enzyme deficiency as central to the pathology of the condition, even
though the precise molecular basis of the enzyme abnormality is not yet
understood. Such conditions are for historical reasons usually referred to as

'inborn errors of metabolism', though in principle this name might well be applied to virtually any inherited disease. Various examples have already been discussed (Ch. 5) and others are listed in Appendix 1 (p. 368).

Typically an unusual but quite distinctive pattern of biochemical changes is observed, characterised by abnormally increased concentrations of certain substances in the body fluids or intracellularly, and by the relative deficiency of others. The detailed character of these changes will in general depend on the role of the particular enzyme in normal metabolism and on its tissue localisation. Their magnitude and, to some extent, their distribution will depend on the degree to which the specific enzyme activity has been reduced. In some conditions it may be effectively absent, but in others the reduction in activity is less extreme.

The clinical abnormalities seen in these various metabolic disorders presumably represent secondary consequences of the distorted biochemical pattern set up by the specific enzyme defect. But the detailed causal relationships are often difficult to discern. Even in such an extensively investigated condition as phenylketonuria (pp. 191–196), where a great deal is now known about the character of the metabolic disturbance, and much information about the altered concentrations of a wide variety of metabolites has been obtained, it has still not proved possible to obtain a satisfactory explanation for the severe mental retardation which is the outstanding clinical feature. The neurones of the developing brain are evidently affected in some way by the biochemical upset, but the details of the process are still quite obscure.

It is perhaps worth commenting on the unexpected character of the biochemical disturbances which have often been discovered to underlie particular clinical syndromes. The point is well illustrated by the condition known as homocystinuria (p. 378). The clinical syndrome is complex and includes among its characteristic features such diverse abnormalities as mental retardation, dislocated lens, a tendency both to arterial and venous thrombosis, and abnormalities in the development of the bones. An abnormal urinary excretion of homocystine was discovered to be associated with this syndrome in the course of a routine screening programme of the aminoacids in the urine of mentally retarded patients. This led to the recognition of a disturbance in methionine metabolism due to a specific deficiency of the enzyme cystathionine synthetase. Although a great deal was known about the metabolic pathways involved in the conversion of methionine to cysteine prior to the discovery of homocystinuria, biochemists would hardly have predicted that a block in this pathway might give rise to the particular complex of clinical abnormalities that are observed. Indeed it is still not known how the various clinical features are brought about, though there is little doubt that they all in some way stem from the primary deficiency of cystathionine synthetase.

The condition known as Pompe's disease (p. 215) illustrates the same general point. This had long been recognised as a quite characteristic disorder in which there was progressive accumulation of glycogen particularly in heart muscle. But although the main pathways in the synthesis and degradation of glycogen had apparently been established, and the various enzymes thought to be concerned had been examined in Pompe's disease, the underlying nature of the condition had remained obscure. Indeed the discovery that it is due to a specific deficiency of an α-(1,4) glucosidase, normally present with many other hydrolases in the intracellular organelles known as lysosomes, was quite unanticipated because this enzyme was not previously thought to have any significant role in glycogen degradation.

When one considers such examples it is perhaps not surprising that there remain very many inherited disorders in which we still have virtually no idea what enzyme or protein may be defective, or indeed what area of metabolism could be involved. The clues may well be present in the symptomatology, but they are certainly in most cases far from obvious and as yet no one appears to have thought of the right biochemical systems to investigate.

9.2. Dominance and recessivity

With the progressive elucidation of the molecular pathology of a disorder one may expect to see more clearly the reasons for its particular mode of inheritance; that is, why it should occur predominantly in the heterozygous state and so be classified as a 'dominant', or why it should be seen only in the homozygous state and so be classed as a 'recessive'.

Sickle-cell disease which is inherited as a 'recessive' characteristic provides a simple illustration of the point. The clinical abnormalities in this condition derive from the fact that the haemoglobin present in the red cells is extremely insoluble when deoxygenated and so causes red cell deformation (sickling) *in vivo*, in those parts of the circulation where the oxygen tension is low. In heterozygotes for the sickle-cell gene, about 65% of the haemoglobin present in the red cells is usually of the normal type and only about 35% is of the sickle-cell type, though the total amount of haemoglobin per cell is not significantly reduced. The mixture considered as a whole shows a reduction in solubility when compared with normal haemoglobin, and sickling of the red cells can be readily demonstrated *in vitro* if the oxygen tension is sufficiently reduced. But the degree of deoxygenation required is greater than normally occurs *in vivo*. So untoward consequences do not usually occur in the heterozygotes who are generally quite healthy.

This example emphasises the important point that the terms 'dominant' and 'recessive' as generally used, only have meaning in reference to a specific characteristic or phenotype. Sickle-cell disease is inherited as a 'recessive' characteristic; but the sickle-cell phenomenon, that is the occurrence of sickling of the red cells when they are subjected to appropriate *in vitro* procedures, is inherited as a 'dominant' characteristic since it occurs both in heterozygotes and homozygotes for the abnormal gene.

One may usefully contrast 'recessively' inherited sickle-cell disease with the 'dominantly' inherited forms of chronic anaemia due to the so-called unstable haemoglobins. In these conditions the heterozygote synthesises in his red cells both the unstable haemoglobin and also the normal haemoglobin, but because of its rapid denaturation the amount of the unstable form and hence the total amount of haemoglobin present in a functionally active state, is progressively reduced as the red cells mature. This causes a chronic anaemic state which is further accentuated by precipitation of the denatured abnormal haemoglobin, which tends to cause the red cells to be more rapidly destroyed and removed from the circulation. Thus unlike the situation in the sickle-cell heterozygote the normal haemoglobin synthesised by these heterozygotes fails to protect the red cells from the adverse effects of the abnormal form. The genes determining the various types of unstable haemoglobin are all extremely rare and so the disorders produced in the homozygous states have not been observed. One may predict however that they would be extremely severe forms of anaemia, often lethal in early life.

In diseases arising from the deficiency of a specific enzyme, the question as to whether the disorder is inherited as a 'recessive' or as a 'dominant' may largely depend on the average activity of the enzyme in the normal homozygous state, and in particular on how far this is in excess of the minimal level actually needed to maintain healthy function. In the heterozygous state the specific enzyme activity is generally intermediate between that present in the normal and in the abnormal homozygotes, and in the extreme case where the abnormal gene leads to complete loss of the enzyme activity, the heterozygote usually shows about one half the average activity found in normal homozygotes (pp. 234–238). So if, as usually appears to be the case, the activity in the normal homozygotes is on average many times the minimum actually required for efficient metabolic function, the reduced activity present in heterozygotes will also usually be in excess, and no untoward consequences will result. Clinical abnormality will then only be seen in the abnormal homozygotes where the activity is presumably so diminished that it is insufficient to maintain healthy function. In fact most of the so-called 'inborn errors of metabolism' which have so far been identified have a recessive pattern of inheritance and one may infer that the normal activity of the specific enzyme involved is much greater than is strictly required. There is, as it were, a considerable built in safety factor.

Dominant inheritance of a disease due to an enzyme deficiency is most likely to occur where the enzyme in question happens to be rate limiting in the metabolic pathway in which it takes part, because the activity of such enzymes in the normal organism will in general be closer to the minimum required to maintain normal function.

However 'dominantly' inherited disease is perhaps more often likely to reflect the occurrence of mutations at loci determining structural or other non-enzymic proteins. The essential point is that most enzyme proteins because they function catalytically are generally present at extremely low molecular concentrations, and even so usually appear to occur in considerable excess of the minimal quantity required for normal function, without placing an undue load on the overall protein synthetic processes in the cell, and without causing dysfunction due to hyperactivity. In contrast other proteins are often present in cells in much greater absolute amounts than enzyme proteins, and also the structure and function of the cell will often require an optimal quantity to be present. Consequently the amount of the protein present in the cells of the normal individual will not, as in the case of many enzymes, be in significant excess of the amount required to maintain normal structure and function.

Thus defective synthesis of say half the total of a particular protein, as might occur in heterozygotes, is often likely to result in markedly defective function and hence overt clinical disease. Such effects could occur if the mutant allele determined an abnormal polypeptide chain with defective properties, and might be particularly marked in polymeric proteins because here more than half of the protein molecules actually formed in the heterozygote might contain structurally defective polypeptides with aberrant functional characteristics. Clinical abnormality might also be expected to occur should the mutant result in a failure in synthesis of the polypeptide, if there were no compensatory increase in the synthesis of the normal polypeptide by the normal allele.

9.3. *Heterogeneity of inherited disease*

Very often a particular syndrome which at first was thought to be a discrete entity and presumed to be determined by a single abnormal gene has turned out on closer analysis to represent a collection of distinct disorders, each the consequence of a quite different mutation and each with its own specific underlying pathology. The degree of genetical heterogeneity that may be uncovered in what at first sight seems a relatively homogeneous condition may be quite remarkable, and there seems no doubt that this will prove to be a very widespread and general phenomenon.

Such heterogeneity can arise in several different ways, and it is convenient to classify them as follows:
a) Different mutant alleles at a single locus;
b) Mutant alleles at different loci,
  i) affecting the same enzyme or protein,
  ii) affecting different enzymes or proteins.

9.3.1. *Heterogeneity due to multiple allelism at a single locus*

A very large number of different alleles, each determining a structurally distinct variant of a particular polypeptide chain, may be generated by separate mutations within the confines of a single gene (p. 278). Some of these mutants may result in clinical abnormality, and frequently the clinical conditions produced by quite different mutations may be very similar to one another and indeed often indistinguishable. For example more than twenty-five unstable variants of haemoglobin, each with a different structural abnormality in the β polypeptide chain (see p. 24 and p. 112) have been identified in different individuals (Stamatoyannopoulos 1972). Each is due to a different mutation at the same gene locus, and each results in a form of chronic haemolytic anaemia. Yet the different anaemias although they vary somewhat in degree are for the most part difficult to distinguish from each other on clinical or pathological grounds alone, and prior to their charac-terisation by structural studies of the haemoglobins would no doubt have been regarded as a single genetic entity with perhaps some superimposed variability, or at most as two or three separate syndromes each determined by a single mutation. The various chronic haemolytic disease due to different variants of the enzyme glucose-6-phosphate dehydrogenase illustrate the same general phenomenon (p. 172). Here more than twenty structural distinct forms of the enzyme each resulting in a form of chronic haemolytic disease, and each determined by a different mutant allele at a single locus have been identified (Yoshida et al. 1971, Beutler and Yoshida 1973). The clinical manifestations in these cases are again very similar to one another.

There is no reason to think that these particular examples are in any way peculiar. In fact as inherited diseases are increasingly studied at the enzyme level, more and more examples of what appear to be essentially the same general phenomenon are coming to light.

An interesting illustration is provided by the work which followed the recognition of the nature of the enzyme deficiency in the Lesch-Nyhan syndrome, a complex and very severe neurological syndrome in children which is characterised by mental retardation, spastic cerebral palsy, choreo-athetosis, and a curious behavioural disorder manifested by self-destructive biting (Lesch and Nyhan 1964). This was found to be associated with hyper-uricaemia due to overproduction of uric acid, and in some cases uric acid renal calculi and signs of gout are present. The condition turned out to arise from what appeared to be a very marked and quite specific deficiency of the enzyme hypoxanthine-guanine phosphoribosyl transferase (Seegmiller et al. 1967), determined by an abnormal gene located on the X chromosome (Nyhan et al. 1967).

Because of its evident relationship to uric acid formation, the enzyme was then examined in a number of adult patients with typical clinical histories of acute gouty arthritis or uric acid nephrolithiasis and with hyperuricaemia shown to be due to uric acid overproduction (Kelley et al. 1967, Kelley 1968). Among this series of patients some were found to have quite marked deficiencies of the enzyme, though not as severe as in the Lesch-Nyhan syndrome. The results of some of the enzyme assays are shown in table 9.1. They suggest that several quite distinct types of defect may occur and give rise to hyperuricaemia, but in any one family the same specific abnormality is present among the affected individuals. Thus in one family (table 9.1, family J) the affected individuals showed levels of activity as measured in red cells of only about 1 % of the normal when either hypoxanthine or guanine was used as substrate. Furthermore, the enzyme protein was found to be significantly more thermolabile than the normal enzyme. In another family (table 9.1, family L) the affected individuals also showed much reduced enzyme activity, but the reduction was considerably more marked with guanine as substrate than with hypoxanthine, a finding which indicates an altered pattern of substrate specificity. Furthermore here the enzyme protein appeared to be less thermolabile than the normal one.

Benke et al. (1973) have reported a remarkable example of hyperuricaemia in a fourteen year old boy, in which the hypoxanthine-guanine phosphoribosyl transferase activity appeared normal by standard assay procedures, but kinetic studies revealed a much reduced affinity of the enzyme for phosphoribosyl-pyrophosphate though the apparent Michaelis constants for hypoxanthine and guanine were evidently not significantly altered.

These findings suggest that several distinct abnormal genes, each producing a structurally altered form of this enzyme protein with abnormal properties, separately occur in the various families. They each evidently lead to marked enzyme deficiency resulting in overproduction of uric acid with hyperuric-

TABLE 9.1

Hypoxanthine-guanine phosphoribosyl transferase activity assayed with hypoxanthine or guanine as substrate in red cells. (Kelley 1968).

	Phosphoribosyl transferase activity (mμmoles/mg protein/hr)		Heat stability relative to normal
	Hypoxanthine	Guanine	
Control subjects (18)	103 ± 18	103 ± 21	—
Patients with Lesch-Nyhan syndrome (9)	<0.01	<0.004	—
Patients with gout due to uric acid overproduction (a) with normal enzyme (10)	99 ± 13	106 ± 21	—
(b) with defective enzyme (10)			
J family			
FJ	1.3	0.6	reduced
RJ	1.5	0.8	reduced
TJ	1.8	0.8	reduced
L family			
FL	11.8	0.5	increased
ML	8.7	0.5	increased
S_1 family			
TS	9.9	9.5	not tested
D family			
AD	12.2	17.3	not tested
G family			
JG	9.4	8.8	not tested
RG	9.2	7.5	not tested
S_2 family			
GS	0.03	0.009	not tested

aemia, and the clinical consequences (gout and nephrolithiasis) are very similar in the different cases. But the clinical picture contrasts very strikingly with that of the Lesch–Nyhan syndrome which is apparently the consequence of a complete or almost complete deficiency of the enzyme. Here a very severe neurological disorder manifesting in childhood occurs.

Although in the Lesch-Nyhan syndrome itself the red cells are usually completely or nearly completely devoid of hypoxanthine-guanine phosphori-bosyl activity, some activity (0.6 to 7% of normal) has been demonstrated in fibroblasts grown in tissue culture (Kelley and Mande 1971). Studies on the properties of the residual enzyme activity in fibroblasts derived from a series of patients showed that there were significant differences between the various cases, in the thermostability and also certain kinetic charac-teristics of the enzyme (Kelley and Arnold 1973). This provides further evidence of genetic heterogeneity.

In the past it was usually assumed that if a metabolic disorder attributable to a specific enzyme deficiency segregated in families in the typical manner of an autosomal recessive characteristic, that the affected individuals were ipso-facto homozygous for a single mutant gene. It will however be apparent from what has been said about heterogeneity arising from multiple allelism that this will often not be the case. Although the condition segregates as an autosomal recessive, it may in many cases be due to heterozygosity for two different mutant alleles each producing an enzyme deficit, but in a different way. Such heterozygous individuals would have received one of the mutants from one parent and one from the other. So elucidation of the situation will often require studies not only on the characteristics of the defective enzyme in the affected patients, but also studies on the two parents and their relatives where usually the separate mutant alleles will be in heterozygous combination with the normal allele. Examples of studies in which heterozygosity for two different mutant alleles has been demonstrated as the cause of an enzyme deficiency resulting in clinical abnormality are provided by investigations of the enzyme defect in certain cases of phosphohexoseisomerase deficiency (Detter et al. 1968, Blume et al. 1972) and of pyruvate kinase deficiency (Paglia et al. 1972). This type of phenomenon is probably by no means uncommon in recessively inherited diseases due to enzyme deficiencies. However it may often be peculiarly difficult to recognise because it involves in the patients the characterisation of the products of two different alleles in a mixture where the total enzyme activity is by the nature of the situation very much reduced, and in the parents it involves characterising the product of the mutant allele in the presence of the product of the normal allele which would of course exhibit much greater enzyme activity.

So far we have been considering genetic heterogeneity in situations where different alleles at a single locus determine different abnormalities which at the clinical level are very similar in their manifestations or indistinguishable. It is however pertinent also to emphasise a somewhat different phenomenon. This is the occurrence of conditions which clinically may appear to be very different and yet are determined by alleles at a single gene locus. The disparate clinical appearances of conditions such as sickle-cell anaemia, haemoglobin C disease, methaemoglobinaemia and chronic haemolytic anaemia due to an unstable haemoglobin, all of which can arise in consequence of different mutations at the haemoglobin β-gene locus, illustrate the point. Another example is the marked difference mentioned earlier in the clinical appearances seen in infants with the Lesch-Nyhan syndrome and in adults with gout due to different mutations at the gene locus determining hypoxanthine guanine phosphoribosyl transferase. In the past such disorders if considered on clinical grounds alone, would probably have been regarded as due to mutations at quite different gene loci.

9.3.2. *Heterogeneity due to mutants at different gene loci*

Some proteins contain two non-identical polypeptide chains coded by genes at different loci. Mutants at either locus may result in structural abnormalities of the protein and cause overt clinical disease. However since the same protein is involved the clinical consequences of mutants at the two separate loci may be very similar. Haemoglobin A ($\alpha_2\beta_2$) provides a simple illustration of the point. Mutant alleles at either the α- or the β-gene loci can result, for example, in unstable forms of the protein which give rise to chronic haemolytic disease (p. 25). The α- and β-thalassaemias (p. 133–146) illustrate the same general point.

More often however heterogeneity arises from mutants at loci determining different and quite distinct enzymes or proteins. That this should be so is not very surprising. Loss of function of one or another of a series of enzymes involved in a sequential series of reactions in a metabolic pathway, or which are associated together in a complex of metabolic relationships, could well result in the same or very similar consequences at the clinical level.

A rather obvious example is provided by the complex network of metabolic reactions which is present in the red cell, and which is necessary for maintaining its functional and structural integrity. Disruption of the normal metabolism of the red cell with its consequent premature destruction and hence chronic haemolytic disease can clearly occur as a result of defects in a number of quite different enzymes. And indeed a series of clinically not very dissimilar forms of genetically determined haemolytic anaemias have already been differentiated and shown to be due to deficiencies of quite different enzymes determined by separate gene loci (table 9.2). These various haemolytic diseases vary considerably in severity. But in practice it seems that they are difficult to differentiate one from another in the absence of information provided by enzyme studies. This difficulty is further enhanced by the fact that at any one of these loci multiple alleles may also occur which differ in the degree of enzyme deficit they produce and hence in the clinical severity (e.g. pyruvate kinase p. 203). So superimposed on the variation due to mutations at separate loci affecting different enzymes, is the added variation due to multiple allelism at single loci.

TABLE 9.2

Genetically determined deficiencies of red cell enzymes resulting in haemolytic disease. (For references see Appendix 1.)

Enzymes of glycolytic pathways	Other enzymes
Hexokinase	Glucose-6-phosphate dehydrogenase
Glucose phosphate isomerase	Glutathione reductase
Phosphofructokinase	Glutathione peroxidase
Triose phosphate isomerase	γ-Glutamyl-cysteine synthetase
Diphosphoglycerate mutase	Glutathione synthetase
Phosphoglycerate kinase	ATPase
Pyruvate kinase	Adenylate kinase

The series of conditions known as the mucopolysaccharidoses illustrate the same general point (p. 217). The syndromes known as Sanfilippo A and B, for example, are evidently indistinguishable clinically, yet A is due to a deficiency of the enzyme heparan sulphate sulphatase (Kresse and Neufeld 1972) and B to a deficiency of N-acetyl-α-glucosaminidase (O'Brien 1972, Figara and Kresse 1972). Both enzymes are concerned in the degradation of the carbohydrate chains of the mucopolysaccharide heparan sulphate, and their separate deficiencies both result in the progressive lysosomal accumulation of partially degraded products of this substance.

The classical Hurler and Hunter syndromes are both severe diseases of early childhood with mental deterioration, stunting of growth and bone deformities, and gross enlargement of the liver and spleen due to the progressive lysosomal accumulation of partially degraded products of the mucopolysaccharides dermatan sulphate and heparan sulphate (p. 220). Although very similar clinically they had long been recognised as genetically distinct because from pedigree studies it appeared that while the gene determining Hunter's syndrome was on the X-chromosome, that determining the Hurler's syndrome was on one of the autosomal chromosomes. Eventually it turned out that Hurler's syndrome is due to a deficiency of α-iduronidase (Matalon and Dorfman 1972, Bach et al. 1972), while Hunter's syndrome is due to a deficiency of sulphoiduronate sulphatase (Bach et al. 1973). Both enzymes are normally concerned in degrading the polysaccharides which accumulate, and it is not therefore surprising that the two syndromes resemble each other so closely. There is however one consistent difference in their clinical manifestations, this is that in Hurler's syndrome the corneae characteristically become cloudy, whereas in Hunter's syndrome they remain clear. This implies that corneal mucopolysaccharides do not require sulphoiduronate sulphatase for their degradation, and are presumably therefore devoid of sulphated iduronate residues (Bach et al. 1973).

9.3.3. Congenital methaemoglobinaemia: a model example

Congenital methaemoglobinaemia provides a simple model example, which illustrates the quite remarkable degree of genetic heterogeneity which may occur even in what is by any standards an exceedingly rare syndrome. The characteristic clinical feature of this condition is a greyish blue cyanotic appearance which is due to the fact that a significant fraction of the iron in the haemoglobin in the circulating red cells is in the ferric state and incapable of transporting oxygen. The abnormality is apparent at birth or shortly after, and persists throughout life usually with very little variation. Most affected individuals are not seriously incapacitated, though occasionally the abnormality is associated with some degree of mental retardation. The condition needs to be distinguished from other causes of chronic cyanosis, such as congenital malformations of the heart, but this can usually be done relatively easily on clinical grounds.

The early studies on the inheritance of the condition showed that at least two sorts of genetical abnormality occurred (Barcroft et al. 1945, Hörlein and Weber 1948). In some cases the disorder appeared to be inherited as an autosomal recessive condition, while in others it was apparently inherited as an autosomal dominant. In the autosomal recessive cases a specific deficiency of the red cell enzyme NADH diaphorase (methaemoglobin reductase) was identified (Gibson 1948, Scott and Griffith 1959). In affected individuals, the level of this enzyme activity is consistently very low and often only barely detectable. Among their parents, children and certain other relatives, who may be presumed to be heterozygotes, a partial reduction in the level of activity of this enzyme is found (Scott 1960), but this is evidently insufficient to lead to any significant degree of methaemoglobinaemia or cyanosis. Electrophoretic studies on the residual enzyme activity in certain affected individuals indicate at least in some cases, that a structural defect in the enzyme protein is probably the cause of the enzyme deficit, but that the structural alteration differs in patients from different families (Kaplan and Beutler 1967, West et al. 1967, Bloom and Zarkowski 1969, Hsieh and Jaffe 1971). Evidently several different abnormal alleles at the gene locus coding for the methaemoglobin reductase enzyme protein can bring about this condition and at least five have been shown to occur.

In the group of 'dominantly' inherited forms of congenital methaemoglobinaemia, NADH diaphorase is not abnormal. However, in most of these cases there is a structural defect of the haemoglobin itself, and a number of distinct types of abnormality have been recognised. They each involve a specific single aminoacid substitution occurring in a region of the molecule where a haem group is attached to a polypeptide chain, and either the α- or the β-chains which are of course coded at separate gene loci may be affected. A minor difference in clinical manifestation is that in the α-chain mutants the cyanotic appearance is present at birth because the α-chain occurs in both foetal ($\alpha_2\gamma_2$) and adult ($\alpha_2\beta_2$) haemoglobin, whereas in the β-chain mutants it only becomes apparent some weeks after birth when adult haemoglobin begins to be the predominant form present.

Thus the syndrome of congenital methaemoglobinaemia can arise because of mutations at at least three distinct gene loci. One is concerned with defining the structure of methaemoglobin reductase, and the others with defining the α- and β-chains of haemoglobin, and at each of these loci several different abnormal alleles causing the syndrome evidently occur. That mutations at yet other gene loci may also result in the same sort of clinical disorder is indicated by the report of a form of congenital methaemoglobinaemia apparently inherited as an autosomal dominant, but showing no abnormality in either haemoglobin or methaemoglobin reductase (Townes and Morrison, 1962).

Congenital methaemoglobinaemia is an extremely rare condition, and probably in most populations occurs with an incidence of only perhaps one in several hundred thousand births. Nevertheless at least ten different abnormal genes causing it have already been identified; two at the Hb α-locus, three at the Hb β-locus and at least five at the locus determining NADH diaphorase. The syndromes they separately produce are difficult if not impossible to distinguish on clinical grounds alone.

One may anticipate that a similar degree of genetic heterogeneity is likely to be found in many other inherited conditions which at present can only be defined in clinical terms, or in terms of some of the secondary biochemical or physiological disturbances originating from the underlying enzyme or protein abnormality.

9.3.4. Therapeutic implications of genetic heterogeneity

The possible occurrence of genetic heterogeneity is of obvious interest if one is trying to understand the fundamental nature of any particular disease. But awareness of this possibility may also, at least in certain instances, be of considerable therapeutic importance.

The point is well illustrated by work on the rare metabolic disorder usually referred to as methylmalonicaciduria (Oberholzer et al. 1967, Stokke et al. 1967, Rosenberg et al. 1968a, b, Lindblad et al. 1968, 1969, Morrow et al. 1969). The condition is characterised by the excretion of large amounts of methylmalonic acid in the urine. The affected children fail to thrive and show pronounced ketoacidosis. They may be severely retarded and often die in early life. The abnormal excretion of methylmalonic acid was shown in a number of cases to be due to deficient activity of the enzyme methylmalonyl-CoA carbonyl mutase which catalyses one of the steps in the reaction sequence by which propionate is converted to succinate (fig. A.11, p. 390). This enzyme requires for its normal activity the cofactor 5′-deoxyadenosyl cobalamin, one of the coenzyme forms of vitamin B_{12}.

Patients with methylmalonicaciduria respond in two quite different ways when large amounts of vitamin B_{12} are administered or if 5′-deoxyadenosyl cobalamin itself is given. In some patients the biochemical disturbances are rapidly corrected and there is a dramatic clinical improvement. In other patients no significant effect is observed. Various studies have shown that in the B_{12} responsive patients the essential defect is in the biosynthetic pathway leading to the formation of the active B_{12} cofactor (Rosenberg et al. 1969, Mahoney and Rosenberg 1970). In the non-responsive patients the defect is probably in the methyl malonyl CoA carbonyl mutase enzyme protein itself. The two defects may be presumed to be determined by mutants at different loci. However they both result in a gross deficiency of the mutase activity and hence the same clinical disturbance. In fact the two types of disorder are most simply differentiated by the therapeutic response or non-response to the administration of vitamin B_{12}.

Homocystinuria (p. 378) is another disorder in which the biochemical abnormalities can be largely corrected by the administration of large amounts of a vitamin B_6, in some cases (Barber and Spaeth 1967, Gaull et al. 1969, Carson and Carré 1969, Mudd et al. 1970), while other cases fail to respond at all (Turner 1967, Hooft et al. 1967, Carson and Carré 1969). In both types of case deficient activity of the enzyme cystathionine synthetase is the cause of the disease (fig. A.5, p. 379). Evidently however at least two different mutants are involved, since responsiveness or non-responsiveness to vitamin B_6 administration runs true to type among affected patients in different sibships. But the exact reason for this difference in response to the vitamin is not known.

Most inborn errors are very rare diseases and are irregularly distributed in different populations. In view of the high degree of genetic heterogeneity that exists, which makes it likely that very often different mutants cause the disease in different families particularly if they come from different populations, it is clearly unwise to assume that if a particular syndrome has failed to respond to a specific line of treatment in some cases it will necessary fail to respond in others.

9.3.5. Other causes of variation in inherited disease

Most inherited diseases show some variability in their manifestations. Much of this variability is often attributable to genetic heterogeneity of the sort which has already been discussed. However the variation may also arise from other causes, both genetic and environmental.

Genetic causes: Variation in the expression of a particular inherited disease in different individuals may derive not only from differences in the particular mutant genes causing the specific enzyme or protein defect which is the basic cause of the disease in question, but also from differences in the rest of the genetical constitution of the individuals in whom the particular abnormal gene occurs. Probably no two individuals, with the exception of monozygotic twins, are exactly alike in this respect, because of the multiplicity of different 'normal' allelic combinations that are present at many gene loci (pp. 297–300). Such differences in so-called 'genetical background', because they define many of the details of the biochemical and physiological milieu against which the effects of the particular abnormal gene are expressed, may well influence the manifestations of a particular disorder. Some combinations of genes at other loci might minimise the pathological consequences, and others accentuate them. Since such different combinations of genes may often result in little or no obvious differences between individuals not carrying the particular abnormal gene, the detailed manner in which they contribute to the variation is generally difficult to define.

In this connexion it is necessary to distinguish between variability due to alleles at loci other than the one where the abnormal gene whose effects are being considered occurs, and variability due to the existence of different so-called 'normal' alleles at the same locus. The latter source of variation can of course only exist where the affected individuals are heterozygous (i.e. in 'dominant' diseases). It has been shown that several distinct alleles having different effects on function without necessarily resulting in overt disease may occur at a particular gene locus, and that sometimes two or more such alleles are each relatively common. In these circumstances the abnormal gene, whose effects one is considering, may be present in heterozygous combination with one or another of these different 'normal' alleles, and this could influence the expression of the clinical disorder that ensues. This effect is sometimes termed 'allelic modification'. It tends to result in affected sibs resembling one another more closely in the manifestations of the particular disease than affected parents and children.

Environmental causes: The clinical manifestations of an enzyme or protein abnormality determined by a particular mutant allele, may depend in some degree on the environmental circumstances to which individuals with the abnormality are exposed. Often the variability in clinical expression may be quite small. But in other cases it can be very wide and may give rise to quite unexpected clinical differences. One very striking example of this is provided by certain mutant alleles which give rise to a deficiency in blood serum of the protein known as α_1-antitrypsin.

This protein which is readily demonstrable in normal serum is originally synthesised in liver. It has the characteristic property of binding to trypsin and certain other proteolytic enzymes and inhibiting their proteolytic activity. Electrophoretic studies of the protein have led to the discovery of a series of variants which can be attributed to different alleles at the gene locus (*Pi*) coding for α_1-antitrypsin (for review see Fagerhol and Laurell 1970). One of these alleles Pi^z gives rise to a marked deficiency of the protein in serum. But the deficiency in this case is evidently largely due to a structural peculiarity of the protein which prevents its release from the liver where it is synthesised, because it has been shown that in homozygotes $Pi^z Pi^z$ that although the amount of the α_1-antitrypsin is greatly reduced in serum (about 15% of normal), it is present in quite large quantities in liver (Lieberman et al. 1972). Another allele Pi^o appears to result in a true deficiency of the protein, and in homozygotes it appears to be virtually absent in serum (Talamo et al. 1973).

It was discovered that certain adults with pulmonary emphysema, particularly among those in which the disease has a relatively early onset, there is a marked deficiency of serum α_1-antitrypsin (Laurell and Eriksson 1963, Eriksson 1964, 1965). The association is sufficiently strong as to indicate a causal relationship, although only a small proportion of all patients with emphysema have α_1-antitrypsin deficiency. Exactly how the deficiency of the protein results in the pulmonary damage is not known, but it is possible that it causes a failure of inhibition of proteolysis in the lung in certain circumstances (for review and discussion see Mittman 1972, Hutchison 1973).

The individuals with pulmonary emphysema and marked deficiency of serum α_1-antitrypsin appear to be mainly homozygotes of the genotype $Pi^z Pi^z$, but some are heterozygotes $Pi^z Pi^o$ and rare patients are $Pi^o Pi^o$. However from studies on their families it became clear that only a proportion of individuals with these genotypes (probably about 15–30%) actually develop emphysema. Others remain quite healthy and may do so to an advanced age. It seems in fact that those individuals with serum α_1-antitrypsin activity who have developed emphysema must have been exposed to some environmental factors or factors which precipitate the development of the disease. The nature of such environmental causes are not as yet understood, though cigarette smoking is thought to be one likely possibility (Hutchison 1973).

Subsequently and rather unexpectedly it was discovered that there is also a highly significant association between serum α_1-antitrypsin deficiency and neonatal hepatitis (Sharp et al. 1969, Porter et al. 1972, Aagenes et al. 1972). The evidence indicates that infants with the genotype $Pi^z Pi^z$ are particularly prone to develop the disorder, though only a proportion of them (perhaps 10–20%) actually do so. Again it seems that there is some, as yet unidentified, environmental factor or factors to which the infants who develop the condition are exposed.

In European populations about 1 in 1,500–3,000 newborn infants have serum α_1-antitrypsin deficiency. A proportion of these develop neonatal hepatitis which may be rapidly fatal or which may lead to hepatic cirrhosis and death in childhood. The majority however survive into adult life, when a further proportion develop pulmonary emphysema. Many however appear to remain healthy. Thus the variation in the clinical consequences of serum α_1-antitrypsin deficiency is quite remarkably wide, and includes two very different types of clinical syndrome, neonatal hepatitis in infants and pulmonary emphysema in adults, which at first sight might appear to be quite unrelated conditions.

9.4. Heredity and environment

The characteristics of an individual depend not only on the genes he carries and the types of enzymes and proteins he consequently makes, but also on the environment or more correctly the series of environments in which he develops and in which he lives. The interactions between what Francis Galton called 'nature' and 'nurture' are often complex and difficult to disentangle. However we implicitly recognise their importance in our ideas of what we mean by inherited disease, and the question can perhaps be examined most simply by looking at a few selected diseases in which the essential nature of the genetical abnormality at the enzyme level has been defined, and also the principle environmental factors affecting its expression are more or less understood.

TABLE 9.3

Genetically determined enzyme deficiencies and environmental factors in the causation of different disorders.

Condition	Enzyme deficiency	Environmental factor
Phenylketonuria	Phenylalanine hydroxylase	Phenylalanine in dietary protein
Galactosaemia	Galactose-1-phosphate uridyl transferase	Lactose in milk
Hereditary fructose intolerance	Liver aldolase (aldolase A)	Fructose, sucrose
Favism, Primaquine sensitivity	Glucose-6-phosphate dehydrogenase	Fava bean, primaquine, sulphonamides, etc.
Scurvy	L-gulonolactone oxidase	Vitamin C deficiency

Consider the series of conditions listed in table 9.3.

Phenylketonuric infants lack the enzyme phenylalanine hydroxylase because they have inherited from each of their parents a mutant form of the gene necessary for the normal formation of this enzyme. In consequence they are unable to metabolise phenylalanine and since this is a necessary constituent of the protein in their diet they become mentally retarded, unless they are fed a phenylalanine restricted diet (p. 191). Similarly galactosaemic infants lack the enzyme galactose-1-phosphate uridyl transferase which is required to metabolise galactose (p. 196). Galactose in the form of the disaccharide lactose is a major carbohydrate constituent of milk, and so in the usual course of events newborn infants will receive large amounts of it. For galactosaemic infants who lack the transferase enzyme, the clinical consequences are severe. However if a galactose-free diet is fed a dramatic improvement occurs, and if no irreversible damage has already occurred the infant will develop in a normal manner.

So in a sense we can regard phenylketonuria and galactosaemia as conditions in which certain infants because of their specific genetical constitutions, are unable to cope with a particular facet of the normal environment, that is the occurrence of phenylalanine in dietary protein or galactose in milk. If their environments are modified appropriately the ill effects of the disability can be prevented. The individual will still lack the particular enzyme, but in his new environment this will not disable him. He will as it were, be predisposed without being clinically affected.

The condition known as hereditary fructose intolerance (pp. 199–203) illustrates the same general point. Here there is a deficiency of the liver enzyme aldolase B, which is necessary for the normal metabolism of fructose. If fructose or sucrose is fed, rather severe symptoms follow. If fructose is excluded from the diet no harmful effects occur. Unlike galactose, fructose is not an obligatory feature of the normal diet of an infant. Consequently the manifestation and severity of the disease are more variable. An important factor appears to be the time when breast feeding is discontinued. If an early change to artificial milk feeds with added sucrose is made, rapid deterioration in the infant's condition occurs and should the nature of the abnormality not be recognised, irreversible damage may take place. If, however, the infant is not weaned until he is several months old he is often able to make a positive rejection of feeds which make him ill, and the peculiarity is likely to be diagnosed. Furthermore such children tend to develop a strong aversion to sugar, sweets and fruit, and thus tend to protect themselves. This illustrates rather clearly how the severity of an inherited disease may depend on quite fortuitous and apparently unconnected factors in the infant's environment. In this case the time when breast feeding is discontinued. In passing one may also note that it also illustrates, in a somewhat dramatic manner, the way in which an individual's taste and food preferences may be a direct consequence of a specific enzyme abnormality.

Favism provides a particularly clear example of the way environmental and genetical factors may interact in the causation of a disease process. This condition has been known for many years as a severe form of recurrent haemolytic anaemia in which the attacks follow the ingestion of the fava bean. It is found particularly in the Middle East and certain parts of Southern Europe where the fava bean is a common feature of the diet. However by no means all people who eat fava beans develop.favism. Those that do have been shown to have a genetically determined deficiency of G-6-PD usually due to the variant known as Gd Mediterranean (p. 165). So to develop the disease an individual must carry the gene and eat the bean. Both genetical and environmental factors can thus be clearly seen as essential to the causation of the disease.

As far as we know, all members of our own species and also other primates are incapable of synthesising L-ascorbic acid (vitamin C). Consequently if for some reason their diet is deficient in this substance they develop scurvy. This is not the case in many other mammalian species which can apparently manage without an independent dietary source of L-ascorbic acid, because they can synthesise it from D-glucose via the reaction sequence:

D-glucose → D-glucuronolactone → L-gulonolactone → L-ascorbic acid.

Evidently in man and other primates the enzyme activity capable of converting L-gulonolactone to L-ascorbic acid is lacking (Burns 1957) presumably because the necessary gene has been lost in the course of evolution. So scurvy with good reason is considered to be a disease caused by an unfavourable environment, namely one in which the individual's diet is deficient in vitamin C. But it can equally well be thought of as due to an inborn error of metabolism which we all happen to possess (Snyder 1959).

We have little hesitation in categorising phenylketonuria, galactosaemia or fructose intolerance as inherited disorders, because virtually all infants with the appropriate genetic constitutions will manifest the clinical disorder in the absence of specific treatment. Similarly we regard scurvy as an environmentally caused condition because it will develop as far as we know in anybody whose diet is deficient in Vitamin C. But it is clear that in each case both genetic and environmental factors are involved. In fact the way we classify these disorders depends essentially in each case on the prevalence in the population of the genetically determined predisposition on the one hand,

and the prevalence of the specific environmental conditions which elicit the abnormality on the other. Where the genetically determined predisposition is uncommon and the environmental factor is universal, as in phenyl-ketonuria or galactosaemia, the disorder is said to be inherited. Whereas if the genetically determined predisposition is universal or nearly so and the specific environmental circumstances relatively uncommon, as in scurvy, we regard the condition as environmental in origin.

It is not surprising that there should be many situations intermediate between these two extremes, where neither the genetic nor the environmental factors are universally present. In such cases only a proportion of the population is genetically predisposed to the condition and only a proportion is exposed to the particular environmental circumstances. This is clearly so in favism. It is evidently also the case with respect to the development of neonatal hepatitis in infants and pulmonary emphysema in adults who have a genetically determined defect of the serum protein, α_1-antitrypsin (p. 361).

If we survey the range of human disease and abnormalities, we find at one extreme conditions such as sickle-cell anaemia, phenylketonuria, haemophilia, muscular dystrophy and so on – which we regard as inherited disorders because all individuals who have the appropriate genes develop the condition. At the other extreme are the typically environmental diseases – such as the severe infections like plague, anthrax, typhus and the vitamin deficiency diseases like scurvy – in which it appears that virtually everyone who is sufficiently exposed to the unfavourable environmental agent develops the disease. But in between there are many disorders, often common ones, in which both genetical and environmental factors are apparently important. Typical examples are schizophrenia, diabetes mellitus peptic ulcer and many congenital abnormalities such as cleft-palate and hare-lip. In each of these conditions there is good evidence for genetical predisposition, and it is also clear that only a fraction of those genetically predisposed actually develop the condition. But we do not know the nature of the primary effects of the genes involved. Nor can we define very clearly the particular features of the environment which cause some individuals but not others among those genetically predisposed to develop the condition.

The elements of the situation can be illustrated by a simple diagram (fig. 9.1). The area encompassed by the outer circle represents a population. The area enclosed by the inner circle represents those individuals in the population genetically predisposed to develop a particular kind of disease. The two lines drawn from the centre to the periphery divide the whole population into those people who happen to be exposed to environmental

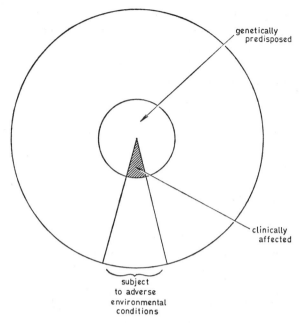

Fig. 9.1. Diagram illustrating in a simple form the roles of genetical predisposition and adverse environmental factors in the causation of disease (see text). Area bounded by outer circle represents a population, and the area enclosed within the inner circle those individuals in the population who are genetically predisposed to develop a particular kind of disease. The two lines from the centre to the periphery enclose a segment of the population subject to some particular environmental conditions that may elicit the disease. The hatched part of the segment represents those individuals who actually develop the disease.

factors which tend to elicit the abnormality (in this case the smaller group), and those who are not so exposed. Only the small segment of the population who are both genetically predisposed and also subject to the unfavourable environmental situation, actually develop the clinical disorder.

The practical significance of this kind of formulation **is** that it emphasises how if by genetical research one could find ways of identifying the individuals in the population who are genetically predisposed to a particular condition, then one would be provided with a powerful tool with which to discover the critical environmental factors. In effect one would be able to ask what significant differences exist or have existed between the environmental circumstances of the predisposed individuals who actually develop the disease, and those who have not. And of course if these environmental circumstances can be recognised, they can hopefully be adjusted in an appropriate manner for the genetically predisposed individuals before the clinical abnormality develops.

Clearly for any particular condition the fraction of the population genetically predisposed may be very small or quite large, and the proportion exposed to the adverse environmental situations can vary similarly. Furthermore one must expect that often many different genes can give rise in one way or another to a particular kind of predisposition, and since their effects will not be identical and also because they will occur in different combinations in different individuals, the degree of predisposition will vary and be graded in severity. Similarly the relevant environmental factors are likely to vary in strength. So the lines shown in the diagram should be regarded as fuzzy rather than sharp.

Nevertheless this simple if diagrammatic approach provides a useful way of thinking about the problems posed by disease states, and particularly about those which are relatively common. It also leads to an interesting paradox. This is that the study of the genetics of many diseases may lead to their prevention or amelioration by purely environmental methods. Indeed it is very probable that one of the most important social and medical applications of genetical research will lie in the control of the environment, since the more it becomes possible to characterise the genetical constitution of an individual precisely, the more likely are we to see how to modify or tailor the environment according to his needs.

Disorders due to specific enzyme deficiencies (inborn errors of metabolism)

A1.1 Disorders of carbohydrate metabolism

Enzyme deficiency

1. Hexokinase (red cell isozyme)
   E.C. 2.7.1.1.

2. Phosphohexose isomerase (glucose phosphate isomerase)
   E.C. 5.3.1.9

3. 6-Phosphofructokinase (muscle type isozyme)
   E.C. 2.7.1.11

Condition

Hexokinase deficiency haemolytic anaemia

Defective phosphorylation of glucose (fig. A1) limits glycolysis and results in premature breakdown of red cells with chronic haemolytic anaemia.

Refs.: Valentine et al. (1967), Keitt (1969), Necheles et al. (1970).

Phosphohexoseisomerase (glucose phosphate isomerase) deficiency

Defective conversion of glucose-6-phosphate to fructose-6-phosphate (fig. A1) in red cells limits glycolysis and results in severe haemolytic anaemia. Enzyme abnormality also present in leucocytes and probably other tissues but this causes no obvious clinical consequences. Several different alleles producing structurally altered forms of the enzyme with deficient functional activity occur, as well as a number of other variants.

Refs.: Baughan et al. (1968), Paglia et al. (1969), Detter et al. (1968), Blume et al. (1972), Arnold et al. (1973), Nakashima et al. (1973).

Phosphofructokinase deficiency (p. 215)

Restriction of glycolysis in muscle due to gross reduction of the enzyme activity leads to glycogen accumulation. Muscular weakness and stiffness develops on prolonged exertion. Partial reduction of activity also present in red cells

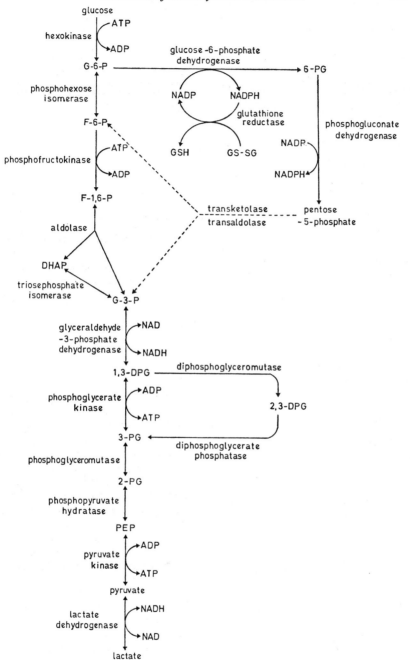

Fig. A1. Enzymes concerned in the metabolism of glucose in red cells (from Valentine 1968).

Enzyme deficiency   Condition

and causes chronic mild haemolysis. The
residual phosphofructokinase activity in red cells
(about 50%) is due to a different isozyme not
affected in this condition.
Refs.: Tarui et al. (1965), Layzer et al. (1967),
    Tarui et al. (1969).

4. Triosephosphate isomerase   *Triosphosphate isomerase deficiency*
   E.C. 5.3.1.1   Enzyme deficiency present in red cells, leucocytes,
                                       muscle, cerebrospinal fluid and probably other
                                       tissues. Results in defective glycolysis with
                                       marked elevation of dihydroxyacetone phos-
                                       phate concentration (fig. A1). Main clinical
                                       features are chronic haemolytic anaemia, and
                                       in some patients who survive the first year or so,
                                       progressive neurological damage affecting both
                                       peripheral nerves and central nervous tissue.
                                       Refs.: Schneider et al. (1965, 1968a, b), Valentine
                                           et al. (1966), Harris et al. (1970).

5. Aldolase B (liver isozyme, see p. 199   *Hereditary fructose intolerance* (pp. 199–203)
   E.C. 4.1.2.7

6. Phosphoglycerate kinase   *Phosphoglycerate kinase deficiency*
   E.C. 2.7.2.3   X-linked. Defective red cell glycolysis (fig. A1)
                                       causing severe haemolytic anaemia in hemizy-
                                       gous males, and mild haemolytic disease in
                                       some heterozygous females. Enzyme deficiency
                                       demonstrable in both red cells and leucocytes.
                                       Refs.: Kraus et al. (1968), Valentine (1968),
                                           Konrad et al. (1973).

7. Pyruvate kinase (red cell isozyme) *Pyruvate kinase deficiency* (pp. 203–204)
   E.C. 2.7.1.40

8. Diphosphoglycerate mutase   *Diphosphoglycerate mutase deficiency*
   E.C. 2.7.5.4   Red cell deficiency results in severe haemolytic
                                       disease.
                                       Ref.: Schröter (1965).

9. Glucose-6-phosphate dehydrogenase *Glucose-6-phosphate dehydrogenase deficiency,
   E.C. 1.1.1.49   primaquine sensitivity, favism, etc.* (pp. 162–174)

Enzyme deficiency

Condition

10. Galactokinase
   E.C. 2.7.1.6

Galactokinase deficiency (p. 198)

11. Galactose-1-phosphate uridyl transferase
   E.C. 2.7.7.12

Galactosaemia (pp. 196–198)

12. Fructokinase
   E.C. 2.7.1.3

Essential fructosuria
Defective phosphorylation of fructose results in high blood fructose levels and abnormal excretion of fructose in the urine after taking foods containing fructose. No pathological consequences occur.
Refs.: Sachs et al. (1942), Lasker (1941), Schapira et al. (1961).

13. Glucose-6-phosphatase
   E.C. 3.1.3.9

von Gierke's disease   (p. 215)

14. Amylo-1,6-glucosidase
   E.C. 3.2.1.33

Forbes' disease (p. 214)

15. Amylo-(1,4→1,6)-transglucosidase
   . (glycogen branching enzyme)
   E.C. 2.4.1.18

Andersen's disease   (pp. 212–214)

16. Phosphorylase (muscle type)
   E.C. 2.4.1.1

a) McArdle's disease (muscle phosphorylase deficiency) (p. 214)
b) Liver phosphorylase deficiency (p. 214).

17. Phosphorylase kinase
   E.C. 2.7.1.38

Glycogen storage disease [*one type*] (p. 214)

18. L-xylulose reductase
   E.C. 1.1.1.10

Congenital pentosuria
L-xylulose (L-xyloketose) is a normal intermediate in the pathway by which D-glucuronic acid is metabolised (fig. A2). Deficiency of the reductase results in accumulation of L-xylulose and its continuous excretion in large amounts in the urine. Increased excretion of L-arabitol has also been found. No pathological consequences occur. The urinary output of L-xylulose is greatly enhanced if glucuronic acid is given, or if

Enzyme deficiency

Condition

certain drugs which stimulate glucuronic acid formation and are themselves excreted as glucuronides are taken.

Refs.: Enklewitz and Lasker (1935), Lasker et al. (1936), Touster (1959), Hiatt (1972), Wang and Van Eis (1970).

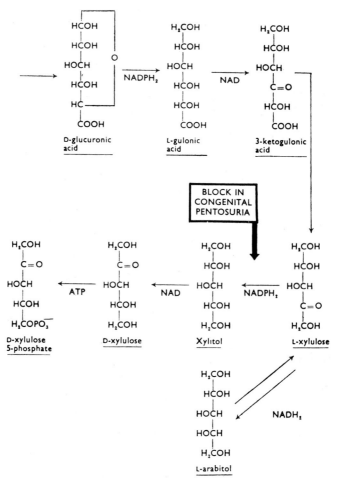

Fig. A2. The glucuronic acid oxidation pathway showing the site of the metabolic block in congenital pentosuria.

Enzyme deficiency   Condition

19. Isomaltase (maltase 1a) and sucrase (maltase 1b)

Sucrose and isomaltose intolerance
Inability to hydrolyse sucrose and isomaltose (formed from starch) in the course of intestinal digestion. On a normal diet there is a persistent chronic diarrhoea with frothy and liquid acid stools. The low *p*H of the stools is due to the presence of lactic and other organic acids formed from undigested carbohydrate by bacterial fermentation in the large intestine.
Refs.: Weijers et al. (1960), Prader and Auricchio (1965), Rey and Frézal (1967), Marshall et al. (1967), Starnes et al. (1970), Dubs et al. (1973).

20. Lactase (intestinal)
    E.C. 3.2.1.23

Congenital lactose intolerance
Failure to hydrolyse ingested lactose in milk. Chronic diarrhoea starting a few days after birth results.
Refs.: Holzel et al. (1959), Lifschitz (1966).

21. Fructose-1,6-diphosphatase
    E.C. 3.1.3.11

Fructose-1,6-diphosphatase deficiency
Deficiency of the enzyme in liver results in severe impairment of gluconeogenesis. Hypoglycaemia and metabolic acidosis occurs on fasting and also after taking fructose or sucrose. There is liver enlargement and biopsy specimens show gross ballooning of hepatic cells with large lipid filled vacuoles.
Refs.: Baker and Winegrad (1970), Pagliara et al. (1972), Melancon et al. (1973).

22. Pyruvate decarboxylase
    E.C. 4.1.1.1

Pyruvate decarboxylase deficiency
Neurological disorder with intermittent ataxia and choreoathetosis. Elevated blood, C.S.F. and urine pyruvate levels. Enzyme deficiency demonstrated in muscle, leucocytes and cultured fibroblasts.
Ref.: Blass et al. (1970, 1971).

A1.2 Lysosomal enzyme deficiencies

Enzyme deficiency   Condition

1. α-L-iduronidase   a) *Hurler syndrome* (pp. 221, 230–231, 255–256)
   E.C. 3.2.1.76   b) *Scheie syndrome* (pp. 221, 230–231)

2. Sulphoiduronate sulphatase   *Hunter syndrome* (pp. 221, 230–231, 355–356)

3. Heparan sulphate sulphatase   *Sanfilippo syndrome-A*
                                                      Progressive accumulation of partially degraded
                                                      heparan sulphate. Severe mental retardation,
                                                      but relatively mild somatic effects.
                                                      Refs.: Sanfilippo et al. (1963), Kresse and
                                                              Neufeld (1972)

4. N-acetyl-α-glucosaminidase   *Sanfilippo syndrome-B*
   E.C. 3.2.1.50   Clinically indistinguishable from Sanfilippo
                                                      syndrome-A.
                                                      Refs.: McKusick (1972), O'Brien (1972), Figura
                                                              and Kresse (1972).

5. β-glucuronidase   *β-glucuronidase deficiency*
   E.C. 3.2.1.31   Progressive accumulation of partially degraded
                                                      dermatan sulphate. Mental retardation, hepatos-
                                                      plenomegaly and bone abnormalities.
                                                      Refs.: McKusick (1972), Hall et al. (1973), Sly
                                                              et al. (1973).

6. GM_1-β-galactosidase   *Generalised gangliosidosis* (pp. 222–223, 224, 225)
   E.C. 3.2.1.23

7. β-N-acetylhexosaminidase   a) *Tay Sachs' disease* (pp. 222–223, 227–230)
   E.C. 3.2.1.30   b) *Sandhoff's disease* (pp. 227–230)

8. Ceramide-lactoside-β-galactosidase   *Ceramide-lactoside lipidosis*
   E.C. 3.2.1.23   Progressive accumulation of ceramide lactoside
                                                      (fig. 6.12, p. 220), associated with brain damage
                                                      and hepatosplenomegaly.
                                                      Ref.: Dawson and Stein (1970).

9. Glucosylceramidase   *Gaucher's disease* (p. 225)
   E.C. 3.2.1.45

Enzyme deficiency	Condition

10. Aryl-sulphatase A
   (cerebroside sulphatase)
   E.C. 3.1.6.8

Metachromatic leucodystrophy
Accumulation of ceramide galactose-3-sulphate (fig. 6.12, p. 222), with progressive degeneration of the central nervous system.
Refs.: Austin et al. (1963), Mehl and Jatzkewitz (1965), Jatzkewitz and Mehl (1969).

11. Galactosylceramidase
   E.C. 3.2.1.46

Krabbe's disease (Globoid cell leucodystrophy)
Accumulation of galactosylceramide (fig. 6.12, p. 222). Rapidly progressive neurological damage with fatal outcome in infancy.
Refs.: Suzuki and Suzuki (1970, 1971, 1972).

12. Ceramide trihexosidase
   (α-galactosidase)
   E.C. 3.2.1.22

Fabry's disease (angiokeratoma corporis diffusum)
Accumulation of ceramide trihexoside. (fig. 6.12, p. 222). Development of skin lesions (small dark purple macules and papules), ocular abnormalities (corneal opacities, cataracts and retinal oedema), and cardiovascular, neurological and gastrointestinal disorders. Severe burning pains in the extremities are a characteristic feature. Inheritance X linked.
Refs.: Wise et al. (1962), Sweeley and Klionsky (1963), Opitz et al. (1965), Brady et al. (1967a, b), Kint (1970), Romeo and Migeon (1970).

13. Spingomyelinase
   E.C. 3.1.4.12

Niemann-Pick disease
Accumulation of sphingomyelin (fig. 6.12, p. 222). Progressive degeneration of central nervous system associated with hepatosplenomegaly.
Refs.: Brady et al. (1966a), Sloan et al. (1969).

14. α-1,4-glucosidase
   E.C. 3.2.1.20

Pompe's disease (pp. 215–216)

15. α-fucosidase
   E.C. 3.2.1.51

α-Fucosidosis
Accumulation of fucose-rich acid mucopolysaccharides and glycolipids. Severe progressive cerebral degeneration in infancy.
Refs.: Durand et al. (1969), Loeb et al. (1969), Van Hoof and Hers (1968).

Enzyme deficiency   Condition

16. *α*-mannosidase   *Mannosidosis*
    E.C. 3.2.1.24   Accumulation of mannose-rich oligosaccharides.
                                                     Brain damage, bony abnormalities and hepa-
                                                     tosplenomegaly.
                                                     Refs.: Ockerman (1967, 1969), Kjellman et al.
                                                         (1969).

17. Acid lipase   *Wolman's disease*
                                                     Deficiency of acid lipase of lysosomal origin
                                                     associated with widespread occurrence in visceral
                                                     organs of 'foamy' cells containing large amounts
                                                     of triglycerides and cholesterylesters. The
                                                     condition is characterised by xanthomatosis,
                                                     calcification of the adrenals, and hepatospleno-
                                                     megaly.
                                                     Refs.: Abramov et al. (1956), Wolman et al.
                                                         (1961), Patrick and Lake (1969), Lake
                                                         and Patrick (1970).

18. 'Acid' phosphatase   *Lysosomal acid phosphatase deficiency*
    E.C. 3.1.3.2   Failure to thrive in infancy with rapid deteriora-
                                                     tion and death.
                                                     Refs.: Nadler and Egan (1970).

19. *β*-aspartyl acetyl glycosaminidase   *Aspartylglycosaminuria*
    E.C. 3.2.2.11   This lysosomal enzyme deficiency appears to
                                                     result in defective breakdown of certain glyco-
                                                     proteins. There is severe mental retardation and
                                                     motor impairment associated with the abnormal
                                                     urinary excretion of large amounts of aspartyl-
                                                     glycosamine (2-acetamido-1-(β^--L-aspartamido)-
                                                     1,2-dideoxyglucose) (fig. A3).
                                                     Refs.: Pollitt et al. (1968), Pollitt and Jenner
                                                         (1969), Palo and Mattson (1970), Aula
                                                         et al. (1973), Autio et al. (1973).

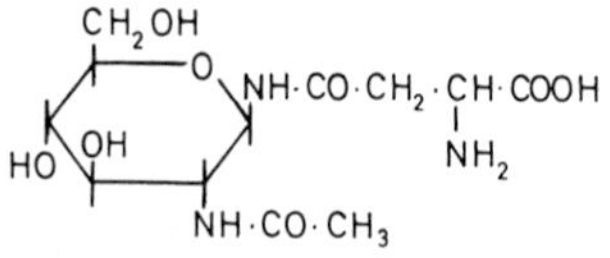

2 - acetamido -1 - (β - L - aspartamido) - 1, 2 - dideoxyglucose

Fig. A3. Aspartylglycosamine.

A1.3 Disorders of aminoacid metabolism

Enzyme deficiency	Condition
1. Phenylalanine 4-hydroxylase E.C. 1.14.16.1	*Phenylketonuria* (pp. 191–196).
2. Homogentic acid oxidase (homogentisate oxygenase) E.C. 1.13.11.5	*Alkaptonuria* (pp. 187–190)
3. Histidase (histidine α-deaminase) E.C. 4.3.1.3	*Histidinaemia* Failure to form urocanic acid from histidine (fig. A4). Elevated blood histidine. Increased urinary excretion of histidine imidazole-pyruvic acid. Absence of formiminoglutamic acid in urine after histidine load. Absence of urocanic acid in sweat. Delayed onset of speech, and speech defects sometimes occur. Refs.: Ghadimi et al. (1961), La Du et al. (1962), Auerbach et al. (1961), Ghadimi and Partington (1967), La Du (1967), Levy (1973).

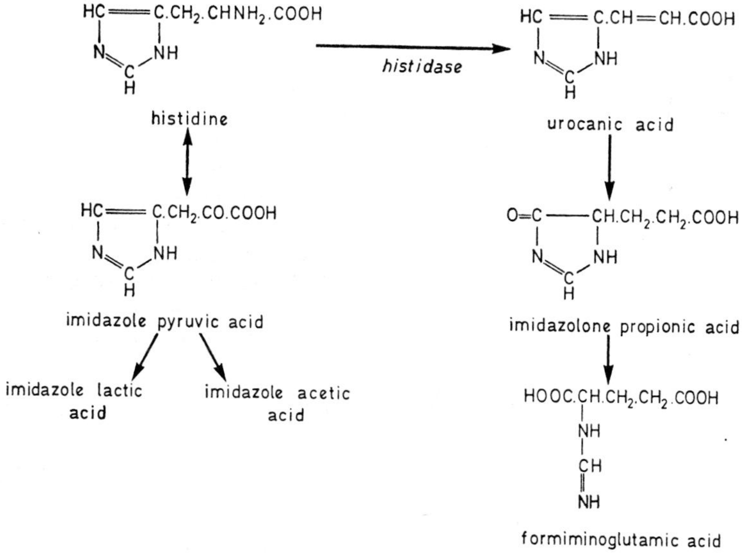

Fig. A4. Pathways in histidine metabolism.

Enzyme deficiency   Condition

4. Urocanase   *Urocanic aciduria*
   E.C. 4.2.1.49   Excessive urinary excretion of urocanic acid,
                                                      exaggerated further by an oral dose of histidine
                                                      (fig. A4). No detectable formimino-glutamic acid
                                                      in urine after oral histidine or intravenous
                                                      urocanate. Associated mental retardation.
                                                      Ref.: Yoshida et al. (1971).

5. Cystathionine β-synthetase   *Homocystinuria*
   E.C. 4.2.1.22   Failure to form cystathionine from homocysteine
                                                      and serine, which is a step in the pathway by
                                                      which cysteine is formed from methionine
                                                      (fig. A5). Abnormal excretion of homocystine
                                                      in urine. Elevated levels of homocystine and of
                                                      methionine in serum. Cystathionine normally
                                                      found in significant quantities in brain tissue is
                                                      virtually absent in this condition. Clinical syn-
                                                      drome includes mental retardation, ectopia
                                                      lentis, skeletal abnormalities, and thromboem-
                                                      bolic phenomena in both arteries and veins.
                                                      Refs.: Gerritsen and Waisman (1964a, b),
                                                             Carson et al. (1965), Brenton et al. (1966),
                                                             Uhlendorf and Mudd (1968), also see
                                                             p. 359.

6. Cystathionase (homoserine dehydratase)   *Cystathioninuria*
   E.C. 4.4.1.1   Failure to cleave cystathionine to give cystine,
                                                      α-ketobutyrate and ammonia. This reaction is a
                                                      step in the major pathway of methionine metab-
                                                      olism (fig. A5). There is a high urinary excretion
                                                      of cystathionine, and an increased concentration
                                                      of cystathionine in the tissues and in serum.
                                                      Mental retardation and psychotic abnormalities
                                                      have been seen in some patients, but it is doubt-
                                                      ful whether these are a consequence of the
                                                      metabolic disorder.
                                                      Refs.: Harris et al. (1959), Frimpter et al. (1963),
                                                             Frimpter (1965, 1967), Finkelstein et al.
                                                             (1966), Tada et al. (1968), Perry et al.
                                                             (1968), Scott et al. (1970).

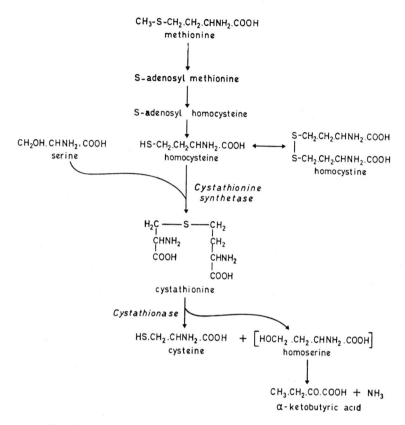

Fig. A5. Pathway for the formation of cysteine from methionine.

Enzyme deficiency

Condition

7. Iodotyrosine deiodinase
   E.C. 1.11.1.8

Goitrous cretinism (one type)
Defective deiodination of mono- and di-iodo-tyrosine during thyroid hormone synthesis. Continuous leakage of these hormone precursors from the thyroid and from the body leads to depletion of iodine stores and thyroid hyperplasia. Clinically this results in gross enlargement of the thyroid associated with severe hypothyroidism.

Refs.: Hutchison and McGirr (1956), Querido et al. (1956), Murray et al. (1965), Stanbury (1972).

Enzyme deficiency

Condition

8. 'Branched chain ketoacid decarboxylase(s)'

(a) *Maple syrup urine disease*
The branched chain aminoacids leucine, iso-leucine and valine are normally degraded via a series of reactions which involve first their conversion to the corresponding ketoacids and then their decarboxylation (fig. A6). The enzyme system (or separate systems) concerned in the decarboxylation step is deficient in this condition. High serum levels of leucine, isoleucine and valine occur; and large amounts of these amino-acids as well as the corresponding ketoacids (isovaleric acid, α-methylbutyric acid and iso-butyric acid) are excreted in the urine which has a characteristic smell. Alloisoleucine probably derived from isoleucine is also present in ab-normal amounts in the serum. Rapidly progres-sive neurological disease with marked cerebral degeneration is usually apparent shortly after birth, and is generally fatal within weeks or months.

Refs.: Dancis et al. (1959), Mackenzie and Woolf (1959), Dancis et al. (1963), Snyderman (1967), Goedde and Keller (1967).

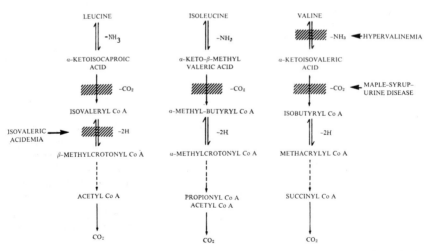

Fig. A6. Pathways in the metabolism of the branched chain aminoacids. Sites of metabolic blocks in maple syrup urine disease, hypervalinaemia and isovaleric acidaemia. (From Budd et al. 1967.)

Enzyme deficiency

Condition

(b) *Intermittent branched-chain ketoaciduria*
Slightly less severe deficiency of the enzyme(s) which are grossly deficient in maple syrup urine disease. Intermittent episodes with increased excretion of branched chain ketoacids and aminoacids occur in urine which may also have the characteristic maple syrup odour. During the episodes, signs and symptoms attributable to neurological disorder (toxic encephalopathy) occur and may be fatal. Between episodes there may be no clinical abnormality, and the urinary excretion of ketoacids and aminoacids is within the normal range. Episodes apparently precipitated by infections or high protein intake.
Refs.: Kiil and Rokkenes (1964), Morris et al.
       (1966), Dancis et al. (1967).
(c) *Mild branched-chain ketoaciduria.*
Partial enzyme deficiency with much milder metabolic upset.
Ref.: Schulman et al. (1970).
(d) *Thiamine responsive branched-chain ketoaciduria.*
Ref.: Scriver et al. (1971).

9. Valine transaminase
   E.C. 2.6.1.32

Hypervalinaemia
Elevated blood valine level and grossly abnormal excretion of valine in the urine, associated with failure to thrive and mental retardation. The defect appears to be specific to the transamination of valine (fig. A6). The transamination of other aminoacids such as leucine or isoleucine is not affected.
Refs.: Wada et al. (1963), Tada et al. (1967),
       Dancis et al. (1967), Goedde et al. (1970).

10. Isovaleryl-coenzyme A dehydrogenase

Isovaleric acidaemia
Grossly abnormal amounts of isovaleric acid in blood and urine, due to defective conversion of isovaleryl-Co A to β-methylcrotonyl-Co A (fig. A6). Periodic bouts of acidosis and coma, apparently precipitated by infections.
Refs.: Tanaka et al. (1966), Budd et al. (1967),
       Levy et al. (1973), Scriver and Rosenberg
       (1973).

Enzyme deficiency   Condition

11. Proline oxidase
    (Pyrroline-5-carboxylate reductase)
    E.C. 1.5.1.2

Hyperprolinaemia
Abnormal elevation of serum proline due to failure of conversion to Δ'-pyrroline-5-carboxylate. Oxidation of hydroxyproline is unaffected. However there is an increased urinary excretion not only of proline but also hydroxyproline and glycine due probably to saturation of the common transport system for these aminoacids in the renal tubules by excess proline (see p. 182). Renal abnormalities and mental retardation have been found in some cases, but how far they are a consequence of the specific metabolic disorder is unclear.
Refs.: Shafer et al. (1962), Efron (1965), Harriesi et al. (1971).

12. Hydroxyproline oxidase

Hydroxyprolinaemia
Abnormal amounts of hydroxyproline in serum and urine due to failure of conversion to Δ'-pyrroline-3-hydroxy-5-carboxylate. No abnormality in collagen metabolism or in peptide-bound urinary hydroxy proline. Proline metabolism apparently normal. Severe mental retardation associated with the defect.
Refs.: Efron et al. (1965), Pelkonen and Kivirikko (1970).

13. Argininosuccinase (argininosuccinate lyase)
    E.C. 4.3.2.1

Arginininosuccinic aciduria (p. 206)

14. Argininosuccinate synthetase
    E.C. 6.3.4.5

Citrullinaemia (pp. 206–208)

15. Ornithine carbamoyl transferase
    (ornithine transcarbamylase)
    E.C. 2.1.3.3

Hyperammonaemia (one type p. 209)

16. Carbamyl phosphate synthetase
    E.C. 2.7.2.5

Hyperammonaemia (one type p. 209)

17. Arginase
    E.C. 3.5.3.1

Argininaemia (p. 209)

Enzyme deficiency   Condition

18. Prolidase (peptidase D)
    E.C. 3.4.13.9

Prolidase deficiency
Massive excretion of di- and tri-peptides all containing proline with some other aminoacid which is amino-terminal.
Ref.: Powell et al. (1973).

19. Carnosinase
    E.C. 3.4.13.3

Carnosinase deficiency
Excessive urinary excretion of carnosine even on a meat-free diet. Associated with mental retardation and progressive neurological disorder.
Ref.s: Perry et al. (1967, 1968); Van Heeswijk et al. (1969).

20. Tyrosine aminotransferase (soluble form)
    E.C. 2.6.1.5

Hypertyrosinaemia (one type).
Elevated blood and urine tyrosine levels. Increased urinary excretion of p-hydroxyphenyl-pyruvic, -lactic and -acetic acids. Associated with syndrome of keratosis palmoplantaris, corneal dystrophy and mental retardation. Mitochondrial tyrosine aminotransferase and also p-hydroxyphenyl pyruvate hydroxylase are not deficient.
Refs.: Fellman et al. (1969), Goldsmith et al. (1972).

21. Lysine: α-ketoglutarate reductase

Persistent hyperlysinaemia
Defective formation of saccharopine from lysine and α-ketoglutarate results in abnormally high levels of blood lysine, and excessive lysine excretion in urine. Some patients were mentally retarded but it is not clear whether this was a consequence of the metabolic disorder.
Refs.: Woody (1964), Woody et al. (1966), Dancis et al. (1969).

22. Kynureninase
    E.C. 3.7.1.3

Xanthurenic aciduria
Kynureninase catalyses the conversion of 3-hydroxykynurenine to 3-hydroxyanthranilic acid (fig. A7) and also kynurenine to anthranilic acid. Vit. B_6 is required as coenzyme. Deficiency of kynureninase results in excessive urinary excretion of xanthurenic acid. This is accentuated after giving an oral tryptophan load, and there is also increased kynurenine and 3-hydroxy-

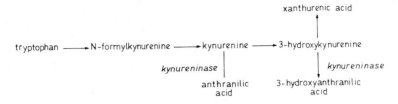

Fig. A7. One pathway in tryptophan metabolism.

Enzyme deficiency	Condition

kynurenine excretion. Several of the patients have been mentally retarded but it is not clear if this was a consequence of the metabolic disorder. Administration of a large amount of Vit. B₆ corrects the metabolic disturbance temporarily. Addition of excess Vit. B₆ to extracts of liver biopsy material increases the level of enzyme activity to near normal levels. It is thought that there is a structural abnormality of the enzyme which reduces its affinity for the co-enzyme.

Refs.: Knapp (1960), O'Brien and Jensen (1963), Tada et al. (1967, 1968).

A1.4 Miscellaneous disorders

1. Xanthine oxidase
   E.C. 1.2.3.2.

Xanthinuria

There is a failure to form uric acid from xanthine so that xanthine effectively replaces uric acid as the end product of purine metabolism. Serum and urinary uric acid levels are extremely low, and there is a grossly abnormal excretion of xanthine in the urine. Due to the low solubility of xanthine, xanthine calculi tend to form in the renal tract.

Refs.: Dent and Philipot (1954), Watts et al. (1963), Engleman et al. (1964), Sperling et al. (1971).

2. Sulphite oxidase
   E.C. 1.8.3.1

Sulphite oxidase deficiency

Inability to convert sulphite to sulphate leading to increased formation of S-sulpho-L-cysteine and thiosulphate. Greatly increased amounts of S-sulpho-L-cysteine, sulphite and thiosulphate

Enzyme deficiency

Condition

in the urine, but virtually no inorganic sulphate present. Progressive neurological abnormalities, dislocation of lenses, and mental retardation. Refs.: Mudd et al. (1967), Irreverre et al. (1967).

3. Catalase
E.C. 1.11.1.6

Acatalasia

Gross deficiency of catalase in all tissues. In some individuals ulceration of the mucosae of the nose and mouth occurs and severe oral gangrene may develop. But other individuals with less than 1% of the normal catalase level remain quite healthy and apparently suffer no ill effects.
Refs.: Takahara (1952), Kaziro et al. (1952), Aebi et al. (1964), Aebi (1967), Takahara (1968), Aebi et al. (1968), Aebi and Suter (1971).

4. *NADH* diaphorase (Methaemoglobin reductase)

Congenital methaemoglobinaemia (pp. 356–357)

5. Orotidine-5'-phosphate pyrophosphorylase
E.C. 2.4.2.10
and
Orotidine-5'-phosphate decarboxylase
(orotate phosphoribosyl transferase)
E.C. 4.1.1.23

Orotic aciduria

A gross deficiency of *two* sequential enzymes, which are concerned in the conversion of orotic acid to uridine-5'-phosphate (fig. A8), is present. There is marked retardation of growth and development, severe megaloblastic anaemia, and

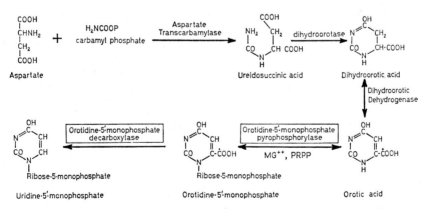

Fig. A8. Pathway of biosynthesis of uridine-5'-monophosphate showing sites of enzyme defects in orotic aciduria. (From Wuu and Krooth 1968.)

Enzyme deficiency   Condition

                                                     abnormal excretion of orotic acid in the urine.
                                                     Refs.: Huguley et al. (1959), Smith et al. (1961),
                                                         Fallon et al. (1964), Krooth (1964),
                                                         Howell et al. (1967).
                                                     b) Similar condition in which there is a deficiency
                                                     of only orotidine-5-phosphate decarboxylase.
                                                     Ref.: Fox et al. (1969).

6. γ-glutamyl-cysteine synthetase   *γ-glutamyl-cysteine synthetase deficiency*
    E.C. 6.3.2.2   Mild chronic haemolytic anaemia associated
                                                     with much reduced levels of red cell glutathione
                                                     (fig. A9).
                                                     Ref.: Konrad et al. (1972).

7. Glutathione synthetase   *Glutathione synthetase deficiency*
    E.C. 6.3.2.3   Mild chronic haemolytic anaemia associated
                                                     with much reduced levels of red cell glutathione
                                                     (fig. A9).
                                                     Ref.: Mohler et al. (1970).

Fig. A9. Biosynthesis of glutathione.

8. Glutathione reductase   *Glutathione reductase deficiency*
    E.C. 1.6.4.2   Chronic haemolytic disease varying consider-
                                                     ably in degree and often precipitated or accen-
                                                     tuated by drugs.
                                                     Refs.: Löhr and Waller (1962), Waller et al.
                                                         (1965), Waller (1968) Staal et al. (1969).

9. Glutathione peroxidase   *Red cell glutathione peroxidase deficiency*
    E.C. 1.11.1.9   Mild chronic haemolytic disease.
                                                     Refs.: Necheles et al. (1969, 1970), Boivin et al.
                                                         (1969).

10. Serum cholinesterase (pseudocholinesterase)   *Suxamethonium sensitivity* (pp. 150–160)
    E.C. 3.1.1.8

Enzyme deficiency

Condition

11. Alkaline phosphatase
    E.C. 3.1.3.1

Hypophosphatasia

Very low levels of serum alkaline phosphatase associated with skeletal abnormalities due to defective ossification. Increased urinary excretion of phosphoryl ethanolamine is a characteristic feature. The enzyme deficiency is present in bone and cartilage and probably in liver and kidney, but not in intestine. The clinical manifestations vary widely in severity in different families, which suggests genetic heterogeneity. In some cases, severe osteodystrophic changes are present at birth or appear shortly after, with death in the first year. In others, early development may appear normal, but after a few months skeletal abnormalities resembling severe rickets develop. The teeth are hypoplastic and shed prematurely. In still other cases, the disorder is not recognised until adult life, the patients presenting with bony deformities and spontaneous fractures, though there is often a history of rickets in childhood.

Refs.: Rathbun (1948), Sobel et al. (1953), Fraser (1957), Currarino et al. (1957), Harris and Robson (1959), Rathbun et al. (1961), Danovitch et al. (1968), Rasmussen (1968), Warshaw et al. (1971).

12. Adenosine triphosphatase
    E.C. 3.6.1.3

Red cell ATPase deficiency

Haemolytic anaemia occurring in apparently heterozygous individuals.

Refs.: Harvald et al. (1964), Hanel et al. (1971).

13. Pancreatic lipase
    E.C. 3.1.1.3

Congenital pancreatic lipase deficiency

Lipase in intestinal juice reduced to about 10% of normal, resulting in marked inability to digest fats. About 50-80% of ingested fats are excreted in the stools which are very bulky and oily.

Refs.: Sheldon (1964), Rey et al. (1966).

14. Hypoxanthine-guanine phosphoribosyl transferase
    E.C. 2.4.2.8

(a) *Lesch–Nyhan syndrome* (pp. 351–353)
(b) *Gout due to uric acid overproduction*
(pp. 351–353).

The role the enzyme is thought to play in the pathways of purine biosynthesis is indicated in fig. A 10.

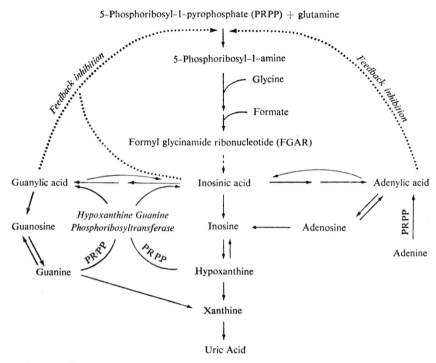

Fig. A10. Pathways in purine biosynthesis and interconversions (Seegmiller et al. 1967).

Enzyme deficiency	Condition
15. 2-oxo-glutarate : glyoxalate carboligase	*Primary hyperoxaluria* (one type) Increased urinary excretion of oxalate, glycolate and glyoxalate. Calcium oxalate calculi form in urinary tract, and calcium oxalate may be deposited within renal parenchyma (nephrocalcinosis) and more widely throughout the body (oxalosis). Refs.: Archer et al. (1957), Hockaday et al. (1964), Koch et al. (1967).
16. D-glycerate dehydrogenase E.C. 1.1.1.29	*Primary hyperoxaluria* (one type) Increased urinary excretion of oxalate and L-glycerate. Oxalate calculi form in renal tract. Ref.: Williams and Smith (1968, 1971).

Enzyme deficiency   Condition

17. Trypsinogen   *Trypsinogen deficiency disease*
                                                 The duodenal juice lacks not only trypsin activity
                                                 but also chymotrypsin and carboxypeptidase
                                                 activity. This is because trypsin, formed from
                                                 trypsinogen, is required to activate the chy-
                                                 motrypsinogen and procarboxypeptidase which
                                                 are formed in normal amounts. Consequently no
                                                 protein digestion occurs in the duodenum and
                                                 small intestine. The affected newborn fails to
                                                 grow and there is a progressive development
                                                 of marked hypoproteinaemia, oedema, and anae-
                                                 mia. The condition responds well to the addition
                                                 of hydrolysed protein to the diet.
                                                 Refs.: Townes (1965), Townes et al. (1967),
                                                        Morris and Fisher (1967).

18. Enterokinase (enteropeptidase)   *Intestinal enterokinase deficiency*
     E.C. 3.4.21.9   Deficiency of enterokinase results in a failure to
                                                 convert trypsinogen, which is present in normal
                                                 amounts, to trypsin. In the absence of trypsin
                                                 activity, chymotrypsinogen and procarboxy-
                                                 peptidase are not converted to chymotrypsin
                                                 and carboxypeptidase. So there is a failure of
                                                 protein digestion, with consequences identical
                                                 to those seen in trypsinogen deficiency disease
                                                 (see above).
                                                 Refs.: Hadorn et al. (1969), Tarlow et al. (1970),
                                                        Polonowski and Bier (1970).

19. Propionyl-CoA carboxylase   *Propionicacidaemia*
     E.C. 6.4.1.3   Ketoacidosis in infancy with propionic acidaemia
                                                 and sometimes hyperglycinaemia (fig. A11).
                                                 May respond to biotin administration.
                                                 Refs.: Hommes et al. (1968), Gompertz et al.
                                                        (1970), Barnes et al. (1970), Hsia et al.
                                                        (1969, 1971).

20. Methylmalonyl-Co A mutase   *Methylmalonic aciduria (one type)* (p. 358)
     E.C. 5.4.99.2

21. Methylmalonyl-Co A racemase   *Methylmalonic aciduria (one type)*
     E.C. 5.1.99.1   Failure to convert d-methylmalonyl-Co A to

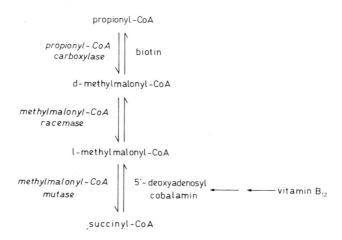

Fig. A11. Metabolic pathway for the conversion of propionyl-CoA to succinyl-CoA.

Enzyme deficiency

Condition

l-methylmalonyl-Co A (fig. A11). Severe infantile ketoacidosis and gross methylmalonic aciduria.
Ref.: Kang et al. (1972).

22. β-methylcrotonyl-CoA carboxylase
E.C. 6.4.1.4

β-methylcrotonyl glycinuria
Excessive excretion of β-methylcrotonylglycine and β-hydroxyisovaleric acid. Metabolic acidosis and failure to thrive.
Refs.: Eldjarn et al. (1970), Gompertz et al. (1971, 1973).

23. Methylenetetrahydrofolate reductase
E.C. 1.1.1.68

Methylenetetrahydrofolate reductase deficiency
Elevated serum and urine homocystine, but no increased serum methionine as in homocystinuria. Enzyme deficiency apparently results in a reduced rate of formation of methionine from homocysteine.
Ref.: Mudd et al. (1972).

24. Lysyl-protocollagen hydroxylase
E.C. 1.14.11.4

Hydroxylysine-deficient collagen disease
Syndrome of severe scoliosis, recurrent joint dislocation and hyperextensible skin, associated with a reduced hydroxylysine content of the collagen.
Refs.: Pinnell et al. (1972), Krane et al. (1972).

Enzyme deficiency   Condition

25. Formiminotransferase   *Formiminotransferase deficiency syndrome*
     E.C. 2.1.2.5   The syndrome includes mental and physical
                                                      retardation, neurological abnormalities and in-
                                                      creased urinary excretion of formaminoglutamic
                                                      acid after an oral histidine load, despite the
                                                      presence of abnormally high serum folate levels.
                                                      Refs.: Arakawa et al. (1965), Arakawa (1970).

26. Uroporphyrinogen synthetase   *Acute intermittent porphyria* (pp. 238–241)
     E.C. 4.3.1.8

27. Uroporphyrinogen III cosynthetase   *Congenital erythropoietic porphyria*
                                                      Excessive production in bone marrow, deposition
                                                      in tissues and excretion in urine and faeces of
                                                      large amounts of uroporphyrin I and copro-
                                                      porphyrin I, which are products of the spontane-
                                                      ous oxidation of uroporphyrinogen I and its
                                                      decarboxylated derivative coproporphyrinogen
                                                      I. The disorder is characterised by photosensiti-
                                                      vity, haemolytic anaemia, and a pinkish dis-
                                                      colouration of the deciduous teeth, associated
                                                      with excessive porphyrinuria. The enzyme de-
                                                      ficiency has been demonstrated in red cells and
                                                      in cultured fibroblasts.
                                                      Refs.: Romeo and Levin (1969), Romeo et al.
                                                      (1970), Marver and Schmid (1972).

28. 25-hydroxycholecalciferol-1-hydroxylase   *Hereditary vitamin-D dependent rickets*
                                                      One form of vitamin-D refractory rickets. There
                                                      is severe rickets which will respond only to very
                                                      large doses of vitamin D (D_2 or D_3 or 25-hydroxy-
                                                      D_3). However healing of the bony rachitic
                                                      condition is obtained by the administration of
                                                      minute amounts (1.0 μg per day) of 1α, 25-
                                                      dihydroxy vitamin D (fig. A12).
                                                      Ref.: Fraser et al. (1973).

29. Adenosine deaminase   *Combined immunodeficiency disease*
     E.C. 3.5.4.4   Gross deficiency of adenosine deaminase (red
                                                      cell and 'tissue' isozymes) associated with a
                                                      severe condition characterised by a failure of
                                                      immune response.
                                                      Refs.: Gibblett et al. (1972), Knudsen and
                                                      Dissing (1973), Hirschhorn and Beratis (1973),
                                                          Pollara and Meuwissen (1973).

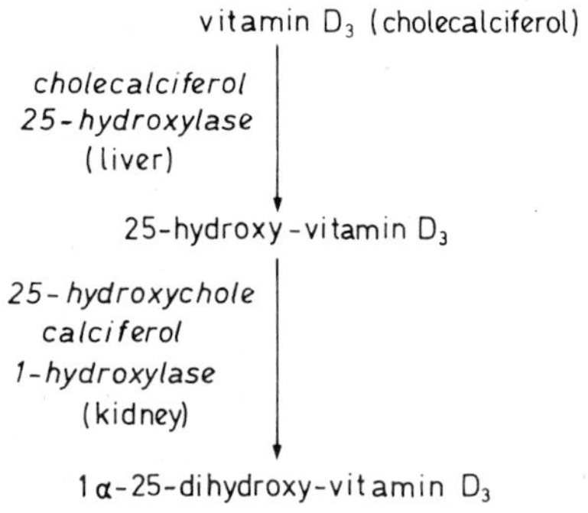

vitamin D₃ (cholecalciferol)

*cholecalciferol
25-hydroxylase*
(liver)

25-hydroxy-vitamin D₃

*25-hydroxychole
calciferol
1-hydroxylase*
(kidney)

1α-25-dihydroxy-vitamin D₃

Fig. A12. Vitamin D₃ metabolism.

Enzyme deficiency	Condition
30. Phytanic acid α-hydroxylase	*Refsum's syndrome* The condition is characterised by accumulation of phytanic acid, a 20-carbon branched-chain fatty acid, in liver and kidney. Clinically there is peripheral neuropathy and ataxia, retinitis pigmentosa and abnormalities in the skin and bones. Phytanic acid may constitute as much as 30% of the total fatty acids of the plasma, whereas it is normally present only in trace amounts. Phytanic acid is dietary in origin. It is normally oxidised through an initial α-oxidation to pristanic acid and then by further successive β-oxidation steps. The enzyme defect in Refsum's disease prevents the conversion of phytanic acid to α-hydroxyphytanic acid which is the initial step in its oxidation to pristanic acid. Refs.: Refsum (1946), Steinberg et al. (1967), Herndon et al. (1969), Steinberg (1972)
31. Adenylate kinase E.C. 2.7.4.3	*Haemolytic disease* (one type) Refs.: Szeinberg et al. (1969), Boivin et al. (1970).
32. Lecithin: cholesterol acyltransferase E.C. 2.3.1.43	*Familial lecithin: cholesterol acyl transferase deficiency* The enzyme (LCAT) catalyses the esterification of cholesterol in plasma. An unsaturated fatty acid is transferred from the 2-position of lecithin

| Enzyme deficiency | Condition |

to the 3-hydroxyl position of cholesterol. The products of the reaction are esterified cholesterol and lysolecithin. The enzyme deficiency results in a decrease in plasma cholesteryl ester and lysolecithin. Plasma free cholesterol, lecithin and triglycerides are usually elevated. The plasma is turbid or milky-white in appearance. Clinically the disorder is characterised by corneal opacities, normochromic anaemia and proteinuria due to renal damage.

Refs.: Norum and Gjone (1967a, b), Gjone and Norum (1968), Hamnström et al. (1969), Norum et al. (1972).

33. Steroid 21-hydroxylase
    E.C. 1.14.99.10

Steroid 21-hydroxylase deficiency

a) *Mild form – simple virilizing adrenal hyperplasia*

The enzyme deficiency results in decreased cortisol synthesis (fig. A13). This induces increased ACTH secretion which in turn results in overproduction of cortisol precursors and sex steroids which do not require 21-hydroxylation for their biosynthesis. There is excessive excretion of pregnanetriol, the metabolite of 17-hydroxyprogesterone and more than twenty other metabolic products lacking 21-OH have been found in the urine. The most prominent clinical feature is virilization. In the female the external genitalia are masculinised, but the internal genitalia are normal (female pseudohermaphroditism). In both sexes there is early fusion of the bony epiphyses and eventual short stature.

b) *Severe form – salt losing form of congenital adrenal hyperplasia.*

In addition to the above features of the 'mild' or 'compensated' form, there is a profound deficiency of aldosterone secretion with consequent salt and water loss.

Refs.: see reviews by Bongiovanni (1972), New and Levine (1973).

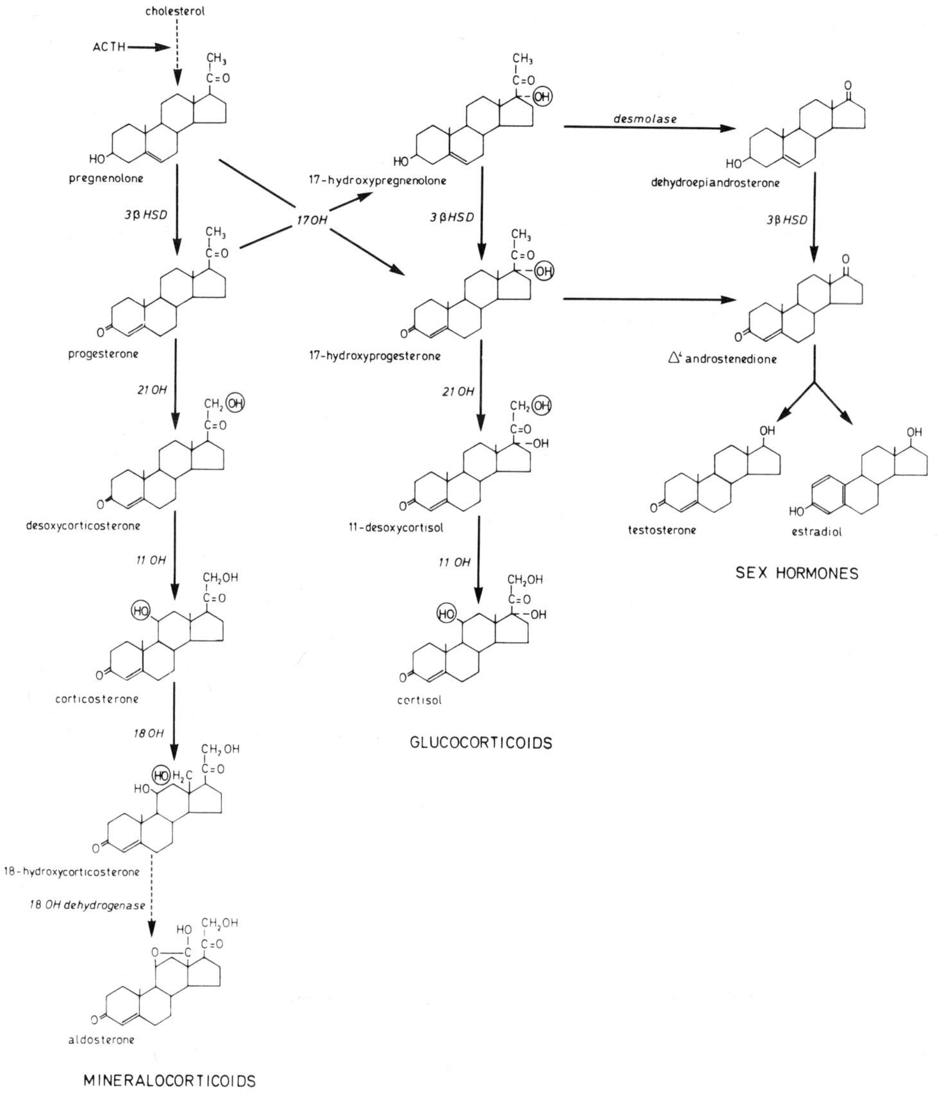

Fig. A13. Simplified scheme for adrenal steroidogenesis. Each hydroxylation step is indicated and the newly added hydroxyl group is circled (from New and Levine 1973). Key: 3β HSD, 3β-hydroxysteroid dehydrogenase; 21 OH, steroid 21-hydroxylase; 11 OH, steroid 11β-hydroxylase; 18 OH, steroid 18-hydroxylase; 17 OH, steroid 17α-hydroxylase.

Enzyme deficiency   Condition

34. Steroid 11β-hydroxylase
    E.C. 1.14.15.4

'Hypertensive' form of congenital adrenal hyper-plasia
The enzyme deficiency results in decreased cortisol synthesis (fig. A13), which induces increased ACTH secretion and this causes overproduction of precursors, particularly desoxycorticosterone (DOC) an active salt retaining hormone and 11-desoxycortisol. The most prominent clinical feature is virilization and in females female pseudohermaphroditism. Many but not all patients show hypertension probably due to excessive DOC production.
Refs.: see reviews by Bongiovanni (1972), New and Levine (1973).

35. 3β-hydroxysteroid dehydrogenase
    E.C. 1.1.1.51

'Non-virilizing' congenital adrenal hyperplasia
The enzyme deficiency results in decreased synthesis of all three classes of adrenal steroid, i.e. mineralcorticoids, glucocorticoids, and sex steroids (fig. A13). In particular there is impaired production of aldosterone, cortisol and testosterone. There are symptoms of severe adrenal insufficiency shortly after birth. In males there is male pseudohermaphroditism.
Refs.: see reviews by Bongiovanni (1972), New and Levine (1973).

36. Steroid 17α-hydroxylase
    E.C. 1.14.99.9

Steroid 17α-hydroxylase deficiency
Diminished secretion of all glucocorticoids and sex steroids, and excessive secretion of mineralcorticoids (fig. A13). Hyperkalaemia and hypertension. Females have sexual infantilism, males male pseudohermaphroditism.
Refs.: see reviews by Bongiovanni (1972), New and Levine (1973).

37. Steroid 18-hydroxylase

Steroid 18-hydroxylase deficiency
Deficient aldosterone formation, but no impairment of cortisol or sex hormone synthesis (fig. A13). Severe salt losing syndrome.
Refs.: see reviews by Bongiovanni (1972), New and Levine (1973).

Enzyme deficiency   Condition

38. Cholesterol desmolase   *Cholesterol desmolase deficiency*
    (cholesterol 20α-hydroxylase)   Deficient conversion of cholesterol to pregnen-
                                           alone (fig. A13). Impaired synthesis of all three
                                           classes of adrenal corticosteroids.
                                           Refs.: see reviews by Bongiovanni (1972), New
                                               and Levine (1973).

Enzyme and protein polymorphisms[*]

A2.1 Enzyme polymorphisms

Enzyme	Method of detection	References
1. Glucose-6-phosphate dehydrogenase	Quantitative assay and electrophoresis	See pp. 162–170
2. Phosphoglucomutase (PGM_1)	Electrophoresis	See pp. 61–63, 293
3. Phosphoglucomutase (PGM_3)	Electrophoresis	See pp. 64–65, 294
4. Placental alkaline phosphatase	Electrophoresis	See pp. 39–41, 303–304
5. Serum cholinesterase (E_1)	Quantitative inhibition tests	See pp. 150–160
6. Serum cholinesterase (E_2)	Electrophoresis	See pp. 160–162
7. Peptidase A	Electrophoresis	See pp. 32–34, 309
8. Liver acetyl-transferase	Metabolic loading test (isoniazid)	see pp. 251–254
9. Red cell acid phosphatase	Electrophoresis	See pp. 179–186
10. Adenylate kinase	Electrophoresis	Fildes and Harris (1966), Rapley et al. (1967), Bowman et al. (1967).
11. Adenosine deaminase	Electrophoresis	Spencer et al. (1968)
12. Phosphogluconate dehydrogenase	Electrophoresis	Fildes and Parr (1963), Parr (1966), Carter et al. (1968).
13. Erythrocyte nicotinamide adenine dinucleotide nucleosidase (NADase)	Quantitative assay	Ng et al. (1968).
14. Galactose-1-phosphate uridyl transferase (Duarte variant)	Quantitative assay and electrophoresis	Beutler et al. (1965, 1966), Mathai and Beutler (1966).
15. Glutathione reductase	Electrophoresis	Long (1967)
16. Pancreatic amylase	Electrophoresis	Kamaryt and Laxova (1965), Merritt et al. (1973).
17. Salivary amylase	Electrophoresis	Merritt et al. (1973)
18. Peptidase D (prolidase)	Electrophoresis	Lewis and Harris (1969b).
19. Peptidase C	Electrophoresis	Povey et al. (1972)
20. Pepsinogens (group I)	Electrophoresis	Samloff and Townes (1970), Samloff et al. (1973).
21. Carbonic anhydrase (CA II)	Immunoelectrophoresis	Moore et al. (1971, 1973).

[*] Situations in which two or more common alleles, each producing a discrete effect on the formation of a protein, have been found in one or several major human populations. A locus is regarded as 'polymorphic' if the frequency of the commonest allele is less than 0.99. (see p. 295).

Enzyme or protein	Method of detection	References
22. Glutamate-oxalate transaminase (mitochondrial form)	Electrophoresis	Davidson et al. (1970), Hackel et al. (1972).
23. Glutamate-pyruvate transaminase (soluble form)	Electrophoresis	Chen and Giblett (1971), Chen et al. (1972).
24. Alcohol dehydrogenase (ADH_2)	pH-activity ratios and electrophoresis	Von Wartburg et al. (1965), Von Wartburg and Schürch (1968), Smith et al. (1971).
25. Alcohol dehydrogenase (ADH_3)	Electrophoresis	Smith et al. (1971, 1973).
26. Malic enzyme (mitchondrial form)	Electrophoresis	Cohen and Omenn (1972).
27. Acetylcholinesterase	Electrophoresis	Coates and Simpson (1972)
28. Esterase D	Electrophoresis	Hopkinson et al. (1973).
29. Glutathione peroxidase	Electrophoresis	Beutler and West (1974)
30. Uridine monophosphate kinase	Electrophoresis	Giblett et al. (1974)
31. Aryl hydrocarbon hydroxylase	Quantitative assay on cultured lymphocytes	Kellerman et al. (1973)

A2.2 'Protein' polymorphisms

Protein	Method of detection	References
1. Haemoglobin (β-locus)	Electrophoresis and haematology	See pp. 5ff, 134ff, 285–288
2. Haemoglobin (α-locus)	Electrophoresis and haematology	See pp. 5ff, 142ff
3. Haptoglobin	Electrophoresis	See pp. 91ff, 290–293
4. Transferrin	Electrophoresis	Smithies (1957), Smithies and Hiller (1959), Giblett (1962), Kirk (1968).
5. Serum α-globulin (Gc)	Immuno-electrophoresis	Hirschfeld (1959), Hirschfeld (1962), Bearn et al. (1964), Rienskou (1968).
6. Immunoglobulins γ-Heavy chains (Gm system)	Immunological (antibody neutralisation)	Grubb (1956), Grubb and Laurell (1956), Steinberg (1967), Grubb (1970).
7. Immunoglobulins κ-Light chains (Inv system)	Immunological (antibody neutralisation)	Ropartz et al. (1961), Ritter and Wendt (1964), Steinberg (1967), Grubb (1970).
8. Immunoglobulins α-Heavy chains (Am system)	Immunological (antibody neutralisation)	Vyas and Fudenberg (1969), Kunkel et al. (1969), Grubb (1970).

Protein	Method of detection	References
9. Serum β-lipoproteins (*Ag* system)	Immuno-diffusion and precipitation	Allison and Blumberg (1961), Blumberg et al. (1962), Hirschfeld and Okachi (1967), Morganti et al. (1967), Hirschfeld (1968).
10. Serum β-lipoproteins (*Lp* system)	Immuno-diffusion and precipitation	Berg (1963, 1965, 1966, 1968), Bundschuh and Vogt (1965), Bütler (1967).
11. Caeruloplasmin	Electrophoresis	Shreffler et al. (1967).
12. Serum α_1 trypsin inhibitor (*Pi* system)	Electrophoresis and immuno-electrophoresis	Fagerhol and Braend (1965), Fagerhol and Laurell (1967), Fagerhol (1967, 1968, 1969).
13. Third component of complement (*C3* system)	Electrophoresis	Wieme and Demeulenaere (1967), Wieme and Segers (1968), Alper and Propp (1968), Azen and Smithies (1968).
14. Glycine-rich-β-glycoprotein (GBG)	Electrophoresis and immunofixation	Alper et al. (1972)
15. Serum α_2-macroglobulin (*Xm* system)	Immuno-diffusion and precipitation	Berg and Bearn (1966, 1968).
16. Serum α_1-acid glycoprotein (oroseromucoid)	Electrophoresis	Schmid et al. (1964, 1965), Johnson et al. (1969).
17. Serum β_2-glycoprotein I	Quantitative immunodiffusion	Cleve (1968), Koppe et al. (1970).
18. α-casein	Electrophoresis	Voglino and Ponzone (1972).
19. β-casein	Electrophoresis	Voglino and Ponzone (1972).
20. Parotid saliva basic protein (*Pb*)	Electrophoresis	Azen (1972).
21. Proline-rich salivary proteins (*Pr*)	Electrophoresis	Azen and Oppenheim (1973).

References

AAGENAES, Ø, MATLARY, A., ELGJO, K., MUNTHE, E., FAGERHOL, M. K. (1972). Neonatal cholestasis in alpha₁-antitrypsin deficient children. Acta Paediat. Scand. *61*, 632.

ABRAMOV, A., SCHORR, S. and WOLMAN, M. (1956). Fatal exfoliative dermatitis following injection of triple antigen and Salk vaccine. Am. J. Diseases Children *91*, 282.

ABRAMSON, R. K., RUCKNAGEL, D. L., SHREFFLER, D. C. and SAAVE, J. J. (1970). Homozygous Hb J Tongariki: Evidence for only one alpha chain structural locus in Melanesians. Science *169*, 194.

AEBI, H. (1967). The investigation of inherited enzyme deficiencies with special reference to acatalasia. Proc. Third Int. Congr. Human Genetics. Johns Hopkins Press, Baltimore, p. 189.

AEBI, H., BAGGIOLINI, M., DEWALD, B., LAUBER, E., SUTER, H., MITCHELI, A. and FREI, J. (1964). Observations in two Swiss families with acatalasia. Enzymol. Biol. Clin. *4*, 121.

AEBI, H. BOSSI, E., CANTZ,M., MATSUBARA, S. and SUTER, H. (1968). Acatalas(em)ia in Switzerland. *In:* 'Hereditary disorders or erythrocyte metabolism', ed. E. Beutler. Grune & Stratton, New York.

AEBI, H. SUTER, H. (1971). Acatalasemia. Advances in Human Genetics *2*, 143. Plenum Press, New York.

AGARWAL, R. P. and PARKS, R. E. (1969). Purine nucleoside phosphorylase from human erythrocytes. IV Crystallisation and some properties. J. Biol. Chem. *244*, 644.

AGER, J. A. M. and LEHMANN, H. (1958). Observations on some 'fast' haemoglobins: K, J, N and 'Bart's'. Brit. Med. J. *1*, 929.

AGER, J. A. M., LEHMANN, H. and VELLA, F. (1958). Haemoglobin 'Norfolk': a new haemoglobin found in an English family. Brit. Med. J. *2*, 529.

ALLAN, J. D., CUSWORTH, D. L., DENT, C. E. and WILSON, V. D. (1958). A disease, probably hereditary, characterised by severe mental deficiency and a constant gross abnormality of aminoacid metabolism. Lancet *1*, 182.

ALLISON, A. C. (1954). Protection afforded by the sickle cell trait against subtertian malarial infection. Brit. Med. J. *1*, 290.

ALLISON, A. C. (1964). Polymorphism and natural selection in human populations. Cold Spring Harbor Symp. Quant. Biol. *29*, 137.

ALLISON, A. C. and BLUMBERG, B. (1961). An isoprecipitation reaction distinguishing human serum protein types. Lancet *1*, 634.

ALPER, C. A., BOENISCH, T. and WATSON, L. (1972). Genetic polymorphism in human glycine-rich-beta-glycoprotein. J. Exp. Med. *135*, 68.

ALPER, C. A. and PROPP, R. P. (1968). Genetic polymorphism of the third component of human complement (C'3). J. Clin. Invest. *47*, 2181.

ALTLAND, K. and GOEDDE, H. W. (1970). Heterogeneity in the silent gene phenotype of pseudocholinesterase of human serum. Biochem. Genetics, 4, 321.

ANANTHAKRISHNAN, R., BLAKE, N. M., KIRK, R. L. and BAXI, A. J. (1970). Further studies on the distribution of genetic variants of lactate dehydrogenase (LDH) in India. Med. J. Austr. 2, 787.

ANDERSEN, D. H. (1956). Familial cirrhosis of the liver with storage of abnormal glycogen. Lab. Invest. 5, 11.

ANDERSON, N. L., PERUTZ, M. F. and STAMATOYANNOPOULOS, G. (1973). Site of amino-acid substitution in haemoglobin Seattle ($\alpha_2{}^A\beta_2{}^{70\mathrm{Asp}}$). Nature New Biol. 243, 274.

ANDERSSON, B., NYMAN, P. O. and STRID, L. (1972). Amino acid sequence of human erythro-cyte carbonic anhydrase B. Biochem. Biophys. Res. Commun. 48, 670.

APPELLA, E. and MARKERT, C. L. (1961). Dissociation of lactate dehydrogenase into subunits with guanidine hydrochloride. Biochem. Biophys. Res. Commun. 6, 171.

ARAKAWA, T. (1970). Congenital defects in folate metabolism. Amer. J. Med. 48, 594.

ARAKAWA, T., OHARA, K. and TAKAHASHI, Y. (1965). Formiminotransferase deficiency syndrome: a new inborn error of folic acid metabolism. Ann. Paediat. 205, 1.

ARCHER, H. E., DORMER, A. E., SCOWEN, E. F. and WATTS, R. W. E. (1957). Primary hyperoxa-luria. Lancet 2, 320.

ARMSTRONG, M. D. and ROBINSON, K. S. (1954). On the excretion of indole derivatives in phenylketonuria. Arch. Biochem. 52, 287.

ARMSTRONG, M. D. and TYLER, F. H. (1955). Studies on phenylketonuria. I. Restricted phenylalanine intake in phenylketonuria. J. Clin. Invest. 34, 565.

ARMSTRONG, M. D., SHAW, K. N. F. and ROBINSON, K. S. (1955). Studies on phenylketonuria. II. The excretion of o-hydroxyphenyl-acetic acid in phenylketonuria. J. Biol. Chem. 213, 797.

ARNOLD, H., BLUME, K-G., ENGELHARDT, R. and LÖHR, G. W. (1973). Glucosephosphate isomerase deficiency: evidence for in vivo instability of an enzyme variant with hemolysis. Blood 41, 691.

ARNOLD, W. J., MEADE, J. C. and KELLEY, W. N. (1972). Hypoxanthine-guanine phospho-ribosyltransferase: characteristics of the mutant enzyme in erythrocytes from patients with the Lesch-Nyhan syndrome. J. Clin. Invest. 51, 1805.

ARNONE, A. (1972). X-ray diffraction study of binding of 2, 3 diphosphoglycerate to human deoxy-haemoglobin. Nature (Lond.) 237, 146.

ARROW, V. K. and WESTALL, R. G. (1958). Aminoacid clearances in cystinuria. J. Physiol. (London) 142, 141.

AUERBACH, V. H., DIGEORGE, A. M., BALDRIDGE, R. C., TOURTELLOTTE, C. D. and BRIGHAM, M. P. (1962). Histidinemia. A deficiency in histidase resulting in the urinary excretion of histidine and of imidazolepyruvic acid. J. Pediat. 60, 487.

AULA, P., NÄNTÖ, V., LAIPIO, M.-L, and AUTIO, S. (1973). Aspartylglucosaminuria: Deficiency of aspartylglucosaminidase in cultured fibroblasts of patients and their heterozygous parents. Clin. Genet. 4, 297.

AUSTIN, J., BALASUBRAMANIAN, A., PATTABIRAMAN, T., SARSWATHI, S., BASU, D. and BACHHA-WAT, B. (1963). A controlled study of enzymic activities in three human disorders of glycolipid metabolism. J. Neurochem. 10, 805.

AUTIO, S. (1972). Aspartylglycosaminuria. J. Mental Deficiency Res. Monograph Series I, 1.

AUTIO, S., VISAKORPI, J. K. and JARVINEN, H. (1973). Aspartylglycosaminuria (AGU). Further aspects on its clinical picture, mode of inheritance and epidemiology based on a series of 57 patients. Ann. Clin. Res. 5, 149.

AYALA, F. J. and ANDERSON, W. W. (1973). Evidence of natural selection in molecular evolution. Nature New Biol. *241*, 274.

AZEN, E. A. (1972). Genetic Polymorphism of basic proteins from parotid saliva. Science *176*, 673.

AZEN, E. A. and OPPENHEIM, F. G. (1973). Genetic polymorphism of proline-rich human salivary proteins. Science *180*, 1067.

AZEN, E. A. and SMITHIES, O. (1968). Genetic polymorphism of C'3 (β1C-globulin) in human serum. Science 162, 905.

AZEVEDO, E., KIRKMAN, H. N., MORROW, A. C. and MOTULSKY, A. G. (1968). Variants of red cell glucose-6-phosphate dehydrogenase among Asiatic Indians. Ann. Hum. Genet. Lond. *31*, 375.

BACH, G., EISENBERG, F., CANTZ, M. and NEUFELD, E. F. (1973). The defect in the hunter syndrome: deficiency of sulfoiduronate sulfatase. Proc. Natl. Acad. Sci. U.S. *70*, 2134.

BACH, G., FRIEDMAN, R., WEISSMANN, B. and NEUFELD, E. F. (1972). The defect in the Hurler and Scheie syndromes: deficiency of a-L-iduronidase. Proc. Natl. Acad. Sci. U.S. *69*, 2048.

BADR, F. M., LORKIN, P. A. and LEHMANN, H. (1973). Haemoglobin P-Nilotic containing a β-ζ chain. Nature New Biol. *242*, 107.

BAGLIONI, C. (1962a). The fusion of two polypeptide chains in haemoglobin Lepore and its interpretation as a genetic deletion. Proc. Natl. Acad. Sci. U.S. *48*, 1880.

BAGLIONI, C. (1962b). A chemical study of hemoglobin Norfolk. J. Biol. Chem. *237*, 69.

BAGLIONI, C. (1963). A child homozygous for persistence of foetal haemoglobin. Nature (Lond.) *198*, 1117.

BAGLIONI, C. and INGRAM, V. M. (1961). Four adult haemoglobin types in one person. Nature (Lond.) *189*, 465.

BAKER, L. and WINEGRAD, A. I. (1970). Fasting hypoglycaemia and metabolic acidosis associated with deficiency of hepatic fructose 1,6 diphosphatase activity. Lancet *2*, 13.

BAKER, W. W. and MINTZ, B. (1969). Subunit structure and gene control of mouse NADP-malate dehydrogenase. Biochem. Genet. *5*, 351.

BANK, A. and MARKS, A. (1966). Excess a chain synthesis relative to β chain synthesis in thalassaemia major and minor. Nature (Lond.) *212*, 1198.

BANK, A., BRAVERMAN, S., O'DONNELL, J. V. and MARKS, P. A. (1968). Absolute rates of globin chain synthesis in thalassaemia. Blood *31*, 226.

BARBER, G. W. and SPAETH, G. L. (1967). Pyridoxine therapy in homocystinuria. Lancet *1*, 337.

BARCROFT, H., GIBSON, Q., HARRISON, D. and MCMURRAY, J. (9145). Familial idiopathic methaemoglobinaemia and its treatment with ascorbic acid. Clin. Sci. *5*, 145.

BARGELLESI, A., PONTREMOLI, S. and CONCONI, F. (1967). Absence of β globin synthesis and excess of a globin synthesis in homozygous β-thalassemia. Europ. J. Biochem. *1*, 73.

BARNABAS, J. and MULLER, C. J. (1962). Haemoglobin Lepore (Hollandia). Nature (Lond.) *194*, 931.

BARNABAS, J., GOODMAN, M. and MOORE, G. W. (1971). Evolution of haemoglobins in primates and other therian mammals. Comp. Biochem. Physiol. *39B*, 455.

BARNES, N. D., HULL, D., BALGOBIN, L. and GOMPERTZ, D. (1970). Biotin responsive propionic-acidaemia. Lancet *2*, 244.

BARON, D. N., DENT, C. E., HARRIS, H., HART, E. W. and JEPSON, J. B. (1956). Hereditary pellagra-like skin rash with temporary cerebellar ataxia, constant renal aminoaciduria, and other bizarre biochemical features. Lancet *2*, 421.

BAUDHUIN, P., HERS, H. G. and LOEB, H. (1964). An electron microscopic and biochemicat study of type II glycogenosis. Lab. Invest. *13*, 1140.

BAUGHAN, M. A., VALENTINE, W. N., PAGLIA, M. D., WAYS, P. O., SIMON, E. R. and DE MARSH Q. B. (1968). Hereditary hemolytic anemia associated with glucosephosphate isomerase (GPI) deficiency – a new enzyme defect of human erythrocytes. Blood *32*, 236.

BEARDMORE, T. D., MEADE, J. C. and KELLEY, W. N. (1973). Increased activity of two enzymes of pyrimidine biosynthesis *de novo* in erythrocytes from patients with the Lesch-Nyhan syndrome. J. Lab. Clin. Med. *81*, 43.

BEARN, A. F., BOWMAN, B. H. and KITCHIN, F. D. (1964). Genetical and biochemical considerations of the serum group-specific component. Cold Spring Harbor Symp. Quant. Biol. *29*, 435.

BEAUCHAMP, C. and FRIDOVITCH, I. (1971). Superoxide dismutase: improved assays and an assay applicable to acrylamide gels. Analyt. Biochem. *44*, 276.

BEAVEN, G. H., HORNABROOK, R. W., FOX, R. H. and HUEHNS, E. R. (1972). Occurrence of heterozygotes and homozygotes for the α-chain haemoglobin variant Hb-J (Tongariki) in New Guinea. Nature (Lond.) *235*, 46.

BECKER, M. A., KOSTEL, P. J., MEYER, L. J. and SEEGMILLER, J. E. (1973b). Human phosphoribosylpyrophosphate synthetase: increased enzyme specific activity in a family with gout and excessive purine synthesis. Proc. Natl. Acad. Sci. U.S. *70*, 2749.

BECKER, M. A. MEYER, L. J., WOOD, A. W. and SEEGMILLER, J. E. (1973). Purine overproduction in man associated with increased phosphoribosylpyrophosphate synthetase activity. Science *179*, 1123.

BECKMAN, G. (1970). Placental alkaline phosphatase, relation between phenotype and enzyme activity. Human Heredity *20*, 74.

BECKMAN, G. (1973). Population studies in northern Sweden. IV. Polymorphism of superoxide dismutase. Hereditas *73*, 305.

BECKMAN, L. and BECKMAN, G. (1967). Individual and organ specific variations of human acid phosphatase. Biochem. Genet. *1*, 145.

BEET, E. A. (1949). The genetics of sickle cell trait in a Bantu tribe. Ann. Eugenics *14*, 279.

BELL, J. C. and RIEMENSNIDER, D. K. (1957). Use of a serum microbiologic assay technique for estimating patterns of isoniazid metabolism. Ann. Rev. Tuberc. *75*, 995.

BELLINGHAM, A. J. and HUEHNS, E. R. (1968). Compensation in haemolytic anaemias caused by abnormal haemoglobins. Nature (Lond.) *218*, 924.

BELSEY, M. A. (1973). The epidemiology of favism. Bull. Wld. Hlth. Org. *48*, 1.

BENESCH, R. and BENESCH, R. E. (1969). Intracellular organic phosphates as regulators of oxygen release by haemoglobin. Nature (Lond.) *221*, 618.

BENESCH, R., BENESCH, R. E. and TYUMA, I. (1966). Subunit exchange and ligand binding. II. The mechanism of the allosteric effect in hemoglobin. Proc. Natl. Acad. Sci. U.S. *56*, 1268.

BENESCH, R. E., RANNEY, H. M., BENESCH, R. and SMITH, M. (1961). The chemistry of the Bohr effect. II. Some properties of haemoglobin H. J. Biol. Chem. *236*, 2926.

BENKE, P. J., HERRICK, N. and HEBERT, A. (1973). Hypoxanthine-guanine phosphoribosyltransferase variant associated with accelerated purine synthesis. J. Clin. Invest. *52*, 2234.

BERATIS, N. G., SEEGERS, W. and HIRSCHHORN, K. (1970). Properties of placental alkaline phosphatase. I Molecular size and electrical charge of the various electrophoretic components of the six common phenotypes. Biochem. Genet. *4*, 689.

BERATIS, N. G., SEEGERS, W. and HIRSCHHORN, K. (1971). Properties of placental alkaline phosphatase. II. Interactions of fast and slow-migrating components. Biochem. Genet. *5*, 367.

BERETTA, A., PRATO, V., GALLO, E. and LEHMANN, H. (1968). Haemoglobin Torino – $\alpha43$ (CD1) phenylalanine → valine. Nature (Lond.) *217*, 1016.

BERG, K. (1963). A new serum type system in man – the Lp system. Acta Pathol. Microbiol. Scand. *59*, 369.

BERG, K. (1965). A new serum type system in man – the Ld system. Vox Sanguinis *10*, 513.

BERG, K. (1966). Further studies on the Lp system. Vox Sanguinis *11*, 419.

BERG, K. (1968). The Lp system. Series Haematologica *1*, 111.

BERG, K. and BEARN, A. G. (1966). An inherited X-linked serum system in man. The Xm System. J. Expl. Med. *123*, 379.

BERG, K. and BEARN, A. G. (1968). Human serum protein polymorphisms. Ann. Rev. Genet. *2*, 341.

BERNINI, L. F., DE JONG, W. W. W. and KHAN, P. M. (1970). Varianti emoglobiniche nella popolazione tribale dell'Andhra Pradesh. Molteplicita dell locus αHb nell'uomo. Atti Associazione Genetica Italiana *15*, 191.

BERRY, H., SUTHERLAND, B. S. and GUEST, G. M. (1957). Phenylalanine tolerance tests on relatives of phenylketonuric children. Am. J. Human Genet. *9*, 310.

BEUTLER, E. and BALUDA, M. C. (1964). The separation of glucose-6-phosphate dehydrogenase deficient erythrocytes from the blood of heterozygotes for glucose-6-phosphate dehydrogenase deficiency. Lancet *1*, 189.

BEUTLER, E., BALUDA, M. C., STURGEON, P. and DAY, R. (1965). A new genetic abnormality resulting in galactose-1-phosphate uridyltransferase deficiency. Lancet *1*, 353.

BEUTLER, E., BALUDA, M. C., STURGEON, P. and DAY, R. (1966). The genetics of galactose-1-phosphate uridyl transferase deficiency. J. Lab. Clin. Med. *64*, 646.

BEUTLER, E., MATHAI, C. K. and SMITH, J. E. (1968). Biochemical variants of glucose-6-phosphate dehydrogenase giving rise to congenital non spherocytic hemolytic disease. Blood *31*, 131.

BEUTLER, E. and WEST, C. (1974). Red cell glutathione peroxidase polymorphism in Afro-Americans. Am. J. Hum. Genet. *26*, 255.

BEUTLER, E., YEH, M. and FAIRBANKS, V. F. (1962). The normal human female as a mosaic of X chromosome activity: studies using the gene for G-6-PD deficiency as a marker. Proc. Natl. Acad. Sci. U.S. *48*, 9.

BEUTLER, E. and YOSHIDA, A. (1973). Human glucose-6-phosphate dehydrogenase variants: a supplementary tabulation. Ann. Hum. Genet. Lond. *37*, 151.

BHENDE, Y., DESHPANDE, C. K., BHATIA, H. M., SANGER, R., RACE, R. R., MORGAN, W. T. J. and WATKINS, W. M. (1952). A 'new' blood group character related to the ABO system. Lancet *1*, 903.

BIANCO, A. and ZINKHAM, W. H. (1963). Lactate dehydrogenases in human testes. Science *139*, 601.

BIANCO, A., ZINKHAM, W. H. and KUPCHYK, L. (1964). Genetic control and ontogeny of lactate dehydrogenase in pigeon testes. J. Exptl. Zool. *156*, 137.

BIANCO, I., MONTALENTI, G. SILVESTRONI, E. and SINISCALCO, M. (1952). Further data on genetics of microcythaemia or thalassaemia minor and Cooley's disease or thalassaemia major. Ann. Eugenics (Lond.) *16*, 299.

BIENZLE, U., AYENI, O., LUCAS, A. O. and LUZZATTO, L. (1972). Glucose-6-phosphate dehydrogenase and malaria. Lancet *1*, 107.

BIGLEY, R. H. and KOLER, R. D. (1968). Liver pyruvate kinase (PK) isozymes in a PK-deficient patient. Ann. Hum. Genet. Lond. *31*, 383.

BIGLEY, R. H., STENZEL, P., JONES, R. T., CAMPOS, J. O. and KOLER, R. D. (1968). Tissue distribution of human pyruvate kinase isozymes. Enzymol. Biol. Clin. *9*, 10.

BLACK, J. A. and DIXON, G. H. (1968). Aminoacid sequence of alpha chains of human haptoglobins. Nature (Lond.) *218*, 736.

BLACK, J. A. and SIMPSON, K. (1967). Fructose intolerance. Brit. Med. J. *4*, 138.

BLAKE, N. M., KIRK, R. L., PRYKE, E. and SINNETT, P. (1969). Lactate dehydrogenase electrophoretic variant in a New Guinea Highland population. Science *163*, 701.

BLASS, J. P., AVIGAN, J. and UHLENDORF, B. W. (1970). A defect in pyruvate decarboxylase in a child with intermittent movement disorder. J. Clin. Invest. *49*, 423.

BLASS, J. P., KARK, R. A. P. and ENGEL, W. K. (1971). Clinical studies of a patient with pyruvate decarboxylase deficiency. Arch. Neurol. *25*, 449.

BLUMBERG, B. S., BERNANKE, D. and ALLISON, A. C. (1962). A human lipoprotein polymorphism. J. Clin. Invest. *41*, 1936.

BLUME, K. G., HRYNIUK, W., POWARS, D., TRINIDAD, F., WEST, C. and BEUTLER, E. (1972). Characterization of two new variants of glucose-phosphate-isomerase deficiency with hereditary nonspherocytic hemolytic anemia. J. Lab. Clin. Med., *79*, 942.

BODANSKY, D., SCHWARTZ, M. K. and NISSELBAUM, J. S. (1966). Isozymes of aspartate aminotransferase in tissues and blood of man. Adv. Enz. Reg. *4*, 299.

BOETTCHER, B. and DE LA LANDE, F. A. (1969). Electrophoresis of human saliva and identification of inherited variants of amylase isozymes. Aust. J. Exptl. Biol. Med. Sci. *47*, 97.

BOIVIN, P. and GALAND, C. (1967). Constant de Michaelis anormale pour le phospho-énol-pyruvate au cours d'un déficit en pyruvate-kinase erythrocitaire. Rev. Franc. Etudes Clin. Biol. *12*, 372.

BOIVIN, P., GALAND, C., HAKIM, J., ROGÉ, J. and GUEROULT, N. (1969). Anémie hemolytique avec déficit en glutathione-peroxydase chez un adulte. Enzym. Biol. Clin. (Basel) *10*, 68.

BOIVIN, P., GALAND, C., HAKIM, J., SIMONY, D. and SELIGMAN, M. (1970). Deficit congenital en adenylat-kinase erythrocitaire. Presse Med. *78*, 1443.

BONAVENTURA, J. and RIGGS, A. (1968). Haemoglobin Kansas, a human haemoglobin with a neutral amino acid substitution and an abnormal oxygen equilibrium. J. Biol. Chem. *243*, 980.

BONGIOVANNI, A. M. (1972). Disorders of adrenocortical steroid biogenesis. *In:* The Metabolic Basis of Inherited Disease. Eds. J. B. Stanbury, J. B. Wyngaarden and D. S. Fredrickson. 3rd Edit. McGraw-Hill, New York. p. 857.

BOOKCHIN, R. M., NAGEL, R. L. and RANNEY, H. M. (1967). Structure and properties of Hb C Harlem, a human haemoglobin variant with aminoacid substitutions in two residues of the β polypeptide chain. J. Biol. Chem. *242*, 248.

BOONE, C. M., CHEN, T. R. and RUDDLE, F. H. (1972). Assignment of LDH-A locus in man to chromosome C-11 using somatic cell hybrids. Proc. Natl. Acad. Sci. U.S. *69*, 510.

BOONE, C. M. and RUDDLE, F. H. (1969). Interspecific hybridization between human and mouse somatic cells: enzyme and linkage studies. Biochem. Genet. *3*, 119.

BOTTINI, E. and MODIANO, G. (1964). Effect of oxidised glutathione on human red cell acid phosphatases. Biochem. Biophys. Res. Commun. *17*, 260.

BOURNE, J. G., COLLIER, H. O. J. and SOMERS, G. F. (1952). Succinylcholine (succinoyl-choline). Muscle relaxant of short action. Lancet *1*, 1225.

BOWMAN, E., FRISCHER, H., AJMAR, F., CARSON, P. and GOWER, M. K. (1967). Population, family and biochemical investigation of human adenylate kinase polymorphism. Nature (Lond.) *214*, 1156.

BOWMAN, H. S. and PROCOPIO, F. (1963). Hereditary non-spherocytic haemolytic anaemia of the pyruvate kinase deficient type. Ann. Internal Med. *58*, 567.

BOYER, S. H. (1961). Alkaline phosphatase in human sera and placentae. Science *134*, 1002.

BOYER, S. H. (1971). Primate haemoglobins: some sequences and some proposals concerning the character of evolution and mutation. Biochem. Genet. *5*, 405.

BOYER, S. H., CHARACHE, S., FAIRBANKS, V. F., MALDONADO, J. E., NOYES, A. and GAYLE, E. E. (1972). Hemoglobin Malmö, β97 (FG4) Histidine $\rightarrow$ Glutamine: a cause of polycythemia. J. Clin. Invest. *51*, 666.

BOYER, S. H., CROSBY, E. F., THURMON, T. F., NOYES, A. N., FULLER, G. F., LESLIE, S. E., SHEPARD, M. K. and HERNDON, C. N. (1969). Hemoglobins A and A_2 in New World primates: comparative variation and its evolutionary implications. Science *166*, 1428.

BOYER, S. H., CROSBY, E. F., NOYES, A. N., FULLER, G. F., LESLIE, S. E., DONALDSON, L. J., VRABLIK, G. R., SCHAEFER, E. W. and THURMON, T. F. (1971). Primate hemoglobins: some sequences and some proposals concerning the character of evolution and mutation. Biochem. Genet. *5*, 405.

BOYER, S. H., FAINER, D. C. and WATSON-WILLIAMS, E. J. 1963. Lactate dehydrogenase variant from human blood: evidence for molecular subunits. Science *141*, 642.

BOYER, S. H., HATHAWAY, P. and GARRICK, M. D. (1964). Modulation of protein synthesis in man: an *in vitro* study of haemoglobin synthesis by heterozygotes. Cold Spring Harbor Symp. Quant. Biol. *29*, 333.

BOYER, S. H., NOYES, A. N., TIMMONS, C. F. and YOUNG, R. A. (1972). Primate hemoglobins: polymorphisms and evolutionary patterns. J. Hum. Evol. *1*, 515.

BOYER, S. H., PORTER, I. H. and WEILBACHER, R. G. (1962). Electrophoretic heterogeneity of glucose-6-phosphate dehydrogenase and its relationship to enzyme deficiency in man. Proc. Natl. Acad. Sci. U.S. *48*, 1868.

BOYLEN, J. B. and QUASTEL, J. H. (1962). Effects of 2-phenylalanine and sodium phenylpyruvate on the formation of melanin from L-tyrosine in melanoma. Nature (Lond.) *193*, 376.

BRADLEY, T. B., BOYER, S. H. and ALLEN, F. H. (1961). Hopkins-2-hemoglobin: a revised pedigree with data on blood and serum groups. Bull. Johns Hopkins Hosp. *108*, 75.

BRADLEY, T. B. JR., BRAWNER, J. N. III and CONLEY, C. L. (1961). Further observations on an inherited anomaly characterised by persistance of fetal hemoglobin. Bull. Johns Hopkins Hosp. *108*, 242.

BRADLEY, T. B. JR., WOHL, R. C. and RIEDER, R. F. (1967). Hemoglobin Gun Hill: deletion of five amino acid residues and impaired heme-globin binding. Science *157*, 1581.

BRADY, R. O. (1968). Enzymatic defects in the sphingolipidoses. Advan. Clin. Chem. *11*, 1.

BRADY, R. O., GAL, A. E., BRADLEY, R. M. and MARTENSSON, E. (1967a). The metabolism of ceramidetrihexosides. I. Purification and properties of an enzyme that cleaves the terminal galactose molecule of galactosylgalactosylglucosylceramide. J. Biol. Chem. *242*, 1021.

BRADY, R. O., GAL, A. E., BRADLEY, R. M., MARTENSSON, E., WARSHAW, A. L. and LASTER, L. (1967b). Enzymatic defect in Fabry's disease. Ceramidetrihexosidase deficiency. New Engl. J. Med. *276*, 1163.

BRADY, R. O., KANFER, J. N., MOCK, M. B. and FREDRICKSON, D. S. (1966a). The metabolism of sphingomyelin. II. Evidence of an enzymatic deficiency in Niemann-Pick disease. Proc. Natl. Acad. Sci. U.S. *55*, 366.

BRADY, R. O., KANFER, J. N. and SHAPIRO, D. (1965). The metabolism of glycocerebrosides. II. Evidence of an enzymatic deficiency in Gaucher's disease. Biochem. Biophys. Res. Commun. *18*, 221.

BRADY, R. O., KANFER, J. N., SHAPIRO, D. and BRADLEY, R. M. (1966b). Demonstration of a deficiency of glucocerebroside-cleaving enzyme in Gaucher's disease. J. Clin. Invest. *45*, 1112.

BRANCATI, C. and BAGLIONI, C. (1966). Homozygous $\beta\delta$ thalassaemia ($\beta\delta$-microcythaemia). Nature (Lond.) *212*, 262.

BRATU, V., LORKIN, P. A., LEHMANN, H. and PREDESCU, C. (1971). Haemoglobin Bucuresti β42 (CDI) Phe → Leu, a cause of unstable haemoglobin haemolytic anaemia. Biochim. Biophys. Acta *251*, 1.

BRAUNITZER, G., HILSE, K., RUDOLFF, V. and HILSCHMANN, N. (1964). The haemoglobins. Advan. Protein Chem. *19*, 1.

BRAVERMAN, A. S. and BANK, A. (1969). Changing rates of globin chain synthesis during erythroid cell maturation in thalassaemia. J. Mol. Biol. *42*, 57.

BREMER, H. J. and NEUMANN, W. (1966). Tolerance of phenylalanine after intravenous administration in phenylketonurics, heterozygous carriers, and normal adults. Nature (Lond.) *209*, 1148.

BRENTON, D. P., CUSWORTH, D. C., DENT, C. E. and JONES, E. E. (1966). Homocystinuria. Clinical and dietary studies. Quart. J. Med. XXXV, 325.

BREWER, G. J. (1967). Achromatic regions of tetrazolium stained starch gels: inherited electrophoretic variation. Am. J. Hum. Genet. *19*, 674.

BREWER, G. J., EATON, J. W., KNUTSEN, C. S. and BECK, C. C. (1967). A strach-gel electrophoretic method for the study of diaphorase isozymes and preliminary results with sheep and human erythrocytes. Biochem. Biophys. Res. Commun. *29*, 198.

BRIMHALL, B., JONES, R. T., BAUR, E. W. and MOTULSKY, A. G. (1969). Structural characterization of hemoglobin Tacoma. Biochemistry *8*, 2125.

BRINKMANN, R., KOOPS, E., KLOPP, O., HEINDL, K. and RÜDIGER, H. W. (1972). Inherited partial deficiency of the PGM_1^1 gene: biochemical and densitometric studies. Ann. Hum. Genet. Lond. *35*, 363.

BRITTEN, R. J. and DAVIDSON, E. H. (1969). Gene regulation for higher cells: a theory. Science *165*, 349.

BRITTEN, R. J. and DAVIDSON, E. H. (1971). Repetitive and non-repetitive DNA sequences and a speculation on the origins of evolutionary novelty. Quart. Rev. Biol. *46*, 111.

BRITTEN, R. J. and KOHNE, D. E. (1968). Repeated sequences in DNA. Science *161*, 529.

BROWN, B. I. and BROWN, D. H. (1966). Lack of an α-1, 4-glucan: α-1, 4-glucan 6-glycosyl transferase in a case of Type IV glycogenosis. Proc. Natl. Acad. Sci. U.S. *56* 725.

BROWN, S. W. and CHANDRA, H. S. (1973). Inactivation system of the mammalian X chromosome. Proc. Natl. Acad. Sci. U.S. *70*, 195.

BUDD, M. A., TANAKA, K., HOLMES, L. B., EFRON, M. L., CRAWFORD, J. D. and ISSELBACHER, K. J. (1967). Isovaleric acidemia. Clinical features of a new genetic defect of leucine metabolism. New Engl. J. Med. *277*, 321.

BUNDSCHUH, G. and VOGT, A. (1965). Die Häufigkeit des Merkmals Lp (x) in der Berliner Bevölkerung. Humangenetik *1*, 379.

BURGERHOUT, W., VAN SOMEREN, H. and BOOTSMA, D. (1973). Cytological mapping of the genes assigned to the human A1 chromosome by use of radiation-induced chromosome breakage in a human-Chinese hamster hybrid cell line. Humangenetik *20*, 159.

BURNS, J. J. (1957). Missing step in man, monkey and guinea-pig required for biosynthesis of L-ascorbic acid. Nature (Lond.) *180*, 553.

BÜTLER, R. (1967). Polymorphisms of the human low-density lipoproteins. Vox Sanguinis *12*, 2.

BYERS, D. A., FERNLEY, H. N. and WALKER, P. G. (1972). Studies on alkaline phosphatase. Catalytic activity of human-placental alkaline-phosphatase variants. Eur. J. Biochem. *29*, 205.

CAHN, R. D., KAPLAN, N. O., LEVINE, L. and ZWILLING, E. (1962). Nature and development of lactic dehydrogenases. Science *136*, 962.

CAMPBELL, A. G. M., ROSENBERG, L. E., SNODGRASS, P. J. and NUZUM, T. C. (1971). Lethal neonatal hyperammonemia due to complete ornithine transcarbamylase deficiency. Lancet *2*, 217.

CAMPBELL, A. G. M., ROSENBERG, L. E., SNODGRASS, P. J. and NUZUM, C. T. (1973). Ornithine transcarbamylase deficiency. A cause of lethal neonatal hyperammonemia in males. New Engl. J. Med. *288*, 1.

CAPP, G. L., RIGAS, D. A. and JONES, R. T. (1970) Evidence for a new haemoglobin chain (ζ-chain). Nature (Lond.) *228*, 278.

CARMODY, P. J., RATTAZZI, M. C. and DAVIDSON, R. G. (1973). Tay-Sachs disease – the use of tears for the detection of heterozygotes. New Engl. J. Med. *289*, 1072.

CARRELL, R. W., LEHMANN, H. and HUTCHISON, H. E. (1966). Haemoglobin Köln (β-98 Valine → Methionine): an unstable protein causing inclusion-body anaemia. Nature (Lond.) *210*, 915.

CARRELL, R. W., LEHMANN, H., LORKIN, P. A., RAIK, E. and HUNTER, E. (1967). Haemoglobin Sydney: β 67 (E11) Valine → Alanine: an emerging pattern of unstable haemoglobins. Nature (Lond.) *215*, 626.

CARRELL, R. W. and OWEN, M. C. (1971). A new approach to haemoglobin variant identification: haemoglobin Christchurch β71 (E15) Phenylalanine → Serine. Biochim. Biophys Acta *236*, 507.

CARROLL, M. and ROBINSON, D. (1973). Immunological Properties of N-acetyl-β-D-glucosaminidase of normal human liver and of GM_2-gangliosidosis liver. Biochem. J. *131*, 91.

CARSON, N. A. and CARRÉ, I. J. (1969). Treatment of homocystinuria with pyridoxine. Arch. Dis. Child. *44*, 387.

CARSON, N. A. J., DENT, C. E., FIELD, C. M. B. and GAULL, G. E. (1965). Homocystinuria. Clinical and pathological review of ten cases. J. Pediat. *66*, 565.

CARSON, P. E., FLANAGAN, C. I., ICKES, C. E. and ALVING, A. S. (1956). Enzymatic deficiency in primaquine sensitive erythrocytes. Science *124*, 484.

CARTER, N. D., FILDES, R. A., FITCH, L. I. and PARR, C. W. (1968). Genetically determined electrophoretic variations of human phosphogluconate dehydrogenase. Acta Genet. *18*, 109.

CASTELLS, S. and BRANDT, I. K. (1968). Phenylketonuria: evaluation of therapy and verification of diagnosis. J. Pediat. *72*, 34.

CAVALLI-SFORZA, L. L. (1966). Population structure and human evolution. Proc. Roy. Soc. Lond. B *164*, 362.

CAVALLI-SFORZA, L. L. (1973). Some current problems of human population genetics. Amer. J. Hum. Genet. *25*, 82.

CEPPELLINI, R. (1955). Nuova interpretazione sulla genetica dei carratteri Lewis eritrocitari e salivari derivante dall'analysi di 87 famiglie. Ric. Sci. Suppl. *25*, 3.

CEPPELLINI, R. (1959). *In:* Biochemistry of human genetics. Ciba Foundation Symposium, ed. G. E. W. Wolstenholme and C. M. O'Connor. p. 133. Churchill, London.

CHAMBERS, R. A. and PRATT, R. T. C. (1956). Idiosyncrasy to fructose. Lancet *2*, 340.

CHARACHE, S. and CONLEY, C. L. (1969). Hereditary persistence of fetal haemoglobin. Ann. New York Acad. Sci. *165*, 37.

CHARACHE, S., OSTERTAG, W. and VON EHRENSTEIN, G. (1971). Clinical studies and physiological properties of Hopkins-2 haemoglobin. Nature New Biol. *234*, 248.

CHARACHE, S., WEATHERALL, D. J. and CLEGG, J. B. (1966). Polycythemia associated with a hemoglobinopathy. J. Clin. Invest. *45*, 813.

CHASE, G. A. and MCKUSICK, V. A. (1972). Founder effect in Tay-Sachs disease. Amer. J. Hum. Genet. *24*, 339.

CHEN, S.-H., ANDERSON, J. E. and GIBLETT, E. R. (1971). 2,3-Diphosphoglycerate mutase: its demonstration by electrophoresis and the detection of a genetic variant. Biochem. Genet. *5*, 481.

CHEN, S.-H., FOSSUM, B. L. G. and GIBLETT, E. R. (1972). Genetic variation of the soluble form of NADP dependent isocitrate dehydrogenase. Amer. J. Hum. Genet. *24*, 325.

CHEN, S.-H. and GIBLETT, E. R. (1971). Polymorphism of soluble glutamic-pyruvic transaminase: a new genetic marker in man. Science *173*, 148.

CHEN, S.-H., GIBLETT, E. R., ANDERSON, J. E. and FOSSUM, B. L. G. (1972). Genetics of glutamic-pyruvic transaminase: its inheritance, common and rare variants, population distribution and differences in catalytic activity. Ann. Hum. Genet. Lond. *35*, 401.

CHEN, T. R., MCMORRIS, F. A., CREAGAN, R., RICCIUTI, F., TISCHFIELD, J. and RUDDLE, F. H. (1973). Assignment of the genes for malate oxidoreductase decarboxylating to chromosome 6 and peptidase B and lactate dehydrogenase B to chromosome 12 in man. Amer. J. Hum. Genet. *25*, 200.

CHESTER, M. A. and WATKINS, W. M. (1969). α-L-Fucosyltransferases in human submaxillary gland and stomach tissues associated with the H, Le^a and Le^b blood group characters and ABH secretor status. Biochem. Biophys. Res. Comm. *34*, 835.

CHILDS, B., ZINKHAM, W., BROWN, E. A., KIMBRO, E. L. and TORBET, J. V. (1958). A genetic study of a defect in glutathione metabolism of the erythrocyte. Bull. Johns Hopkins Hosp. *102*, 21.

CLARK, S. W., GLAUBIGER, G. A. and LA DU, B. N. (1968). Properties of plasma cholinesterase variants. Ann. N.Y. Acad. Sci. *151*, 710.

CLARKE, B. (1970). Darwinian evolution of proteins. Science *168*, 1009.

CLARKE, B. (1970). Selective constraints on amino-acid substitutions during the evolution of proteins. Nature (Lond.) *228*, 159.

CLEGG, J. B. and WEATHERALL, D. J. (1967). Haemoglobin synthesis in α-thalassaemia, (Haemoglobin H. disease). Nature (Lond.) *215*, 1241.

CLEGG, J. B. and WEATHERALL, D. J. (1974). Haemoglobin Constant Spring, an unusual α-chain variant involved in the aetiology of haemoglobin H disease. Ann. New York Acad. Sci. (in press).

CLEGG, J. B., WEATHERALL, D. J. and MILNER, P. F. (1971). Haemoglobin Constant Spring –

A chain terminating mutant? Nature (Lond.) *234*, 337.

CLEGG, J. B., WEATHERALL, D. J., BOON, W. H. and MUSTAFA, D. (1969). Two new human haemoglobin variants involving proline substitutions. Nature (Lond.) *222*, 379.

CLEGG, J. B., WEATHERALL, D. J., NA-NAKORN, S. and WASI, P. (1968). Haemoglobin synthesis in β thalassaemia. Nature (Lond.) *220*, 664.

CLEVE, H. (1968). Genetic studies on the deficiency of β_2-glycoprotein I of human serum. Humangenetik *5*, 294.

CLEVE, H. and HERZOG, P. (1969). Phenotypic variants of haptoglobin Johnson types. Humangenetik *7*, 218.

CLEVE, H., GORDON, S., BOWMAN, B. H. and BEARN, A. G. (1967). Comparison of the tryptic peptides and amino acid composition of the beta polypeptide chains of the three common haptoglobin phenotypes. Am. J. Human Genet. *19*, 713.

COATES, P. M. and SIMPSON, N. E. (1972). Genetic variation in human erythrocyte acetyl cholinesterase. Science *175*, 1466.

COHEN, P. T. W. and OMENN, G. S. (1972). Genetic variation of the cytoplasmic and mitochondrial malic enzymes in the monkey *Macaca nemestrina*. Biochem. Genet. *7*, 289.

COHEN, P. T. W. and OMENN, G. S. (1972). Human malic enzyme: high frequency polymorphism of the mitochondrial form. Biochem. Genet. *7*, 303.

COHEN, P. T. W., OMENN, G. S., MOTULSKY, A. G., CHEN, S.-H. and GIBLETT, E. R. (1973). Restricted variation in the glycolytic enzymes of human brain and erythrocytes. Nature New Biol. *241*, 229.

COMINGS, D. E. and MOTULSKY, A. G. (1966). Absence of cis delta chain synthesis in (δβ) thalassaemia (F-talassaemia). Blood *28*, 54.

CONLEY, C. L., WEATHERALL, D. J. RICHARDSON, S. N., SHEPHERD, M. K. and CHARACHE, C. (1963). Hereditary persistence of fetal hemoglobin: a study of 79 affected persons in 15 Negro families in Baltimore. Blood *21*, 261.

CONNELL, G. E., DIXON, G. H. and SMITHIES, O. (1962). Subdivision of the three common haptoglobin types based on hidden differences. Nature (Lond.) *193*, 505.

CONNELL, G. E., SMITHIES, O. and DIXON, G. H. (1966). Geneaction in the human haptoglobins. II. Isolation and physical characterisation of alpha polypeptide chains. J. Mol. Biol. *21*, 225.

COOK, P. J. L., GRAY, J. E., BRACK, R. A., ROBSON, E. B. and HOWLETT, R. M. (1969). Data on haptoglobin and the D group chromosomes. Ann. Hum. Genet. Lond. *33*, 125.

CORBEEL, L. M., COLOMBO, J. P., VAN SANDE, M. and WEBER, A. (1969). Periodic attacks of lethargy in a baby with ammonia intoxication due to a congenital defect in ureogenesis. Arch. Dis. Child. *44*, 681.

CORI, G. T. (1954). Glycogen structure and enzyme deficiencies in glycogen storage diseases. Harvey Lectures Ser. *48*, p. 145.

CORI, G. T. (1957). Biochemical aspects of glycogen deposition diseases. Mod. Probl. Pediat. *3*, 344.

CORI, G. T. and CORI, C. F. (1952). Glucose-6-phosphatase of the liver in glycogen storage disease. J. Biol. Chem. *199*, 661.

COTE, R. H. and MORGAN, W. T. J. (1956). Some nitrogen-containing disaccharides isolated from human blood-group A substances. Nature (London) *178*, 1171.

COURT-BROWN, W. M. and SMITH, P. G. (1969). Human population cytogenetics. Brit. Med. Bull *25*, 74.

COX, R. P., KRAUSS, M. R., BALIS, M. E. and DANCIS, J. (1970). Evidence for transfer of enzyme product as the basis of metabolic co-operation between tissue culture fibroblasts of Lesch-Nyhan disease and normal cells. Proc. Natl. Acad. Sci. U.S. *67*, 1573.

CRICK, F. H. C. (1967). The genetic code. Proc. Roy. Soc. B *167*, 331.

CROW, J. F. (1973). The dilemma of nearly neutral mutations: How important are they for evolution and human welfare? J. Hered. *63*, 306.

CURRARINO, G., NEWHAUSER, E., REYERSBACK, G. and SOBEL, E. (1957). Hypophosphatasia. Am. J. Roentgenol. *78*, 392.

CURTAIN, C. C. (1964). A structural study of abnormal haemoglobins occurring in New Guinea. Australian J. Exptl. Biol. Med. Sci. *42*, 89.

DACIE, J. V., SHINTON, N. K., GAFFNEY, P. J., CARRELL, R. W. and LEHMANN, H. (1967). Haemoglobin Hammersmith (β42 (CD1) Phe → Ser). Nature (Lond.) *216*, 663.

DACREMONT, G. and KINT, J. A. (1968). Gm$_1$ – ganglioside accumulation and β-galactosidase deficiency in a case of Gm$_1$ gangliosidosis (Landing disease). Clin. Chim. Acta *21*, 421.

DANCIS, J., HUTZLER, J. and LEVITZ, M. (1963). The diagnosis of maple syrup urine disease (branch chain ketoaciduria) by the *in vitro* study of the peripheral leucocyte. Pediatrics *32*, 234.

DANCIS, J., HUTZLER, J. and LEVITZ, M. (1965). Detection of the heterozygote in maple syrup urine disease. J. Pediat. *66*, 595.

DANCIS, J., HUTZLER, J. and ROKKONES, T. (1967). Intermittent branched-chain ketonuria. New Engl. J. Med. *276*, 84.

DANCIS, J., HUTZLER, J., COX, R. P. and WOODY, N. C. (1969). Familial hyperlysinemia with lysine-ketoglutarate reductase insufficiency. J. Clin. Invest. *48*, 1447.

DANCIS, J., LEVITZ, M., MILLER, S. and WESTALL, R. G. (1959). Maple syrup urine disease. Brit. Med. J. *1*, 91.

DANCIS, J., HUTZLER, J., TADA, K., WADA, Y., MORIKAWA, T. and ARAKAWA, T. (1967). Hypervalinemia. A defect in valine transamination. Pediatrics *39*, 813.

DANES, B. S. and BEARN, A. G. (1967). Hurler's syndrome: a genetic study of clones in cell culture with particular reference to the Lyon hypothesis. J. Exptl. Med. *126*, 509.

DANOVITCH, S. H., BAER, P. N. and LASTER, L. (1968). Intestinal alkaline phosphatase activity in familial hypophosphatasia. New Engl. J. Med. *278*, 1253.

DAS, S. R., MUKHERJEE, B. N., DAS, S. K. and MALHOTRA, K. C. (1972). Four types of genetic variants of LDH found in India. Hum. Hered., *22*, 264.

DAVIDSON, R. G. and CORTNER, J. A. (1967a). Genetic variant of human erythrocyte malate dehydrogenase. Nature (Lond.) *215*, 761.

DAVIDSON, R. G. and CORTNER, J. A. (1967b). Mitochondrial malate dehydrogenase: a new genetic polymorphism in man. Science *157*, 1569.

DAVIDSON, R. G., CORTNER, J. A., RATTAZZI, M. C. RUDDLE, F. H. and LUBS, H. A. (1970). Genetic polymorphisms of human mitochondrial glutamic oxaloacetic transaminase. Science *169*, 391.

DAVIDSON, R. G., FILDES, R. A., GLEN-BOTT, A. M.., HARRIS, H., ROBSON, E. B. and CLEGHORN, T. E. (1965). Genetical studies on a variant of human lactate dehydrogenase (subunit A). Ann. Hum. Genet. Lond. *29*, 5.

DAVIDSON, R. G., NITOWSKY, J. M. and CHILDS, B. (1963). Demonstration of two populations of cells in the human female heterozygous for glucose-6-phosphate dehydrogenase variants. Proc. Natl. Acad. Sci. U.S. *50*, 481.

DAVIES, R. O., MARTON, A. V. and KALOW, W. (1960). The action of normal and atypical cholinesterase of human serum upon a series of esters of choline. Can. J. Biochem. Physiol. *38*, 545.

DAWSON, G. and STEIN, A. D. (1970). Lactosylceramidosis: catabolic enzyme defect of glycosphingolipid metabolism. Science *170*, 556.

DEAN, G. (1963). The porphyrias: a story of inheritance and environment. Pitman, London.

DEAN, G. (1969). The porphyrias. Brit. Med. Bull. *25*, 48.

DE DUVE, C. (1963). The lysosome concept. *In:* CIBA Foundation Symposium on Lysosomes. Churchill, London.

DE JONG, W. W. W., WENT, L. N. and BERNINI, L. F. (1968). Haemoglobin Leiden: deletion of β6 or 7 glutamic acid. Nature (Lond.) *220*, 788.

DE LA LANDE, F. A. and BOETTCHER, B. (1969). Electrophoretic examination of human serum amylase isozymes. Enzymologia *37*, 335.

DENT, C. E. and PHILPOT, G. R. (1954). Xanthimuria, an inborn error (or deviation) of metabolism. Lancet *1*, 182.

DENT, C. E. and ROSE, G. A. (1951. Aminoacid metabolism in cystinuria. Quart. J. Med. *20*, 205.

DENT, C. E. and SENIOR, B. (1955). Studies on the treatment of cystinuria. Brit. J. Urol. *27*, 317.

DENT, C. E., SENIOR, B. and WALSHE, J. M. (1954). The pathogenesis of cystinuria. II. Polarographic studies of the metabolism of sulphur-containing aminoacids. J. Clin. Invest. *33*, 1216.

DERN, R. J. (1966). A new hereditary quantitative variant of G-6-PD characterised by a marked increase in enzyme activity. J. Lab. Clin. Med. *68*, 560.

DERN, R. J., MCCURDY, P. R. and YOSHIDA, A. (1969). A new structural variant of glucose-6-phosphate dehydrogenase with a high production rate (G6PD Hektoen). J. Lab. Clin. Med. *73*, 283.

DERN, R. J., WEINSTEIN, I. M., LEROY, G. V., TALMAGE, D. W. and ALVING, A. S. (1954). The hemolytic effect of primaquine. I. The localization of the drug-induced hemolytic defect in primaquine-sensitive individuals. J. Lab. Clin. Med. *43*, 303.

DETTER, J. C., WAYS, P. O., GIBLETT, E. R., BAUGHAN, M. A., HOPKINSON, D. A., POVEY, S. and HARRIS, H. (1968). Inherited variations in human phosphohexose isomerase. Ann. Hum. Genet. Lond. *31*, 329.

DEVADATTA, S., GANGADHARAM, P. R., ANDREWS, R. H., FOX, W., REMAKRISHNAN, C. V., SELKON, J. B. and VELU, S. (1960). Peripheral neuritis due to isoniazid. Bull. World Health Org. *23*, 587.

DE VERDIER, C.-H. and GARBY, L. (1969). Low binding of 2,3-diphosphoglycerate to haemoglobin F. A contribution to the knowledge of the binding site and explanation for the high oxygen affinity of foetal blood. Scand. J. Clin. Lab. Invest. *23*, 149.

DINGLE, J. T. and ,ELL, H. B. (1969). Lysosomes in Biology and Pathology. North-Holland, Amsterdam.

DE SANT'AGNESE, P. A., ANDERSEN, D. H. and MASON, H. (1950). Glycogen storage disease of the heart. II. Critical review of the literature. Pediatrics *6*, 607.

DIXON, G. H. (1966). Mechanisms of protein evolution. Essays in Biochemistry *2*, 147. Academic Press, New York.

DOENICKE, A., GARTNER, T., KRENZBERG, G., REMES, I., SPIESS, W. and STEINBEREITHNER, K. (1963). Serum cholinesterase anenzymia. Acta Anaesthesiol. Scand. *7*, 59.

DONALD, L. J. and ROBSON, E. B. (1973). Rare variants of placental alkaline phosphatase. Ann. Hum. Genet. Lond. *37*, 303.

DONNELL, G. N., BERGREN, W. R., BRETTHAUER, R. K. and HANSEN, R. G. (9160). The enzymatic expression of heterozygosity in families of children with galactosemia. Pediatrics *25*, 572.

DOOLAN, P. D., HARPER, H. A., HUTCHIN, M. E. and ALPEN, E. L. (1957). Renal clearance of lysine in cystinuria. Am. J. Med. *23*, 416.

DORFMAN, A. and MATALON, R. (1972). The mucopolysaccharidoses. P. 1218 In: The metabolic basis of inherited disease. Eds. J. B. Stanbury, J. B. Wyngaarden and D. S. Fredrickson. 3rd Edition: McGraw Hill, New York.

DOUGLAS, G. R., MCALPINE, P. J. and HAMERTON, J. L. (1973). Regional localisation of loci for human *PGM*₁ and *6PGD* on human chromosome one by use of hybrids of chinese hamster-human somatic cells. Proc. Natl. Acad. Sci. U.S. *70*, 2737.

DOWDLE, E. B., MUSTARD, P. and EALES, L. (1967). δ-aminolaevulinic acid synthetase activity in normal and porphyric human livers. S. African Med. J. *41*, 1093.

DREYER, W. J. and BENNETT, J. C. (1965). The molecular basis of antibody formation: a paradox. Proc. Natl. Acad. Sci. U.S. *54*, 864.

DUBS, R., STEINMANN, B. and GITZELMANN, R. ((1973). Demonstration of an inactive enzyme antigen in sucrase-isomaltase deficiency. Helv. Paediat. Acta., *28*, 187.

DUFOUR, A. P., KNIGHT, R. A. and HARRIS, H. W. (1964). Genetics of isoniazid metabolism in Caucasian, Negro and Japanese populations. Science *145*, 391.

DUMA, H., EFREMOV, G., SADIKARIO, A., TEODOSIJEV, D., MLADENOVSKI, B., VLASKI, R. and ANDREEVA, M. (1968). Study of nine families with haemoglobin-Lepore. Brit. J. Haematol. *15*, 161.

DURAND, P., BARRONE, C. and DELLA CELLA, G. (1969). Fucosidosis. J. Paediat. *75*, 665.

EDELMAN, G. M. and GALLY, J. A. (1967). Somatic recombination of duplicated genes, an hypothesis on the origin of antibody diversity. Proc. Natl. Acad. Sci. U.S. *57*, 353.

EDELSTEIN, S. J., TELFORD, J. N. and CREPEAU, R. H. (1973). Structure of fibers of sickle cell hemoglobin. Proc. Natl. Acad. Sci. U.S. *70*, 1104.

EDINGTON, G. M. and LEHMANN, H. (1954). Haemoglobin G: a new haemoglobin found in a West African. Lancet *2*, 173.

EDINGTON, G. M. and LEHMANN, H. (1955). Expression of the sickle cell gene in Africa. Brit. Med. J. *1*, 1308, and *2*, 1328.

EDINGTON, G. M. and WATSON-WILLIAMS, E. J. (1964). Sickling, haemoglobin C, glucose-6-phosphate dehydrogenase deficiency and malaria in Western Nigeria. *In:* Abnormal haemoglobins in Africa, ed. J. H. P. Jonxis. Blackwell, Oxford.

EDWARDS, Y. H., EDWARDS, P. A. and HOPKINSON, D. A. (1973). A trimeric structure for mammalian purine nucleoside phosphorylase. FEBS Letters *32*, 235.

EDWARDS, Y. H., HOPKINSON, D. A. and HARRIS, H. (1971). Inherited variants of human nucleoside phosphorylase. Ann. Hum. Genet. Lond. *34*, 395.

EFRON, M. L. (1965). Familial hyperprolinemia. Report of a second case, associated with congenital renal malformations, hereditary hematuria and mild mental retardation, with demonstration of an enzyme defect. New Engl. J. Med. *272*, 1243.

EFRON, M. L. (1965). Hydroxyprolinemia. II. A rare metabolic disease due to a deficiency of the enzyme 'hydroxyproline oxidase'. New Engl. J. Med. *272*, 1299.

ELDJARN, L., JELLUM, E., STOKKE, O., PANDE, H. and WALLER, P. E. (1970). β-hydroxyisovaleric aciduria & β-methylcrotonylglycinuria: a new inborn error of metabolism. *Lancet 2*, 521.

ELDER, G. H., GRAY, C. H. and NICHOLSON, D. C. (1972). The porphyrias: a review. J. Clin. Pathol. *25*, 1013.

ELLARD, G. A., ABER, V. R., GAMMON, P. T., MITCHISON, D. A., LAKSHMINARAYAN, S., CITRON, K. M., FOX, W. and TALL, R. (1972). Pharmacology of some slow release preparations of isoniazid of potential use in intermittent treatment of tuberculosis. Lancet *1*, 340.

ELLARD, G. A., GAMMON, P. T., HELMY, H. S. and REES, R. J. W. (1972). Dapsone acetylation and the treatment of Leprosy. Nature (Lond.) *239*, 159.

ENCKLEWITZ, M. and LASKER, M. (1935). The origin of L-xyloketose (urine pentose). J. Biol. Chem. *110*, 443.

ENGLEMAN, K., WATTS, R. W. E., KLINENBERG, J. R., SJOERDSMA, A. and SEEGMILLER, J. E. (1964). Clinical, physiological and biochemical studies of a patient with xanthinuria and pheochromocytoma. Am. J. Med. *37*, 839.

EPSTEIN, C. J. and SCHECHTER, A. N. (1968). An approach to the problem of conformational isozymes. Ann. New York Acad. Sci. *151*, 85.

ERIKSSON, S. (1964). Pulmonary emphysema and alpha$_1$-antitrypsin deficiency. Acta. Med. Scand. *175*, 197.

ERIKSSON, S. (1965). Studies in α_1-antitrypsin deficiency. Acta. Med. Scand. *Suppl. 432*.

EVANS, D. A. P. (1963). Pharmacogenetics. Am. J. Med. *34*, 639.

EVANS, D. A. P. and CLARKE, C. A. (1961). Pharmacogenetics. Brit. Med. Bull. *17*, 234.

EVANS, D. A. P., DAVIDSON, K. and PRATT, R. T. C. (1965). The influence of acetylator phenotype on the effect of treating depression with phenelzine. Clin. Pharmacol. Therap. *6*, 430.

EVANS, D. A. P., MANLEY, K. E. and MCKUSICK, V. A. (1960). Genetic control of isoniazid metabolism in man. Brit. Med. J. *2*, 485.

EVANS, D. A. P., STOREY, P. B. and MCKUSICK, V. A. (1961). Further observations on the determination of the isoniazid inactivator phenotype. Bull. Johns Hopkins Hosp. *108*, 60.

EVANS, D. A. P. and WHITE, T. A. (1964). Human acetylation polymorphism. J. Lab. Clin. Med. *63*, 394.

EVANS, F. T., GRAY, P. W. S., LEHMANN, H. and SILK, E. (1952). Sensitivity to succinyl-choline in relation to serum cholinesterase. Lancet *1*, 1229.

EZE, L. C., TWEEDIE, M. C. K., BULLEN, M. F., WREN, P. J. J. and EVANS, D. A. P. (1973). Quantitative genetics of human red-cell acid phosphatase. Ann. Hum. Genet. Lond. *37*, 333.

FAGERHOL, M. K. (1967). Serum Pi types in Norwegians. Acta Pathol. Microbiol. Scand. *70*, 421.

FAGERHOL, M. K. (1968). The Pi system: genetic variants of serum α_1 antitrypsin. Series Haematologica *1*, 153.

FAGERHOL, M. K. (1969). Quantitative studies on the inherited variants of serum α-antitrypsin. Scand. J. Clin. Lab. Invest. *23*, 97.

FAGERHOL, M. K. and BRAEND, M. (1965). Serum prealbumin: polymorphism in man. Science *149*, 986.

FAGERHOL, M. K. and LAURELL, C.-B. (1967). The polymorphism of 'prealbumins' and α_1-antitrypsin in human sera. Clin. Chim. Acta *16*, 199.

FAGERHOL, M. K. and LAURELL, C.-B. (1970). The Pi system: inherited variant of serum

α_1-antitrypsin. *In:* Progress in Medical Genetics 7, ed. by A. G. Steinberg and A. G. Bearn.

FALLON, H. J., SMITH, L. H., GRAHAM, J. B. AND BURNETT, C. H. (1964). A genetic study of hereditary orotic aciduria. New Engl. J. Med. *270*, 878.

FELLMAN, J. H., VANBELLINGHEN, P. J., JONES, R. T. and KOLER, R. D. (1969). Soluble and mitochondrial forms of tyrosine aminotransferase. Relationship to human tyrosinemia. Biochemistry *8*, 615.

FELLOUS, M., BENGTSSON, B., FINNEGAN, D. and BODMER, W. F. (1973). Expression of the Xg^a antigen on cells in culture and its segregation in somatic cell hybrids. Ann. Hum. Genet. Lond. *37*, (in press).

FESSAS, P. and STAMATOYANNOPOULOS, G. (1964). Hereditary persistence of foetal haemoglobin in Greece. A study and comparison. Blood *24*, 223.

FILDES, R. A. and HARRIS, H. (1966). Genetically determined variation of adenylate kinase in man. Nature (Lond.) *209*, 261.

FILDES, R. A. and PARR, C. W. (1963). Human red cell phosphogluconate dehydrogenase. Nature (Lond.) *200*, 890.

FIGURA, K. and KRESSE, H. (1972). The Sanfilippo B corrective factor: An N-acetyl-α-D-glucosaminidase. Biochem. Biophys. Res. Comm. *48*, 262.

FINCH, J. T., PERUTZ, M. F., BERTLES, J. F. and DOBLER, J. (1973). Structure of sickled erythrocytes and of sickle-cell hemoglobin fibers. Proc. Natl. Acad. Sci. U.S. *70*, 718.

FINKELSTEIN, J. D., MUDD, S. H., IRREVERRE, F. and LASTER, L. (1966). Deficiencies of cystathionase and homoserine dehydratase activities in cystathioninuria. Proc. Natl. Acad. Sci. U.S. *55*, 865.

FISHER, R. A. (1930). The genetical theory of natural selection. Clarendon Press, Oxford.

FISHER, R. A. and HARRIS, H. (1969). Studies on the purification and properties of the genetic variants of red cell acid phosphohydrolase in man. Ann. New York Acad. Sci. *166* 380.

FISHER, R. A. and HARRIS, H. (1971). Studies on the separate isozymes of red cell acid phosphatase phenotypes A and B. I. Chromatographic separation of the isozymes. Ann. Hum. Genet. Lond. *34*, 431.

FISHER, R. A. and HARRIS, H. (1971). Studies on the separate isozymes of red cell acid phosphatase phenotypes A and B. II. Comparison of kinetics and stabilities of the isozymes. Ann. Hum. Genet. Lond. *34*, 439.

FISHER, R. A. and HARRIS, H. (1971). Further studies on the molecular size of red cell acid phosphatase. Ann. Hum. Genet. Lond. *34*, 449.

FISHER, R. A. and HARRIS, H. (1972). 'Secondary' isozymes derived from the three *PGM* loci. Ann. Hum. Genet. Lond. *36*, 69.

FISHER, R. A., TURNER, B. M., DORKIN, H. L. and HARRIS, H. (1974). Studies on human erythrocyte inorganic pyrophosphatase. Ann. Hum. Genet. Lond. *37*, 341.

FITCH, W. M. (1971). The nonidentity of invariable positions in the cytochromes *c* of different species. Biochem. Genet. *5*, 231.

FITCH, W. M. (1973). Are human hemoglobin variants distributed randomly among the positions? J. Molec. Evolution *2*, 181.

FITCH, W. M. and MARGOLIASH, E. (1967). Construction of phylogenetic trees. Science *155*, 279.

FITCH, W. M. and MARKOWITZ, E. (1970). An improved method for determining codon variability in a gene and its application to the rate of fixation of mutations in evolution. Biochem. Genet. *4*, 579.

FLATZ, G., KINDERLERER, J. L., KILMARTIN, J. V. and LEHMANN, H. (1971). Haemoglobin Tak: a variant with additional residues at the end of the β-chains. Lancet *1*, 732.

FÖLLING, A. (1934). Über Ausscheidung von Phenylbrenztraubensäure in den Harn als Stoffwechselanomalie in Verbindung mit Imbezillität. Hoppe-Seyelers Z. Physiol. Chem. *227*, 169.

FORBES, G. B. (1953). Glycogen disease. Report of a case with abnormal glycogen structure in liver and skeletal muscle. J. Pediat. *42*, 645.

FORGET, B. G., BENZ, E. J., SKOULTCHI, A., BAGLIONI, C. and HOUSMAN, D. (1974). Absence of messenger RNA for beta globin chain in $β^0$-thalassaemia. Nature (Lond.) *247*, 379.

FOX, R. M., O'SULLIVAN, W. J. and FIRKIN, B. G. (1969). Orotic aciduria. Differing enzyme patterns. Am. J. Med. *47*, 332.

FRASER, D. (1957). Hypophosphatasia. Am. J. Med. *22*, 730.

FRASER, D., KOOH, S. W., KIND, H. P., HOLICK, M. F., TANAKA, Y. and DELUCA, H. F. (1973). Vitamin-D dependent rickets: an inborn error of vitamin D metabolism. New Engl. J. Med. *289*, 817.

FRATANTONI, J. C., HALL, C. W. and NEUFELD, E. F. (1968). Hurler and Hunter syndromes: mutual correction of the defect in cultured fibroblasts. Science *162*, 570.

FRATANTONI, J. C., HALL, C. W. and NEUFELD, E. F. (1968). The defect in Hurler's and Hunter's syndromes: faulty degradation of mucopolysaccharide. Proc. Natl. Acad. Sci. U.S. *60*, 699.

FRATANTONI, J. C., HALL, C. W. and NEUFELD, E. F. (1969). The defect in Hurler and Hunter syndromes: II Deficiency of specific factors involved in mucopolysaccharide degradation. Proc. Natl. Acad. Sci. U.S. *64*, 360.

FREDRICKSON, D. S. (1966). Sphingomyelin lipidosis: Niemann-Pick disease. *In:* The metabolic basis of inherited disease, ed. J. B. Stanbury, J. B. Wyngaarden and D. S. Fredrickson. 2nd ed. McGraw-Hill, New York.

FREDRICKSON, D. S. and SLOAN, H. R. (1972). Glucosyl ceramide lipidoses: Gaucher's disease. P. 730 *In:* The metabolic basis of inherited disease, eds. J. B. Stanbury, J. B. Wyngaarden and D. S. Fredrickson. 3rd Edit. McGraw Hill, New York.

FREEMAN, J. M., NICHOLSON, J. F., SCHIMKE, R. T., ROWLAND, L. P. and CARTER, S. (1970). Congenital hyperammonemia: association with hyperglycinemia and decreased levels of carbamyl phosphate synthetase. Arch. Neurol. *23*, 430.

FRICK, P. G., HITZIG, W. H. and BETKE, K. (1962). Hemoglobin Zurich I. A new hemoglobin anomaly associated with acute hemolytic episodes with inclusion bodies after sulfonamide therapy. Blood *20*, 261.

FRIEDMAN, P. A., FISHER, D. B., KANG, E. S. and KAUFMAN, S. (1973). Detection of hepatic phenylalanine 4-hydroxylase in classical phenylketonuria. (genetic disease/tetrahydrobiopterin/lysolecithin). Proc. Natl. Acad. Sci. U.S. *70*, 552.

FRIMPTER, G. W. (1965). Cystathioninuria: nature of the defect. Science *149*, 1095.

FRIMPTER, G. W. (1967). Cystathioninuria. *In:* Aminoacid metabolism and genetic variation, ed. W. L. Nyhan. McGraw-Hill, New York.

FRIMPTER, G. W., HAYMOVITZ, A. and HORWITH, M. (1963). Cystathioninuria. New Engl. J. Med. *268*, 333.

FRISCH, E. R. (1972). Essential fructosuria and hereditary fructose intolerance. *In:* The metabolic basis of inherited disease, eds. J. B. Stanbury, J. B. Wyngaarden and D. S. Fredrickson. 3rd Edit. McGraw Hill, New York.

FRITZ, P. J., VESELL, E. S., WHITE, E. L. and PRUITT, K. M. (1969). The roles of synthesis and degradation in determining tissue concentrations of lactate dehydrogenase-5. Proc. Natl. Acad. Sci. U.S. *62*, 558.

FRITZ, P. J., WHITE, L. E., VESELL, E. L. and PRUITT, K. M. (1971). New theory of the control of protein concentrations in animal cells. Nature New Biol. *230*, 119.

FROESCH, E. R., PRADER, A., LABHART, A., STUBER, H. W. and WOLF, H. P. (1957). Die hereditäre Fructoseintoleranz, einer bisher nicht bekannte kongenitaler Stoffwechselstörung. Schweiz. Med. Wschr. *87*, 1168.

FROESCH, E. R., WOLF, H. P., BAITSCH, H., PRADER, A. and LABHART, A. (1963). Hereditary fructose intolerance. An inborn defect of hepatic fructose-1-phosphate splitting aldolase. Am. J. Med. *34*, 151.

FUDENBERG, H. H. and WARNER, N. L. (1970). Genetics of immunoglobulins. Advances in Human Genetics *1*, 131. Eds. H. Harris and K. Hirschhorn, Plenum Press.

FUKUI, M. and WAKASUGI, C. (1972). Liver alcohol dehydrogenase in a Japanese population. Jap. J. Legal. Med. *26*, 46.

FUNAKOSHI, S. and DEUTSCH, H. F. (1969). Human carbonic anhydrases. II Some physicochemical properties of native isozymes and of similar isozymes generated *in vitro*. J. Biol. Chem. *244*, 3438.

FUNAKOSHI, S. and DEUTSCH, H. F. (1971). Human carbonic anhydrases. VI. Levels of isozymes in old and young erythrocytes and in various tissues. J. Biol. Chem. *246*, 1088.

GAFFNEY, P. J. and LEHMANN, H. (1969). Residual enzyme activity in the serum of a homozygote for the silent pseudocholinesterase gene. Human Heredity *19*, 234.

GALLY, J. A. and EDELMAN, G. M. (1972). The genetic control of immunoglobulin synthesis. Ann. Rev. Genet. *6*, 1.

GARROD, A. E. (1909). Inborn errors of metabolism. Oxford University Press.

GARTLER, S. M., CHEN, S-H., FIALKOW, P. J., GIBLETT, E. R. and SINGH, S. (1972). X chromosome inactivation in cells from an individual heterozygous for two X-linked genes. Nature New Biol. *236*, 149.

GARTLER, S. M., LISKAY, R. M., CAMPBELL, B. K., SPARKES, R. and GANT, N. (1972). Evidence for two functional X-chromosomes in human oocytes. Cell Differentiation *1*, 215.

GARTLER, S. M., SCOTT, R. C., GOLDSTEIN, J. L., CAMPBELL, B. and SPARKES, R. (1971). Lesch-Nyhan syndrome: rapid detection of heterozygotes by use of hair follicles. Science *172*, 572.

GASKILL, P. and KABAT, D. (1971). Unexpectedly large size of globin messenger ribonucleic acid. Proc. Natl. Acad. Sci. U.S. *68*, 72.

GAULL, G., RASSIN, D. K. and STURMAN, J. A. (1969). Enzymatic and metabolic studies of homocystinuria: effects of pyridoxine. Neuropediatrie *1*, 199.

GELBER, R., PETERS, J. H., GORDON, G. R., GLAZKO, A. J. and LEVY, L. (1971). The polymorphic acetylation of dapsone in man. Clin. Pharmacol. Therap. *12*, 225.

GELEHRTER, T. D. and SNODGRASS, P. J. (1974). Lethal neonatal deficiency of carbamyl phosphate synthetase. New Engl. J. Med. *290*, 430.

GERALD, P. S. and DIAMOND, L. K. (1958). A new hereditary haemoglobinopathy (the Lepore trait) and its interaction with the thalassaemia trait. Blood *12*, 835.

GERALD, P. S. and EFRON, M. L. (1961). Chemical studies of several varieties of Hb-M. Proc. Natl. Acad. Sci. U.S. *47*, 1758.

GERRITSEN, T. and WAISMAN, H. A. (1964a). Homocystinuria: absence of cystathionine in the brain. Science *145*, 588.

GERRITSEN, T. and WAISMAN, H. A. (1964b). Homocystinuria: an error in the metabolism of methionine. Pediatrics *33*, 413.

GHANDIMI, H. and PARTINGTON, M. W. (1967). Salient features of histidinemia. Am. J. Diseases Children *113*, 83.

GHANDIMI, H., PARTINGTON, M. W. and HUNTER, A. (1961). A familial disturbance of histidine metabolism. New Engl. J. Med. *265*, 221.

GIBLETT, E. R. (1962). The plasma transferrins. *In:* Progress in medical genetics, Vol. 2, ed. A. G. Bearn and A. G. Steinberg. Grune and Stratton, New York.

GIBLETT, E. R. (1964). Variant haptoglobin phenotypes. Cold Spring Symp. Quant. Biol. *29*, 321.

GIBLETT, E. R. (1968). The haptoglobin system. Series Haematol. *1*, 3.

GIBLETT, E. R. (1969). Genetic markers in human blood. Blackwell, Oxford.

GIBLETT, E. R., ANDERSON, J. E., COHEN, F., POLLARA, B. and MEUWISSEN, H. J. (1972). Adenosine-deaminase deficiency in two patients with severely impaired cellular immunity. Lancet *II*, 1067.

GIBLETT, E. R., ANDERSON, J. A., CHEN, S.-H., TENG, Y.-S. and COHEN, F. (1974). Uridine monophosphate kinase: a new genetic polymorphism with possible clinical implications. Am. J. Hum. Genet. *26*, 627.

GIBLETT, E. R. and SCOTT, N. M. (1965). Red cell acid phosphatase: racial distribution and report of a new phenotype. Am. J. Hum. Genet. *17*, 425.

GIBLETT, E. R. and STEINBERG, A. G. (1960). The inheritance of serum haptoglobin types of American Negroes: evidence for a third allele Hp^{2M}. Am. J. Hum. Genet. *12*, 160.

GIBSON, Q. (1948). The reduction of methaemoglobin in red blood cells and studies on the cause of idiopathic methaemoglobinaemia. Biochem. J. *42*, 13.

GILBERT, W. and MÜLLER-HILL, B. (1966). Isolation of the lac repressor. Proc. natl. Acad. Sci. U.S. *56*, 1891.

GILLES, H. M., FLETCHER, K. A., HENDRICKSE, R. G., LINDNER, R., REDDY, S. and ALLAN, N. (1967). Glucose-6-phosphate dehydrogenase deficiency, sickling, and malaria in African children in South Western Nigeria. Lancet *1*, 138.

GITZELMANN, R. (1967). Hereditary galactokinase deficiency, a newly recognised cause of juvenile cataracts. Pediat. Res. *1*, 14.

GITZELMANN, R., CURTIUS, H.-C. and SCHNELLER, I. (1967). Galactitol and galactose-1-phosphate in the lens of a galactosemic infant. Exptl. Eye Res. 6, 1.

GJONE, E. nad NORUM, K. R. (1968). Familial serum cholesterol ester deficiency. Acta Med. Scand. *183*, 107.

GOEDDE, H. W., GEHRING, D. and HOFMANN, R. A. (1965). On the problem of a 'silent gene' in pseudocholinesterase polymorphism. Biochim. Biophys. Acta *107*, 391.

GOEDDE, H. W. and KELLER, W. (1967). Metabolic pathways in maple syrup urine disease. *In:* Amino acid metabolism and genetic variation, ed. W. L. Nyhan. McGraw-Hill, New York.

GOEDDE, H. W., LANGENBECK, V., BRACKERZ, D., KELLER, W., ROKKONES, T., HALVORSEN, S., KIIL, R. and MERTON, B. (1970). Clinical and biochemical-genetical aspects of intermittent branched-chain ketoaciduria. Acta Paed. Scand. *59*, 83.

GOEDDE, H. W., RICHTER, E., HÜFNER, M. and VON ZUR MÜHLEN, A. (1964a). Untersuchungen zur Ahornsirupkrankheit an zwei Familien. Humangenetik *1*, 163.

GOEDDE, H. W., RICHTER, E., HÜFNER, M. and SIXEL, B. (1964b). Arbeitsvorschrift eines vereinfachten Heterozygotentestes für die Ahornsirup-Krankheit. Klin. Wschr. *15*, 818.

GOLD, R. J. M., MAAG, V. R., NEAL, J. L. and SCRIVER, C. R. (1973). The use of biochemical data in screening for mutant alleles and in genetic counselling. Ann. Hum. Genet. (Lond). *37*, 315.

GOLDBERG, E. (1963). Lactic and malic dehydrogenases in human spermatozoa. Science *139*, 602.

GOLDBERG, E. (1965). Lactate dehydrogenases in spermatozoa: subunit interactions *in vitro*. Arch. Biochem. Biophys. *109*, 134.

GOLDSMITH, L., KANG, E., BIENFANG, D. and BADEN, H. (1972). Tyrosinemia with phenolicaciduria in the Richter-Hanhart syndrome. Amer. J. Hum. Genet. *24*, 25a.

GOLDSTEIN, J. L., CAMPBELL, B. K. and GARTLER, S. M. (1973). Homocystinuria: heterozygote detection using phytohemagglutinin-stimulated lymphocytes. J. Clin. Invest. *52*, 218.

GOMPERTZ, D., DRAFFAN, G. H., WATTS, J. L. and HULL, D. (1971). Biotin responsive β-methylcrotonyl-glycinuria. Lancet *2*, 22.

GOMPERTZ, D., GOODEY, P. A. and BARTLETT, K. (1973). Evidence for the enzymic defect in methylcrotonylglycinuria. Febs Letters *32*, 13.

GOMPERTZ, D., STORRS, L. N., BAU, D. C. K., PETERS, T. J. and HUGHES, E. A. (1970). Localisasation of enzymic defect in propionicacidemia. Lancet *1*, 1140.

GOODMAN, M., BARNABAS, J., MATSUDA, G. and MOORE, G. W. (1971). Molecular evolution in the descent of man. Nature *233*, 604.

GOODMAN, S. I., MCINTYRE, C. A. and O'BRIEN, D. (1967). Impaired intestinal transport of proline in a patient with familial iminoaciduria. J. Pediat. *71*, 246.

GRANICK, J. (1966). The induction *in vitro* of the synthesis of δ-aminolaevulinic acid synthetase in chemical porphyria: a response to certain drugs, sex hormones and foreign chemicals. J. Biol. Chem. *241*, 1359.

GREENE, M. L., BAYLES, J. R. and SEEGMILLER, J. E. (1970). Substrate stabilisation: genetically controlled reciprocal relationship of two human enzymes. Science *167*, 887.

GRIMES, A. J., MEISLER, A. and DACIE, J. V. (1964). Hereditary non-spherocytic haemolytic anaemia. A study of red cell carbohydrate metabolism in twelve cases of pyruvate kinase deficiency. Brit. J. Haematol. *10*, 403.

GRUBB, R. (1948). Correlation between Lewis blood group and secretor character in man. Nature (Lond.) *162*, 933.

GRUBB, R. (1951). Observations on the human group system Lewis. Acta Pathol. Microbiol Scand. *28*, 61.

GRUBB, R. (1956). Agglutination of erythrocytes coated with 'incomplete' anti-Rh by certain rheumatoid arthritis sera and some other sera. The existence of human serum groups. Acta Pathol. Microbiol. Scand. *39*, 195.

GRUBB, R. (1970). The genetic markers of human immunoglobulins. Chapman and Hall, London.

GRUBB, R. and LAURELL, A. B. (1956). Hereditary serological human serum groups. Acta Pathol. Microbiol. Scand. *39*, 390.

GRUMBACH, M. M., MARKS, P. A. and MOROSHIMA, A. (1962). Eryhrocyte glucose-6-phosphate dehydrogenase activity and X chromosome polysomy. Lancet *1*, 1330.

GUIDOTTI, G., KONIGSBERG, W. and CRAIG, L. C. (1963). On the dissociation of normal adult human haemoglobin. Proc. Natl. Acad. Sci. U.S. *50*, 774.

GUTSCHE, B. B., SCOTT, E. M. and WRIGHT, R. C. (1967). Hereditary deficiency of pseudo-cholinesterase in Eskimos. Nature (Lond.) *215*, 322.

HACKEL, E., HOPKINSON, D. A. and HARRIS, H. (1972). Population studies on mitochondrial glutamate-oxaloacetate transaminase. Ann. Hum. Genet. Lond. *35*, 491.

HADORN, B., TARLOW, M. J., LLOYD, J. K. and WOLFF, O. H. (1969). Intestinal enterokinase deficiency. Lancet *1*, 812.

HAHN, E. V. and GILLESPIE, E. B. (1927). Sickle-cell anaemia: report of a case greatly improved by splenectomy; experimental study of sickle-cell formation. Arch. Int. Med. *39*, 233.

HAIGH, J. and MAYNARD SMITH, J. (1972). Population size and protein variation in man. Genet. Res., *19*, 73.

HAKOMORI, S. and STRYCHARZ, G. D. (1968). Investigations on cellular blood-group substances. I. Isolation and chemical composition of blood-group ABH and Le^h iso-antigens of sphingoglycolipid nature. Biochem. *7*, 1279.

HALL, C. W., CANTZ, M. and NEUFELD, E. F. (1973). A β-glucuronidase deficiency mucopolysaccharidosis: studies in cultured fibroblasts. Arch. Bioch. Biophys. *155*, 32.

HALL CRAGGS, M., MARSDEN, P. D., RAPER, A. B., LEHMANN, H. and BEALE, D. (1964). Homozygous sickle cell anaemia arising from two different haemoglobins S. Brit. Med. J. *2*, 87.

HAMILTON, H. B., IUCHI, I., MIYAJI, T. and SHIBATA, S. (1969). Hemoglobin Hiroshima (β^{143}-Histidine aspartic acid): a newly identified fast moving beta chain variant associated with increased oxygen affinity and compensatory erythremia. J. Clin. Invest. *48*, 525.

HAMNSTROM, B., GJONE, E. and NORUM, K. R. (1969). Familial plasma lecithin: cholesterol acyltransferase deficiency. Brit. Med. J. *2*, 283.

HANEL, H. K., COHN, J. and HARVALD, B. (1971). Adenosine-triphosphatase deficiency in a family with non-spherocytic haemolytic anaemia. Human Heredity *21*, 313.

HARRAP, G. J. and WATKINS, W. M. (1964). Characterisation of the enzyme from *T. foetus* that destroys the serological specificity of blood-group A substance. Biochem. J. *93*, 9P

HARRIES, J. T., PIESOWICZ, A. T., SEAKINS, J. W. T., FRANCIS, D. E. M. and WOLFF, O. W. (1971). Low proline diet in Type 1 hyperprolinemia. Arch. Dis. Child. *46*, 72.

HARRIS, H. (1966). Enzyme polymorphisms in man. Proc. Roy. Soc. B *164*, 298.

HARRIS, H. (1969). Genes and isozymes. Proc. Roy. Soc. B *174*, 1.

HARRIS, H. (1969). Enzyme and protein polymorphism in human populations. Brit. Med. Bull. *25*, 5. (New Aspects of Human Genetics).

HARRIS, H. and HOPKINSON, D. A. (1972). Average heterozygosity per locus in man: an estimate based on the incidence of enzyme polymorphism. Ann. Hum. Genet. Lond. *36*, 9.

HARRIS, H., HOPKINSON, D. A., LUFFMAN, J. E. and RAPLEY, S. (1968). Electrophoretic variation in red cell enzymes. *In:* Hereditary disorders of erythrocyte metabolism, ed. E. Beutler. Grune and Stratton, New York.

HARRIS, H., HOPKINSON, D. A. and ROBSON, E. B. (1962). Two dimensional electrophoresis of pseudo-cholinesterase components in normal human serum. Nature (Lond.) *196*, 1296.

HARRIS, H., HOPKINSON, D. A. and ROBSON, E. B. (1973). The incidence of rare alleles determining electrophoretic variants: data on 43 enzyme loci in man. Ann. Hum. Genet. Lond. *37*, 237.

HARRIS, H., HOPKINSON, D. A., ROBSON, E. B. and WHITTAKER, M. ó1963a). Genetical studies on a new variant of serum cholinesterase detected by electrophoresis. Ann. Hum. Genet. Lond. *26*, 359.

HARRIS, H., HOPKINSON, D. A., SPENCER, N., COURT-BROWN, W. M. and MANTLE, D. (1963b). Red cell glucose-6-phosphate dehydrogenase activity in individuals with abnormal numbers of X-chromosomes. Ann. Hum. Genet. Lond. *27*, 59.

HARRIS, H., MITTWOCH, U., ROBSON, E. B. and WARREN, F. L. (1955a). Pattern of aminoacid excretion in cystinuria. Ann. Hum. Genet. Lond. *19*, 196.

HARRIS, H., MITTWOCH, U., ROBSON, E. B. and WARREN, F. L. (1955b). Pehnotypes and genotypes in cystinuria. Ann. Hum. Genet. Lond. *20*, 57.

HARRIS, H., PENROSE, L. S. and THOMAS, D. H. H. (1959). Cystathioninuria. Ann. Hum. Genet. Lond. *23*, 442.

HARRIS, H. and ROBSON, E. B. (1959). A genetical study of ethanolamine phosphate excretion in hypophosphatasia. Ann. Hum. Genet. Lond. *23*, 421.

HARRIS, H. and ROBSON, E. B. (1963). Fractionation of human serum cholinesterase components by gel filtration. Biochim. Biophys. Acta *73*, 649.

HARRIS, H., ROBSON, E. B., GLÉN-BOTT, A. M. and THORNTON, J. A. (1963c). Evidence for non-allelism between genes affecting human serum cholinesterase. Nature (Lond.) *200*, 1185.

HARRIS, H., ROBSON, E. B. and SINISCALCO, M. (1958). Atypical segregation of haptoglobin types in man. Nature (Lond.) *182*, 1324.

HARRIS, H. and WARREN, F. L. (1953). Quantitative studies on the urinary cystine in patients with cystine stones and their relatives. Ann. Eugen. Lond. (now Ann. Hum. Genet.) *18*, 125.

HARRIS, H. and WHITTAKER, M. (1961). Differential inhibition of human serum cholinesterase with fluoride. Recognition of two new phenotypes. Nature (Lond.) *191*, 496.

HARRIS, H. and WHITTAKER, M. (1962). The serum cholinesterase variants: a study of twenty-two families selected via the 'intermediate' phenotype. Ann. Hum. Genet. Lond. *26*, 73.

HARRIS, H., WHITTAKER, M., LEHMANN, H. and SILK, E. (1960). The pseudocholinesterase variants. Esterase levels and dibucaine numbers in families selected through suxamethonium sensitive individuals. Acta Genet. Stat. Med. *10*, 1.

HARRIS, J. W. (1950). Studies on the destruction of red blood cells. VIII. Molecular orientation in sickle cell haemoglobin solutions. Proc. Soc. Exp. Biol. N.Y. *75*, 197.

HARRIS, S. R., PAGLIA, D. E., JAFFE, E. R., VALENTINE, W. N. and KLEIN, R. L. (1970). Triosephosphate isomerase deficiency in an adult. Clin. Res. *18*, 529.

HARTMAN, G. (1941). Group antigens in human organs. Munksgaard, Copenhagen.

HARVALD, B., HANEL, K. H., SQUIRES, R. and TRAP-JENSEN, J. (1964). Adenosine-triphosphatase deficiency in patients with non-spherocytic haemolytic anaemia. Lancet *1*, 18.

HAYASHI, A., STAMATOYANNOPOULOS, G., YOSHIDA, A. and ADAMSON, J. (1971). Haemoglobin Rainer: β145(HC2)tyrosine → cysteine and haemoglobin Bethesda: β145(HC2) tyrosine → histidine. Nature New Biol. *230*, 264.

HEARN, V. M., SMITH, Z. G. and WATKINS, W. M. (1968). An α-N-acetyl-D-galactosaminyl-transferase associated with the human blood-group A character. Biochem. J. *109*, 315.

HENDERSON, L. E., HENRIKSSON, D. and NYMAN, P. O. (1973). Amino acid sequence of human erythrocyte carbonic anhydrase C. Biochem. Biophys. Res. Commun. *52*, 1388.

HERBICH, J., FISHER, R. A. and HOPKINSON, D. A. (1970). Atypical segregation of human red cell acid phosphatase phenotypes: evidence for a rare 'silent' allele P^0. Ann. Hum. Genet. Lond. *34*, 145.

HERNDON, J. H. JR., STEINBERG, D., UHLENDORF, B. W. and FALES, H. M. (1969). Refsum's disease: characterization of the enzyme defect in cell culture. J. Clin. Invest. *48*, 1017.

HERRICK, J. B. (1910). Peculiar elongated and sickle shaped red corpuscles in a case of severe anaemia. Arch. Intern. Med. *6*, 517.

HERS, H, G. (1959). Etudes enzymatiques sur fragments hépatiques. Application à la classification des glycogenoses. Rev. Intern. Hépatol. *9*, 35.

HERS, H. G. (1962). α-glucosidase deficiency in generalised glycogen-storage disease (Pompe's disease). Biochem. J. *86*, 11.

HERS, H. G. (1965). Inborn lysosomal diseases. Gastroenterology *48*, 625.

HERS, H. G. and JOASSIN, G. (1961). Anomalie de l'aldolase hépatique dans 'intolerance' au fructose. Enzymol. Biol. Clin. *1*, 4.

HERS, H. G. and VAN HOOF, F. (Eds.) (1973). Lysosomes and storage diseases. Academic Press, New York.

HIATT, H. H. (1972). Pentosuria. *In:* The metabolic basis of inherited disease, eds. J. B. Stanbury, J. B. Wyngaarden and D. S. Fredrickson. 3rd Edition. McGraw-Hill, New York.

HIRSCHFELD, J. (1959). Immunoelectrophoretic demonstration of qualitative differences in normal human sera and their relation to haptoglobins. Acta Pathol. microbiol. Scand. *47*, 160.

HIRSCHFELD, J. (1962). The Gc system. Immunoelectrophoretic studies of normal human sera with special reference to a new genetically determined serum system (Gc). Progr. Allergy *6*, 155.

HIRSCHFELD, J. (1968). The Ag system – a comparison of different isoprecipitation sera. Series Haematologica *1*, 38.

HIRSCHFELD, J. and OKACHI, K. (1967). Distribution of Ag(x) and Ag(y) antigens in some populations. Vox Sang. *13*, 1.

HIRSCHHORN, R. and BERATIS, N. G. (1973). Severe combined immunodeficiency and adenosine-deaminase deficiency. Lancet *2*, 1217.

HOCKADAY, R. D. R., CLAYTON, J. E., FREDERICK, E. W. and SMITH, L. H. JR. (1964). Primary hyperoxaluria. Medicine *43*, 315.

HOCKWALD, R. S., ARNOLD, J., CLAYMAN, B. and ALVING, S. A. (1952). Toxicity of primaquine in Negroes, J. Am. Med. Ass. *149*, 1568.

HODGKIN, W. E., GIBLETT, E. R., LEVINE, H., BAUER, W. and MOTULSKY, A. G. (1965). Complete pseudocholinesterase deficiency: genetic and immunologie characterization. J. Clin. Invest. *44*, 486.

HOLLAN, S. R., SZELENYI, J. G., BRIMHALL, B., DUERST, M., JONES, R. T., KOLER, R. D. and STOCKLEN, Z. (1972). Multiple alpha chain loci for human haemoglobins: Hb J-Buda and Hb G-Pest. Nature (Lond.) *235*, 47.

HOLLENDER, A., LORKIN, P. A., LEHMANN, H. and SVENSSON, B. (1969). New unstable haemoglobin Borås: $\beta88$ (F4) leucine → arginine. Nature (Lond.) *222*, 953.

HOLTON, J. B. (1965). Skin L-histidine ammonia-lyase activity in the family of a child with histidinaemia. Clin. Chim. Acta *11*, 193.

HOLZEL, A. and KOMROVER, G. M. (1955). A study of the genetics of galactosaemia. Arch. Disease Childhood *30*, 155.

HOLZEL, A., SCHWARTZ, V. and SUTCLIFFE, K. W. (1959). Defective lactose absorption causing malnutrition in infancy. Lancet 1, 1126.

HOMMES, F. A., DE GROOT, C. J., WILMINK, C. W. and JONXIS, J. H. P. (1969). Carbamyl-phosphate synthetase deficiency in an infant with severe cerebral damage. Arch. Diseases Children *44*, 688.

HOMMES, F. A., KUIPERS, J. R. G., ELEMA, J. D., JANSEN, J. F. and JONXIS, J. H. P. (1968). Propionicacidemia, a new inborn error of metabolism. Pediat. Res. *2*, 519.

HOOD, L. and TALMAGE, D. W. (1970). Mechanism of antibody diversity: germ line basis for variability. Science *168*, 325.

HOOFT, C., CARTON, D. and SAMYN, W. (1971). Pyridoxine treatment in homocystinuria. Lancet *1*, 1384.

HOOK, E. B.., STAMATOYANNOPOULOS, G., YOSHIDA, A. and MOTULSKY, A. N. (1968). Glucose-6-phosphate dehydrogenase Madrona: a slow electrophoretic glucose-6-phosphate dehydrogenase variant with kinetic characteristics similar to those of normal type. J. Lab. Clin. Med. *72*, 404.

HOPKINS, I. J., CONNELLY, J. F., DAWSON, A. G., HIRD, F. J. R. and MADDISON, T. G. (1969). Hyperammonemia due to ornithine transcarbamylase deficiency. Arch. Diseases Children *44*, 143.

HOPKINSON, D. A., CORNEY, G., COOK, P. J. L., ROBSON, E. B. and HARRIS, H. (1970). Genetically determined electrophoretic variants of human red cell NADH diaphorase. Ann. Hum. Genet. Lond. *34*, 1.

HOPKINSON, D. A. and HARRIS, H. (1965). Evidence for a second 'structural' locus determining human phosphoglucomutase. Nature (Lond.) *208*, 410.

HOPKINSON, D. A. and HARRIS, H. (1966). Rare phosphoglucomutase phenotypes. Ann Hum. Genet. Lond. *30*, 167.

HOPKINSON, D. A. and HARRIS, H. (1968). A third phosphoglucomutase locus in man. Ann. Hum. Genet. Lond. *31*, 359.

HOPKINSON, D. A. and HARRIS, H. (1969). Red cell acid phosphatase, phosphoglucomutase and adenylate kinase. *In:* Biochemical methods in red cell genetics, ed. G. Yunis. Academic Press, New York.

HOPKINSON, D. A. and HARRIS, H. (1971). Recent work on isozymes in man. Ann. Rev. Genet. *5*, 5.

HOPKINSON, D. A., MESTRINER, M. A., CORTNER, J. and HARRIS, H. (1973). Esterase D: a new human polymorphism. Ann. Hum. Genet. Lond. *37*, 119.

HOPKINSON, D. A., PETERS, J. and HARRIS, H. (1974). Rare electrophoretic variants of glycerol-3-phosphate dehydrogenase: evidence for two structural gene loci (GPD_1 and GPD_2). Ann. Hum. Genet. Lond. *37*, 477.

HOPKINSON, D. A., SPENCER, N. and HARRIS, H. (1963). Red cell acid phosphatase variants; a new human polymorphism. Nature (Lond.) *199*, 969.

HOPKINSON, D. A., SPENCER, N. and HARRIS, H. (1964). Genetical studies on human red cell acid phosphatase. Am. J. Hum. Genet. *16*, 141.

HÖRLEIN, H. and WEBER, G. (1948). Über chronische familiäre Methämoglobinämia und eine neue Modifikation des Methämoglobins. Deutsch. Med. Wschr. *73*, 476.

HORNER, F. A., STREAMER, C. W., ALEJANDRINO, L. L., REED, L. H. and IBBOTT, F. (1962). Termination of dietary treatment of phenylketonuria. New Engl. J. Med. *266*, 79.

HORTON, B. F. and HUISMAN, T. H. J. (1963). Linkage of the β-chain and γ-chain structural genes of human haemoglobins. Am. J. Hum. Genet. *15*, 394.

HOUSMAN, D., FORGET, B. G., SKOULTCHI, A. and BENZ, E. J. (1973). Quantitative deficiency of chain-specific globin messenger ribonucleic acids in the thalassemia syndromes. Proc. Natl. Acad. Sci. U.S. *70*, 1809.

HOWELL, R., KLINENBERG, J. R. and KROOTH, R. S. (1967). Enzyme studies on diploid cell strains developed from patients with hereditary orotic aciduria. Johns Hopkins Med. J. *120*, 81.

HOWELL, R. R. and STEVENSON, R. E. (1971). The offspring of phenylketonuric women. Soc. Biol. *18*, 519.

HSIA, D. Y. Y. (1970), Phenylketonuria and its variants. Progress in Medical Genetics 7, 29. Grune and Stratton, New York.

HSIA, D. Y. Y., PAINE, R. S. and DRISCOLL, K. W. (1957). Phenylketonuria: Detection of the heterozygous carrier. J. Mental Deficiency Res. *1*, 53.

HSIA, D. Y. Y., DRISCOLL, K., TROLL, W. and KNOX, W. E. (1956). Detection by phenylalanine tolerance tests of heterozygous carriers of phenylketonuria. Nature (Lond.) *178*, 1239.

HSIA, Y. E., SCULLY, K. J. and ROSENBERG, L. E. (1969). Defective propionate carboxylation in ketotic hyperglycinaemia. Lancet *1*, 757.

HSIA, Y. E., SCULLY, K. J. and ROSENBERG, L. E. (1971). Inherited propionyl-CoA carboxylase deficiency in 'ketotic hyperglycinemia'. J. Clin. Invest., *50*, 127.

HUBBY, J. L. and LEWONTIN, R. C. (1966). A molecular approach to the study of genic heterozygosity in natural populations. I. The number of alleles at different loci in Drosophila pseudoobscura. Genetics *54*, 577.

HUEHNS, E. R., DANCE, N., BEAVEN, G. H., HECHT, F. and MOTULSKY, G. (1964). Human embryonic haemoglobins. Cold Spring Harbor Symp. Quant. Biol. *29*, 327.

HUEHNS, E. R., DANCE, N., BEAVEN, G. H., KEIL, J. V., HECHT, F. and MOTULSKY, A. G. (1964). Human embryonic haemoglobins. Nature (Lond.) *201*, 1095.

HUEHNS, E. R. and MODELL, C. B. (1967). Haemoglobin synthesis in thalassaemia. Trans. Roy. Soc. Trop. Med. Hyg. *61*, 157.

HUG, G., SCHUBERT, W. K. and CHUCK, G. (1966). Phosphorylase kinase of the liver: deficiency in a girl with increased hepatic glycogen. Science *153*, 1534.

HUGHES, H. P., BIEHL, J. P., JONES, A. P. and SCHMIDT, L. H. (1954). Metabolism of isoniazid in man as related to the occurrence of peripheral neuritis. Am. Rev. Tuberc. *70*, 266.

HUGH-JONES, K., NEWCOMB, A. L. and HSIA, D. Y. Y. (1960). The genetic mechanism of galactosaemia. Arch. Disease Childhood *35*, 521.

HUGULEY, C. M. JR., BAIN, J A., RIVERS, S. and SCOGGINS, R. (1959). Refractory megaloblsatic anaemia associated with excretion of orotic acid. Blood *14*, 615.

HUIJING, F. (1967). Phosphorylase kinase in normal subjects and patients with glycogen storage disease. Biochem. Biophys. Acta. *148*, 601.

HUIJING, F., OBBINK, H. J. K. and VAN CREVELD, S. (1968). The activity of the debranching-enzyme system in leucocytes. A genetic study of glycogen storage disease type III. Acta Genet. Basel *18*, 128.

HUISMAN, T. H. J., BROWN, A. K., EFREMOV, G. D., WILSON, J. B., REYNOLDS, C. A., UY, R. and SMITH, L. L. (1971). Hemoglobin Savannah (B6(24)β glycine → valine): an unstable variant causing anemia with inclusion bodies. J. Clin. Invest. *50*, 650.

HUISMAN, T. J. H., SCHROEDER, W. A., DOZY, A. M., SHELTON, J. R., SHELTON, J. B., BOYD, E. M. and APELL, G. (1969). Evidence for multiple structural genes for the γ-chains of human fetal haemoglobin in hereditary persistence of fetal hemoglobin. Ann. New York Acad. Sci. *165*, 320.

HUISMAN, T. J. H., SCHROEDER, W. A., STAMATOYANNOPOULOS, G., BOUVER, N., SHELTON,

References

J. R., and APELL, G. (1970). Nature of fetal hemoglobin in the Greek type of hereditary persistence of fetal haemoglobin with and without concurrent β-thalassaemia. J. Clin. Invest. *49*, 1035.

HUISMAN, T. H. J., WRIGHTSTONE, R. N., WILSON, J. B., SCHROEDER, W. A. and KENDALL, A. G. (1972). Haemoglobin Kenya, the product of fusion of γ and β polypeptide chains. Arch. Biochem. Biophys. *153*, 850.

HUNT, J. A. and INGRAM, V. M. (1958). Allelomorphism and the chemical differences of the human haemoglobins A, S and C. Nature (Lond.) *184*, 1062.

HUNT, J. A. and LEHMANN, H. (1959). Haemoglobin Bart's: a foetal haemoglobin without α-chains. Nature (Lond.) *184*, 872.

HUNTER, R. L. and MARKERT, C. L. (1957). Histochemical demonstration of enzymes separated by zone electrophoresis in starch gels. Science *125*, 1294.

HUNTSMAN, R. G., HALL, M., LEHMANN, H. and SUKUMARAN, P. K. (1963). A second and a third abnormal haemoglobin in Norfolk. Brit. Med. J. *1*, 720.

HUTCHISON, D. C. S. (1973). Alpha-1-antitrypsin deficiency and pulmonary emphysema: the role of proteolytic enzymes and their inhibitors. Brit. J. Dis. Chest. *67*, 171.

HUTCHISON, J. H. and MCGIRR, E. M. (1956). Sporadic non-endemic goitrous cretinism. Lancet *1*, 1035.

ILLINGWORTH, B. and CORI, G. T. (1952). Structure of glycogens and amylopectins. III. Normal and abnormal human glycogen. J. Biol. Chem. *199*, 653.

ILLINGWORTH, B., CORI, G. T. and CORI, C. F. (1956). Amylo 1, 6 glucosidase activity in muscle tissue in generalised glycogen storage disease. J. Biol. Chem. *218*, 123.

IMAMURA, K. and TANAKA, T. (1972). Multimolecular forms of pyruvate kinase from rat and other mammalian tissues. I. Electrophoretic studies. J. Biochem. *71*, 1043.

IMAMURA, K., TANIUCHI, K. and TANAKA, T. (1972). Multimolecular forms of pyruvate kinase from Yoshida ascites 130 cells and comparative studies on the enzymological and immunological properties of the three types of pyruvate kinases, L, M_1, and M_2. J. Biochem., *72*, 1001.

IMAMURA, T., FUJITA, S., OHTA, Y., HANADA, M. and YANASE, T. (1969). Hemoglobin Yoshizuka (G10(108)β asparagine $\rightarrow$ aspartic acid): A new variant with a reduced oxygen affinity from a Japanese family. J. Clin. Invest. *48*, 2341.

INGRAM, V. M. (1957). Gene mutations in human haemoglobin: the chemical difference between normal and sickle cell haemoglobin. Nature (Lond.) *180*, 326.

INGRAM, V. M. (1959). Abnormal human haemoglobins. III. The chemical difference between normal and sickle cell haemoglobins. Biochim. Biophys. Acta *36*, 402.

INGRAM, V. M. (1961). Gene evolution and the haemoglobins. Nature (Lond.) *189*, 704.

INGRAM, V. M. and STRETTON, A. O. W. (1959). Genetic basis of the thalassemia diseases. Nature (Lond.) *184*, 903.

IRREVERRE, F., MUDD, S. H., HEIZER, W. D. and LASTER, L. (1967). Sulfite oxidase deficiency: studies of a patient with mental retardation, dislocated ocular lenses and abnormal urinary excretion of S-sulfo-L-cysteine, sulfite and thiosulfate. Biochem. Med., *1*, 187.

ISEKI, S., FUROKAWA, K. and YAMAMOTO, S. (1959). B substance-decomposing enzyme produced by an anaerobic bacterium. II. Chemical action of the B-decomposing enzyme. Proc. Japan Acad. *35*, 513.

ISEKI, S. and MASAKI, S. (1953). Transformation of blood group substance by bacterial enzyme. Proc. Japan Acad. *29*, 460.

ISSELBACHER, K. J., ANDERSON, E. P., KURAHASHI, K. and KALCKAR, H. M. (1956). Congenital galactosaemia: a single enzymatic block in galactose metabolism. Science *123*, 635.

ITANO, H. A. (1957). The human haemoglobins: their properties and genetic control. Advan. Protein Chem. *12*, 216.

ITANO, H. A. (1965). The synthesis and structure of normal and abnormal haemoglobins. *In:* Abnormal haemoglobins in Africa, ed. J. H. P. Jonxis. Blackwell, Oxford.

ITANO, H. A. (1966). Genetic regulation of peptide synthesis in haemoglobins. J. Cell Physiol. *67*, suppl. 1, 65.

ITANO, H. A. and NEEL, J. V. (1950). A new inherited abnormality of human haemoglobin. Proc. Natl. Acad. Sci. U.S. *36*, 613.

ITANO, H. A. and ROBINSON, E. A. (1960). Genetic control of the α-chain and β-chains of hemoglobin. Proc. Natl. Acad. Sci. U.S. *46*, 1492.

JACOB, F. and MONOD, J. (1961). Genetic regulatory mechanisms in the synthesis of proteins. J. Mol. Biol. *3*, 318.

JARKOVSKY, Z., MARCUS, D. M. and GROLLMAN, A. P. (1970). Fucosyltransferase found in human milk. Product of the Lewis blood group gene. Biochemistry *9*, 1123.

JATZKEWITZ, H. and MEHL, E. (1969). Cerebroside-sulphatase and arylsulphatase A deficiency in metachromatic leukodystrophy (ML). J. Neurochem. *16*, 19.

JENKINS, T. and CORFIELD, V. (1972). The red cell acid phosphatase polymorphism in Southern Africa: population data and studies on R, RA and RB phenotypes. Ann. Hum. Genet. Lond. *35*, 379.

JENNE, J. W. (1965). Partial purification and properties of the isoniazid trans-acetylase in the human liver. Its relationship to the acetylation of p-aminosalicylic. J. Clin. Invest. *44*, 1992.

JEPSON, J. B. (1966). Hartnup disease. *In:* The metabolic basis of inherited disease, ed. J. B. Stanbury, J. B. Wyngaarden and D. S. Fredrickson. McGraw-Hill, New York.

JERNE, N. K. (1971). The somatic generation of immune recognition. Eur. J. Immunol. *1*, 1.

JERVIS, G. A. (1950). Excretion of phenylalanine and derivatives in phenylpyruvic oligophrenia. Proc. Soc. Exptl. Biol. Med. *75*, 83.

JERVIS, G. A. (1953). Phenylpyruvic oligophrenia: deficiency of phenylalanine oxidising system. Proc. Soc. Exptl. Biol. Med. *82*, 514.

JERVIS, G. A. (1960). Detection of heterozygotes for phenylketonuria. Clin. Chim. Acta *5*, 471.

JOHNSON, A. M., SCHMID, K. and ALPER, C. A. (1969). Inheritance of human α_1-acid glycoprotein (oroseromucoid) variants. J. Clin. Invest., *48*, 2293.

JONES, R. T., SCHROEDER, W. A., BALOG, J. E. and VINOGRAD, J. R. (1959). Gross structure of hemoglobin H. J. Am. Chem. Soc. *81*, 3161.

JONES, R. T., BRIMHALL, B., HUISMAN, T. H. J., KLEIHAUER, E. and BETKE, K. (1966). Hemoglobin Freiburg: abnormal hemoglobin due to deletion of a single aminoacid residue. Science *154*, 1024.

JONGSMA, A., VAN SOMEREN, H., WESTERVELD, A., HAGEMEIJER, A. and PEARSON, P. (1973). Localization of genes on human chromosomes by studies of human-chinese hamster somatic cell hybrids. Humangenetik *20*, 195.

JUSTICE, P., O'FLYNN, M. E. and HSIA, D. Y. Y. (1967). Phenylalanine hydroxylase activity in hyperphenylalanemia. Lancet *1*, 928.

KABACK, M. M. and ZEIGER, R. S. (1972). Heterozygote detection in Tay-Sachs disease: a prototype community screening program for the prevention of recessive genetic disorders. *In:* Sphingolipids, spingolipidoses and allied disorders, Eds. B. W. Volk and S. M. Aronson. Plenum Press, New York.

KABAT, E. A. (1956). Blood group substances; their chemistry and immunochemistry. Academic Press, New York.

KABAT, E. A. and LESKOWITZ, S. (1955). Immunochemical studies on blood groups. XVII. Structural units involved in blood group A and specificity. J. Am. Chem. Soc. 77, 5159.

KACIAN, D. L., GAMBINO, R., DOW, L. W., GROSSBARD, E., NATTA, C., RAMIREZ, F., SPIEGELMAN, S., MARKS, P. A. and BANK, A. (1973). Decreased globin messenger RNA in thalassemia detected by molecular hybridization. Proc. Natl. Acad. Sci. U.S. 70, 1886.

KALCKAR, H. M., ANDERSON, E. P. and ISSELBACHER, K. J. (1956). Galactosemia, a congenital defect in a nucleotide transferase. Biochim. Biophys. Acta 20, 262.

KALOW, W. (1959). Cholinesterase types. In: Biochemistry of human genetics, Ciba Foundation Symp., ed. G. E. W. Wolstenholme and C M. O'Connor. Churchill, London.

KALOW, W. and DAVIES, R. O. (1958). The activity of various esterase inhibitors towards atypical human serum cholinesterase. Biochem. Pharmacol. 1, 183.

KALOW, W. and GENEST, K. (1957). A method for the detection of atypical forms of human serum cholinesterases. Determination of dibucaine numbers. Canad. J. Biochem.

KALOW, W. and STARON, N. (1957). On the distribution and inheritance of atypical forms of human serum cholinesterase as indicated by dibucaine numbers. Canad. J. Biochem. Physiol. 35, 1305.

KAMARYT, J. and LAXOVA, R. (1965). Amylase heterogeneity. Some genetic and clinical aspects. Humangenetik 1, 579.

KANG, E. S., SOLLEE, N. D. and GERALD, P. S. (1970). Results of treatment and termination of the diet in phenylketonuria. Pediatrics 46, 881.

KAPLAN, J.-C. and BEUTLER, E. (1967). Electrophoresis of red cell NADH- and NADPH-diaphorases in normal subjects and patients with congenital methemoglobinemia. Biochem. Biophys. Res. Commun. 29, 605.

KAPLAN, N. O. (1968). Nature of multiple molecular forms of enzymes. Ann. N.Y. Acad. Sci. 151, 382.

KAPLAN, N. O., EVERSE, J. and ADMIRAAL, J. (1968). Significance of substrate inhibition of dehydrogenases. Ann. N.Y. Acad. Sci. 151, 400.

KARN, R. C., SHULKIN, J. D., MERRITT, A. D. and NEWELL, R. C. (1973). Evidence for post-transcriptional modification of human salivary amylase (Amy_1) isozymes. Biochem. Genet. 10, 341.

KARP, G. W. JR. and SUTTON, H. E. (1967). Some new phenotypes of human red cell acid phosphatase. Am. J. Hum. Genet. 19, 54.

KAUFMAN, S. (1958). Phenylalanine hydroxylation co-factor in phenylketonuria. Science 128, 1506.

KAZIRO, K., KIKUCHI, G., NAKAMURA, H. and YOSHIDA, M. (1952). Die Frage nach der physiologischen Funktion der Katalase im menschlichen Organismus: Notiz über die Entdeckung einer Konstitutionsanomalie 'Anenzymia catalasea'. Chem. Ber. 85, 866.

KEITT, A. S. (1966). Pyruvate kinase deficiency and related disorders of red cell glycolysis. Am. J. Med. 41, 742.

KEITT, A. S. (1969). Haemolytic anemia with impaired hexokinase activity. J. Clin. Invest. 48, 1997.

KELLER, P. J., KAUFFMAN, D. L., ALLAN, B. J. and WILLIAMS, B. L. (1971). Further studies on the structural differences between the isoenzymes of human parotid amylase. Biochemistry 10, 4867.

KELLERMAN, G., LUYTEN-KELLERMAN, M. and SHAW, C. R. (1973). Genetic variation of aryl hydrocarbon hydroxylase in human lymphocytes. Am. J. Hum. Genet. *25*, 327.

KELLEY, W. N. (1968). Hypoxanthine-guanine phosphoribosyl transferase deficiency in the Lesch-Nyhan syndrome and gout. Federation Proc. *27*, 1047.

KELLEY, W. N. and ARNOLD, W. J. (1973). Human hypoxanthine-guanine phosphoribosyl-transferase: studies on the normal and mutant forms of the enzyme. Fed. Proc. *32*, 1656.

KELLEY, W. N. and MEADE, J. C. (1971). Studies on hypoxanthine-guanine phosphoribosyl transferase in fibroblasts from patients with the Lesch-Nyhan syndrome. Evidence for genetic heterogeneity. J. Biol. Chem. *246*, 2953.

KELLEY, W. N., ROSENBLOOM, F. M., HENDERSON, J. F. and SEEGMILLER, J. E. (1967). A specific enzyme defect in gout associated with overproduction of uric acid. Proc. Natl. Acad. Sci. U.S. *57*, 1735.

KENDALL, A. G., OJWANG, P. J., SCHROEDER, W. A. and HUISMAN, T. H. J. (1973). Hemoglobin Kenya, the product of a γ-β fusion gene: studies of the family. Am. J. Hum. Genet. *25*, 548.

KENNEDY, J. L., WERTELECKI, W., GATES, L., SPERRY, B. P. and CASS, V. M. (1967). The early treatment of phenylketonuria. Am. J. Diseases Children *113*, 16.

KIIL, R. and ROKKONES, T. (1964). Late manifesting variant of branched-chain keto-aciduria (maple syrup urine disease). Acta Pedlat. *53*, 356.

KIM, Y. S., PERDOMO, J., BELLA, A. JR. and NORDBERG, J. (1971). N-acetyl-D-galactosaminyl-transferase in human serum and erythrocyte membranes. Proc. Natl. Acad. Sci. U.S. *68*, 1753.

KIMURA, M. (1968). Evolutionary rate at the molecular level. Nature (Lond.) *217*, 624.

KIMURA, M. (1968). Genetic variability maintained in a finite population due to mutational production of neutral and nearly neutral isoalleles. Genet. Res. Camb. *11*, 247.

KIMURA, M. (1969). The rate of molecular evolution considered from the standpoint of population genetics. Proc. Natl. Acad. Sci. U.S. *63*, 1181.

KIMURA, M. and CROW, J. F. (1964). The number of alleles that can be maintained in a finite population. Genetics *49*, 725.

KIMURA, M. and OHTA, T. (1971). Theoretical aspects of population genetics. Princeton University Press, Princeton.

KIMURA, M. and OHTA, T. (1971a). Protein polymorphism as a phase of molecular evolution. Nature (Lond.) *229*, 467.

KIMURA, M. and OHTA, T. (1971b). On the rate of molecular evolution. J. Molec. Evolution *1*, 1.

KIMURA, M. and OHTA, T. (1973). Mutation and evolution at the molecular level. Genetics *Suppl. 73*, 19.

KING, J. and DIXON, R. I. (1969). A further factor contributing to inherited suxamethonium sensitivity. Brit. J. Anaesth. *41*, 1023.

KING, J. and MORGAN, H. G. (1970). The temperature activity relationships of serum cholinesterases. J. Clin. Pathol. *23*, 730.

KING, J. and MORGAN, H. G. (1971). Temperature activity of serum cholinesterases. J. Clin. Pathol. *24*, 182.

KING, J. L. and JUKES, T. H. (1969). Non-Darwinian evolution. Science *164*, 788.

KING, M. A. R., WILTSHIRE, B. G., LEHMANN, H. and MORIMOTO, H. (1972). An unstable hae-moglobin with reduced oxygen affinity: haemoglobin Peterborough, β111(G13) valine → phenylalanine, its interaction with normal haemoglobin and with haemoglobin Lepore. Brit. J. Haematol. *22*, 125.

KINGSBURY, K. J. (1971). Relation of ABO blood-groups to atherosclerosis. Lancet *I*, 199.

KINT, J. A. (1970). Fabry's disease: alpha-galactosidase deficiency. Science *167*, 1268.

KIRK, R. L. (1968). The haptoglobin groups in man. Karger, Basel.

KIRK, R. L. (1968). The world distribution of transferrin variants and some unsolved problems. Acta. Genet. Med. Gemell. *17*, 613.

KIRKMAN, H. N. (1959). Characteristics of glucose-6-phosphate dehydrogenase from normal and primaquine sensitive erythrocytes. Nature (Lond.) *184*, 1291.

KIRKMAN, H. N. (1968). Glucose-6-phosphate dehydrogenase variants and drug-induced hemolysis. Ann. N.Y. Acad. Sci. *151*, 753.

KIRKMAN, H. N. and BYNUM, E. (1959). Enzymic evidence of a galactosaemic trait in parents of galactosaemic children. Ann. Hum. Genet. Lond. *23*, 117.

KIRKMAN, H. N. and HENDRICKSON, E. M. (1963). Sex-linked electrophoretic difference in glucose-6-phosphate dehydrogenase. Am. J. Hum. Genet. *15*, 241.

KIRKMAN, H. N. and RILEY, H. D. JR. (1961). Congenital non-spherocytic haemolytic anaemia. Studies on a family with a qualitative defect in glucose-6-phosphate dehydrogenase. Am. J. Diseases Children *102*, 313.

KIRKMAN, H. N., MCCURDY, P. R. and NAIMAN, J. L. (1964a). Functionally abnormal G6PD. Cold Spring Harbor. Symp. Quant. Biol. *29*, 391.

KIRKMAN, H. N., RILEY, H. D. JR. and CROMWELL, B. B. (1960). Different enzymic expressions of mutants of human glucose-6-phosphate dehydrogenase. Proc. Natl. Acad. Sci. N.Y. *46*, 938.

KIRKMAN, H. N., SCHETTINI, F. and PICKARD, B. M. (1964c). Mediterranean variant of glucose-6-phosphate dehydrogenase. J. Lab. Clin. Med. *63*, 726.

KIRKMAN, H. N., SIMON, E. R. and PICKARD, B. M. (1965). Seattle variant of glucose-6-phosphate dehydrogenase. J. Lab. Clin. Med. *66*, 834.

KIRKMAN, H. N., ROSENTHAL, I. M., SIMON, E. R., CARSON, P. E. and BRINSON, A. G. (1964b). 'Chicago 1' variant of glucose-6-phosphate dehydrogenase in congenital hemolytic disease. J. Lab. Clin. Med. *63*, 715.

KITAMURA, A., IIJIMA, N., HASHIMOTO, F. and HIRATSUKA, A. (1971). Hereditary deficiency of subunit H of lactate dehydrogenase. Clin. Chim. Acta *34*, 419.

KITTO, G. B., STOLZENBACH, F. E. and KAPLAN, N. O. (1970). Mitochondrial malate dehydrogenase: further studies on multiple electrophoretic forms. Biochem. Biophys. Res. Commun. *38*, 31.

KITTO, G. B., WASSARMAN, P. G. and KAPLAN, N. O. (1966). Enzymatically active conformers of mitochondrial malate dehydrogenase. Proc. Natl. Acad. Sci. U.S. *56*, 578.

KJELLMAN, B., GAMSTORP, I., BRUN, A., OCKERMAN, P.-Å. and PALMGREN, B. (1969). Mannosidosis: a clinical and histopathologic study. J. Paediat. *75*, 366.

KLEIHAUER, E. F., REYNOLDS, C. A., DOZY, A. M., WILSON, J. B., MOORES, R. R., BERENSON, M. P., WRIGHT, C. S. and HUISMAN, T. H. I. (1968). Haemoglobin Bibba or $\alpha_2{}^{)36 \text{ Pro}} \beta_2$, an unstable α chain abnormal haemoglobin. Biochim. Biophys. Acta *154*, 220.

KNAPP, A. (1960). Über einer neuer Hereditäre von Vitamin-B$_6$ abhängige Störung im Tryptophan-stoff Wechsel. Clin. Chim. Acta *5*, 6.

KNIGHT, R. A., SELIN, M. J. and HARRIS, H. W. (1959). Genetic factors influencing isoniazid blood levels in humans. Trans. Conf. Chemotherap. Tuberc. (St. Louis) *18*, 52.

KNOX, W. E. (1972). Phenylketonuria. *In:* The Metabolic Basis of Inherited Disease, eds.

J. B. Stanbury, J. B. Wyngaarden and D. S. Fredrickson. 3rd Edition. McGraw Hill, New York.

KNOX, W. E. and MESSINGER, E. (1958). The detection of the metabolic effect of the recessive gene for phenylketonuria. Am. J. Hum. Genet. *10*, 53.

KNUDSEN, B. B. and DISSING, J. (1973). Adenosine deaminase deficiency in a child with severe combined immunodeficiency. Clin. Genetics *4*, 344.

KNUDSON, A. G. JR. (1973). Founder effect in Tay-Sachs disease. Am. J. Hum. Genet. *25*, 108.

KNUDSON, A. G. and KAPLAN, W. D. (1962). Genetics of the sphingolipidoses. *In:* Cerebral Sphingolipidoses: A symposium on Tay-Sachs' disease and allied disorders. eds.: S. M. Aronson and B. W. Volk. Academic Press, New York.

KOBATA, A. and GINSBERG, V. (1970). Uridine diphosphate-N-acetyl-D-galactosamine: D-galactose α-3-N-acetyl-D-galactosaminyl transferase, a product of the gene that determines blood type A in man. J. Biol. Chem. *245*, 1484.

KOBATA, A., GROLLMAN, E. F. and GINSBURG, V. (1968a). An enzymatic basis for blood type A in humans. Arch. Biochem. Biophys. *124*, 609.

KOBATA, A., GROLLMAN, E. F. and GINSBURG, V. (1968b). An enzymatic basis for blood type B in humans. Biochem. Biophys. Res. Commun. *32*, 272.

KOCH, J., SKOKSTAD, E. L. R., WILLIAMS, H. E. and SMITH, L. H. JR. (1967). Deficiency of 2-oxo-glutarate-glyoxylate carboligase activity in primary hyperoxaluria. Proc. Natl. Acad. Sci. U.S. *57*, 1123.

KOLODNY, E. H., BRADY, R. O. and VOLK, B. W. (1969). Demonstration of an alteration of ganglioside metabolism in Tay-Sachs disease. Biochem. Biophys. Res. Comm. *37*, 526.

KÖMPF, J., RITTER, H. and SCHMITT, J. (1971). Genetic polymorphism of glycerol-3-phosphate dehydrogenase (E.C.: 1.1.1.8). I Transspecific variability of G-3-PD subunit *β* in primates. Humangenetik *13*, 75.

KÖMPF, J., RITTER, H. and SCHMITT, J. (1972). Genetic polymorphism of glycerol-3-phosphate dehydrogenase (E.C. 1.1.1.8). II Transspecific variability of G-3-PD subunit A in primates. Formal genetics and population genetics. Humangenetik *14*, 103.

KOMROWER, G. M., SCHWARZ, V., HOLZEL, A. and GOLBERG, L. (1956). A clinical and biochemical study of galactosaemia. Arch. Disease Childhood *31*, 254.

KONOTEY-AHULU, F. I. D., GALLO, E., LEHMANN, H. and RINGELHAHN, B. (1968). Haemoglobin Korle-Bu (*β*73 aspartic acid → asparagine). J. Med. Genet. *5*, 107.

KONRAD, P. N., RICHARDS, F., VALENTINE, W. N. and PAGLIA, D. E. (1972). *γ*-Glutamyl-cysteine synthetase deficiency. New. Engl. J. Med., *286*, 558.

KONRAD, P. N., MCCARTHY, D. J., MAUER, A. M., VALENTINE, W. N. and PAGLIA, D. E. (1973). Erythrocyte and leukocyte phosphoglycerate kinase deficiency with neurologic disease. J. Pediat. *82*, 456.

KOPPE, A. L., WALTER, H., CHOPRA, V. P. and BAJATZADEH, M. (1970). Investigations on the genetics and population genetics of the *β*$_2$-glycoprotein I polymorphism. Humangenetik *9*, 164.

KOSCIELAK, J. (1967). Isolation of ABO antigens from red cells. *In:* Methods in immunology and immunochemistry, Vol. 1, ed. C. A. Williams and M. W. Chase. Academic Press, New York.

KÓSCIELAK, J., PIASEK, A. and GÓRNIAK, H. (1970). Studies on the chemical structure of blood group A specific glycolipids from human erythrocytes. In: Blood and tissue antigens, ed. D. Aminoff. Academic Press, New York.

KOSTER, J. F., FERNANDES, J., SLEE, R. G., VAN BERKEL, TH. J. C. and HULSMÄNN, W. C. (1973). Hepatic phosphorylase deficiency: a biochemical study. Biochem. Biophys. Res. Comm. *53*, 282.

KRANE, S. M., PINNELL, S. R. and EBBE, R. W. (1972). Decreased lysyl-protocollagen hydroxylase activity in fibroblasts from a family with a newly recognized disorder: hydroxylysine deficient collagen. J. Clin. Invest. *51*, 52a.

KRAUS, A. P., LANGSTON, M. F. and LYNCH, B. L. (1968). Red cell phosphoglycerate kinase deficiency. Biochem. Biophys. Res. Commun. *30*, 173.

KRAUS, A. P. and NEELY, C. L. JR. (1964). Human erythrocyte lactate dehydrogenase: four genetically determined variants. Science *145*, 595.

KRESSE, E. and NEUFELD, E. F. (1972). The Sanfilippo A corrective factor. J. Biol. Chem. *247*, 2164.

KRESSE, H., WIESMANN, U., CANTZ, M., HALL, C. W. and NEUFELD, E. F. (1971). Biochemical heterogeneity of the Sanfilippo syndrome: preliminary characterization of two deficient factors. Biochem. Biophys. Res. Comm. *42*, 892.

KROOTH, R. S. (1964). Properties of diploid cell strains developed from patients with an inherited abnormality of uridine biosynthesis. Cold Spring Harbor Symp. Quant. Biol. *29*, 189.

KUMAHARA, Y., FEINGOLD, D. S., FREEDBERG, I. M. and HIATT, H. H. (1961). Studies of pentose metabolism in normal subjects and in patients with pentosuria and pentosuria trait. J. Clin. Endocr. *21*, 887.

KUNKEL, H. G., CEPPELLINI, R., MULLER-EBERHARD, U. and WOLF, J. (1957). Observations on the minor basic haemoglobin component in blood of normal individuals and patients with thalassaemia. J. Clin. Invest. *36*, 1615.

KUNKEL, H. G., SMITH, W. K., JOSLIN, F. G., NATWIG, J. B. and LITWIN, S. D. (1969). Genetic marker of the γA2 subgroup of γA immunoglobulins. Nature *223*, 1247.

KURACHI, S., HERMODSON, M., HORNUNG, S. and STAMATOYANNOPOULOS, G. (1973). Structure of haemoglobin Seattle. Nature New Biol. *243*, 275.

KYRIAKIDES, E. C., PAUL, B. and BALINT, J. A. (1972). Lipid accumulation and acid lipase deficiency in fibroblasts from a family with Wolman's disease, and their apparent correction *in vitro*. J. Lab. Clin. Med. *80*, 810.

LABERGE, C. (1969). Hereditary tyrosinemia in a French Canadian isolate. Am. J. Hum. Genet. *21*, 36.

LABIE, D., ROSA, J., BELKHODJA, O. and BIERME, R. (1971). Hemoglobin Toulouse α_2 $\beta_2^{66(\mathrm{E})10\mathrm{Lys} \rightarrow \mathrm{Glu}}$. Structure and consequences in molecular pathology. Biochim. Biophys. Acta *236*, 201.

LABIE, D., SCHROEDER, W. A. and HUISMAN, T. H. J. (9166). The aminoacid sequence of the $\delta\beta$ chains of haemoglobin Lepore Augusta = Lepore Washington. Biochim. Biophys. Acta *127*, 428.

LABRIE, F. (1969). Isolation of an RNA with the properties of haemoglobin messenger. Nature (Lond.) *221*, 1217.

LA DU, B. N. (1966). Alcaptonuria. *In:* The metabolic basis of inherited diseases, ed. J. B. Stanbury, J. B. Wyngaarden and D. S. Fredrickson. 2nd edition. McGraw-Hill, New York.

LA DU, B. N. (1967). The enzymatic deficiency in tyrosinemia. Am. J. Diseases Children *113*, 54.

LA DU, B. N. (1967). Histidinemia. Am. J. Diseases Children *113*, 88.

LA DU, B. N., HOWELL, R. R., JSCOBY, G. A., SEEGMILLER, J. E. and ZANNONI, V. G. (1962). The enzymatic effect in histidinemia. Biochem. Biophys. Res. Comm. *7*, 398.

LA DU, B. N., ZANNONI, V. G., LASTER, L. and SEEGMILLER, J. E. (1958). The nature of the defect in tyrosine metabolism in alkaptonuria. J. Biol. Chem. *230*, 251.

LAKE, B. D. and PATRICK, A. D. (1970). Wolman's disease: deficiency of E 600-resistant acid esterase activity with storage of lipids in lysosomes. J. Pediat. *76*, 262.

LAMBOTTE-LEGRAND, J. and LAMBOTTE-LEGRAND, C. (1958). Notes complémentaires sur la drépanocytes. II. Sicklémie et malaria. Ann. Soc. Belge. Med. Trop. *38*, 45.

LAMM, L. U. (1970). Family, population and mother-child studies of two phosphogluco-mutase loci (PGM$_1$ and PGM$_3$). Lack of close linkage between the two loci ($\omega > 0.33$). Human Heredity *20*, 292.

LA MOTTA, R. V., WORONICK, C. L. and REINFRANK, R. F. (1970). Multiple forms of serum cholinesterase: molecular weights of isoenzymes. Arch. Biochem. and Biophys. *136*, 448.

LANDING, R. H., SILVERMAN, F. N., CRAIG, J. M., JACOBY, M. D., LAHEY, M. E. and CHADWICK, D. L. (1964). Familial neurovisceral liposis. An analysis of eight cases of a syndrome previously reported as 'Hurler variant', 'pseudo-Hurler disease' and Tay-Sachs disease with visceral involvement. Am. J. Diseases Children *108*, 503.

LANDSTEINER, K. (1901). Über Agglutinationserscheinungen normalen menschlichen Blutes. Wien. Klin. Wschr. *14*, 1132.

LANGLEY, C. H. and FITCH, W. M. (1973). The constancy of evolution: a statistical analysis of the α and β hemoglobins, cytochrome c and fibrinopeptide A. *In:* Genetic Structure of Populations, ed. N. E. Morton. University Press of Hawaii, Honolulu.

LANGLEY, C. H. and FITCH, W. M. (1974). An examination of the constancy of the rate of molecular evolution. J. Mol. Evol. (in press).

LARIZZA, P., BRUNETTI, P., GRIGNANI, F. and VENTURE, S. (1958). L'individualita bioenzima-tica dell'eritrocite 'fabico' sopra alcune anomalie biochemiche ed enzimatiche della emazie nei pazienti affetti da favismo e nei loro familiari. Haematologia *43*, 205.

LASKER, M. (1941). Essential fructosuria. Human Biol. *13*, 51.

LASKER, M., ENKLEWITZ, M. and LASKER, G. W. (1936). The inheritance of L-xyloketosuria essential pentosuria). Human Biol. *8*, 243.

LASTER, L., MUDD, S. H., FINKELSTEIN, J. D. and IRREVERRE, F. (1965). Homocystinuria due to cystathionine synthase deficiency: the metabolism of L-methionine. J. Clin. Invest. *44*, 1708.

LAURELL, C.-B. and ERIKSSON, S. (1963). The electrophoretic α_1-globulin pattern of serum in α_1-antitrypsin deficiency. Scand. J. Clin. Lab. Invest. *15*, 132.

LAYZER, R. B., ROWLAND, L. P. and RANNEY, H. M. (1967). Muscle phosphofructokinase deficiency. Arch. Neurol. *17*, 512.

LEBHERZ, H. G. and RUTTER, W. J. (1969). Distribution of fructose diphosphate aldolase variants in biological systems. Biochemistry *8*, 109.

LECAM, L. M., NEYMAN, J. and SCOTT, E. L. (Eds.) (1972). 'Darwinian, neo-Darwinian and non-Darwinian evolution'. Proc. Sixth Berkeley Symposium on mathematical statistics and probability. V. Univ. California Press, Berkeley.

LEGUM, C. P. and NITOWSKY, H. M. (1969. Studies on leucocyte brancher enzyme activity in a family with type IV glycogenosis. J. Pediat. *74*, 84.

LEHMANN, H. (1970). Different types of alpha-thalassaemia and significance of haemo-globin Bart's in neonates. Lancer *2*, 78.

LEHMANN, H. and CARRELL, R. W. (1968). Differences between α- and β-chain mutants of human haemoglobin and between α- and β- thalassaemia. Possible duplication of the α-chain gene. Brit. Med. J. IV, 748.

LEHMANN, H. and CARRELL, R. W. (1969). Variations in the structure of human haemoglobin. Brit. Med. Bull. *25*, 14.

LEHMANN, H. and HUNTSMAN, R. G. (1966). Man's haemoglobins: North-Holland Publishing Co., Amsterdam.

LEHMANN, H. and HUNTSMAN, R. G. (1972). The haemoglobinopathies. *In:* The metabolic basis of inherited disease, ed. J. B. Stanbury, J. B. Wyngaarden and D. S. Fredrickson. McGraw-Hill, New York. 3rd edition.

LEHMANN, H. and RYAN, E. (1956). The familial incidence of low pseudocholinesterase level. Lancet 2, 124.

LEHRS, H. (1930). Über die gruppenspezifische Eigenschaften des menschlichen Speichels. Z. Immun. Forsch. *66*, 175.

LESCH, M. and NYHAN, W. L. (1964). A familial disorder of uric acid metabolism and central nervous system function. Am. J. Med. 36, 561.

LEVIN, B. (1967). Arginino succinic aciduria. Am. J. Diseases Children *113*, 162.

LEVIN, B., ABRAHAM, T. J., OBERHOLZER, V. G. and BURGESS, E. A. (1969a). Hyperammonemia: a deficiency of liver ornithine transcarbamylase. Occurrence in mother and child. Arch. Dis. Child. *44*, 152.

LEVIN, B., DOBBS, R. H., BURGESS, E. A. and PALMER, T. (1969b). Hyperammonemia: a variant type of deficiency of liver ornithine transcarbamylase. Arch. Diseases Children *44*, 152.

LEVINE, P., ROBINSON, E., CELANO, M., BRIGGS, O. and FALKINBURG, L. (1955). Gene interaction resulting in suppression of blood group substance. B. Blood *10*, 1100.

LEVY, H. L. (1973). Genetic screening. Advances in Human Genetics, *4*, 1, eds.: H. Harris and K. Hirschhorn. Plenum Press, New York.

LEVY, H. L., ERICKSON, A. M., LOTT, I. T. and KURTZ, D. J. (1973). Isovaleric acidemia: results of family study and dietary treatment. Pediatrics *52*, 83.

LEVY, H. L., SHIH, V. E., KAROLKEWICZ, E., FRERICH, W. A., CARR, J. R., CASS, V., KENNEDY, J. L. JR. and MACCREADY, R. A. (1971). Persistent mild hyperphenylalaninaemia in the untreated state. A prospective study. New Engl. J. Med. *285*, 424.

LEWIS, E. B. (1951). Pseudo allelism and gene evolution. Cold Spring Harbor Symp. Quant. Biol. *16*, 159.

LEWIS, W. H. P. (1973). Common polymorphism of Peptidase A. Electrophoretic variants associated with quantitative variation of red cell levels. Ann. Hum. Genet. Lond. *36*, 267.

LEWIS, W. H. P., CORNEY, G. and HARRIS, H. (1968). Pep A 5–1 and Pep A 6–1: two new variants of peptidase A with features of special interest. Ann. Hum. Genet. Lond. *32*, 35.

LEWIS, W. H. P. and HARRIS, H. (1967). Human red cell peptidases. Nature (Lond.) *215*, 351.

LEWIS, W. H. P. and HARRIS, H. (1969a). *In vitro* 'hybridisation' of peptidase A subunits between different human phenotypes, and between human and monkey. Ann. Hum. Genet. Lond. *33*, 89.

LEWIS, W. H. P. and HARRIS, H. (1969b). Peptidase D (prolidase) variants in man. Ann. Hum. Genet. Lond. *32*, 317.

LEWONTIN, R. C. (1967). An estimate of average heterozygosity in man. Amer. J. Hum. Genet. *19*, 681.

LEWONTIN, R. C. and HUBBY, J. L. (1966). A molecular approach to the study of genetic heterozygosity in natural populations. II. Amount of variation and degree of heterozygosity in natural population of Drosophila pseudoobscura. Genetics *54*, 595.

LEWONTIN, R. C. and KRAKAUER, J. (1973). Distribution of gene frequency as a test of the theory of the selective neutrality of polymorphisms. Genetics *74*, 175.

LIDDELL, J. BROWN, D., BEALE, D., LEHMANN, H. and HUNTSMAN, R. G. (1964). A new haemoglobin – J^a Oxford found during a survey of an English population. Nature (Lond.) *204*, 269.

LIDDELL, J., LEHMANN, H. and DAVIES, D. (1963). Harris and Whittaker's pseudocholinesterase variant with increased resistance to fluoride. A study of four families and the identification of the homozygote. Acta Genet. *13*, 95.

LIDDELL, J., LEHMANN, H. and SILK, E. (1962). A 'silent' pseudocholinesterase gene. Nature (Lond.) *193*, 561.

LIEBERMAN, J., MITTMAN, C. and GORDON, H. W. (1972). Alpha1-antitrypsin in the livers of patients with emphysema. Science *175*, 63.

LIE-INJO, L. E., GANESAN, J., CLEGG, J. B. and WEATHERALL, D. J. (1974). Homozygous state for Hb Constant Spring (slow moving X components). Blood *43*, 251.

LIE-INJO, L. E., LIE, H. G., AGER, J. A. M. and LEHMANN, H. (1962). α-thalassaemia as a cause of hydrops foetalis. Brit. J. Haematol. *8*, 1.

LIFSCHITZ, F. (1966). Congenital lactase deficiency. J. Pediat. *69*, 229.

LINDBLAD, B., LINDSTRAND, K., SVANBERG, G. and ZETTERSTRÖM, R. (1969). The effect of cobamide coenzyme in methylmalonic acidemia. Acta Paediat. Scand. *58*, 178.

LINDBLAD, B., OLIN, P., SVANBERG, B. and ZETTERSTRÖM, R. (1968). Methylmalonic acidemia. Acta Paediat. Scand. *57*, 417.

LINDER, D. and GARTLER, S. M. (1965a). Distribution of glucose-6-phosphate dehydrogenase electrophoretic variants in different tissues of heterozygotes. Am. J. Hum. Genet. *17*, 212.

LINDER, D. and GARTLER, S. M. (1965b). Glucose-6-phosphate dehydrogenase mosaicism: utilisation as a cell marker in the study of leiomyomas. Science *150*, 67.

LINES, J. G. and MCINTOSH, R. (1967). Oxygen binding by haemoglobin J-Cape Town ($\alpha_2$92 Arg → Gln). Nature (Lond.) *215*, 297.

LIVINGSTONE, F. B. (1967). Abnormal hemoglobins in human populations. Aldine Publishing Co., Chicago.

LIVINGSTONE, F. B. (1969). The founder effect and deleterious genes. Am. J. Phys. Anthropol. *30*, 55.

LIVINGSTONE, F. B. (1971). Malaria and human polymorphisms. Ann. Rev. Genet. *5*, 33.

LLOYD, K. O. and KABAT, E. A. (1968). Immunochemical studies of blood groups XLI. Proposed structure for the carbohydrate portions of blood groups A, B, H, Lewis[a] and Lewis[b] substances. Proc. Natl. Acad. Sci. U.S. *61*, 1470.

LLOYD, K. O., KABAT, E. A. and LICERIO, E. (1968). Immunochemical studies on blood groups. 38. Structure and activities of oligosaccharides produced by alkaline degradation of blood group Lewis[a] substance. Proposed structure of the carbohydrate chains of human blood group A, B, H, Le[a] and Le[b] substances. Biochemistry *7*, 2976.

LLOYD, K. O., KABAT, E. A., LAYUG, E. J. and GRUEZO, F. (1966). Immunochemical studies on blood groups. 34. Structure of some oligosaccharides produced by alkaline degradation of blood group A, B, and H substances. Biochemistry *5*, 1489.

LOEB, H., TONDEUR, M., JONNIAUX, G., MOCKELPOHL, S. and VAMOS-HURWITZ, E. (1969).

Biochemical and ultracentrifugal studies in a case of mucopolysaccharide F (Fucosidosis). Helv. Paed. Acta. *24*, 519.

LÖHR, G. W. and WALLER, H. D. (1962). Eine neue enzymopenische hämolytische Anämie mit Glutathion-reductase-Mangel. Med. Klin. *57*, 1521.

LONG, W. K. (1967). Glutathione reductase in red blood cells: variant associated with gout. Science *155*, 712.

LONG, W. K., KIRKMAN, H. N., SUTTON, H. E. (1965). Electrophoretically slow variants of glucose-6-phosphate dehydrogenase from red cells of Negroes. J. Lab. Clin. Med. *65*, 81.

LORKIN, P. A. (1973). Fetal and embryonic haemoglobins. J. Med. Genet. *10*, 50.

LORKIN, P. A., LEHMANN, H., FAIRBANKS, V. F., BERGLUND, G. and LEONHARD, T. (1970). Two new pathological haemoglobins: Olmstead β141(H19) Leu → Arg and Malmo β97(FG4) His → Gln. Biochem. J. *119*, 68P.

LUFFMAN, J. E. and HARRIS, H. (1967). A comparison of some properties of human red cell acid phosphatase in different phenotypes. Ann. Hum. Genet. Lond. *30*, 387.

LUZZATTO, L. and ALLAN, N. C. (1965). Different properties of glucose-6-phosphate dehydrogenase from human erythrocytes with normal and abnormal enzyme levels. Biochem. Biophys. Res. Commun. *21*, 547.

LUZZATTO, L., USANGA, E. A. and REDDY, S. (1969). Glucose-6-phosphate dehydrogenase deficient red cells: resistance to infection by malarial parasites. Science *164*, 839.

LYON, M. F. (1962). Sex chromatin and gene action in the mammalian X-chromosome. Am. J. Hum. Genet. *14*, 135.

LYON, M. F. (1971). Possible mechanisms of X-chromosome inactivation. Nature New Biol. *232*, 229.

MABRY, C. C., DENNISTON, J. C. and COLDWELL, J. G. (1966). Mental retardation in children of phenylketonuric mothers. New Engl. J. Med. *275*, 1331.

MACBRINN, M., OKADA, S., WOOLLACOTT, M., PATEL, V., HO, M. W., TAPPELL, A. L. and O'BRIEN, J. S. (1969). Beta-galactosidase deficiency in the Hurler syndrome. New Engl. J. Med. *281*, 338.

MACIVER, J. E., WENT, L. N. and IRVINE, R. A. (1961). Hereditary persistence of foetal haemoglobin: a family study suggesting allelism of the F gene to the S and C haemoglobin genes. Brit. J. Haematol. *7*, 373.

MACKENZIE, D. Y. and WOOLF, L. I. (1959). Maple syrup urine disease – an inborn error of the metabolism of valine, leucine and isoleucine associated with gross mental deficiency. Brit. med. J. *1*, 90.

MAHONEY, M. J. and ROSENBERG, L. E. (1970). Inherited defects of B_{12} metabolism. Am. J. Med. *48*, 548.

MARCUS, D. M. and CASS, L. E. (1969). Glycosphingolipids with Lewis blood group activity: uptake by human erythrocytes. Science *164*, 553.

MARKERT, C. L. (1963). Lactate dehydrogenase isozymes: dissociation and recombination of subunits. Science *140*, 1329.

MARKERT, C. L. (1968). The molecular basis for isozymes. Ann. N.Y. Acad. Sci. *151*, 14.

MARKERT, C. L. and MOLLER, F. (1959). Multiple forms of enzymes: tissue ontogenetic, and species specific patterns. Proc. Natl. Acad. Sci. U.S. *45*, 753.

MARKS, P. A. (1958). Red cell glucose-6-phosphate and 6-phosphogluconic dehydrogenases and nucleoside phosphorylase. Science *127*, 1338.

MARKS, P. A., BANKS, J. and GROSS, R. T. (1962). Genetic heterogeneity of glucose-6-phosphate dehydrogenase deficiency. Nature (Lond.) *194*, 454.

MARKS, P. A., GROSS, R. T. and HURWITZ, R. E. (1959). Gene action in erythrocyte deficiency of glucose-6-phosphate dehydrogenase deficiency: tissue enzyme levels. Nature (Lond.) *183*, 1266.

MARKS, P. A., SZEINBERG, A. and BANKS, J. (1961). Erythrocyte glucose-6-phosphate dehydrogenase of normal and mutant human subjects. Properties of purified enzymes. J. Biol. Chem. *236*, 10.

MARR, A. M. S., DONALD, A. S. R., WATKINS, W. M. and MORGAN, W. T. J. (1967). Molecular and genetic aspects of human blood-group Le^b specificity. Nature (Lond.) *215*, 1345.

MARSHALL, W. C., LLOYD-STILL, J. and SEAKINS, J. W. T. (1967). Congenital sucrase and isomaltase deficiency with temporary lactose intolerance. Acta. Paediat. Scand. *56*, 211.

MARVER, H. S. and SCHMID, R. (1972). The porphyrias. *In:* The metabolic basis of inherited disease, eds. J. B. Stanbury, J. B. Wyngaarden and D. S. Fredrickson. 3rd Edition. McGraw Hill, New York.

MATALON, R. and DORFMAN, A. (1972). Hurler's syndrome, an α-L-iduronidase deficiency. Biochem. Biophys. Res. Commun. *47*, 959.

MATHAI, C. K. and BEUTLER, E. (1966). Electrophoretic variation of galactose-1-phosphate uridyltransferase. Science *154*, 1179.

MATSUDA, I., ARASHIMA, S., NAMBU, J., TAKEKOSHI, Y. and ANAKURA, M. (1971). Hyperammonemia due to a mutant enzyme of ornithine transcarbamylase. Pediatrics *48*, 595.

MATSUNAGA, E. (1962). An inert allele Hp^0 at the Hp locus. Jap. J. Hum. Genet. *7*, 133.

MCALPINE, P. J. (1974). Isozyme analysis in somatic cell hybrids. Proc. IIIrd International Isozyme Conference. Academic Press, N.Y. (in press).

MCALPINE, P. J., HOPKINSON, D. A. and HARRIS, H. (1970). Thermostability studies on the isozymes of human phosphoglucomutase. Ann. Hum. Genet. Lond. *34*, 61.

MCALPINE, P. J., HOPKINSON, D. A. and HARRIS, H. (1970). The relative activities attributable to the three phosphoglucomutase loci, (PGM_1, PGM_2 and PGM_3) in human tissues. Ann. Hum. Genet. Lond. *34*, 169.

MCALPINE, P. J., HOPKINSON, D. A. and HARRIS, H. (1970). Molecular size estimates of the human phosphoglucomutase isozymes by gel filtration chromatography. Ann. Hum. Genet. Lond. *34*, 177.

MCARDLE, R. B. (1951). Myopathy due to a defect in muscle glycogen breakdown. Clin. Sci. *10*, 13.

MCCARTHY, C. F., BORLAND, J. L. JR., LYNCH, H. J. JR., OWEN, E. E. and TYOR, M. P. (1964). Defective uptake of basic aminoacids and L-cystine by intestinal mucosa of patients with cystinuria. J. Clin. Invest. *43*, 1518.

MCCONNELL, R. B. (1966). The genetics of gastro-intestinal disorders. Oxford University Press.

MCCURDY, P. R., KIRKMAN, H. N., NAIMAN, J. L., JIM, R. T. S. and PICKARD, B. M. (1966). A chinese variant of glucose-6-phosphate dehydrogenase. J. Lab. Clin. Med. *67*, 374.

MCCURDY, P. R., PEARSON, H. and GERALD, P. S. (1961). A new haemoglobinopathy of unusual genetic significance. J. Lab. Clin. Med. *58*, 86.

MCKUSICK, V. A. (1971). Mendelian inheritance in man. Catalogs of autosomal Dominant, Autosomal recessive, and X linked phenotypes. 3rd Edition. Johns Hopkins Press, Baltimore.

MCKUSICK, V. A. (1972). Hereditable disorders of connective tissue. 4th Edition. C. V.

Mosby, Saint Louis.

MCKUSICK, V. A., HOWELL, R. R., HUSSELS, I. E., NEUFELD, E. F. and STEVENSON, R. E. (1972). Allelism, non-allelism and genetic compounds among the mucopolysaccharidoses. Lancet *1*, 993.

MCMURRAY, W. C., RATHBUN, J. C., MOHYUDDIN, F. and KOEGLER, S. J. (1963). Citrullinuria. Pediatrics *32*, 347.

MEDALIE, J. H., LEVENE, C., PAPIER, C., GOLDBOURT, U., DREYFUSS, F., ORON, D., NEUFELD, H. and RISS, E. (1971). Blood groups, myocardial infarction and angina pectoris among 10,000 adult males. New Engl. J. Med. *285*, 192.

MEERA KHAN, P., WESTERVELD, A., GRZESCHIK, K. H., DEYS, B. F., GARSON, O. M. and SINISCALCO, M. (1971). X-Linkage of human phosphoglycerate kinase confirmed in man-mouse and man-chinese hamster somatic cell hybrids. Am. J. Hum. Genet. *23*, 614.

MEHL, E. and JATZKEWITZ, M. (1965). Evidence for a genetic block in metachromatic leukodystrophy. Biochem. Biophys. Res. Commun. *19*, 407.

MELANCON, S. B., KHACHADURIAN, A. K., NADLER, H. L. and BROWN, B. I. (1973). Metabolic and biochemical studies in fructose 1,6-diphosphatase deficiency. Pediatrics *82*, 650.

MELARTIN, L. and BLUMBERG, B. S. (1966). Albumin Naskapi: a new variant of serum albumin. Science *153*, 3744.

MERRITT, A. D., RIVAS, M. L. and WARD, J. C. (1972). Human amylase loci: evidence for close linkage. Nature *239*, 243.

MERRITT, A. D., LOVRIEN, E. W., RIVAS, M. L. and CONNEALLY, P. M. (1973b). Human amylase loci: genetic linkage with the Duffy blood group locus and assignment to linkage group I. Am. J. Hum. Genet. *25*, 523.

MERRITT, A. D., RIVAS, M. L., BIXLER, D. and NEWELL, R. (1973). Salivary and pancreatic amylase: electrophoretic characterizations and genetic studies. Am. J. Hum. Genet. *25*, 510.

MEYER, V. A., STRAND, J., DOSS, M., REES, A. C. and MARVER, H. S. (1972). Intermittent acute porphyria – demonstration of a genetic defect in porphobilinogen metabolism. New Engl. J. Med. *286*, 1277.

MIDELFORT, C. F. and MEHLER, A. H. (1972). Deamidation *in-vivo* of an asparagine residue of rabbit muscle aldolase. Proc. Natl. Acad. Sci. U.S. *69*, 1816.

MIGEON, B. R. (1971). Studies of skin fibroblasts from 10 families with HGPRT deficiency, with reference to X-chromosomal inactivation. Am. J. Human Genet. *23*, 199.

MIGEON, B. R., DER KALOUSTIAN, V. M., NYHAN, W. L., YOUNG, W. J. and CHILDS, B. (1968). X-linked hypoxanthine-guanine phosphoribosyl transferase deficiency: heterozygote has two clonal populations. Science *160*, 425.

MILLER, A. L. and MCLEAN, P. (1967). Urea cycle enzymes in the liver of a patient with argininosuccinic aciduria. Clin. Sci. *32*, 385.

MILNE, M. D., ASATOOR, A. M., EDWARDS, K. G. D. and LOUGHBRIDGE, L. W. (1961). The intestinal absorption defect in cystinuria. Gut *2*, 323.

MILNER, P. F., CLEGG, J. B. and WEATHERALL, D. J. (1971). Haemoglobin-H disease due to a unique haemoglobin variant with an elongated α-chain. Lancet *1*, 729.

MISHU, M. K. and NANCE, W. E. (1969). Further evidence for close linkage of the Hb^β and Hb^δ loci in man. J. Med. Genet. *6*, 190.

MITOMA, C., AULD, R. M. and UDENFRIEND, S. (1957). On the nature of enzymatic defect in phenylpyruvic oligophrenia. Proc. Soc. Exptl. Biol. N.Y. *94*, 634.

MITTMAN, C. (Ed.) (1972). Pulmonary Emphysema and Proteolysis. Academic Press, New York.

MIYAJI, T., IUCHI, I., SHIBATA, S., TAKEDA, I. and TAMURA, A. (1963). Possible aminoacid substitution in the α chain (α 87 Tyr) of Hb M Iwata. Acta Haematol. Japan *26*, 538.

MIYAMOTO, M. and FITZPATRICK, T. B. (1957). Competitive inhibition of mammalian tyrosinase by phenylalanine and its relationship to hair pigmentation in phenylketonuria. Nature (Lond.) *179*, 199.

MIZRAHI, O. and SER, I. (1963). Essential pentosuria. *In:* Genetics of migrant and isolate populations, ed: E. Goldschmidt, Williams and Wilkins Baltimore.

MODIANO, G., FILIPPI, G., BRUNELLI, F., FRATTAROLI, W. and SINISCALCO, G. M. (1967). Studies on red cell acid phosphatase in Sardinia and Rome. Absence of correlation with past malarial morbidity. Acta. Genet. Basel. *17*, 17.

MODIANO, G., SCOZZARI, R., GIGLIANI, F., SANTOLAMAZZA, C., SPENNATI, G. F. and SAINI, P. (1970). Enzyme activity in two red cell adenylate kinase phenotypes. Am. J. Hum. Genet. *22*, 292.

MOHLER, D. N., MAJERUS, P. W., MINNICH, V., HESS, C. E. and GARRICK, M. D. (1970). Glutathione synthetase deficiency as a cause of hereditary hemolytic disease. New. Engl. J. Med. *283*, 1253.

MOHYUDDIN, F., RATHBUN, J. C. and MCMURRAY, W. C. (1967). Studies on aminoacid metabolism in citrullinuria. Amer. J. Diseases Children *113*, 152.

MONN, E. (1969a). Relation between blood cell phosphoglucomutase isoenzymes and age of cell population. Scand. J. Haematol. *6*.

MONN, E. (1969b). Chromatographic studies on human red cell phosphoglucomutase. Protein Res. *1*, 1.

MOORE, M. J., DEUTSCH, H. F. and ELLIS, F. R. (1973). Human carbonic anhydrases. IX. Inheritance of variant erythrocyte forms. Am. J. Hum. Genet. *25*, 29.

MOORE, M. J., FUNAKOSHI, S. and DEUTSCH, H. F. (1971). Human carbonic anhydrases. VII. A new C type isozyme in erythrocytes of American Negroes. Biochem. Genet. *5*, 497.

MORGAN, W. T. J. (1967). Soluble b.ood group specific substances. *In:* Methods in immunology and immunochemistry p. 75, ed. C. A. Williams and M. W. Chase. Vol. 1, Academic Press, New York.

MORGAN, W. T. J. and VAN HEYNINGEN, R. (1944). The occurrence of A, B and O blood group substances in pseudomucinous ovarian cyst fluids. Brit. J. Exptl. Pathol. *25*, 5.

MORGAN, W. T. J. and WATKINS, W. M. (1948). The detection of a product of the blood group O gene and the relationship of the so-called O substance to the agglutinogens A and B. Brit. J. Exptl. Pathol. *29*, 159.

MORGAN, W. T. J. and WATKINS, W. M. (1953). The inhibition of the haemagglutinins in plant seeds by human blood-group substances and simple sugars. Brit. J. Exptl. Pathol. *34*, 94.

MORGAN, W. T. J. and WATKINS, W. M. (1956). The product of the human blood group A and B genes in individuals belonging to group AB. Nature (Lond.) *177*, 521.

MORGAN, W. T. J. and WATKINS, W. M. (1969). Genetic and biochemical aspects of human blood-group A-, B-, H-, Le[a]- and Le[b]- specificity. Brit. Med. Bull. *25*, 30.

MORGANTI, G., BEOLCHINI, P. E., VERUCCI, A. and BÜTLER, R. (9167). Contributions to the genetics of the serum β-lipoproteins in man. I. Frequency, transmission and penetrance of factors Ag(x) and Ag(y). Humangenetik *4*, 262.

MORIMOTO, H., LEHMANN, H. and PERUTZ, M. F. (1971). Molecular pathology of human haemoglobin: stereo-chemical interpretation of abnormal oxygen affinities. Nature (Lond.) *232*, 408.

MORIN, C. L., THOMPSON, M. W., JACKSON, S. H. and SASS-KORTSAK, A. (1971). Biochemical and genetic studies in cystinuria, observations on double heterozygotes of Genotype I/II. J. Clin. Invest. *50*, 1961.

MORRIS, M. D., FISHER, D. A. and FISER, R. (1966). Late onset branched chain ketoaciduria (maple syrup urine disease). Journal-Lancet *86*, 149.

MORRIS, M. D. and FISHER, D. A. (1967). Trypsinogen deficiency disease. Am. J. Diseases Children *114*, 203.

MORROW, G. (1967). Citrullinemia. Am. J. Diseases Children *113*, 157.

MORROW, III. G., BARNES, L. A., CARDINALE, G. J., ABELES, R. H. and FLAKS, J. G. (1969). Congenital methylmalonic acidemia: enzymatic evidence for two forms of the disease. Proc. Natl. Acad. Sci. U.S. *63*, 191.

MORTON, N. E. (1964). Genetic studies of Northeastern Brazil. Cold Spring Harbor Symp. Quant. Biol. *29*, 69.

MOSER, H. W., EFRON, M. L., BROWN, H., DIAMOND, R. and NEUMANN, L. G. (1967). Argininosuccinicaciduria: Report of two new cases and demonstration of intermittent elevation of blood ammonia. Am. J. Med. *42*, 9.

MOTULSKY, A. G. (1964). Current concepts of the genetics of the thalassaemias. Cold Spring Harbor Symp. Quant. Biol. *29*, 399.

MOURANT, A, E. (1946). A 'new' human blood group antigen of frequent occurrence. Nature (Lond.) *158*, 237.

MOURANT, A. E., KOPEC, A. C. and DOMANIEWSKA-SOBCZAK. (1971). Blood groups and blood clotting. Lancet *1*, 223.

MOWBRAY, S., WATSON, B. and HARRIS, H. (1972). A search for electrophoretic variants of human adenine phosphoribosyl transferase. Ann. Hum. Genet. Lond. *36*, 153.

MUDD, S. H., IRREVERRE, F. and LASTER, L. (1967). Sulfite oxidase deficiency in man: demonstration of the enzymatic defect. Science *156*, 1599.

MUDD, S. H., FINKELSTEIN, J. D. IRREVERRE, F. and LASTER, L. (1964). Homocystinuria: an enzymatic defect. Science *143*, 1443.

MUDD, S. H., EDWARDS, W. A., LOEB, P. M., BROWN, M. S. and LASTER, L. (1970). Homocystinuria due to cystathionine synthase deficiency: the effect of pyridoxine. J. Clin. Invest. *49*, 1762.

MUDD, S. H., UHLENDORF, B. W., FREEMAN, J. M., FINKELSTEIN, J. D. and SHIH, V. E. (1972). Homocystinuria associated with decreased methylenetetrahydrofolate reductase activity. Biochem. Biophys. Res. Comm. *46*, 905.

MULLER, C. J. and KINGMA, S. (1961). Haemoglobin Zurich $\alpha_2{}^A \beta_2{}^{63\mathrm{Arg}}$. Biochim. Biophys. Acta *50*, 595.

MURRAY, P., THOMSON, J. A., MCGIRR, E. M., WALLACE, T. J., MACDONALD, E. M. and MACCABAE, H. J. (1965). Absent and defective iodotyrosine deiodination. Lancet *1*, 183.

MYRIANTHOPOULOS, N. C. and ARONSON, S. M. (1966). Population dynamics of Tay-Sachs disease. I. Reproductive fitness and selection. Am. J. Hum. Genet. *18*, 313.

MYRIANTHOPOULOS, N. C. and ARONSON, S. M. (1972). Population dynamics of Tay-Sachs disease. II. What confers the selective advantage upon the Jewish heterozygote? *In:* Sphingolipids, sphingolipidoses and allied disorders, eds.: B. W. Volk and S. M. Aronson. Plenum Press, New York.

MYRIANTHOPOULOS, N. C., NAYLOR, A. F. and ARONSON, S. M. (1972). Founder effect in Tay-Sachs disease unlikely. Am. J. Hum. Genet. *24*, 341.

NABHOLZ, M., MIGGIANO, V. and BODMER, W. (1969). Genetic analysis with human-mouse somatic cell hybrids. Nature (Lond.) *223*, 358.

NADLER, H. L. and EGAN, T. J. (1970). Deficiency of lysosomal acid phosphatase. A new familial metabolic disorder. New. Engl. J. Med. *282*, 302.

NAGEL, R. L., RANNEY, H. M., BRADLEY, T. B., JACOBS, A. and UDEM, L. (1969). Hemoglobin L Ferrara in a Jewish family associated with a hemolytic state in the propositus. *Blood 34*, 157.

NAKAJIMA, H. (1963). Studies on heme α-methenyl oxygenase, a new enzyme which is capable of transforming haemoglobin-haptoglobin to a possible precursor of biliverdin. Proc. 9th Congr. Eur. Soc. Haematology, Lisbon. p. 840. Karger, Basel.

NAKAJIMA, H., TAKAMURA, T., NAKAJIMA, O. and YAMAOKA, K. (1963). Studies on heme α-methenyl oxygenase. J. Biol. Chem. *238*, 3784.

NAKAO, K., WADA, O., KITAMURA, T. and UONO, M. (1966). Activity of amino-laevulinic acid synthetase in normal and porphyric human livers. Nature (Lond.) *210*, 838.

NAKASHIMA, K., MIWA, S., ODA, S., ODA, E., MATSUMOTO, N., FUKUMOTO, Y. and YAMADA, T. (1973). Electrophoretic and kinetic studies of glucosephosphate isomerase (GPI) in two different Japanese families with GPI deficiency. Am. J. Hum. Genet. *25*, 294.

NA-NAKORN, S. and WASI, P. (1970). Alpha-thalassaemia in Northern Thailand. Am. J. Hum. Genet. *22*, 645.

NANCE, W. E. (1967). Genetic studies of human serum and erythrocyte polymorphisms. Ph. D. Thesis. University of Wisconsin.

NANCE, W. E. and SMITHIES, O. (1963). New haptoglobin alleles: a prediction confirmed. Nature (Lond.) *198*, 869.

NANCE, W. E. and UCHIDA, I. (1964). Turner's syndrome, twinning, and an unusual variant of glucose-6-phosphate dehydrogenase. Am. J. Hum. Genet. *16*, 380.

NANCE, W. E., CLAFLIN, A. and SMITHIES, O. (1963). Lactic dehydrogenase: genetic control in man. Science *142*, 1075.

NAVON, R., PADEH, B. and ADAM, A. (1973). Apparent deficiency of hexosaminidase A in healthy members of a family with Tay-Sachs disease. Am. J. Hum. Genet. *25*, 287.

NECHELES, T. F., RAI, U. S. and CAMERON, D. (1970). Congenital non-spherocytic haemolytic anaemia associated with an unusual erythrocyte hexokinase abnormality. J. Lab. Clin. Med., *76*, 593.

NECHELES, T. F., STEINBERG, M. H. and CAMERON, D. (1970). Erythrocytic glutathione-peroxidase deficiency. Brit. J. Haematol. *19*, 605.

NECHELES, T. F., MALDONADO, N., BARQUET-CHEDIAK, A. and ALLEN, D. M. (1969). Homozygous erythrocyte glutathione-peroxidase deficiency: clinical and biochemical studies. Blood *33*, 164.

NEEB, H., BEIBOER, J. L., JONXIS, J. H. P., KAARS SIJPESTEIJN, J. A. and MULLER, C. J. (1961). Homozygous Lepore haemoglobin disease appearing as thalassaemia major in two Papuan siblings. Trop. Geogr. Med. *13*, 207.

NEEL, J. V. (1949). The inheritance of sickle cell anaemia. Science *110*, 64.

NEEL, J. V. (1951). The inheritance of the sickling phenomenon, with particular reference to sickle cell disease. Blood *6*, 389.

NEEL, J. V. and SALZANO, F. M. (1967). Further studies on the Xavante Indians. X. Some hypotheses – generalizations resulting from these studies. Am. J. Hum. Genet. *19*, 554.

NEUFELD, E. F. and CANTZ, M. J. (1971). Corrective factors for inborn errors of metabolism. Ann. New York Acad. Sci. *179*, 580.

NEUFELD, E. F. and FRATANTONI, J. C. (1970). Inborn errors of mucopolysaccharide metabolism. Science *169*, 141.

NEVO, E. (1973). Test of selection and neutrality in natural populations. Nature (Lond.) *244*, 573.

NEW, M. I. and LEVINE, L. S. (1973). Congenital adrenal hyperplasia. Advances in Human Genetics *4*, 251. Eds. H. Harris, K. Hirschhorn. Plenum Press, New York.

NG, W. G., DONNELL, G. N. and BERGREN, W. R. (1968). Deficiency of erythrocyte nicotinamide adenine dinucleotide nucleosidase (NADase) activity in the Negro. Nature (Lond.) *217*, 64.

NICHOLS, E. A., CHAPMAN, V. M. and RUDDLE, F. H. (1973). Polymorphism and linkage for mannosephosphate isomerase in *Mus musculus*. Biochem. Genet. *8*, 47.

NISHIMURA, E. T., HAMILTON, H. B., KOBARA, T. Y., TAKAHARA, S., OGURA, Y. and DOI, K. (1959). Carrier state in human acatalasemia. Science *130*, 3371.

NIXON, A. D. and BUCHANAN, J. G. (1967). Haemolytic anaemia due to pyruvate kinase deficiency. New Zealand Med. J. *66*, 859.

NOLAN, C. and MARGOLIASH, E. (1968). Comparative aspects of primary structures of proteins. Ann. Rev. Biochem. *37*, 727.

NORDMANN, Y., SCHAPIRA, F. and DREYFUS, J. C. (1968). A structurally modified liver aldolase in fructose intolerance: immunochemical and kinetic evidence. Biochem. Biophys. Res. Commun. 884.

NORUM, K. R. and GJONE, E. (1967a). Familial serum-cholesterol esterification failure. A new inborn error of metabolism. Biochim. Biophys. Acta *144*, 698.

NORUM, K. R. and GJONE, E. (1967b). Familial plasma lecithin: cholesterol acyltransferase deficiency. Biochemical study of a new inborn error of metabolism. Scand. J. Clin. Lab. Invest. *20*, 231.

NORUM, K. R., GLOMSET, J. A. and GJONE, E. (1972). Familial lecithin: cholesterol acyltransferase deficiency. *In:* The Metabolic Basis of Inherited Disease, eds. J. B. Stanbury, J. B. Wyngaarden and D. S. Fredrickson. 3rd Edition. McGraw-Hill.

NOVY, M. J., EDWARDS, M. J. and METCALFE, J. (1967). Haemoglobin Yakima: II High blood oxygen affinity associated with compensatory erythrocytosis and normal haemodynamics. J. Clin. Invest. *46*, 1848.

NUTE, P., STAMATOYANNOPOULOS, G. and FUNK, D. (1972). Hemoglobin Olympia (β20 val → met): an electrophoretically silent variant associated with high oxygen affinity and erythrocytosis. J. Clin. Invest. *51*, 70a.

NYHAN, W. H., BAKAY, B., CONNOR, J. D., MARKS, J. F. and KEELE, D. K. (1970). Hemizygous expression of glucose-6-phosphate dehydrogenase in erythrocytes of heterozygotes for the Lesch-Nyhan syndrome. Proc. Natl. Acad. Sci. U.S. *65*, 214.

NYHAN, W. L., PESEK, J., SWEETMAN, L., CARPENTER, D. G. and CARTER, C. H. (1967). Genetics of an X-linked disorder of uric acid metabolism and cerebral function. Pediat. Res. *1*, 5.

NYMAN, M. (1959). Serum haptoglobin. Methodological and clinical studies. Scand. J. Clin. Lab. Invest. *11*, suppl. *39*.

OATES, J. A., NIRENBERG, P. Z., JEPSON, J. B., SJOERDSMA, A. and UDENFRIEND, S. (1963). Conversion of phenylalanine to phenylethylamine in patients with phenylketonuria. Proc. Soc. Exptl. Biol. Med. *112*, 1078.

OBERHOLZER, V. G., LEVIN, B., BURGESS, E. A. and YOUNG, W. F. (1967). Methylmalonic aciduria. Inborn error of metabolism leading to metabolic acidosis. Arch. Dis. Childhood *42*, 492.

O'BRIEN, D. and JENSEN, C. B. (1963). Pyridoxine dependency in two mentally retarded subjects. Clin. Sci. *24*, 179.

O'BRIEN, J. S. (1972). Ganglioside storage disease. Advances in human Genetics *3*, 39. Plenum Press, New York.

O'BRIEN, J. S. (1972). Sanfilippo syndrome: profound deficiency of alpha-acetylglucosaminidase activity in organs and skin fibroblasts from Type-B patients. Proc. Natl. Acad. Sci. U.S. *69*, 1720.

O'BRIEN, J. S., OKADA, S., CHEN, A. and FILLERUP, D. L. (1970). Tay-Sachs disease: detection of heterozygotes and homozygotes by serum hexosaminidase assay. New Engl. J. Med., *283*, 15.

ÖCKERMAN, P.-A. (1967). A generalised storage disease resembling Hurler's syndrome. Lancet *2*, 239.

ÖCKERMAN, P.-A. (1969). Mannosidosis isolation of oligosaccharides storage material from brain. J. Pediat. *75*, 360.

ÖCKERMAN, P.-A. and KÖHLIN, P. (1968). Tissue acid hydrolase activities in Gaucher's disease. Scand. J. Clin. Lab. Invest. *22*, 62.

O'FLYNN, M. E., TILLMAN, P. and HSIA, D. Y. Y. (1967). Hyperphenylalaninemia without phenylketonuria. Am. J. Dis. Child. *113*, 22.

OHTA, T. (1973). Slightly deleterious mutant substitutions in evolution. Nature (Lond.) *246*, 96.

OHTA, T. and KIMURA, M. (1971). On the constancy of the evolutionary rate of cistrons. J. Molec. Evol. *1*, 18.

OHTA, Y., YAMAOKA, K., SUMIDA, I. and YANASE, T. (1971). Haemoglobin Miyada, a β-δ fusion peptide (anti-Lepore) type discovered in a Japanese family. Nature New Biol. *234*, 218.

OKADA, S., MCCREA, M. and O'BRIEN, J. S. (1972). Sandhoff's disease (GM_2 gangliosidosis Type 2): Clinical, chemical, and enzyme studies in five patients. Pediat. Res. *6*, 606.

OKADA, S. and O'BRIEN, J. S. (1968). Generalized gangliosidosis: β-galactosidase deficiency. Science *160*, 1002.

OKADA, S. and O'BRIEN, J. S. (1969). Tay-Sachs disease: generalised absence of a β-N-acetylhexosaminidase component. Science *165*, 698.

OKADA, S., VEATH, M. L., LEROY, J. and O'BRIEN, J. S. (1971). Ganglioside GM_2 storage diseases: hexosaminidase deficiencies in cultured fibroblasts Am. J. Hum. Genet. *23*, 55.

OKADA, S., VEATH, M. L. and O'BRIEN, J. S. (1970). Juvenile GM_2 gangliosidosis: partial deficiency of hexosaminidase A. J. Pediat. *77*, 1063.

OPFELL, R. W., LORKIN, P. A. and LEHMANN, H. (1968). Hereditary non-spherocytic haemolytic anaemia with post-splenectomy inclusion bodies and pigmenturia caused by an unstable haemoglobin Santa Ana-β88 (F4) leucine $\rightarrow$ proline. J. Med. Genet. *5*, 292.

OPITZ, J. M., STILES, F. C., WISE, D., RACE, R. R., SANGER, R., VON GEMMINGEN, G. R., KIERLAND, R. R., CROSS, E. G. and DE GROTT, W. P. (1965). The genetics of angiokeratoma corporis diffusum (Fabry's disease) and its linkage relations with the Xg locus. Amer. J. Hum. Genet. *17*, 325.

OSTERMAN, J. and FRITZ, P. J. (1973). Pyruvate kinase isozymes: a comparative study in tissues of various mammalian species. Comp. Biochem. Physiol., *44B*, 1077.

OSTERTAG, W. and SMITH, E. W. (1969). A third type of a ζ-β crossover (ζ50, β86). Europ. J. Biochem. *10*, 371.

PAGLIA, D. E., VALENTINE, W. N. and RUCKNAGEL, D. L. (1972). Defective erythrocyte pyruvate kinase with impaired kinetics and reduced optimal activity. Brit. J. Haematol. *22*, 651.

PAGLIA, D. E., HOLLAND, P., BAUGHAN, M. A. and VALENTINE, W. N. (1969). Occurrence of defective hexosephosphate isomerization in human erythrocytes and leucocytes. New Engl. J. Med. *280*, 66.

PAGLIA, D. E., VALENTINE, W. N., BAUGHAN, M. A., MILLER, D. R., REED, C. F. and MCINTYRE. O. R. (1968). An inherited molecular lesion of erythrocyte pyruvate kinase. J. Clin. Invest. *47*, 1929.

PAGLIARA, A. S., KARL, I. E., KEATING, J. P., BROWN, B. I. and KIPNIS, D. M. (1972). Hepatic fructose-1, 6-diphosphatase deficiency. A cause of lactic acidosis and hypoglycemia in infancy. J. Clin. Invest. *51*, 2115.

PAINTER, T. J., WATKINS, W. M. and MORGAN, W. T. J. (1962). Isolation of a B-specific di-saccharide from human blood-group B substance. Nature (Lond.) *193*, 1042.

PALO, J. and MATTSON, K. (1970). Eleven new cases of aspartylglycosaminuria. J. Ment. Defic. Res. *14*, 168.

PARE, C. M. B., SANDLER, M. and STACEY, R. C. (1957). 5-hydroxytryptamine deficiency in phenylketonuria. Lancet *1*, 511.

PARR, C. W. (1966). Erythrocyte phosphogluconate dehydrogenase polymorphism. Nature (Lond.) *210*, 487.

PARR, C. W. and FITCH, L. I. (1967). Inherited quantitative variations of human phospho-gluconate dehydrogenase. Ann. Hum. Genet. Lond. *30*, 339.

PARRINGTON, J. M., CRUICKSHANK, G., HOPKINSON, D. A., ROBSON, E. B. and HARRIS, H. (1968). Linkage relationships between the three phosphoglucomutase loci *PGM*₁, *PGM*₂ and *PGM*₃. Ann. Hum. Genet. Lond. *32*, 27.

PATRICK, A. D. (1965). A deficiency of glucocerebrosidase in Gaucher's disease. Biochem. J. *97*, 17c.

PATRICK, A. D. and LAKE, B. D. (1969). Deficiency of an acid lipase in Wolman's disease. Nature (Lond.) *222*, 1067.

PAULING, L., ITANO, H. A., SINGER, S. J. and WELLS, I. C. (1949). Sickle cell anaemia, a molecular disease. Science *110*, 543.

PEHLKE, D. M., MCDONALD, J. A. and HOLMES, E. W. (1972). Inosinic acid dehydrogenase activity in the Lesch-Nyhan syndrome. J. Clin. Invest. *51*, 1398.

PELKONEN, R. and KIVIRIKKO, K. I. (1970). Hydroxyprolinemia. New Engl. J. Med., *283*, 451.

PEMBERTON, R. E. and BAGLIONI, C. (1972). Duck haemoglobin messenger RNA. J. Mol. Biol. *65*, 531.

PEMBERTON, R. E., HOUSMAN, D., LODISH, H. and BAGLIONI, C. (1972). Isolation of duck haemoglobin messenger RNA and its translation by rabbit reticulocyte cell free system. Nature New Biol. *235*, 99.

PENHOET, E. E., KOCHMAN, M. and RUTTER, W. J. (1969). Molecular and catalytic properties of aldolase C. Biochemistry *8*, 4396.

PENHOET, E. E., RAJKUMAR, T. V. and RUTTER, W. J. (1966). Multiple forms of fructose diphosphate aldolase in mammalian tissues. Proc. Natl. Acad. Sci. U.S. *56*, 1275.

PENHOET, E. E., KOCHMAN, M., VALENTINE, R. and RUTTER, W. J. (1967). The subunit structure of mammalian fructose diphosphate aldolase. Biochemistry *6*, 2940.

PENROSE, L. S. (1954). Quelques principes sur la fréquence des gènes et sa stabilité dans les populations humaines. J. Génét. Hum. *3*, 159.

PENROSE, L. S. (1963). Outline of human genetics. 2nd edition. Heinemann, London.

PERLROTH, M. G., TSCHUDY, D. P., MARVER, H. S., BERARD, C. W., ZEIGEL, R. F., RECHCIGL, M. and COLLINS, A. (1966). Acute intermittent porphyria. New morphologic and biochemical findings. Amer. J. Med. *41*, 149.

PERRY, H. M. JR., TAN, E. M., CARMODY, S. and SAKAMOTO, A. (1970). Relationship of acetyl transferase activity to antinuclear antibodies and toxic symptoms in hypertensive patients treated with hydralazine. J. Lab. Clin. Med. *76*, 114.

PERRY, T. L., HANSEN, S. and LOVE, D. L. (1968). Serum carnosinase deficiency in carnosinemia. Lancet *1*, 1229.

PERRY, T. L., HANSEN, S., TISCHLER, B. and BUNTING, R. (1967). Determination of heterozygosity for phenylketonuria on the aminoacid analyzer. Clin. Chim. Acta *18*, 51.

PERRY, T. L., HANSEN, S., TISCHLER, B., BUNTING, R. and BERRY, K. (1967). Carnosinemia: a new metabolic disorder associated with neurologic disease and mental defect. New. Engl. J. Med., *277*, 1219.

PERRY, T. L., HARDWICK, T. F., HANSEN, S., LOVE, D. L. and ISRAELS, S. (1968). Cystathioninuria in two healthy siblings. New. Engl. J. Med., *278*, 590.

PERUTZ, M. F. and LEHMANN, H. (1968). Molecular pathology of human haemoglobin. Nature (Lond.) *219*, 902.

PERUTZ, M. F. and MITCHISON, J. M. (1950). State of haemoglobin in sickle-cell anaemia. Nature (Lond.) *166*, 677.

PERUTZ, M. F., MUIRHEAD, H., COX, J. M. and GOAMAN, L. C. G. (1968). Three-dimensional Fourier synthesis of horse oxyhaemoglobin at 2.8 Å resolution: the atomic model. Nature (Lond.) *219*, 131.

PERUTZ, M. F., PULSINELLI, P. DEL, EYCK, L. TEN, KILMARTIN, J. V., SHIBATA, S., IUCHI, I., MIYAJI, T. and HAMILTON, H. B. (1971). Haemoglobin Hiroshima and the mechanism of the alkaline Bohr effect. Nature New Biol. *232*, 147.

PETERS, J. H. and LEVY, L. (1971). Dapsone acetylation in man: another example of polymorphic acetylation. Ann. New York Acad. Sci. *179*, 660.

PETERS, J. H., GORDON, G. R. and BROWN, P. (1965). The relationship between the capacities of human subjects to acetylate isoniazid, sulfanilamide and sulfamethazine. Life Sci. *4*, 99.

PINK, R., WANG, A.-C. and FUDENBERG, H. H. (1971). Antibody variability. Ann. Rev. Med. *22*, 145.

PINTO, P. V. C., NEWTON, W. A. JR. and RICHARDSON, K. E. (1966). Evidence for four types of erythrocyte glucose-6-phosphate dehydrogenase from G-6-PD deficient human subjects. J. Clin. Invest. *45*, 823.

PIOMELLI, S., CORASH, L. M., DAVENPORT, D. D., MIRAGLIA, J. and AMBROSI, E. L. (1968). Glucose-6-phosphate dehydrogenase deficiency (types Gd^A and $Gd^{Mediterranean}$); the result of *in vivo* instability of the mutant enzymes. J. Clin. Invest. *47*, 940.

POLLARA, B. and MEUWISSEN, H. J. (1973). Combined immunodeficiency disease and A.D.A. deficiency. Lancet *2*, 1324.

POLLITT, R. J. and JENNER, F. A. (1969). Enzymatic cleavage of 2-acetamido-1-(β'-L-aspartamido)-1,2-dideoxy-β-D-glucose by human plasma & seminal fluid. Failure to detect the heterozygous state for aspartylglycosaminuria. Clin. Chim. Acta. *25*, 413.

POLLITT, R. J., JENNER, F. A. and MERSKEY, H. (1968). Aspartylglycosaminuria. An inborn error associated with mental defect. Lancet *2*, 253.

POLONOWSKI, C. and BIER, H. (1970). Pseudo-trypsinogen deficiency due to lack of intestinal enterokinase. Acta. Paediat. Scand. *59*, 458.

POOTRAKUL, S., WASI, P. and NA-NAKORN, S. (1967a). Haemoglobin Bart's hydrops foetalis in Thailand. Ann. Hum. Genet. Lond. *30*, 293.

POOTRAKUL, S., WASI, P. and NA-NAKORN, S. (1967b). Studies on haemoglobin Bart's (Hb-γ_4) in Thailand: the incidence and the mechanism of occurrence in cord blood. Ann. Hum. Genet. Lond. *31*, 149.

PORETZ, R. D. and WATKINS, W. M. (1972). Galactosyltransferases in human submaxillary glands and stomach mucosa associated with the biosynthesis of blood group B specific glycoproteins. Eur. J. Biochem. *25*, 455.

PORTER, C. A., MOWAT, A. P., COOK, P. J. L., HAYNES, D. W. G., SHILKIN, K. B. and WILLIAMS, R. (1972). α_1-Antitrypsin deficiency and neonatal hepatitis. Brit. Med. J. *3*, 435.

PORTER, I. H., BOYER, S. H., WATSON-WILLIAMS, E. J., ADAM, A., SZEINBERG, A. and SINISCALCO, M. (1964). Variation of glucose-6-phosphate dehydrogenase in different populations. Lancet *1*, 895.

PORTER, M. T., FLUHARTY, A. L. and KIHARA, H. (1971). Correction of abnormal cerebroside sulphate metabolism in cultured metachromatic leukodystrophy fibroblasts. Science *172*, 1263.

POVEY, S., CORNEY, G., LEWIS, W. H. P., ROBSON, E. B., PARRINGTON, J. M. and HARRIS, H. (1972). The genetics of peptidase C in man. Ann. Hum. Genet. Lond. *35*, 455.

POWELL, G. F., RASCO, M. and MANISCALCO, R. M. (1973). A prolidase deficiency in man with iminopeptiduria. Genetics *Suppl.*, 216.

PRADER, A. and AURICCHIO, S. (1965). Defects of intestinal disaccharide absorption. Ann. Rev. Med. *16*, 345.

PRAKASH, S., LEWONTIN, R. C. and HUBBY, J. L. (1969). A molecular approach to the study of genic heterozygosity in natural populations. IV. Patterns of genic variation in central, marginal and isolated populations of *Drosophila pseudoobscura*. Genetics *61*, 841.

PRATO, V., GALLO, E., RICCO, G., MAZZA, U., BIANCO, G. and LEHMANN, H. (1970). Haemolytic anaemia due to haemoglobin Torino. Brit. J. Haematol. *19*, 105.

PRICE, P. M., CONOVER, J. H. and HIRSCHHORN, K. (1972). Chromosomal localization of human haemoglobin genes. Nature (Lond.) *237*, 340.

PROKOP, O. and DIETRICH, A. (1968). Proof of the existence of the Hp^0 gene. Dtsch. Z. Ges. Gerichth. Med. *63*, 111.

PTASHNE, M. (1967). Isolation of the λ phage repressor. Proc. Natl. Acad. Sci. U.S. *57*, 306.

PULSINELLI, P. D., PERUTZ, M. F. and NAGEL, R. L. (1973). Structure of haemoglobin M Boston, a variant with a five-coordinated ferric heme. Proc. Natl. Acad. Sci. U.S. *70*, 3870.

PUSZTAI, A. and MORGAN, W. T. J. (1963). The aminoacid composition of the human blood-group A, B, H and Le^a specific substances. Biochem. J. *88*, 546.

PUTKONEN, T. (1930). Über die gruppenspezifischen Eigenschaften verschiedener Körper-flüssigkeiten. Acta Soc. Med. Fenn. 'Duodecim' A, *14*, No. 12.

QUATTRIN, N., BIANCHI, P., CIMINO, R., DE ROSA, L., DINI, E. and VENTRUTO, V. (1967). Studies on nine families with haemoglobin Lepore in Campania. Acta Haematol. *37*, 266.

QUERIDO, A., STANBURY, J. B., KASSENAAS, A. A. H. and MEIJER, I. W. A. (1956). The metabolism of iodotyrosines. III. Di-iodotyrosine deshalogenating activity of human thyroid tissue. J. Clin. Endocr. *16*, 1096.

QUICK, C. B., FISHER, R. A. and HARRIS, H. (1974). A kinetic study of the isozymes determined by the three phosphoglucomutase loci, PGM_1, PGM_2 and PGM_3. Europ. J. of Biochem. *42*, 511.

RACE, C., ZIDERMAN, D. and WATKINS, W. M. (1968). An a-D-galactosyltransferase associated with the blood group B character. Biochem. J. *107*, 733.

RACE, R. R. (1971). Is the Xg blood group locus subject to inactivation? Proc. Fourth International Congress of Human Genetics, p. 311, Excerpta Medica, Amsterdam.

RACE, R. R. and SANGER, R. (1968). Blood groups in man. 5th edition. Blackwell, Oxford.

RAMOT, B., BEN-BASSAT, I. and SHCHORY, J. (1969). New glucose-6-phosphate dehydrogenase variants observed in Israel and their association with congenital non-spherocytic haemolytic anaemia. J. Lab. Clin. Med. *74*, 895.

RAMOT, B. and BROK, F. (1964). A new glucose-6-phosphate dehydrogenase mutant (Tel-Hashomer mutant). Ann. Hum. Genet. Lond. *28*, 167.

RAMOT, B., FISHER, S., SZEINBERG, A., ADAM, A., SHEBA, C. and GAFNI, D. (1959). A study of subjects with glucose-6-phosphate dehydrogenase deficiency. II. Investigation of leucocyte enzymes. J. Clin. Invest. *38*, 2234.

RANNEY, H. M., JACOBS, A. S., UDEM, L. and ZALUSKY, R. (1968). Hemoglobin Riverdale-Bronx, an unstable hemoglobin resulting from the substitution of arginine for glycine at helical residue B6 of the β polypeptide chain. Biochim. Biophys. Res. Comm. *33*, 1004.

RANNEY, H. M., JACOBS, A. S., BRADLEY, T. B. and CORDOVA, F. A. (1963). A 'new' variant of haemoglobin A_2 and its segregation in a family with haemoglobin S. Nature (Lond.) *197*, 614.

RAPER, A. B. (1956). Sickling in relation to morbidity from malaria and other diseases. Brit. Med. J. *1*, 965.

RAPER, A. B., GAMMACK, D. B., HUEHNS, E. R. and SHOOTER, E. M. (1960). Four haemoglobins in one individual. A study of the genetic interaction of Hb G and Hb C. Brit. Med. J. *2*, 1257.

RAPLEY, S., ROBSON, E. B. and HARRIS, H. (1967). Data on the incidence, segregation and linkage relations of the adenylate kinase (AK) polymorphism. Ann. Hum. Genet. Lond. *31*, 237.

RASMUSSEN, K. (1968). Phosphorylethanolamine and hypophosphatasia. Danish Med. Bull. *15* (supplement).

RATHBUN, J. C. (1948). Hypophosphatasia, a new development anomaly. Amer. J. Diseases Children *75*, 822.

RATHBUN, J. C., MACDONALD, J. W., ROBINSON, H. M. and WANKLIN, J. M. (1961). Hypophosphatasia: a genetic study. Arch. Disease Childhood *36*, 540.

RATTAZZI, M. C., MARKS, J. S. and DAVIDSON, R. G. (1973). Electrophoresis of arylsulfatase from normal individuals and patients with metachromatic leukodystrophy. Am. J. Hum. Genet. *25*, 310.

REED, C. S., HAMPSON, R., GORDON, S., JONES, R. T., NOVY, M. J., BRIMHALL, B., EDWARDS, M. J. and KOLER, R. D. (1968). Erythrocytosis secondary to increased oxygen affinity of a mutant hemoglobin, hemoglobin Kempsey. Blood *31*, 623.

REED, T. E. (1968). Research on blood groups and selection from the Child Health and Development Studies, Oakland, California. III. Couple mating type and reproductive performance. Am. J. Hum. Genet. *20*, 129.

REFSUM, S. (1946). Heredopathia atactica polyneuritiformis. Acta Psychiat. Scand., *Suppl. 38*, 9.

REGE, V. P., PAINTER, T. J., WATKINS, W. M. and MORGAN, W. T. J. (1964a). Isolation of serologically active fucose-containing oligosaccharides from human blood-group H substance. Nature (Lond.) *203*, 360.

REGE, V. P., PAINTER, T. J., WATKINS, W. M. and MORGAN, W. T. J. (1964b). Isolation of a serologically active, fucose-containing trisaccharide from human blood-group Le^a substance. Nature (Lond.) *204*, 740.

REY, J. and FRÉZAL, J. (1967). Les anomalies des disaccharidases. Arch. Franc. Pediat. *24*, 65.

REY, J., FRÉZAL, J., ROYER, P. and LAMY, M. (1966). L'absence congénitale de lipase pancréatique. Arch. Franc. Pediat. *23*, 5.

REYS, L. and YOSHIDA, A. (1971). Chromatographic separation of variant enzyme components of human pseudocholinesterase. J. Génét. Hum. *19*, 261.

RICHMOND, R. C. (1970). Non-Darwinian evolution: a critique. Nature (Lond.) *225*, 1025.

RIEDER, R. F. and BRADLEY, T. B. (1968). Hemoglobin Gun Hill: an unstable protein associated with chronic hemolysis. Blood *32*, 355.

RIEDER, R. F., OSKI, F. A. and CLEGG, J. B. (1969). Hemoglobin Philly (β35 tyrosine → phenylalanine): Studies in molecular pathology of hemoglobin. J. Clin. Invest. *48*, 1627.

RIENSKOU, T. (1968). The Gc system. Series Haematologica *1*, 1, 21.

RITTER, H. and WENDT, G. G. (1964). Untersuchung von 223 Familien zur Formalen Genetik des INV-Polymorphismus. Humangenetik *1*, 123.

ROBINSON, D. and CARROLL, M. (1972). Tay-Sachs disease: interrelation of hexosaminidases A and B. Lancet *1*, 322.

ROBINSON, D. and STIRLING, J. L. (1968). N-acetyl-β-glucosaminidases in human spleen. Biochem. J. *107*, 321.

ROBSON, E. B. and HARRIS, H. (1965). Genetics of the alkaline phosphatase polymorphism of the human placenta. Nature (Lond.) *207*, 1257.

ROBSON, E. B. and HARRIS, H. (1966). Further data on the incidence and genetics of the serum cholinesterase phenotype C_5+. Ann. Hum. Genet. Lond. *29*, 403.

ROBSON, E. B. and HARRIS, H. (1967). Further studies on the genetics of placental alkaline phosphatase Ann. Hum. Genet. Lond. *30*, 219.

ROMEO, G. and LEVIN, E. Y. (1969). Uroporphyrinogen III cosynthetase in human congenital erythropoietic porphyria. Proc. Natl. Acad. Sci. U.S. *63*, 856.

ROMEO, G. and MIGEON, B. R. (1970). Genetic inactivation of the α-galactosidase locus in carriers of Fabry's disease. Science *170*, 180.

ROMEO, G., KABACK, M. M. and LEVIN, E. Y. (1970). Uroporphyrinogen III cosynthetase activity in fibroblasts from patients with congenital erythropoietic porphyria. Biochem. Genet., *4*, 659.

ROPARTZ, C., LENOIR, J. and RIVAT, L. (1961). A new inheritable property of human sera: the InV factor. Nature (Lond.) *189*, 586.

ROSENBERG, L. E. (1966). Cystinuria: genetic heterogeneity and allelism. Science *154*, 1341.

ROSENBERG, L. E. (1967). Genetic heterogeneity in cystinuria. *In:* Aminoacid metabolism and genetic variation, Ed. W. L. Nyhan. McGraw Hill, New York.

ROSENBERG, L. E. (1972). Disorders of propionate, methylmalonate and vitamin B$_{12}$ metabolism. *In:* The metabolic basis of inherited disease. Eds. J. B. Stanbury, J. B. Wyngaarden and D. S. Fredrickson. 3rd Edition. McGraw-Hill, New York.

ROSENBERG, L. E., ALBRECHT, I. and SEGAL, S. (1967). Lysine transport in human kidney: evidence for two systems. Science *155*, 1426.

ROSENBERG, L. E., DURANT, J. L. and ELSAS, L. J. (1968). Familial iminoglycinuria. An inborn error of renal tubular transport. New Engl. J. Med. *278*, 1407.

ROSENBERG, L. E., LILLJEQVIST, A.-CH. and HSIA, Y. E. (1968). Methylmalonic aciduria. New Engl. J. Med. *278*, 1319.

ROSENBERG, L. E., DOWNING, S., DURANT, J. L. and SEGAL, S. (1966). Cystinuria: biochemical evidence for three genetically distinct diseases. J. Clin. Invest. *45*, 365.

ROSENBERG, L. E., LILLJEQVIST, A.-CH., HSIA, Y. E. and ROSENBLOOM, F. M. (1969). Vitamin B_{12} dependent methylmalonicaciduria: Defective B_{12} metabolism in cultured fibroblasts. Biochem. Biophys. Res. Commun. *37*, 607.

ROSENBLOOM, F. M., KELLEY, W. N., HENDERSON, J. F. and SEEGMILLER, J. E. (1967). Lyon hypothesis and X-linked disease. Lancet *2*, 305.

ROTHMAN, M. C. and RANNEY, H. M. (1971). Double heterozygosity for hemoglobin G ($\alpha_2^{68\,\mathrm{Lys}}\beta_2^{A}$) and hemoglobin D ($\alpha_2^{A}\beta_2^{121\,\mathrm{Gln}}$). Blood *37*, 177.

RUBIN, C. S., BALIS, M. E., PIOMELLI, S., BERMAN, P. H. and DANCIS, J. (1969). Elevated AMP pyrophosphorylase activity in congenital IMP pyrophosphorylase deficiency (Lesch-Nyhan disease). J. Lab. Clin. Med. *74*, 732.

RUBIN, C. S., DANCIS, J., YIP, L. C., NOWINSKY, R. C. and BALIS, M. E. (1971). Purification of IMP: pyrophosphate phosphoribosyltransferases catalytically incompetent enzymes in Lesch-Nyhan disease. Proc. Natl. Acad. Sci. U.S. *68*, 1971.

RUBINSTEIN, H. M., DIETZ, A. A., HODGES, L. K., LUBRANO, T. and CZEBOTAR, V. (1970). Silent cholinesterase gene: variations in the property of serum enzyme in apparent homozygote. J. Clin. Invest. *44*, 486.

RUCKNAGEL, D. L., BRANDT, N. J. and SPENCER, H. H. (1971). α-chain mutants of human hemoglobin contributing to the genetics of the α-locus. In: Genetical, functional, and physical studies of hemoglobins. First Inter-American Symposium on Hemoglobin. Eds.: T. Arends, G. Bemski and R. L. Nagel. S. Karger, New York.

RUDDLE, F. H. (1972). Linkage analysis using somatic cell hybrids. Advances in Human Genetics *3*, 173. Edit. by H. Harris and K. Hirschhorn, Plenum Press, New York.

RUDDLE, F. H. (1973). Linkage analysis in man by somatic cell genetics. Nature (Lond.) *242*, 165.

RUSSELL, J. D. and DEMARS, R. (1967). UDP-glucose: α-D-galactose-1-phosphate uridylyltransferase activity in cultured human fibroblasts. Biochem. Genet. *1*, 11.

RUSSELL, A., LEVIN, B., OBERHOLZER, V. G. and SINCLAIR, L. (1962). Hyperammonemia. A new instance of an inborn enzymatic defect of the biosynthesis of urea. Lancet *2*, 699.

RUTTER, W. J., RAJKUMAR, T., PENHOET, E. and KOCHMAN, M. (1968). Aldolase variants: structure and physiological significance. Ann. N.Y. Acad. Sci. *151*, 102.

SACHS, B., STERNFELD, L. and KRAUS, G. (1942). Essential fructosuria: its pathophysiology. Am. J. Diseases Children *63*, 252.

SALZMANN, J., DEMARS, R. and BENKE, P. (1968). Single-allele expression at an X-linked hyperuricemia locus in heterozygous human cells. Proc. Natl. Acad. Sci. U.S. *60*, 545.

SAMLOFF, I. M., LIEBMAN, W. M., GLOBER, G. A., MOORE, J. O. and INDRA, D. (1973). Population studies of pepsinogen polymorphism. Am. J. Hum. Genet. *25*, 178.

SANDHOFF, K. (1969). Variation of β-N-acetylhexosaminidase pattern in Tay-Sachs disease. FEBS Letters *4*, 351.

References

SANDHOFF, K. and JATZKEWITZ, H. (1972). The chemical pathology of Tay-Sachs disease. *In:* Sphingolipids, sphingolipidoses and allied disorders. Ed. B. W. Volk and S. M. Aronson. Plenum Press, New York.

SANDHOFF, K. and WÄSSLE, W. (1972). Anreichung und Characterisierung zweier Formen der menschlichen N-acetyl-β-D-hexosaminidase. Hoppe-Seylers Z. Physiol. Chem. *352*, 1119.

SANFILIPPO, S. J., PODOSIN, R., LANGER, L. O. JR. and GOOD, R. A. (1963). Mental retardation associated with acid mucopolysacchariduria (heparitin sulphate type). J. Pediat. *63*, 837.

SANSONE, G. and PIK, C. (1965). Familial haemolytic anaemia with erythrocyte inclusion bodies, bilifuscinuria and abnormal haemoglobin (haemoglobin Galliera Genova). Brit. J. Haematol. *11*, 511.

SANSONE, G., CARRELL, R. W. and LEHMANN, H. (1967). Haemoglobin Genova: β28 (B10) leucine → proline. Nature (Lond.) *214*, 877.

SANTACHIARA, A. S. B. and MODIANO, G. (1969). Ultracentrifuge studies of red cell phosphoglucomutase. Nature (Lond.) *233*, 625.

SARTORI, E. (1971). On the pathogenesis of favism. J. Med. Genet. 8, 462.

SAUNDUBRAY, J.-M., CATHELINEAU, L., CHARPENTIER, C., BOISSE, J., ALLANEAU, C., LE BONT, H. and LESAGE, B. (1973). Hereditary ornithine-carbamoyl-transferase defect with qualitative enzyme abnormality. Report of a case with neonatal appearance and early lethal evolution in a boy. Arch. Fr. Pediat. *30*, 15.

SCHACHTER, H., MICHAELS, M. A., TILLEY, C. A., CROOKSTON, M. C. and CROOKSTON, J. H. (1973). Qualitative differences in the N-acetyl-D-galactosaminyltransferases produced by human A^1 and A^2 genes. Proc. Natl. Acad. Sci. U.S. *70*, 220.

SCHAPIRA, F., SCHAPIRA, G. and DREYFUS, J.-C. (1961). La lésion enzymatique de la fructosurie benigne. Enzymol. Biol. clin. *1*, 170.

SCHECHTER, A. N. and EPSTEIN, C. J. (1968). Mitochondrial malate dehydrogenase: reversible denaturation studies. Science *159*, 997.

SCHIFF, F. (1927). Über den serologische Nachweiss der Blutgruppeneigenschaft O. Klin. Wschr. *6*, 303.

SCHIFF, F. and SASAKI, H. (1932). Der Ausscheidungstypus, ein auf serologischen Wege nachweisbares Mendelndes Merkmal. Klin. Wschr. *11*, 1426.

SCHIFFMAN, G., KABAT, E. A. amd LESKOWITZ, S. (1962). Immunochemical studies on blood groups. XXVI. The isolation of oligosaccharides from human ovarian cyst blood group A substance including two disaccharides and a trisaccharide involved in the specificity of the blood group A antigenic determinant. J. Am. Chem. Soc. *84*, 73.

SCHIFFMAN, G., KABAT, E. A. and THOMPSON, W. (1964). Immunochemical studies on blood groups. XXX. Cleavage of A, B and H blood-group substances by alkali. Biochemistry *3*, 113.

SCHIFFMAN, G., KABAT, E. A. and THOMPSON, W. (1964). Immunochemical studies on blood groups. XXXII. Immunochemical properties of an possible partial structures for the blood group A, B and H antigenic determinants. Biochemistry *3*, 587.

SCHIMKE, R. N., MCKUSICK, V. A., HUANG, T. and POLLACK, A. D. (1965). Homocystinuria: studies of 20 families with 38 affected members. J. Am. Med. Ass. *193*, 711.

SCHMID, K., BINETT, J. P., TOKITA, K., MOROZ, L. and YOSHIZAKI, H. (1964). The polymorphic forms of α_1-acid glycoprotein of normal Caucasian individuals. J. Clin. Invest. *43*, 2347.

SCHMID, K., TOKITA, K. and YOSHIZAKI, H. (1965). The α_1-acid glycoprotein variants of normal Caucasian and Japanese individuals. J. Clin. Invest. *44*, 1394.

SCHMID, R., ROBBINS, P. W. and TAUT, R. R. (1959). Glycogen synthesis in muscle lacking phosphorylase. Proc. Natl. Acad. Sci. U.S. *45*, 1236.

SCHNEIDER, A. S., DUNN, I., IBSEN, K. H. and WEINSTEIN, I. M. (1968a). Inherited triosephosphate isomerase deficiency. Erythrocyte carbohydrate metabolism and preliminary studies of the erythrocyte enzyme. *In:* Hereditary disorders of erythrocyte metabolism, ed. E. Beutler. Grune and Stratton, New York.

SCHNEIDER, A. S., VALENTINE, W. N., HATTORI, M. and HEINS, H. L. (1965). Hereditary hemolytic anaemia with triosephosphate isomerase deficiency. New Engl. J. Med. *272*, 229.

SCHNEIDER, A. S., VALENTINE, W. N., BAUGHAN, M. A., PAGLIA, D. E., SHORE, N. A. and HEINS, H. L. JR. (1968b). Triose phosphate isomerase deficiency. A multi system inherited disorder. Clinical and genetical aspects. *In:* Hereditary disorders of erythrocyte metabolism, ed. E. Beutler. Grune and Stratton, New York.

SCHNEIDER, R. G., UEDA, S., ALPERIN, J. B., BRIMHALL, B. and JONES, R. T. (1969). Hemoglobin Sabine Beta 91(F7) Leu → Pro. An unstable variant causing severe anemia with inclusion bodies. New Engl. J. Med. *280*, 739.

SCHREIER, K. and FLAIG, H. (1956). Urinary excretion of indolepyruvic acid in normal conditions and Folling's disease. Klin. Wschr. *34*, 1213.

SCHRÖDER, H. and EVANS, D. A. P. (1972). The polymorphic acetylation of sulphapyridine in man. J. Med. Genet. *9*, 168.

SCHROEDER, W. A., HUISMAN, T. H. J., BROWN, A. K., UY, R., BOUVER, N., LERCH, P. O., SHELTON, J. R., SHELTON, J. B. and APELL, G. (1971). Postnatal changes in the heterogeneity of human fetal haemoglobin. Pediatric Res. *5*, 493.

SCHROEDER, W. A., HUISMAN, T. H. J., SHELTON, J. R., SHELTON, J. B., KLEIHAUER, E. F., DOZY, A. M. and ROBBERSON, B. (1968). Evidence for multiple structural genes for the γ chain of human fetal hemoglobin. Proc. Natl. Acad. Sci. U.S. *60*, 537.

SCHROEDER, W. A., SHELTON, J. R., SHELTON, J. B., APELL, G., HUISMAN, T. H. J. and BOUVER, N. G. (1972). World-wide occurrence of nonallelic genes for the γ-chain of human foetal haemoglobin in newborns. Nature New Biol. *240*, 273.

SCHRÖTER, W. (1965). Kongenitale nichtspäracytäre hämolytische Anämie bei 2, 3-Diphosphoglyceratmutase Mangel der Erythrocyten im frühen Säuglingsalter. Klin. Wschr. *43*, 1147.

SCHULMAN, J. D., LUSTBERG, T. J., KENNEDY, J. L., MUSELES, M. and SEEGMILLER, J. H. (1970). A new variant of maple syrup urine disease (branch-chain ketoaciduria). Am. J. Med. *49*, 118.

SCHWARTZ, E., KAN, Y. W. and NATHAN, D. G. (1969). Unbalanced globin chain synthesis in alpha-thalassaemia heterozygotes. Ann. New York Acad. Sci. *165*, 288.

SCHWARZ, V., GOLBERG, L., KOMROWER, G. M. and HOLZEL, A. (1956). Some disturbances of erythrocyte metabolism in galactosaemia. Biochem. J. *62*, 34.

SCOTT, E. M. (1960). The relation of diaphorase of human erythrocytes to inheritance of methaemoglobinemia. J. Clin. Invest. *39*, 1176.

SCOTT, E. M. (1966). Kinetic comparison of genetically different acid phosphatases of human erythrocytes. J. Biol. Chem. *241*, 3049.

SCOTT, E. M. (1973). Inheritance of two types of deficiency of human serum cholinesterase. Ann. Hum. Genet. Lond. *37*, 139.

SCOTT, E. M. and GRIFFITH, I. (1959). The enzymic defect of hereditary methaemoglobin-aemia: diaphorase. Biochim. Biophys. Acta *34*, 584.

SCOTT, E. M. and POWERS, R. F. (1972). Human serum cholinesterase, a tetramer. Nature New Biol. *236*, 83.

SCOTT, E. M., WEAVER, D. D. and WRIGHT, R. C. (1970). Discrimination of phenotypes in human serum cholinesterase deficiency. Am. J. Hum. Genet. *22*, 363.

SCOTT, E. M., WRIGHT, R. C. and WEAVER, D. D. (1969). The discrimination of phenotypes for rate of disappearance of isonicotinoyl hydrazide from serum. J. Clin. Invest. *48*, 1173.

SCOTT, R. C., DASSELI, S. W., CLARK, S. H., CHIANG-TENG, C. and SWEDBERG, K. R. (1970). Cystathioninuria: a benign genetic condition. J. Pediat. *76*, 571.

SCRIVER, C. R. (1967). Aminoacid transport in mammalian kidney. *In:* Aminoacid metabolism and genetic variation Ed. W. L. Nyhan. McGraw Hill, New York.

SCRIVER, C. R. (1968). Renal tubular transport of proline hydroxyproline and glycine. III. Genetic basis for more than one mode of transport in human kidney. J. Clin. Invest. *47*, 823.

SCRIVER, C. R. and HECHTMEN, P. (1970). Human genetics of membrane transport with emphasis on aminoacids. *In:* Advances in human genetics, ed. H. Harris and K. Hirschhorn. vol. 1. Plenum Press, New York.

SCRIVER, C. R. and ROSENBERG, L. E. (1973). Aminoacid metabolism and its disorders. Saunders, Philadelphia.

SCRIVER, C. R., LAROCHELLE, J. and SILVERBERG, M. (1967). Hereditary tyrosinemia and tyrosyluria in a French Canadian geographic isolate. Am. J. Diseases Children *113*, 41.

SCRIVER, C. R., MACKENZIE, S., CLOW, C. L. and DELVIN, E. (1971). Thiamine-responsive maple syrup urine disease. Lancet *1*, 310.

SEEGMILLER, J. E., ROSENBLOOM, F. M. and KELLEY, W. N. (1967). Enzyme defect associated with a sex-linked human neurological disorder and excessive purine synthesis. Science *155*, 1682.

SEID-AKHARAN, M., WINTER, W. P., ABRAMSON, R. K. and RUCKNAGEL, D. (1972). Hemoglobin Wayne: A frameshift variant occurring in two distinct forms. Blood *40*, 927.

SELANDER, R. K. and KAUFMAN, D. W. (1973). Genic variability and strategies of adaptation in animals. Proc. Natl. Acad. Sci. U.S. *70*, 1875.

SHAFER, I. A., SCRIVER, C. R. and EFRON, M. L. (1962). Familial hyperprolinaemia, cerebral dysfunction, and renal anomalies occurring in a family with hereditary nephropathy and deafness. New Engl. J. Med. *267*, 51.

SHARP, H. L., BRIDGES, R. A., KRIVIT, W. and FREIER, E. F. (1969). Cirrhosis associated with alpha-1-antitrypsin deficiency: a previously unrecognized inherited disorder. J. Lab. Clin. Med. *73*, 934.

SHAW, C. R. (1965). Electrophoretic variation in enzymes. Science *149*, 936.

SHAW, C. R., SYNER, F. N. and TASHIAN, R. E. (1962). New genetically determined molecular form of erythrocyte esterase in man. Science *138*, 31.

SHELDON, W. (1964). Congenital pancreatic lipase deficiency. Arch. Disease Childhood. *39*, 268.

SHEN, L., GROLLMAN, E. F. and GINSBURG, V. (1968). An enzymatic basis for secretor status and blood group substance specificity in humans. Proc. Natl. Acad. Sci. U.S. *59*, 224.

SHENKEL-BRUNNER, H. and TUPPY, H. (1970). Enzymes from human gastri cmucosa conferring blood group A and B specificities upon erythrocytes. Eur. J. Biochem. *17*, 218.

SHENKEL-BRUNNER, H. and TUPPY, H. (1973). Enzymatic conversion of human blood-group-O erythrocytes into A_2 and A_1 cells by a a-N-acetyl-D-galactosaminyl transferase of blood-group A individuals. Eur. J. Bioch. *34*, 125.

SHIBATA, S., MIYAJI, T., KARITA, K., IUCHI, I., OHBA, Y. and YAMAMOTO, K. (1967). A new type of hereditary nigremia discovered in Akita – hemoglobin $M_{Hyde\ Park}$ disease. Proc, Japan Acad. *43*, No. 1.

SHIBATA, S., MIYAJI, T., VEDA, S., MATSUOKA, M., IUCHI, I., YAMADA, K. and SHINKAI, N. (1970). Hemoglobin Tochingi (β56–59 deleted). A new unstable haemoglobin discovered in a Japanese family. Proc. Jap. Acad. *46*, 440.

SHIH, V. E. and EFRON, M. L. (1972). Urea cycle disorders. *In:* The metabolic basis of inherited disease, eds. J. B. Stanbury, J. B. Wyngaarden and D. S. Fredrickson. 3rd Edition. McGraw Hill, New York.

SHIM, B. S. and BEARN, A. G. (1964). The distribution of haptoglobin subtypes in various populations, including subtype patterns in some non-human primates. Am. J. Hum. Genet. *16*, 477.

SHINODA, T. (1967). Red cell acid phosphatase types in a Japanese population. Jap. J. Hum. Genet. *11*, 252.

SHORT, E. M., CONN, H. O., SNODGRASS, P. J., CAMPBELL, A. G. M. and ROSENBERG, L. E. (1973). Evidence for X-linked dominant inheritance of ornithine transcarbamylase deficiency. New Engl. J. Med. *288*, 7.

SHOWS, T. B. and RUDDLE, F. H. (1968). Malate dehydrogenase: evidence for tetrameric structure in *Mus musculus*. Science *160*, 1356.

SHOWS, T. B., TASHIAN, R. E. and BREWER, G. J. (1964). Erythrocyte glucose-6-phosphate dehydrogenase in Caucasians: new inherited variant. Science *145*, 1056.

SHREFFLER, D. C., BREWER, G. J., GALL, J. C. and HONEYMAN, M. S. (1967). Electrophoretic variation in human serum ceruloplasmin: a new genetic polymorphism. Biochem. Genet. *1*, 101.

SICK, K., BEALE, D., IRVINE, D., LEHMANN, H., GOODALL, P. T. and MACDOUGALL, S. (1967). Haemoglobin $G_{Copenhagen}$ and Haemoglobin $J_{Cambridge}$. Two new β-chain variants of haemoglobin A. Biochim. Biophys. Acta *140*, 231.

SIEGEL, W., COX, R., SCHROEDER, W. A., HUISMAN, T. H. J., PENNER, O. and ROWLEY, P. T. (1970). An adult homozygous for persistent fetal haemoglobin. Ann. Int. Med. *72*, 533.

SILVERS, D. N., COX, R. P., BALIS, M. E. and DANCIS, J. (1972). Detection of the heterozygote in Lesch-Nyhan disease by hair-root analysis. New Engl. J. Med. *286*, 390.

SIMPSON, E. (1966). Factors influencing cholinesterase activity in a Brazilian population. Am. J. Hum. Genet. *18*, 243.

SIMPSON, N. E. and KALOW, W. (1964). The 'silent' gene for serum cholinesterase. Am. J. Hum. Genet. *16*, 180.

SINGER, J. D., COTLIER, E. and KRIMMER, R. (1973). Hexosaminidase A in tears and saliva for rapid identification of Tay-Sachs disease and its carriers. Lancet *2*, 1116.

SINHA, K. P., LEWIS, W. H. P., CORNEY, G. and HARRIS, H. (1970). Studies on the quantitative variation of human red cell peptidase A activity. Ann. Hum. Genet. Lond. *34*, 153.

SLOAN, H. R., UHLENDORF, B. W., KANFER, J. N., BRADY, R. O. and FREDRICKSON, D. S. (1969). Deficiency of sphingomyelin-cleaving enzyme activity in tissue cultures derived from patients with Niemann-Pick disease. Biochem. Biophys. Res. Commun. *34*, 582.

SLY, W. S., QUINTON, B. A., MCALLISTER, W. H. and RIMOIN, D. L. (1973). Isolated β-glucuronidase deficiency: clinical report of a new mucopolysaccharidosis. J. Pediat. *82*, 249.

SMITH, D. H., CLEGG, J. B., WEATHERALL, D. J. and GILLES, H. M. (1973). Hereditary persistence of foetal haemoglobin associated with a $\gamma\beta$ fusion variant, Haemoglobin Kenya. Nature New Biol. *246*, 184.

SMITH, E. W. and TORBERT, J. V. (1958). Two abnormal haemoglobins with evidence for a new genetic locus for haemoglobin formation. Bull. Johns Hopkins Hosp. *102*, 38.

SMITH, J. M. (1968). 'Haldane's dilemma' and the rate of evolution. Nature (Lond.) *219*, 1114.

SMITH, J. M. (1970). Population size, polymorphism and the rate of non-Darwinian evolution. Am. Naturalist *104*, 231.

SMITH, J. M. (1970). The causes of polymorphism. Symp.: Zool. Soc. Lond. No. 26, 371.

SMITH, L. F. (1966). Species variation in the aminoacid sequence of insulin. Am. J. Med. *40*, 662.

SMITH, L. H., SULLIVAN, M. and HUGULEY, C. M. JR. (1961). Pyrimidine metabolism in man. IV. The enzymatic defect of oroticaciduria. J. Clin. Invest. 40, 656.

SMITH, L. H., HUGULEY, C. M. JR. and BAIN, J. A. (1972). Hereditary orotic aciduria. *In:* The Metabolic Basis of Inherited Disease. eds. J. B. Stanbury, J. B. Wyngaarden and D. S. Fredrickson. 3rd Edition. McGraw-Hill, New York.

SMITH, M., HOPKINSON, D. A. and HARRIS, H. (1971). Developmental changes and polymorphisms in human alcohol dehydrogenase. Ann. Hum. Genet. Lond. *34*, 251.

SMITH, M., HOPKINSON, D. A. and HARRIS, H. (1972). Alcohol dehydrogenase isozymes in adult human stomach and liver; evidence for activity of the ADH_3 locus. Ann. Hum. Genet. Lond. *35*, 243.

SMITH, M., HOPKINSON, D. A. and HARRIS, H. (1973). Studies on the subunit structure and molecular size of the human alcohol dehydrogenase isozymes determined by the different loci, ADH_1, ADH_2 and ADH_3. Ann. Hum. Genet. Lond. *36*, 401.

SMITH, S. E. and KYI, T. (1968). Inactivation of isoniazid in Burmese subjects. Nature (Lond.) *217*, 1273.

SMITHIES, O. (1955). Zone electrophoresis in starch gels: group variations in the serum proteins of normal human adults. Biochem. J. *61*, 629.

SMITHIES, O. (1957). Variations in human serum β-globulins. Nature (Lond.) *180*, 1482.

SMITHIES, O. (1964). Chromosomal rearrangements and protein structure. Cold Spring Harbor Symp. Quant. Biol. *29*, 309.

SMITHIES, O. (1967). Antibody variability. Science *157*, 267.

SMITHIES, O. and CONNELL, G. E. (1959). Biochemical aspects of the inherited variations in human serum haptoglobins and transferrins. *In:* Ciba Foundation Symposium on Biochemistry of Human Genetics, ed. G. E. Wolstenholme and C. M. O'Connor, Churchill, London.

SMITHIES, O., CONNELL, G. E. and DIXON, G. H. (1962a). Inheritance of haptoglobin subtypes. Am. J. Hum. Genet. *14*, 14.

SMITHIES, O., CONNELL, G. E. and DIXON, G. H. (1962b). Chromosomal rearrangements and the evolution of haptoglobin genes. Nature (Lond.) *196*, 232.

SMITHIES, O., CONNELL, G. E. and DIXON, G. H. (1966). Gene action in the human haptoglobins. I. Dissociation into constituent polypeptide chains. J. Mol. Biol. *21*, 213.

SMITHIES, O. and HILLER, O. (1959). The genetic control of transferrin in humans. Biochem. J. *72*, 121.

SMITHIES, O. and WALKER, N. F. (1955). Genetic control of some serum proteins in normal humans. Nature (Lond.) *176*, 1265.

SMITHIES, O. and WALKER, N. F. (1956). Notation for serum protein groups and the genes controlling their inheritance. Nature (Lond.) *178*, 694.

SNEATH, J. S. and SNEATH, P. H. A. (1955). Transformation of Lewis groups of human red cells. Nature (Lond.) *176*, 172.

SNYDER, L. H. (1959). Fifty years of medical genetics. Science *129*, 7.

SNYDERMAN, S. E. (1967). Maple syrup urine disease. *In:* Amino acid metabolism and genetic variation, ed. W. L. Nyhan. McGraw-Hill, New York.

SNYDERMAN, S. E., NORTON, P. and HOLT, L. E. JR. (1955). Effect of tyrosine administration in phenylketonuria. Federation Proc. *14*, 450.

SOBEL, E. H., CLARK, L. C., FOX, R. P. and ROBINOV, M. (1953). Rickets, deficiency of alkaline phosphatase activity and premature loss of teeth in childhood. Pediatrics *11*, 309.

SOLOMONS, G., KELESKE, L. and OPITZ, E. (1966). Evaluation of the effects of terminating the diet in phenylketonuria. J. Pediat. *69*, 596.

SOYAMA, K., SHIMADA, N., KUSUNOKI, T. and NAKAMURA, T. (1973). The diagnostic value of thrombocytic glucose-6-phosphatase in patients with Von Gierke's disease and its heterozygotes. Clin. Chim. Acta. *44*, 327.

SPENCER, N., HOPKINSON, D. A. and HARRIS, H. (1964a). Phosphoglucomutase polymorphism in man. Nature (Lond.) *204*, 742.

SPENCER, N., HOPKINSON, D. A. and HARRIS, H. (1964b). Quantitative differences and gene dosage in the human red cell acid phosphatase polymorphism. Nature (Lond.) *201*, 299.

SPENCER, N., HOPKINSON, D. A. and HARRIS, H. (1968). Adenosine deaminase polymorphism in man. Ann. Hum. Genet. Lond. *32*, 9.

SPERLING, O., LIBERMAN, U. A., FRANK, M. and DE VRIES, A. (1971). Xanthinuria. An additional case with demonstration of xanthine oxidase deficiency. Am. J. Clin. Pathol. *55*, 351.

SRIVASTAVA, S. K. and BEUTLER, E. (1972). Antibody against purified human hexosaminidase B cross-reacting with human hexosaminidase A. Biochem. Biophys. Res. Comm. *47*, 753.

SRIVASTAVA, S. K. and BEUTLER, E. (1973). Hexosaminidase-A and hexosaminidase-B: 'studies in Tay-Sachs' and Sandhoff's disease. Nature (Lond.) *241*, 463.

STAAL, G. E. J., HELLEMAN, P. W., DE WAEL, J. and VEEGER, C. (1969). Purification and properties of an abnormal glutathione reductase from human erythrocytes. Biochim. Biophys. Acta. *185*, 63.

STAMATOYANNOPOULOS, G. (1972). The molecular basis of haemoglobin disease. Ann. Rev. Genet. *6*, 47.

STAMATOYANNOPOULOS, G., BELLINGHAM, A. J., LENFANT, C. and FINCH, C. A. (1971). Abnormal haemoglobins with high and low oxygen affinity. Ann. Rev. Med. *22*, 221.

STAMATOYANNOPOULOS, G., FRASER, G. R., MOTULSKY, A. G., FESSAS, P., AKRIVAKIS, A. and PAPAYANNOPOULOU, T. (1966). On the familial predisposition to favism. Am. J. Hum. Genet. *18*, 253.

STAMATOYANNOPOULOS, G., PARER, J. T. and FINCH, C. A. (1969). Physiologic implications of a hemoglobin with decreased oxygen affinity (hemoglobin Seattle). New Engl. J. Med. *281*, 915.

STAMATOYANNOPOULOS, G., YOSHIDA, A., BACOPOULOS, C. and MOTULSKY, A. G. (1967). Athens variant of glucose-6-phosphate dehydrogenase. Science *157*, 831.

STANBURY, J. B. (1972). Familial goiter. *In:* The metabolic basis of inherited disease. Eds. J. B. Stanbury, J. B. Wyngaarden and D. S. Fredrickson. 3rd Edition. McGraw-Hill, New York.

STARNES, C. W. and WELSH, J. D. (1970). Intestinal sucrase-isomaltase deficiency and renal calculi. New. Engl. J. Med., *282*, 1023.

STEADMAN, J. H., YATES, A. and HUEHNS, E. R. (1970). Idiopathic Heinz body anaemia: Hb Bristol (β67 (E11) Val → Asp). Brit. J. Haematol. *18*, 435.

STEIN, W. H. (1951). Excretion of aminoacids in cystinuria. Proc. Soc. Exptl. Biol. *78*, 705.

STEINBERG, A. G. (1967). Genetic variations in human immunoglobulins. The Gm and Inv types. *In:* Advances in immunogenetics, ed. T. J. Greenwalt. Lippincott, Philadelphia.

STEINBERG, D. (1972). Phytanic acid storage disease: Refsum's syndrome. *In:* The Metabolic Basis of Inherited Disease. eds. J. B. Stanbury, J. B. Wyngaarden and D. S. Fredrickson. 3rd Edition. McGraw-Hill, New York.

STEINBERG, D., HERNDON, J. H. JR., UHLENDORF, B. W., MIZE, C. E., AVIGAN, J. and MILNE, G. W. A. (1967). Refsum's disease: nature of the enzyme defect. Science *156*, 1740.

STETSON, C. A. JR. (1966). The state of haemoglobin in sickled erythrocyte. J. Exptl. Med. *123*, 341.

STOKKE, O., ELDJARN, L., NORUM, K. R., STEIN-JOHNSON, J. and HALVORSEN, S. (1967). Methylmalonic aciduria: a new inborn metabolic error which may cause fatal acidosis in the neonatal period. Scand. J. Clin. Lab. Invest. *20*, 313.

STRAND, L. J., FELSHER, B. F., REDEKER, A. G. and MARVER, H. S. (1970). Heme biosynthesis in intermittent acute porphyria: decreased hepatic conversion of porphobilinogen to porphyrins and increased delta aminolevulinic acid synthetase activity. Proc. Natl. Acad. Sci. U.S. *67*, 1315.

SUBAK-SHARPE, H., BURK, R. R. and PITTS, J. D. (1969). Metabolic cooperation between biochemically marked mammalian cells in tissue culture. J. Cell. Sci. *4*, 353.

SUNAHARA, S., URANO, M. and OGAWA, M. (1961). Genetical and geographical studies on isoniazid inactivation. Science *134*, 1530.

SUNSHINE, P., LINDENBAUM, J. E. and LEVY, H. L. (1972). Hyperammonemia due to a defect in ornithine transcarbamylase. Pediatrics *50*, 100.

SUTTON, H. E. (1970). The haptoglobins. Progress in Medical Genetics *7*, 163. Grune and Stratton, New York.

SUZUKI, K. (1968). Cerebral GM$_1$ gangliosidosis: chemical pathology of visceral organs. Science *159*, 1471.

SUZUKI, K. and SUZUKI, Y. (1970). Globoid cell leukodystrophy (Krabbe's disease): deficiency of galactocerebroside β-galactosidase. Proc. Natl. Acad. Sci. U.S. *66*, 302.

SUZUKI, K. and SUZUKI, Y. (1972). Galactosylceramide lipidosis: globoid cell leucodystrophy (Krabbe's disease). *In:* The Metabolic basis of Inherited Disease, eds. J. B. Stanbury, J. B. Wyngaarden and D. S. Fredrickson. 3rd Edition. McGraw-Hill, New York.

SUZUKI, Y. and SUZUKI, K. (1970). Partial deficiency of hexosaminidase component A in juvenile GM$_2$-gangliosidosis. Neurology *20*, 848.

SUZUKI, Y. and SUZUKI, K. (1971). Krabbe's globoid cell leukodystrophy: deficiency of galactocerebrosidase in serum, leukocytes and fibroblasts. Science *171*, 73.

SUZUKI, Y., SUZUKI, K. and KAMOSHITA, S. (1969). Chemical pathology of GM_1 gangliosidosis. J. Neuropathol. Exptl. Neurol. *28*, 25.

SVED, J. A., REED, T. E. and BODMER, W. F. (1967). The number of balanced polymorphisms that can be maintained in a natural population. Genetics *55*, 469.

SWALLOW, D. M. and HARRIS, H. (1972). A new variant of the placental acid phosphatases; its implications regarding their subunit structures and genetical determination. Ann. Hum. Genet. Lond. *36*, 141.

SWALLOW, D. M., POVEY, S. and HARRIS, H. (1973). Activity of the 'red cell' acid phosphatase locus in other tissues. Ann. Hum. Genet. Lond. *37*, 31.

SWEELEY, C. C. and KLIONSKY, B. (1963). Fabry's disease: classification as a sphingolipidosis and partial characterization of a novel glycolipid. J. Biol. Chem. *238*, PC 3148.

SZEINBERG, A., KAHANA, D., GAVENDO, S., ZAIDMAN, J. and BEN-EZZER, J. (1969). Hereditary deficiency of adenylate kinase in red blood cells. Acta Haematol. *42*, 111.

SZEINBERG, A., SHEBA, C. and ADAM, A. (1958). Enzymatic abnormality in erythrocytes of a population sensitive to *Vicia faba* or drug induced haemolytic anaemia. Nature (Lond.) *181*, 1256.

SZULMAN, A. E. (1964). The histological distribution of the blood group substances in man as disclosed by immunofluorescence. Part III. The A, B and H antigens in embryos and fetuses from 18 mm in length. J. Exptl. Med. *119*, 503.

SZULMAN, A. E. (1966). Chemistry, distribution, and function of blood group substances.

TADA, K., WADA, Y. and ARAKAWA, T. (1967). Hypervalinemia.
113, 64.

TADA, K., YOKOYAMA, Y., NAKAGAWA, H. and ARAKAWA, T. (1968). Vitamin B_6 dependent xanthurenic aciduria. The second report. Tohoku J. Exptl. Med *95*, 107.

TADA, K., YOKOYAMA, Y., NAKAGAWA, H., YOSHIDA, T. and ARAKAWA, T. (1967). Vitamin B_6 dependent xanthurenic aciduria. Tohoku J. Exptl. Med. *93*, 115.

TADA, K., YOSHIDA, T., YOKOYAMA, Y., SATO, T., NAGAKAWA, H. and ARAKAWA, T. (1968). Cystathioninuria not associated with vitamin B_6 dependency: a probably new type of cystathioninuria. Tohoku J. Exptl. Med. *95*, 235.

TAKAHARA, S. (1952). Progressive oral gangrene, probably due to a lack of catalase in the blood (acatalasaemia). Lancet *2*, 1101.

TAKAHARA, S. (1968). Acatalasemia in Japan. *In:* Hereditary disorders of erythrocyte metabolism, ed. E. Beutler. Grune and Stratton, New York.

TALAMO, R. C., LANGLEY, C. E., REED, C. E. and MAKINO, S. (1973). α_1-Antitrypsin deficiency: a variant with no detectable α_1-antitrypsin. Science *181*, 70.

TALLMAN, J. F., JOHNSON, W. G. and BRADY, R. O. (1972). The metabolism of Tay-Sachs ganglioside: catabolic studies with lysosomal enzymes from normal and Tay-Sachs brain tissue. J. Clin. Invest. *51*, 2339.

TANAKA, K. R. and BEUTLER, E. (1969). Hereditary hemolytic anemia due to glucose-6-phosphate dehydrogenase Torrance: a new variant. J. Lab. Clin. Med. *73*, 657.

TANAKA, K. R. and VALENTINE, W. N. (1968). Pyruvate kinase deficiency. *In:* Hereditary disorders of erythrocyte metabolism, ed. E. Beutler. Grune and Stratton, New York.

TANAKA, K. R., VALENTINE, W. N. and MIWA, S. (1962). Pyruvate kinase (PK) deficiency: hereditary non-spherocytic hemolytic anemia. Blood *19*, 267.

TANAKA, K. R., BUDD, M. A., EFRON, M. L. and ISSELBACHER, K. J. (1966). Isovaleric acidemia: a new genetic defect of leucine metabolism. Proc. Natl. Acad. Sci. U.S. *56*, 236.

TANIGUCHI, K. and GJESSING, L. R. (1965). Studies on tyrosinosis: 2. Activity of the transaminase, para-hydroxyl-phenyl pyruvate oxidase and homogentisic acid oxidase. Brit. Med. J. *1*, 968.

TARLOW, M. J., HADORN, B., ARTHURTON, M. W. and LLOYD, J. K. (1970). Intestinal enterokinase deficiency. Arch. Dis. Child. *45*, 651.

TARUI, S., KONO, N., NASU, T. and NISHIKAWA, M. (1969). Enzymatic basis for the coexistence of myopathy and hemolytic disease in inherited muscle phosphofructokinase deficiency. Biochem. Biophys. Res. Commun. *34*, 77.

TARUI, S., OKUNO, G., IKURA, Y., TANAKA, T., SUDA, M. and NISHIKAWA, M. (1965). Phosphofructokinase deficiency in skeletal muscle. A new type of glycogenosis. Biochem. Biophys. Res. Commun. *19*, 517.

TASHIAN, R. E. (1959). Phenylpyruvic acid as a possible precursor of o-hydroxyphenylacetic acid in man. Science *129*, 1553.

TASHIAN, R. E. (1969). The esterases and carbonic anhydrases of human erythrocytes. *In:* Biochemical Methods in Red Cell Genetics. ed. J. J. Yunis, Academic Press, New York.

TASHIAN, R. E. and SHAW, M. W. (1962). Inheritance of an erythrocyte acetylesterase variant in man. Amer. J. Hum. Genet. *14*, 295.

TASHIAN, R. E., PLATO, C. C. and SHOWS, T. B. (1963). Inherited variant of erythrocyte carbonic anhydrase in Micronesians from Guam and haipan. Science *140*, 53.

TASHIAN, R. E., SHREFFLER, D. C. and SHOWS, T. B. (1968). Genetic and phylogenetic variation in the different molecular forms of mammalian erythrocyte carbonic anhydrases. Ann. N.Y. Acad. Sci. *151*, 64.

TATESON, R. and BAIN, A. D. (1971). GM_2 gangliosidoses: consideration of the genetic defects. Lancet *2*, 612.

TEDESCO, T. A. (1972). Human galactose 1-phosphate uridyltransferase. Purification, antibody production and comparison of the wild type, Duarte variant, and galactosemic gene products. J. Biol. Chem. *247*, 6631.

TEDESCO, T. A. and MELLMAN, W. J. (1967). Argininosuccinate synthetase activity and citrulline metabolism in cells cultured from a citrullinemic subject. Proc. Natl. Acad. Sci. U.S. *57*, 829.

TERHEGGEN, H. G., LAVINHA, F., COLOMBO, J. P., VAN SANDE, M. and LOWENTHAL, A. (1972). Familial hyperargininemia. J. Génét. Hum. *20*, 69.

TERHEGGEN, H. G., SCHWENK, A., LOWENTHAL, A., VAN SANDE, M. and COLOMBO, J. P. (1969). Argininemia with arginase deficiency. Lancet *2*, 748.

THIER, S. O., FOX, M., SEGAL, S. and ROSENBERG, L. E. (1964). Cystinuria: *in vitro* demonstration of an intestinal transport defect. Science *143*, 482.

THIER, S. O., SEGAL, S., FOX, M., BLAIR, A. and ROSENBERG, L. E. (1965). Cystinuria: defective intestinal transport of dibasic aminoacids and cystine. J. Clin. Invest. *44*, 442.

THOMAS, D. M. and HARRIS, H. (1971). Comparison of thermostabilities of different human placental alkaline phosphatase phenotypes. Ann. Hum. Genet. Lond. *35*, 221.

TOBARI, Y. N. and KOJIMA, K.-I. ((1972). A study of spontaneous mutation rates at ten loci detectable by starch gel electrophoresis in *Drosophila melanogaster*. Genetics *70*, 397.

TODD, D., LAI, M. C. S., BEAVEN, G. H. and HUEHNS, E. R. (1970). The abnormal haemoglobins in homozygous α-thalassaemia. Brit. J. Haematol. *19*, 27.

TOMLINSON, S. and WESTALL, R. G. (1964). Argininosuccinic aciduria. Argininosuccinase and arginase in human blood cells. Clin. Sci. *26*, 261.

TÖNZ, O. (1968). The congenital methamoglobinemias. Bibliotheca Haematologia No. 28. Karger, Basel.

TOUSTER, O. (1959). Pentose metabolism and pentosuria. Am. J. Med. *26*, 724.

TOWNES, P. L. (1965). Trypsinogen deficiency disease. J. Pediat. *66*, 275.

TOWNES, P. L. and MORRISON, M. (1962). Investigation of the defect in a variant of hereditary methemoglobinemia. Blood *19*, 60.

TOWNESS, P. L., BRYSON, M. F. and MILLER, G. (1967). Further observations on trypsinogen deficiency disease: report of a second case. J. Pediat. *71*, 220.

TOWNSEND, E. H., MASON, H. H. and STRONG, P. S. (1951). Galactosemia and its relation to Laennec's cirrhosis: review of literature and presentation of six additional cases. Pediatrics *7*, 760.

TSCHUDY, D. P., PERLROTH, M. G., MARVER, H. S., COLLINS, A., HUNTER, G. and RECHCIGL, M. (1965). Acute intermittent porphyria: the first 'overproduction disease' localized to a specific enzyme. Proc. Natl. Acad. Sci. U.S. *53*, 841.

TUBERCULOSIS CHEMOTHERAPY CENTRE MADRAS (1970). A controlled comparison of a twice-weekly and three once-weekly regimens in the initial treatment of pulmonary tuberculosis. Bull. Wld. Hlth. Org. *43*, 143.

TURNER, B. (1967). Pyridoxine treatment in homocystinuria. Lancet *2*, 1151.

TURNER, B. M., FISHER, R. A., GARTHWAITE, E., WHALE, R. J. and HARRIS, H. (1973). An account of two new ICD-S variants not detectable in red blood cells. Ann. Hum. Genet. Lond. *37*, 469.

TURNER, B. M., FISHER, R. A. and HARRIS, H. (1971). An association between the kinetic and electrophoretic properties of human purine-nucleoside-phosphorylase isozymes. Eur. J. Biochem. *24*, 288.

TURNER, B. M., FISHER, R. A. and HARRIS, H. (1974). A comparison of the soluble and mitochondrial forms of human isocitrate dehydrogenase with an examination of the secondary isozymes derived from the soluble form. Ann. Hum. Genet. Lond. *37*, 455.

TYUMA, I. and SHIMIZU, K. (1970). Effect of organic phosphates on the difference in oxygen affinity between foetal and adult human haemoglobin. Fed. Proc. *29*, 1112.

UHLENDORF, B. W. and MUDD, S. H. (1968). Cystathionine synthase in tissue culture from human skin: enzyme defect in homocystinuria. Science *160*, 1007.

VALENTINE, W. N. (1968). Hereditary hemolytic anemias associated with specific erythrocyte enzymopathies. Calif. Med. *108*, 280.

VALENTINE, W. N., HSIEH, H.-S., PAGLIA, D. E., ANDERSON, H. M., BAUGHAM, M. A., JAFFÉ, E. R. and GARSON, O. M. (1969). Hereditary hemolytic anaemia associated with phosphoglycerate kinase deficiency in erythrocytes and leukocytes. New. Engl. J. Med., *280*, 528.

VALENTINE, W. N., OSKI, F. A., PAGLIA, D. E., BAUGHAN, M. A., SCHNEIDER, A. S. and NAIMAN, J. L. (1967). Hereditary hemolytic anemia with hexokinase deficiency. New Engl. J. Med. *276*, 1.

VALENTINE, W. N. and PAGLIA, K. R. (1972). Pyruvate kinase deficiency and other enzyme deficiency hereditary hemolytic anaemias. *In:* The metabolic basis of inherited disease, eds. J. B. Stanbury, J. B. Wyngaarden, and D. S. Fredrickson. 3rd Edition. McGraw Hill, New York.

VALENTINE, W. N., SCHNEIDER, A. S., BAUGHAN, M. A., PAGLIA, D. E. and HEINS, H. L. JR. (1966). Hereditary hemolytic anemia with triose-phosphate isomerase deficiency. Am. J. Med. *41*, 27.

VALENTINE, W. N., TANAKA, K. R. and MIWA, S. (1961). A specific erythrocyte glycolytic enzyme defect (pyruvate kinase) in three subjects with congenital non-spherocytic hemolytic anemia. Trans. Ass. Amer. Physicians *74*, 100.

VAN CONG, N., BILLARDON, C., PICARD, J. Y., FEINGOLD, J. and FRÉZAL, J. (1971). Liaison probable (linkage) entre les locus PGM₁ et peptidase C chez l'homme. C. R. Acad. Sci. (D) Paris *272*, 485.

VANDEPITTE, J. M. (1959). The incidence of haemoglobinoses in the Belgian Congo. *In:* Abnormal haemoglobins, ed. J. H. P. Jonxis and J. F. Delafresnaye. Blackwell, Oxford.

VANDEPITTE, J. M., ZUELZER, W. W., NEEL, J. V. and COLAERT, J. (1955). Evidence concerning the inadequacy of mutation as an explanation of the frequency of the sickle cell gene in the Belgian Congo. Blood *10*, 341.

VAN DER ZEE, S. P. M., TRIJBELS, J. M. F., MONNENS, L. A. H., HOMMES, F. A. and SCHRETLEN, E. D. A. M. (1971). Citrullinemia with rapidly fatal neonatal course. Arch. Dis. Child. *46*, 847.

VAN HEESWIJK, P. J., TRIJBELS, J. M. F., SCHRETLEN, E. D. A. M., VAN MUNSTER, P. J. J. and MONNENS, L. A. H. (1969). A patient with a deficiency of serum-carnosinase activity. Acta Paediat. Scand. *58*, 584.

VAN HEYNINGEN, V., CRAIG, I. and BODMER, W. (1973). Genetic control of mitochondrial enzymes in human-mouse somatic cell hybrids. Nature (Lond.) *242*, 509.

VAN HOOF, F. (1967). Amylo-1, 6-glucosidase activity and glycogen content of the erythrocytes of normal subjects. Patients with glycogen storage disease and heterozygotes. Eur. J. Biochem. *2*, 271.

VAN HOOF, F. and HERS, H. G. (1968). The abnormalities of lysosomal enzymes in mucopolysaccharidoses. Eur. J. Biochem. *7*, 34.

VESELL, E. S. (1965a). Formation of human lactate dehydrogenase isozyme patterns *in vitro*. Proc. Natl. Acad. Sci. U.S. *54*, 111.

VESELL, E. S. (1965b). Genetic control of isozyme patterns in human tissues. *In:* Progress in medical genetics, ed. A. G. Steinberg and A. G. Bearn. Grune and Stratton, New York.

VESELL, E. S. (1968). Introduction. Multiple molecular forms of enzymes. Ann. N.Y. Acad. Sci. *151*, 5.

VIDGOFF, J. BUIST, N. R. M. and O'BRIEN, J. S. (1973). Absence of β-N-Acetyl-D-hexosaminidase A activity in a healthy woman. Am. J. Hum. Genet. *25*, 372.

VOGEL, F. (1970). ABO Blood groups and disease. Am. J. Hum. Genet. *22*, 464.

VOGLINO, G. F, and PONZONE, A. (1972). Polymorphism in human casein. Nature New Biol. *238*, 149.

VOLK, B. W. and ARONSON, S. M. (1972). Sphingolipids, sphingolipidoses and allied disorders. Plenum Press, New York.

VON GIERKE, E. (1929). Hepato-nephromegalia glykogenia. Beitr. Pathol. Anat. *82*, 497.

VON WARTBURG, J. P., PAPENBERG, J. and AEBI, H. (1965). An atypical human alcohol dehydrogenase. Can. J. Biochem. *43*, 889.

VON WARTBURG, J. P. and SCHÜRCH, P. M. (1968). Atypical human liver alcohol dehydrogenase. Ann. N.Y. Acad. Sci. *151*, 936.

VYAS, G. N. and FUDENBERG, H. H. (1969). Am (1), the first genetic marker of human immunoglobulin A. Proc. Natl. Acad. Sci. *64*, 1211.

WADA, Y., TAKA, K., MINEGAWA, A., YOSHIDA, T., MORIKAWA, T. and OKAMURA, T. (1963). Idiopathic hypervalinemia. Probably a new entity of inborn error of valine metabolism. Tohoku J. Exptl. Med. *81*, 46.

WAJCMAN, H., LABIE, D. and SCHAPIRA, G. (1973). Two new hemoglobin variants with deletion. Hemoglobin Tours: Thr $\beta 87$ (F3) deleted and hemoglobin St Antoine: Gly-Leu $\beta 74$-75 (E18-19) deleted. Consequences for oxygen affinity and protein stability. Biochim. Biophys. Acta. *295*, 495.

WALDENSTRÖM, J. (1937). Studien über Porphyrie. Acta Med. Scand. Suppl. 82.

WALDENSTRÖM, J. (1957). The porphyrias as inborn errors of metabolism. Amer. J. Med. *22*, 758.

WALDENSTRÖM, J. and HAEGER-ARONSEN, B. (1967). The porphyrias: a genetic problem. *In:* Progress in medical genetics, ed. A. G. Steinberg and A. G. Bearn. Vol. 5. Grune and Stratton, New York.

WALLACE, H. W., MOLDAVE, K. and MEISTER, A. (1957). Studies on conversion of phenylalanine to tyrosine in phenylpyruvic oligophrenia. Proc. Soc. Exptl. Biol. *94*, 632.

WALLER, H. D. (1968). Glutathione reductase deficiency. *In:* Hereditary disorders of erythrocyte metabolism, ed. E. Beutler. Grune and Stratton, New York.

WALLER, H. D., LOHR, G., ZYSNO, E., GEROK, W., VOSS, D. and STRAUSS, G. (1965). Glutathionreductasemangel mit hämatologischen und neurologischen Störungen. Klin. Wschr. *43*, 413.

WANG, Y. M. and VAN EYS, J. (1970). The enzymatic defect in essential pentosuria. New. Engl. J. Med. *282*, 892.

WARD, J. C., MERRITT, A. D. and BIXLER, D. (1971). Human salivary amylase: genetics of electrophoretic variants. Am. J. Hum. Genet. *23*, 403.

WARSHAW, J. B., LITTLEFIELD, J. W., FISHMAN, W. H., INGLIS, N. R. and STOLBACH, L. L. (1971). Serum alkaline phosphatase in hypophosphatasia. J. Clin. Invest. *50*, 2137.

WASI, P. (1973). Annotation: Is the human globin α-chain locus duplicated. Brit. J. Haematol. *24*, 267.

WASI, P., NA-NAKORN, S., POOTRAKUL, S., SOOKANEK, M., DISTHASONGCHAN, P., PORNPATKUL, M. and PANICH, V. (1969). Alpha and beta-thalassaemia in Thailand. Ann. New York Acad. Sci. *165*, 60.

WATERBURY, L. and FRENKEL, E. P. (1969). Phosphofructokinase deficiency in congenital non-spherocytic haemolytic anaemia. Clin. Res. *17*, 347.

WATKINS, W. M. (1956). The appearance of H specificity following the enzymic inactivation of blood-group B substance. Biochem. Soc. *64*, 21P.

WATKINS, W. M. (1966). Blood group substances. Science *152*, 172.

WATKINS, W. M., KOSCIELAK, J. and MORGAN, W. T. J. (1964). The relationship between the specificity of the blood-group A and B substances isolated from erythrocytes and from secretions. Proc. 9th Congr. Int. Soc. Blood Transfusion, Mexico City, 1962. p. 213.

WATKINS, W. M. and MORGAN, W. T. J. (1952). Neutralization of the anti-H agglutinin in eel serum by simple sugars. Nature (Lond.) *169*, 825.

WATKINS, W. M. and MORGAN, W. T. J. (1955a). Inhibition by simple sugars of enzymes which decompose the blood-group substances. Nature (Lond.) *175*, 676.

WATKINS, W. M. and MORGAN, W. T. J. (1955b). Some observations on the O and H characters of human blood and secretions. Vox Sang. (O.S.) 5, 1.

WATKINS, W. M. and MORGAN, W. T. J. (1957a). Specific inhibition studies relating to the Lewis blood-group system. Nature (Lond.) 180, 1038.

WATKINS, W. M. and MORGAN, W. T. J. (1957b). The A and H character of the blood-group substances secreted by persons belonging to group A₂. Acta Genet. Statist. Med. 6, 521.

WATKINS, W. M. and MORGAN, W. T. J. (1962). Further observations on the inhibition of blood-group specific serological reactions by simple sugars of known structure. Vox. Sang. 7, 129.

WATKINS, W. M., ZARNITZ, M. L. and KABAT, E. A. (1962). Development of H activity by human blood-group B substance treated with coffee bean α-galactosidase. Nature (Lond.) 195, 1204.

WATSON, J. D. and CRICK, F. H. C. (1953). Genetical implications of the structure of deoxyribosenucleic acid. Nature (Lond.) 171, 964.

WATTS, R. L. and WATTS, D. C. (1968). The implications for molecular evolution of possible mechanisms of primary gene duplication. J. Theoret. Biol. 20, 227.

WATTS, R. W. E., ENGLEMAN, K., KLINENBERG, J. R., SEEGMILLER, J. E. and SJOERDSMA (1963). Enzyme defect in a case of xanthinuria. Nature (Lond.) 201, 395.

WEATHERALL, D. J. (1969). Genetics of the thalassaemias. Brit. Med. Bull. 25, 24.

WEATHERALL, D. J. and CLEGG, J. B. (1972). The thalassaemia syndromes. 2nd Edition. Blackwell, Oxford.

WEATHERALL, D. J., CLEGG, J. B. and BOON, W. H. (1970). The haemoglobin constitution of infants with the haemoglobin Bart's hydrops foetalis syndrome. Brit. J. Haematol. 18, 357.

WEATHERALL, D. J., CLEGG, J. B. and NAUGHTON, M. A. (1965). Globin synthesis in thalassaemia: an in vitro study. Nature (Lond.) 208, 1061.

WEATHERALL, D. J., SIGLER, A. T. and BAGLIONI, C. (1962). Four hemoglobins in each of three brothers. Genetic and biochemical significance. Bull. Johns Hopkins Hosp. III, 143.

WEIJERS, H. A., VAN DE KAMER, J. H., MOSSELL, D. A. A. and DICKE, W. K. (1960). Diarrhoea caused by deficiency of sugar splitting enzymes. Lancet 2, 296.

WEINREICH, J., BUSCH, D., GOTTSTEIN, V., SCHAEFER, J. and ROHR, J. (1968). Über zwei neue Fälle von hereditärer nichtsphärocytärer hämolytischer Anämie bei glucose-6-phosphate dehydrogenase – defekt in einer norddeutschen Familie. Klin. Wochenschr. 46, 146.

WEITKAMP, L. B., SALZANO, F. M., NEEL, J. V., PORTA, F., GEERDINK, R. A. and TARNOKY, A. L. (1973). Human serum albumin: twenty three genetic variants and their population distribution. Ann. Hum. Genet. Lond. 36, 381.

WEITKAMP, L. R., SHREFFLER, D. C., ROBBINS, J. L., DRACHMANN, O., ADNER, P. L., WIEME, R. J., SIMON, N. M., COOKE, K. B., SANDOR, G., WUHRMANN, F., BRAEND, M. and TARNOKY, A. L. (1967). An electrophoretic comparison of serum albumin variants from nineteen unrelated families. Acta Genet. Basel 17, 399.

WELLS, I. C. and ITANO, H. A. (1951). The ratio of sickle cell anemia haemoglobin to normal haemoglobin in the sicklemics. J. Biol. Chem. 188, 65.

WELLS, W. W., PITTMAN, T. A. and EGANT, T. J. (1964). The isolation and identification of galactitol from the urine of patients with galactosaemia. J. Biol. Chem. 239, 3192.

WELLS, W. W., PITTMAN, T. A., WELLS, H. J. and EGAN, T. J. (1965). The isolation and identification of galactitol from the brains of galactosemic patients. J. Biol. Chem. *240*, 1002.

WEST, C. A., GOMPERTZ, B. D., HUEHNS, E. R., KESSEL, I. and ASHBY, J. R. (1967). Demonstration of an enzyme variant in a case of congenital methaemoglobinaemia. Brit. Med. J. *4*, 212.

WESTALL, R. G. (1960). Argininosuccinic aciduria: identification and reactions of the abnormal metabolites in a newly described form of mental disease. With some preliminary metabolic studies. Biochem. J. *77*, 135.

WESTHAMER, S., FREIBERG, A. and AMARAL, L. (1973). Quantitation of lactate dehydrogenase isoenzyme patterns of the developing human fetus. Clin. Chim. Acta. *45*, 5.

WHEELER, J. T. and KREVANS, J. R. (1961). Homozygous state of persistent fetal haemoglobin and interaction of persistent haemoglobin with thalassaemia. Bull. Johns Hopkins Hosp. *109*, 217.

WHITE, J. M., BRAIN, M. C., LORKIN, P. A., LEHMANN, H. and SMITH, M. (1970). Mild 'unstable haemoglobin haemolytic anaemia' caused by haemoglobin Shepherds Bush (β74 (E18) Gly → Asp). Nature (Lond.) *225*, 939.

WHITTAKER, M. (1964). The pseudocholinesterase variants: esterase levels and increased resistance to fluoride. Acta Genet. *14*, 281.

WHITTAKER, M. (1967). The pseudocholinesterase variants. A study of fourteen families selected via the fluoride resistant phenotype. Acta Genet. *17*, 1.

WHO SCIENTIFIC GROUP (1967). Standardisation of procedures for the study of glucose-6-phosphate dehydrogenase. Wld. Hlth. Org. Tech. Rep. Ser. No. 366 World Health Organisation – Geneva.

WICK, H., BACHMANN, C., BAUMGARTNER, R., BRECHBÜHLER, T., COLOMBO, J. P., WIESMANN, U., MIHATSCH, M. J. and OHNACKER, H. (1973). Variants of citrullinaemia. Arch. Diseases Children, *48*, 636.

WIEME, R. J. and DEMEULENAERE, L. (1967). Genetically determined electrophoretic variant of the human complement component C'3. Nature (Lond.) *214*, 1042.

WIEME, R. J. and SEGERS, J. (1968). Genetic polymorphism of the complement component C'3 in a Bantu population. Nature (Lond.) *220*, 176.

WIESMANN, U. and NEUFELD, E. F. (1970). Scheie and Hurler syndromes: apparent identity of the biochemical defect. Science *169*, 72.

WILLIAMS, H. E. and SMITH, L. H. JR. (1968). L-glyceric aciduria. A new genetic variant of primary hyperoxaluria. New Engl. J. Med. *278*, 233.

WILLIAMS, H. E. and SMITH, L. H. JR. (1971). Hyperoxaluria in L-glyceric aciduria: possible pathogenic mechanism. Science *171*, 390.

WILLIAMS, H. E., KENDIG, E. M. and FIELD, J. B. (1963). Leukocyte debranching enzyme in glycogen storage disease. J. Clin. Invest. *42*, 656.

WILSON, A. C., CAHN, R. D. and KAPLAN, N. O. (1963). Functions of the two forms of lactic dehydrogenase in the breast muscle of birds. Nature (Lond.) *197*, 331.

WINSLOW, R. B. and INGRAM, V. M. (1966). Peptide chain synthesis of human hemoglobins A and A$_2$. J. Biol. Chem. *241*, 1144.

WINTERHALTER, K. H., ANDERSON, N. M., AMICONI, G., ANTONINI, E. and BRUNORI, M. (1969). Functional properties of hemoglobin Zürich. Eur. J. Biochem. *11*, 435.

WISE, D., WALLACE, H. J. and JELLINEK, E. H. (1962). Angiokeratoma corporis diffusum: a clinical study of eight affected families. Quart. J. Med. *31*, 177.

WOESE, C. R. (1967). The genetic code. Harper and Row, New York.

References

WOLMAN, M., STERK, V. V., GATT, S. and FRENKEL, M. (1961). Primary familial xanthematosis with involvement and calcification of the adrenals: report of 2 more cases of siblings of apreviously described infant. Pediatrics *28*, 742.

WOODY, N. C. (1964). Hyperlysinemia. Am. J. Diseases Children *108*, 543.

WOODY, N. C., HUTZLER, J. and DANCIS, J. (1966). Further studies of hyperlysinemia. Am. J. Diseases Children *112*, 577.

WOOLF, L. I. (1951). Excretion of conjugated phenylacetic acid in phenylketonuria. Biochem. J. *49*, ix.

WORLD HEALTH, ORGANISATION (1967). Technical Report Series No. 366. Standardisation of procedures for the study of glucose-6-phosphate dehydrogenase. Geneva.

WRIGHT, S. (1966). Polyallelic random drift in relation to evolution. Proc. Natl. Acad. Sci. U.S. *55*, 1074.

WRIGHTSTONE, R. N. and HUISMAN, T. H. J. (1968). Qualitative and quantitative studies of sickle cell hemoglobin in homozygotes and heterozygotes. Clin. Chim. Acta *22*, 593.

WUNTCH, T., CHEN, R. F. and VESELL, E. S. (1970). Lactate dehydrogenase isozymes: kinetic properties at high enzyme concentrations. Science *167*, 63.

WUNTCH, T., CHEN, R. F. and VESELL, E. S. (1970). Lactate dehydrogenase isozymes: Further kinetic studies at high enzyme concentrations. Science *169*, 480.

WUU, K.-D. and KROOTH, R. S. (1968). Dihydroorotic acid dehydrogenase activity of human diploid cell strains. Science *160*, 539.

YAMAOKA, K. (1971). Hemoglobin Hirose: $\alpha_2\beta_2 37(C3)$ tryptophan yielding serine. Blood *38*, 730.

YAMAZAKI, T. and MURAYAMA, T. (1972). Evidence for the neutral hypothesis of protein polymorphism. Science *178*, 56.

YCAS, M. (1969). The biological code. North-Holland Publishing Co., Amsterdam.

YOSHIDA, A. (1967). A single amino acid substitution (asparagine to aspartic acid) between normal (B+) and the common Negro variant (A+) of human glucose-6-phosphate dehydrogenase. Proc. Natl. Acad. Sci. U.S. *57*, 835.

YOSHIDA, A. (1970). Aminoacid substitution (histidine to tyrosine) in a glucose-6-phosphate dehydrogenase variant (G6PD Hektoen) associated with overproduction. J. Mol. Biol. *52*, 483.

YOSHIDA, A. (1973). Haemolytic anaemia and G6PD deficiency. Science *179*, 532.

YOSHIDA, A., BEUTLER, E. and MOTULSKY, A. G. (1971). Table of human glucose-6-phosphate dehydrogenase variants. Bull. Wld. Hlth. Org. *45*, 243.

YOSHIDA, A. and LIN, M. (1973). Regulation of glucose-6-phosphate dehydrogenase activity in red blood cells from hemolytic and non-hemolytic variant subjects. Blood *41*, 877.

YOSHIDA, A. and MOTULSKY, A. G. (1969). A pseudocholinesterase variant (E. Cynthiana) associated with elevated plasma enzyme activity. Am. J. Hum. Genet. *21*, 486.

YOSHIDA, A., STAMATOYANNOPOULOS, G. and MOTULSKY, A. G. (1967a). Negro variant of glucose-6-phosphate dehydrogenase deficiency (A⁻) in man. Science *155*, 97.

YOSHIDA, A., STEINMANN, L. and HARBART, P. (1967b). *In vitro* hybridization of normal and variant human glucose-6-phosphate dehydrogenase. Nature (Lond.) *216*, 275.

YOSHIDA, T., TATA, K., HONDA, Y. and ARAKAWA, T. (1971). Urocanic–aciduria: a defect in the urocanase activity in the liver of a mentally retarded. Tohoky J. Exptl. Med. *104*, 305.

ZELLER, E. A. (1943). Isolierung von Phenylmilchsäure und Phenyltraubensäure aus Harn bei Imbecillitas Phenylpyruvica. Helv. Chim. Acta *26*, 1614.

ZIDERMAN, D., GOMPERTZ, S., SMITH, Z. G. and WATKINS, W. M. (1967). Glycosyl transferases in mammalian gastric mucosal linings. Biochem. Biophys. Res. Commun. *29*, 56.

ZINKHAM, W. H. (1968). Lactate dehydrogenase isozymes of testis and sperm: biological and biochemical properties and genetic control. Ann. N.Y. Acad. Sci. *151*, 598.

ZINKHAM, W. H., ISENSEE, H. and RENWICK, J. H. (1969). Linkage of lactate dehydrogenase *B* and *C* loci in pigeons. Science *164*, 185.

ZINKHAM, W. H., LENHARD, R. E. JR. and CHILDS, B. (1958). A deficiency of glucose-6-phosphate dehydrogenase activity in erythrocytes of patients with favism. Bull. Johns Hopk. Hosp. *102*, 169.

Subject index

467